INTERNATIONAL SEMINAR ON NUCLEAR WAR AND PLANETARY EMERGENCIES
45th Session

THE ROLE OF SCIENCE IN THE THIRD MILLENNIUM

Lectio Magistralis Global Warming Emergency / Why Science is Needed for The Culture of The Third Millennium / Mitigation of Terrorist Acts – Improvised Nuclear Devices – Gap Analysis of EU – Preventing Nuclear Explosive Terrorism – Public Perception / Climate & Climate Economics / Global Nuclear Energy Issues – Outlook in Japan, China, Europe and USA – Sustainable Nuclear Energy Systems – Economics of Nuclear Power / Energy & Sustainability in Cities – Global Sustainability – Smart Grids – Complexity / Global Food Production & Forest Dynamics / Water, Pollution & Terrorism / Brain Aging & Behavior / Frontiers in Fast Computing & Informatics / Information Security

THE SCIENCE AND CULTURE SERIES

Nuclear Strategy and Peace Technology*

Series Editor: Antonino Zichichi

2009 — International Seminar on Nuclear War and Planetary Emergencies — 42nd Session: World Energy Crisis—Energy & Pollution: Essential Technologies for Managing the Coupled Challenges of Climate Change and Energy Security, Energy, Water, Climate, Pollution & Limits of Development in Asian Countries; Global Monitoring of the Planet—Sensitivity of Climate to Additional CO2 as Indicated by Water Cycle Feed-back Issues, Climate Uncertainties Addressed by Satellites, The Basic Mathematics Needed for all Models; Pollution and Medcine—The Revolution in the Environmental Health Sciences and the Emergence of Green Chemistry; Information Security—Cyber Conflict and Cyber Stability: Finding a Path to Cyber Peace; Cultural Pollution—The Erice Science for Peace Award Scientific Session

2010 — International Seminar on Nuclear War and Planetary Emergencies — 43rd Session: The Role of Science in the Third Millennium Energy: Nuclear Power Plants—Sustainable Systems / Pollution: Oil Spills – Air, Water and Waste Criteria / Climate: Paleoclimate Natural Variations and Implications – Sensitivity, Climategate, USA Legislation / Cosmic Objects: Countermeasures / Mitigation of Terrorist Acts: Scientific Contributions to Biosecurity / Lectio Magistralis: Science in Everyday Life / Information Security: Cyber Conflicts and Stability / The Erice Science for Peace Prize Award

2011 — International Seminar on Nuclear War and Planetary Emergencies — 44th Session: The Role of Science in the Third Millennium: World Energy Crisis - Global Nuclear Energy Issues after Fukushima - Energy Efficiency / Energy and Pollution - Unconventional Natural Gas: Benefits and Risks / Climate - Cosmic Rays and Climatic Processes / Water and Pollution - Water Scarcity and Pollution - Contami nants of Emerging Concern / Information Security - The Role of Science in Information Technologies and Internet Tools in Developing Countries Food, Soil & Medicine - Greenhouse Gases Consequences - Evidence-Based Third Millennium Medicine / Lectio Magistralis - The Role of Science in the Third Millennium

2012 — International Seminar on Nuclear War and Planetary Emergencies — 45th Session: Lectio Magistralis Global Warming Emergency/Why Science is Needed for the Culture of the Third Millennium/ - Mitigation of Terrorist Acts – Improvised Nuclear Devices – Gap Analysis of EU – Precenting Nuclear Explosive Terrorism - Public Perception/Climate & Climate Economics/Global Nuclear Energy Issues – Outlook in Japan, China, Europe and USA – Sustainable Nuclear Energy Systems – Economics of Nuclear Power/Energy & Sustainability in Cities – Global Sustainability - Smart Grids – Complexity/Global Food Production & Forest Dynamics/Water, Pollution & Terrorism/Brain Aging & Behavior/ Frontiers in Fast Computing & Informatics/ Information Security

*For the complete list of titles in this series, please go to:
http://worldscientific.com/series/scsnsp

THE SCIENCE AND CULTURE SERIES

Nuclear Strategy and Peace Technology

INTERNATIONAL SEMINAR ON NUCLEAR WAR AND PLANETARY EMERGENCIES

45th Session

THE ROLE OF SCIENCE IN THE THIRD MILLENNIUM

Lectio Magistralis Global Warming Emergency / Why Science is Needed for The Culture of The Third Millennium / Mitigation of Terrorist Acts – Improvised Nuclear Devices – Gap Analysis of EU – Preventing Nuclear Explosive Terrorism – Public Perception / Climate & Climate Economics / Global Nuclear Energy Issues – Outlook in Japan, China, Europe and USA – Sustainable Nuclear Energy Systems – Economics of Nuclear Power / Energy & Sustainability in Cities – Global Sustainability – Smart Grids – Complexity / Global Food Production & Forest Dynamics / Water, Pollution & Terrorism / Brain Aging & Behavior / Frontiers in Fast Computing & Informatics / Information Security

"E. Majorana" Centre for Scientific Culture

Erice, Italy, 19 – 24 August 2012

Series Editor and Chairman: A. Zichichi

Edited by R. Ragaini

World Scientific

NEW JERSEY • LONDON • SINGAPORE • BEIJING • SHANGHAI • HONG KONG • TAIPEI • CHENNAI

Published by

World Scientific Publishing Co. Pte. Ltd.

5 Toh Tuck Link, Singapore 596224

USA office: 27 Warren Street, Suite 401-402, Hackensack, NJ 07601

UK office: 57 Shelton Street, Covent Garden, London WC2H 9HE

British Library Cataloguing-in-Publication Data
A catalogue record for this book is available from the British Library.

INTERNATIONAL SEMINAR ON PLANETARY EMERGENCIES — 45TH SESSION: THE ROLE OF SCIENCE IN THE Third Millennium: LECTIO MAGISTRALIS GLOBAL WARMING EMERGENCY/WHY SCIENCE IS NEEDED FOR THE CULTURE OF THE THIRD MILLENNIUM/ - MITIGATION OF TERRORIST ACTS – IMPROVISED NUCLEAR DEVICES – GAP ANALYSIS OF EU – PREVENTING NUCLEAR EXPLOSIVE TERRORISM – PUBLIC PERCEPTION/CLIMATE & CLIMATE ECONOMICS/GLOBAL NUCLEAR ENERGY ISSUES – OUTLOOK IN JAPAN, CHINA, EUROPE AND USA – SUSTAINABLE NUCLEAR ENERGY SYSTEMS – ECONOMICS OF NUCLEAR POWER/ ENERGY & SUSTAINABILITY IN CITIES – GLOBAL SUSTAINABILITY – SMART GRIDS – COMPLEXITY/GLOBAL FOOD PRODUCTION & FOREST DYNAMICS/ WATER, POLLUTION & TERRORISM/BRAIN AGING & BEHAVIOR/FRONTIERS IN FAST COMPUTING & INFORMATICS/INFROMATION SECURITY

ISBN 978-981-4531-77-1

In-house Editor: Rhaimie Wahap

Printed in Singapore by B & Jo Enterprise Pte Ltd

CONTENTS

7. WATER, POLLUTION & TERRORISM

8. BRAIN AGING & BEHAVIOR

SPECIAL SESSION

LECTIO MAGISTRALIS

THE MANMADE CONTRIBUTION TO ONGOING GLOBAL WARMING IS NOT A PLANETARY EMERGENCY*

PRESIDENT VÁCLAV KLAUS
President of the Czech Republic
Prague, Czech Republic

Many thanks for the invitation to attend your conference and to speak here. I appreciate that a mere politician, a former economist, has been invited to address this well-known gathering of highly respected scientists. If I understand it correctly, this year's seminar is devoted to the discussion of the role of science and of "planetary emergencies".

To the first topic, **I want to say very clearly that I don't see a special role for science which would be different from doing science.** I have, of course, in mind "normal science", not a "post-normal science" whose ambitions are very often connected with political activism. The role of scientists is not in speculating on the probabilities of events that cannot be directly measured and tested, nor in promoting a pseudo-scientific "precautionary principle", nor in engaging in activities which are the proper function not of scientists but of risk managers.

To the second topic, I have to say that as a conservatively-minded person, I am unaware of any forthcoming "planetary emergency", with the exception of those potential situations which would be the consequences of human failures—of human fanaticism, of false pride, and of lack of modesty. But these are problems of political systems and of ideologies.

This brings me to the topic of my speech. **I will try to argue that current as well as realistically foreseeable global warming, and especially Man's contribution to it, is not a planetary emergency which should bother us.**

I am not a climatologist, but the IPCC and its leading spokespersons are not climatologists either. I am content to be a consumer of climatology and its related scientific disciplines. In this respect, I am located—in the economic jargon—on the demand side of climatology, not on the supply side.

There are many distinguished scientists here, and some of them are on the other side. I have no intention to break into their fields of study. By expressing my doubts about a simple causal relationship between human CO_2 emissions and climate, I do not have the slightest ambition to support one or another competing scientific hypothesis concerning the factors leading to global warming (or eventually cooling).

Nevertheless, my reading both of the available data and of conflicting scientific arguments and theories allows me to argue that it is not global warming caused by human activity that is threatening us.

My views about this issue have been expressed in a number of speeches and articles in the last couple of years all over the world. The book **"Blue Planet in Green Shackles"**[1] has already been published in 18 languages, last month even in Indonesian.

---

[1] Klaus, V.: *Modrá, nikoli zelená planeta: Co je ohroženo, klima nebo svoboda?*, Praha, Dokořán, 2007; English version: *Blue Planet in Green Shackles*, Competitive Enterprise Institute, Washington DC, 2008. Italian version: *Pianeta blu, non verde*, IBL Libri, Torino, 2009.

The subtitle of the book asks, **"What is Endangered: Climate or Freedom?"** The real problem is not climate or global warming, but the **Global Warming Doctrine** and its consequences. They may eventually bring us close to a real planetary emergency. Absolutely unnecessarily, without any connection with global temperature.

This doctrine, **as a set of beliefs, is an ideology, if not a religion.**[2] It lives independently on the science of climatology. Its disputes are not about temperature, but are part of the "conflict of ideologies". Temperature is used and misused in these disputes. The politicians, the media and the public—misled by the very aggressive propaganda produced by the adherents of the global warming doctrine—do not see this. It is our task to help them to distinguish between what is science and what is ideology.

The global warming doctrine believers have not yet presented its authoritative text, its manifesto. One of the reasons is that no one wants to be explicitly connected with it. Another is that to put such a text together would be difficult because this doctrine is not a monolithic concept which can be easily summarized. Its subject-matter does not belong to any single science. It presents itself as a flexible, rather inconsistent, **loosely connected cascade of arguments**, which is why it has quite successfully escaped the scrutiny of science. It comfortably dwells in the easy and self-protecting world of false interdisciplinarity which is really a non-disciplinarity, it is an absence of discipline.

My reading of this **new incarnation of environmentalism** can be summarized in the following way:

1. It starts with the claim that there is **an undisputed and undisputable, empirically confirmed, statistically significant, global, not local, warming;**
2. It continues with the argument that the time series of global temperature exhibit a growing trend which dominates their cyclical and random components. This trend is supposed to be non-linear, perhaps exponential;
3. This trend is declared to be dangerous for the people (in the eyes of "soft" environmentalists) or for the planet (by "deep" environmentalists);
4. This temperature growth is postulated as a solely or chiefly man-made phenomenon attributable to growing emissions of CO_2 from industrial activity and the use of fossil fuels;
5. The sensitivity of global temperature to even small variations in CO_2 concentration in the atmosphere is supposed to be very high;
6. The GWD exponents promise us a solution: **the ongoing temperature increase can be reversed by radical reduction in CO_2 emissions;**[3]
7. They also know how to bring about their solution: they want to organize emissions reduction by means of the institutions of "global governance". They forget to tell us that this is not possible without undermining democracy, the independence of individual countries, human freedom, economic prosperity and a chance to eliminate poverty in the world;
8. They rely on the undefined and undefinable "precautionary principle". Cost-benefit analysis is not relevant to them.

---

[2] I was recently in California. In my hotel room, Al Gore´s book *An Inconvenient Truth* was next to the Bible.

[3] This is what Ray Evans calls, "The Theory of Climate Control", *Quadrant*, No. 3, 2008.

This simple scheme can be, undoubtedly, improved, extended, supplemented or corrected in many ways, but I believe its basic structure is fair and correct.

I do not believe in any one of these eight articles of faith and I am not alone. There are many natural scientists and also social scientists, especially economists, who do not believe in them either. The problem is that most genuine scientists do science and are not willing to discuss this doctrine in the public space.

An additional problem is that natural scientists and social scientists do not talk to each other. They only come into contact with self-proclaimed interdisciplinarists who are very often mere dealers in second-hand ideas. Social scientists, in particular, tend to be silenced by seemingly authoritative statements that "the science is settled", while natural scientists assume a priori that there is nothing "hard" in the social sciences.

Politicians—after having abandoned other ideologies—heartily welcomed this new one. They became rapidly convinced that playing the global warming card is an easy game to play, at least in the short or medium run. They hoped the voters would appreciate their caring about issues more serious than the next elections. The problem is that the politicians (to say nothing about the media) do not take into consideration the long-term consequences and costs of measures demanded by this doctrine.

How to make a change? I dare say that science itself will not make the change, regardless of its achievements. The Global Warming Doctrine is not based on science. Accordingly, scientific debate itself cannot bring it into disrepute. The course of the world-wide global warming debate more or less confirms this elementary methodological argument. Serious scientific research continues to bring us new pieces of knowledge almost on a daily basis, but it has not brought and will not bring us any decisive breakthrough in the public debate on this topic. Climate is a complex system. In spite of the dreams of believers in general systems theory, any scientific discovery concerning this topic will always be only a partial one.

Can a decisive change come as a result of new empirical data? I doubt it. It is evident that the current temperature data confirm neither the alarmist and apocalyptic views of the believers in the GWD, nor their quasi-scientific hypotheses about the exclusivity of the relationship between CO_2 and temperature. The world has not warmed for the last 15 years, but that is too short to shatter the whole carefully built edifice of the global warming doctrine. Moreover, it should not be forgotten that some of us have been arguing that a century in climatology is too short to prove the ongoing global warming as a new long-term trend. That is why, for the sake of symmetry, we must accept that a decade is not sufficient to do the opposite.

Discussing technicalities in more and more depth will not help us, because the supporters of the global warming doctrine are not interested in them. We are not dealing with people who are authentically interested in science, in objective truth, in identifying the causes of incremental changes in temperature. For them, the temperature data are just an instrument in their plans to change the world, to suppress human freedom, to bring people back to underdevelopment. **Their ideas are the ideas of ideologues**, not of scientists or climatologists. Data and theories, however sophisticated, will not change their views.

I mentioned my economic background. Let me turn attention to what the field of economics—with all of its internal disagreements—says to that:

1. Economists believe in the rationality and efficiency of the spontaneous decisions of millions of individuals. They believe in "the wisdom of people" rather than in the wisdom of governments and of their scientific advisors. They do not deny that market failures happen, but they have many reasons to argue that government failures are bigger and much more dangerous than market failures. They consider that jumping on the bandwagon of the global warming doctrine is an example of a serious government failure which undermines markets, human freedom and human prosperity;
2. Economists, at least since Frederic Bastiat, have considered it their duty to warn policymakers against unintended consequences and against failing to differentiate between what is seen and what is not seen;
3. Economists have at their disposal a rather developed subdiscipline called "energy economics". They know something about scarcity, as well as about prices, and they have to warn governments against playing with them.
4. Economists believe in rational risk-aversion, not in the precautionary principle;
5. Economists are aware of externalities and have worked with them for a long time. It is their own concept: it was not discovered by environmentalists. They consider it dangerous in unqualified hands. After decades of studying it, they do not see the world as full of negative externalities a priori;
6. Economists base their thinking about intertemporal events on a rather sophisticated concept of discounting. It was the misunderstanding of discounting in the climatologic modeling that brought me into the subject of global warming some years ago;
7. Economists have some undeniable experience with the analysis of time series. Statistical and econometric methods used in economic analysis are full of sophisticated models not used in natural sciences, because these are based mostly on the analysis of cross-section data samples. They know something about the problems with the imperfect quality of data, about measurement errors, about data mining, about the precariousness of all kinds of averages and other statistical characteristics. They also have some experience with computer modelling in complex systems, with pseudocorrelations, with the sensitivity of parameter adjustments, etc. For that reason they are convinced they have the right to comment on the statistical analyses of climatologists.[4]

Based on all that:

1. Economists do not see the outcome of the cost-benefit comparisons of CO_2 emission reductions as favourably as the adherents of the global warming doctrine. They know that energy demand and supply patterns change only slowly. They see the very high degree of stability of the relationship between man-made carbon dioxide emissions, economic activity and emissions intensity, and possess no hypothesis for expecting a radical shift in this relationship. **Emissions intensity (as a macro-phenomenon) moves**

---

[4] I would like to mention at least R. McKitrick and S. McIntyre and their attack on the bastion of the GWD, on the so called "hockey stick".

only very slowly and does not make miracles. The very robust relationship between CO_2 emissions and the rate of economic growth is here, and is here to stay.

If somebody wants to reduce CO_2 emissions, he must either expect a revolution in economic efficiency (which determines emissions intensity) or start organizing a world-wide economic decline. Revolutions in economic efficiency—at least in relevant time horizons—have never been realized in the past and will not happen in the future either. It was the recent financial and economic crisis, not a technological miracle or preaching by the IPCC, that brought about a slight—and probably temporary—reduction in CO_2 emissions. The GWD adherents should explain to the people worldwide that to achieve their plans economic decline is inevitable.

2. The relationships studied in natural sciences are not influenced by subjective valuations of the variables in question, nor by any rational (or irrational) behaviour, nor by the fact that people make choices. In social or behavioural sciences, it is more difficult. To make rational choices means to pay attention to intertemporal relationships and to look at opportunity costs. It is evident that **by assuming a very low, near-zero discount rate the proponents of the global warming doctrine neglect the issue of time and of alternative opportunities.**

A low discount rate used in global warming models means harming current generations (vis-à-vis future generations). Undermining current economic development harms future generations as well. Economists representing very different schools of thought, from W. Nordhaus at Yale[5] to K. M. Murphy at Chicago,[6] tell us convincingly that the discount rate—indispensable for any intertemporal calculations—should be around the market rate, around 5%, and that it should be close to the real rate of return on capital, because only that rate reflects the true opportunity cost of climate mitigation.

We should not accept claims that by adopting low discount rates we "protect the interests of future generations",[7] or that opportunity costs are irrelevant because in the case of global warming "the problem of choice does not exist" (p. 104). This uneconomic or perhaps anti-economic way of thinking must never be accepted.

3. As someone who personally experienced central planning and attempts to organize the whole of society from one place, I feel obliged to warn against the arguments and ambitions of the believers in the global warming doctrine. Their arguments and ambitions are very similar to those we used to live with decades ago under Communism. The arrogance with which the global-warming alarmists and their fellow-travellers in politics and the media present their views is appalling. They want to suppress the market, they want to control the whole of society, they want to dictate prices (directly or indirectly by means of various interventions, including taxes), they want to "use" the market. I agree with Ray Evans that we experience the "Orwellian use of the words 'market' and 'price' to persuade people to accept a control over their lives".[8] All the

---

[5] *A Question of Balance: Weighing the Options on Global Warming Policies*, Yale University Press, June 2008.

[6] Some Simple Economics of Climate Changes, paper presented to the MPS General Meeting in Tokyo, September 8, 2008.

[7] M. Dore: "A Question of Fudge", *World Economics*, January–February 2009, p. 100.

[8] "The Chilling Costs of Climate Catastrophism", *Quadrant*, June 2008.

standard economic arguments against such attempts should be repeated. It is our duty to do it.

To conclude, I agree with many serious climatologists who say that the warming we may expect will be very small. I agree with Bob Carter and other scientists that it is difficult "to prove that the human effect on the climate can be measured" because "this effect is lost in the variability of natural climate changes".[9] **Provided that there are no irrational attempts to mitigate the human effect on global temperature, the economic losses connected with the warming we may expect will be very small.** The loss generated as a result of the completely useless fight against global warming would be far greater.

---

[9] Heartland Institute's International Conference on Climate Change, New York City, March 2009, p. 23. Professor Carter's arguments are more developed in his book "Climate: The Counter Consensus", Stacey International, London, 2010.

OVERCOMING THE PLANETARY EMERGENCIES IS POSSIBLE BECAUSE SCIENCE, TECHNOLOGY AND CULTURE ARE FOUNDED ON LIBERTY, RIGHT AND TRUTH

HON. ALESSANDRO PAGANO
Chamber of Deputies, Italian Parliament
Rome, Italy

Once again the International Seminar on Planetary Emergencies, held in Erice since 1981 thanks to the Centre for Scientific Culture "Ettore Majorana", has shown the way.

The aim of these Seminars is to apply the scientific know how in accordance with the concept of science advocated by the Centre for Scientific Culture "Ettore Majorana", presided by Professor Antonino Zichichi, president of the World Federation of Scientists (WFS), which is "science without secrets or frontiers".

This year, besides the participation of 110 scientists from over 100 Laboratories and University Centres of 40 countries, the Seminar offers the presence of a few political figureheads from various fields of experience, to contribute to congressional activities. Professor Vàclav Klaus, president of the Czech Republic, the Interior Minister Anna Maria Cancellieri, two members of the House of Parliament, Alfred Mantovano, previous member of various national governments and Alessandro Pagano, previous member of various regional governments and of course the Secretary General of the Republic, Donato Marra. All have alternatively taken part in the 5 daily sessions, to debate on the various courses along which the world is moving.

The basic assumption on which the debate was started was the following: Facts are proving that science, technology and culture alone are not enough for the solution of the multiple problems rooted in the political and economic violence of the past century, which are still afflicting the society of this new millennium. And this is due, as explained authoritatively by Professor Zichichi, to *the culture so-called modern and afflicted with ideologies to the point of negating the evidence and reasoning as was done in pre-Aristotelian times.*

In fact, modern society is lacking in the rigorous evaluation of the problems afflicting our life and our planet. Energy, water, environmental pollution, climate, terrorism, demographic decline in industrialised countries, economic crisis and so on, are only a few of the 71 problems identified and studied by the WFS scientists. This abnormal number of emergencies cannot be a coincidence.

In other, more precise terms, the decline which our society is facing has its origin in the deep interior conflict and the lack of values which permeate and dominate modern man. The recent global economic crisis, just to mention one that is still particularly actual, is the reflection of a much larger, deep and complex crisis which has affected modern man: an existential, ethical and anthropological crisis. This crisis, which has bewildered man, is *universal* since all people have been affected in a major or minor way; *deep* because it strikes at the essence of man's soul and all domains of man's action; and *complex* because it renders chaotic and inextricable the tangle of all these apparently insane forces and is a *process* that, more or less rapidly, could erode and destroy all the positive potentialities.

But why has this crisis exploded into being? This is the major question posed by scientists who, beyond their uncontested scientific competence, are aware of the existence of "something" that conditions negatively the behaviour of this form of living matter to which we belong.

Alfredo Mantovano tried to provide an answer by stating: *"With realism it is reminded that the essence of totalitarianism coincides with the transformation of a lie into truth through the will exercised by one man unto another. One would outright think about the license taken by those who kill another being, weaker species (…). Science and technology are still being put at the service of death with instruments of mass destruction, but also—simultaneously—for sterile medicine, away from cameras and inconvenient testimonies".*

Finally, all the invited politicians and the emissary of the President of the Italian Republic found themselves to be in agreement with Professor Zichichi's concrete proposal and the support the WFS project *"Science, Technology and Culture in the Third Millennium: a Project for Mankind"*, that aims at combining, beyond the three most important components for the development of society which are science, technology and culture, also the three essential moral components of man, which are liberty, right and above all truth, so as to ensure that new technologies are always "pro" and never "against" life.

As taught by the ancient philosophers, that find in Aristotle their supreme component, only *Truth*, which is the "correspondence of matter and intellect", can bring order to the situation. And, in the Immanent, the most genuine source of truth is Science which—it should never be forgotten—was born from an act of faith in the existence of the Author of the Logic which rules the world, from the subnuclear universe to that made of stars and galaxies. If the culture of our times, so-called modern but in fact is pre-Aristotelian, was in stride with the great conquests of Science we wouldn't be facing the 71 Planetary Emergencies which must be fought and resolved.

SESSION 1

OPENING SESSION

«ETTORE MAJORANA» FOUNDATION AND CENTRE FOR SCIENTIFIC CULTURE

INTERNATIONAL SEMINARS
ON PLANETARY EMERGENCIES
45th Session

THE ROLE OF SCIENCE IN THE THIRD MILLENNIUM

Erice 19 - 25 August 2012

**SOTTO L'ALTO PATRONATO
DEL PRESIDENTE DELLA REPUBBLICA ITALIANA**

OPENING SESSION

Erice, 20 August 2012

WHY SCIENCE IS NEEDED FOR THE CULTURE OF THE THIRD MILLENNIUM

Antonino Zichichi
INFN and University of Bologna, Italy
CERN, Geneva, Switzerland
Enrico Fermi Centre, Rome, Italy
World Federation of Scientists, Beijing, Geneva, Moscow, New York

Paul A.M. Dirac Lecture Hall – Patrick M.S. Blackett Institute
Erice, 20 August 2012

WHY SCIENCE IS NEEDED FOR THE CULTURE OF THE THIRD MILLENNIUM

ANTONINO ZICHICHI
INFN and University of Bologna, Italy, CERN, Geneva, Switzerland, Enrico Fermi Centre, Rome, Italy, World Federation of Scientists, Beijing, Geneva, Moscow, New York

THE CULTURE OF OUR TIME AND THE ONE NEEDED

The Culture of our time is based on the first great achievement of our intellect: Language. Language means Poetry, Literature, Music, Arts, Theatre, Economy, Politics and other manifestations of the human intellect such as Philosophy; all intellectual activities that could exist even if neither rigorous Logic nor Science had ever been discovered.

Rigorous Logic and Science must be brought into the cultural patrimony of the third millennium, since the number one enemy of humanity is Ignorance.

The first monumental construction of rigorous Logic is Euclidean geometry—more than 2,000 years ago. Rigorous Logic includes arithmetic (theory of numbers), algebra (theory of variables), analysis (theory of functions), and topology (theory of domains where the functions exist).

None of these activities could be discovered if our intellectual activities were limited in the domain of Language. The next point to realize is that all activities in the domain of rigorous Logic would exist even if Science had never been discovered. In fact, Science is the only rigorous Logic which has been used in order to create the world as it is.

There are many logical theoretical structures which are not found in Nature, e.g., a space with infinite dimensions does not lead to any self-contradiction. The space with infinite number of dimensions exists from a rigorous theoretical logic point of view, but the space where we live has a finite number of dimensions apparently four (three of space and one of time), but probably 43, if the Superworld exists.

If we physicists, using the most powerful Subnuclear Physics collider—LHC (Large Hadron Collider)—with its high precision technological instruments will be able to have the experimental evidence for the existence of the Superworld, new horizons will be opened to our vision of the world where we leave.

I hope this remark allows us to make clear the distinction between rigorous Theoretical Logic (mathematics) and rigorous Experimental Logic (Science). Out of all possible Rigorous Theoretical Logic structures only one has been chosen to build the world as it is. This Logic is Science.

The reason why Science can be an instrument for peace is the fact that Science is the most rigorous way which allows one to distinguish a great "pure" achievement of the human intellect from the use we can make of it.

Science is neither good nor evil. People currently use the word "science" to mean the "use of science" (i.e., technology) which is no longer "science", just as the "use of language" is no longer "language" and the "use of logic" is no longer "logic". Let me elaborate further on this point since the Culture of our time, called Modern Culture, has never devoted the necessary attention to distinguish Science from Technology.

An example of great achievement in "pure" Science is the discovery of the "Standard Model", i.e., the superb synthesis that can explain all phenomena of our world in terms of three Fundamental Structures (called the three families of elementary particles) and three Fundamental Forces of Nature (the electroweak, the strong subnuclear and the gravitational). See Addenda I.a–II.b.

An example of great achievement in "pure" Logic is the invention of the rigorous and formidable logical structure called the Infinite.

Examples of great achievement in "pure" Language are the *Pietà* by Michelangelo, the *Primavera* by Botticelli, the *Ninth Symphony* by Beethoven, the *Fifth Symphony* by Mahler, the *Divina Commedia* by Dante. All these achievements in "pure" Science, Logic and Language are for mankind.

Applied Science, Logic and Language do exist.

These "applications" can be for and against mankind. For example the use of Science against mankind, corresponds to war technology.

The use of Logic against mankind corresponds to the invention of computer technology for the control of billions of people, thus destroying individual freedom.

The use of Language against mankind corresponds to the invention of ideologies like Stalinism and Nazism.

Examples that are for mankind do indeed exist in the domain of Applied Science, Language and Logic.

Peace technology, instruments to help mankind in Medicine, Agriculture and Meteorology, all correspond to Applied Science for mankind.

Applied Logic for mankind corresponds to the invention of new robots to avoid dangerous and unpleasant work for man, of specialized computer-based applications for medicine, for weather forecasting, etc.

Applied Language for mankind corresponds to the performance of pieces of poetry, of music and of art, plus all other activities designed to help increase the quality of everyday life in all sectors that would exist even if Logic and/or Science had never been discovered.

The components of Applied Logic for mankind and Applied Science for mankind contribute enormously to improve the quality of life.

The three pillars (Language, Logic, Science) in their applied part for man, have a common task and reach the same goal, i.e., to improve the quality of everyday life.

Science, in the Immanent (therefore without appealing to any existential topic connected with the Sphere of the Transcendent), is the source of a new hope, well rooted in the Logic of Nature with its Fundamental Laws, which all aim at good and never at evil.

But for this to happen also in Applied Science, it is necessary that the technological applications of the great scientific discoveries be entrusted not to political bodies, but to the scientific community itself.

This has never been the case because Applied Science implies decision-making actions and this means political activity, which belongs to Applied Language.

This can be for mankind (Democracy) and against mankind (Dictatorship). It is Dictatorship that produces weapons and war technology, nevertheless everybody considers us scientists responsible for Applied Science against mankind.

Worse still, people think that Science was born of Technology. Some would even say that we scientists would not be here, if it were not for military technological development research work.

The claim, by the fellows responsible for the Culture of our time, that Technology precedes Basic Science is due to the fact that the "fire" and the "wheel" were invented before Galilei discovered Science: i.e., the Logic of Nature.

They never say that the "fire" and the "wheel" were understood after Basic Science had been discovered by Galilei. To "invent" a new instrument does not necessarily mean understanding "why" it works.

The spectacular successes in the construction of Pyramids and other masterpieces of Architecture, all over the world, did not give rise to the discovery of the first pieces of the Logic of Nature, such as the Principle of Inertia and the other two laws of Mechanics. A discovery in Basic Science corresponds to "understanding" all possible instruments that can be invented.

For example, the discovery of a Fundamental Force of Nature, the electromagnetic force, has enabled us to understand that all our senses (sight, hearing, smell, taste and touch) are manifestations of the same fundamental force.

This force originates from a unique entity, called "the electric charge". If we could switch off this "charge" our five senses would cease to exist. Thus all present and future "inventions" connected with sight, hearing, smell, taste and touch are understood even before they are really implemented.

Before Basic Science was discovered, thanks to Galileo Galilei, it was the other way around; i.e.,: since the Fundamental Laws of Nature had not been discovered, the technological inventions were always rotating around the same two original ones, the "fire" and the "wheel".

Since neither the "fire" nor the "wheel" were "understood", the technological development could not produce anything really new.

And this went on until Galilei in the 17th Century started to discover the first Fundamental Laws of Nature.

This is how, in just four centuries, we have implemented the enormous number of technological inventions that are a part of our daily life: telephone, TV, computers, internet, and all such instruments that would take pages and pages to list. This list is the proof that the motor of progress is scientific discovery.

If the quality of life in the industrialized world has reached the level never realized before, this is due to the enormous number of scientific discoveries in four centuries of Science. If fundamental Science stopped discovering new phenomena, the quality of life would stop improving.

Language, Logic and Science, all contribute to improve the quality of our life, but the decisive role is due to Science, since Applied Science for mankind is the real **motor for progress**.

An example: when Galilei started to discover the Logic of Nature, mankind was measuring the Time using the sundial, with errors of at least a few seconds per day. Galilei having discovered the laws governing the oscillation of the pendulum made great advancements in the precision of time measurement and this enabled us in just four centuries to reach the present day precision of one second per Universe lifetime, i.e., 20 billion years.

During the many millennia preceding the discovery of Science, all civilizations were measuring time with the same error of one second per day. If Science had never been discovered we could never have reached the vision of the Universe as it looks through our instruments, from the structure of the proton to the farthest Galaxy.

This is synthetically represented in Figures 1 and 2 shown here.

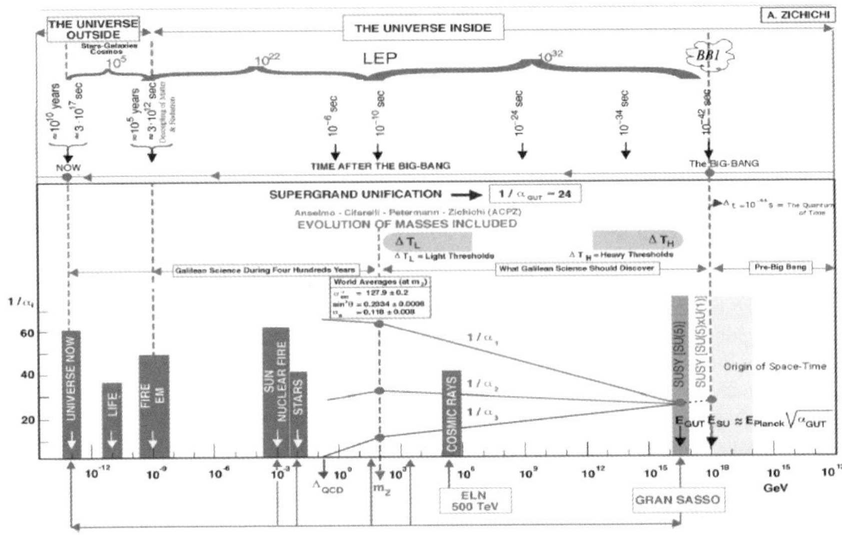

Fig. 1.

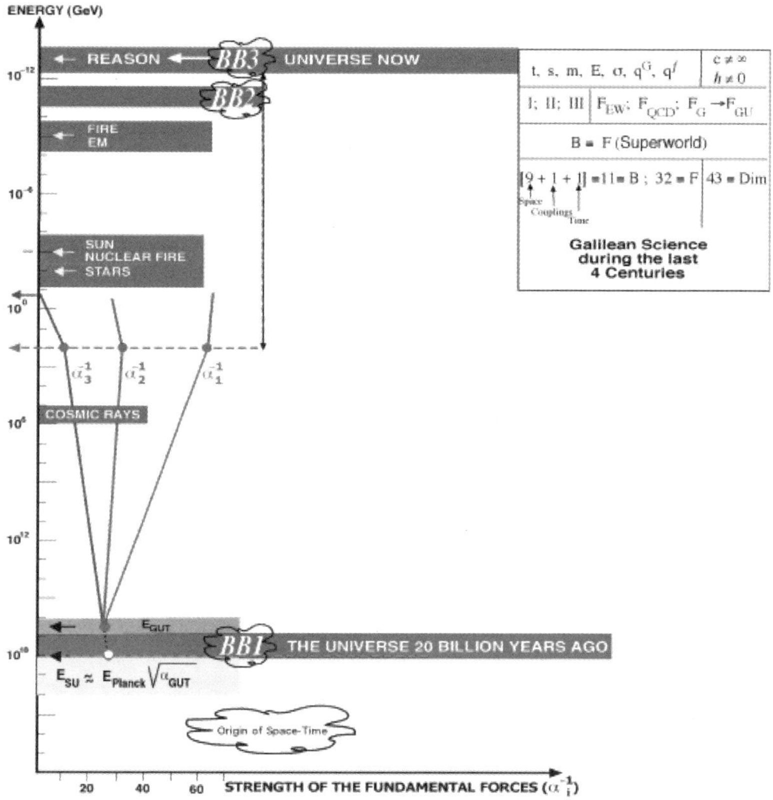

Fig. 2.

Figure 1: The convergence of the Fundamental Forces of Nature are reported by the three lines of the Figure. For this convergence it is necessary the hypothesis that the Superworld exists. We see that the "Universe" illustrated in this figure consists of many important details. The "Universe outside" is the one which comes after the decoupling of protons, electrons and photons, when atoms started their formation, 380 thousand years after Big Bang-1. The mathematical basis, which includes the hypothesis of the Superworld, is reported in Figure 3.

Figure 3: The lines are the result of calculations executed with a supercomputer using the following system of three weakly coupled differential non-linear equations:

$$\mu \frac{d\alpha_i}{d\mu} = \frac{b_i}{2\pi} \alpha_i^2 + \sum_j \frac{bij}{8\pi^2} \alpha_j \alpha_i^2$$

This system describes the evolution of all phenomena including the Superworld, from the maximum level of energy, EGUT, to our world which is at the minimum energy level.

Fig. 3.

To achieve this superb synthesis of human knowledge Reductionism has been the Basic Logic: to solve one problem and only one at the time. The latest news coming from the culture of our time, called Modern Culture, is that all Fields of Science need HOLISM not REDUCTIONISM.

Modern Culture claims that from proteins up to History **everything interacts with the environment**.

In fact **History** needs us, the living form of matter, L, **i.e., human being**, whose roots are in the most elementary form of life, the so called "minimal life cell", C, for which genes and proteins are needed: **everything interacts with the environment**.

But proteins and all other pieces needed for L are made with molecules and from molecules down to the Basic Constituents, the interaction with the environment is zero: the greatest achievements of Science are coming from the study of our Constituents which do not interact with the environment.

The Modern Culture neglects what is shown in Figure 4: i.e., from molecules, down to the Basic Quantities there is no interaction with the environment and all achievements of our knowledge come from Reductionism, started four centuries ago.

I INTERACTIONS WITH ENVIRONMENT FROM BASIC QUANTITIES TO LIVING MATTER

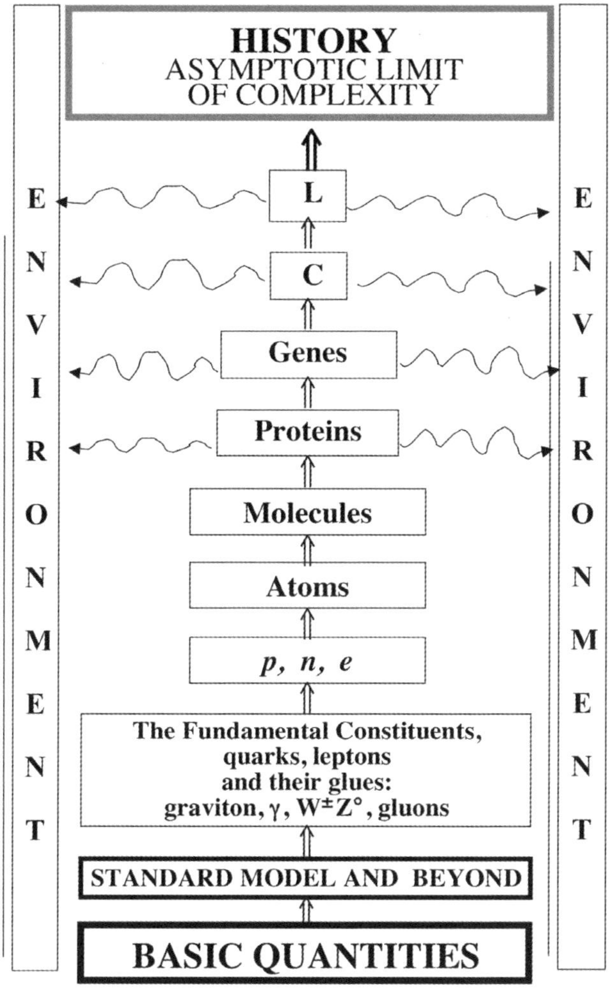

Fig. 4.

The Culture of our Time neglects that Holism has been dominant for 10,000 years, without ever being able to discover any Fundamental Law of Nature, as reported in Figure 5.

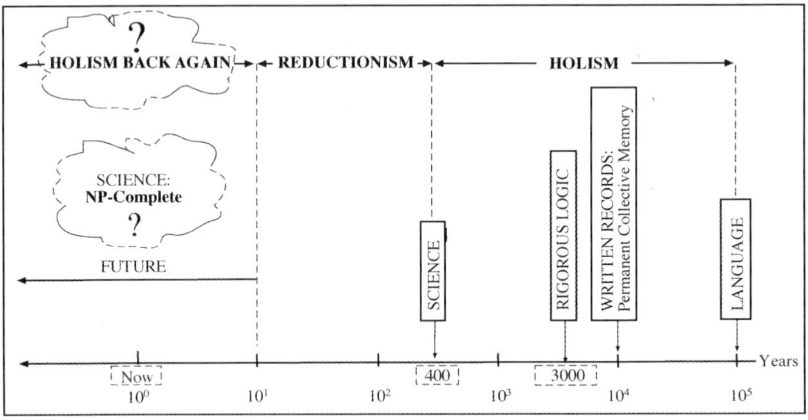

Fig. 5.

It turns out that Complexity in the real world exists, no matter the mass-energy and space-time scales considered, as illustrated in the Figure 6. This result is coming thanks to REDUCTIONISM.

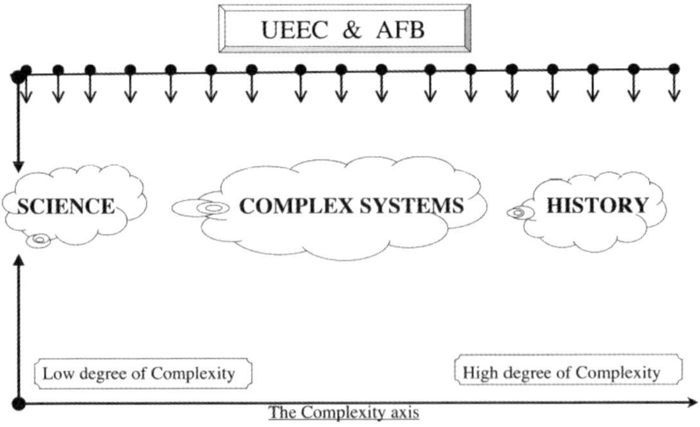

Fig. 6.

To the great achievements of Reductionism have contributed many distinguished scientists, some of them are here: T.D. Lee the Co-chair of our Seminars; Dick Garwin, Dick Wilson and many other scientists who have been during many decades here in Erice at the Subnuclear Physics Courses.

To these great achievements of Science correspond the 71 Planetary Emergencies to which we will return later. Let me explain how these achievements of Science are related to our Project. This is in fact the logic of our Project.

II THE LOGIC OF OUR PROJECT

The Logic of our Project is that Basic Science and Technology should be dedicated to the fight against Planetary Emergencies by promoting knowledge transfer and education, which in turn will allow the increase the standards of living of a Nation. Let me quote a statement by Abdus Salam a co-founder of WFS(1): "On science and technology depend the standards of living of a nation." This Statement explains how are the achievements of Science, as it the Standard Model and Beyond, related to our Project.

Planetary Emergencies cannot be solved without having with us the best fellows from developing Countries.

1) New projects within the Basic Science and Technology program of the PROJECT for MANKIND.

The ICSC-World Laboratory works on the principle that one of the better ways of helping Developing Countries is to support the participation of their scientific elite in projects aimed at the solution of their particular problems, working in collaboration with their peers in Developed Countries and contributing to the advancement of science and human knowledge as a whole.

This program is closely related with CERN, the LAA project and R&D works for advanced experiments and technology.

Hundreds of bright young scientists from developing countries were given training in CERN, to help in the development of highly sophisticated detectors for the LAA, the ALICE and the R&D for Advanced Technologies experiments.

Specific new pilot projects should now be considered and implemented to promote research and innovation using technology as a tool for international development. For this I refer to the great success of the LAA project at CERN, where the recruitment of technicians, engineers, and physicists, dedicated to basic research, innovation and development, allow them to invent new detectors and initiate technologies, such as advanced electronics. Even today, some 25 years later, technological spin-offs triggered by the LAA project find spectacular new applications ([2])

2) Use the advances in communications technology, the informatics and computing frontiers.

Peoples training and the use of advanced computing and informatics technologies will help the economy and the development of a country.

A similar training effort was part of the very successful programs of the ICSC-World Laboratory in the past when supporting hundreds of students and fellows from Developing Countries to participate in top research experiments at CERN.

Today—in the times of distributed computing with GRIDS and CLOUDS—training can be organised remotely but requires highly developed informatics infrastructures and competent areal or local teams in each participating institution and country.

This has been demonstrated by the outstanding success of the LHC experimental teams of students when considering the speed with which the groups identified the Higgs signal within the Tsunami of data received via the computer grids.

Effective monitoring of Planetary Emergencies is helped by real-time collection and distribution of information. This in turn is the basis for prevention, response and recovery from emergencies.

3) The QGCW experiment proposal will allow the study of new accelerator techniques, imaging and prototype new detector concepts and find solutions for ultra fast time synchronization.

Examples of the classifical applications: fast timing synchronization of TOF detectors for medical treatments with PET detectors. The development of CRAB cavities for future colliders and energy recovery linear accelerators are only examples of planned global efforts to improve accelerator performance.

New ideas—such as plasma acceleration for example—need more efforts and bright students to get involved.

All these techniques are part of the accelerator R&D plans, which will undoubtedly lead to applications outside basic research experiments.

An important part of these global collaborations is also the work on improved fabrication methods and investment in superconducting magnets and cryogenic systems.

Another excellent project would be training of technicians and engineers in magnetic field measurements and in cold testing of magnets, as well demonstrated by a recent successful project of this type, launched for LHC between CERN and India.

4) Use the ongoing general progress in the domain of microelectronics, new sensors, and radiation detector development for use in health care, for the protection of environment and/or for energy saving measures.

Dedicated projects on focused training and education of students will help Developing Countries to master and improve techniques, which are essential for improved domestic health care efforts to avoid future emergencies.

5) A Permanent Monitoring Panel (PMP) on Technology and Innovation has been established in the basic field of new research developments in technological instruments and informatics. The PMP will be chaired by Professor Horst Wenninger.

His activities will start with the EPS-TIG workshop, scheduled at Erice in October 2012 aiming to *highlight - review - evaluate* technologies and devices, developed for use in basic research and identified as candidates for potential applications in health care and in production of innovative products. The invited keynote speakers are themselves involved in the development of innovative technologies or experts with good knowledge and overview of ongoing activities and speakers with experience in best practices of technology transfer.

All these projects are ideal training fields for new talents helping them to acquire competences in most advanced technologies.

Working on basic sciences motivated by curiosity results in competences required for applied sciences, designed to answer specific questions.

This is the reason why we have launched last year a competition to encourage our fellows, the world over, to think about Totally Unexpected Discoveries (TUD).

6) A competition the world over for young fellows to think about Totally Unexpected Discoveries (TUD).

Proposal: a competition worldwide for young fellows to start thinking about new problems.

Why? Because the Culture of our Time, called modern, is in fact pre-Aristotelian: neither **rigorous theoretical logic** (**Math**) nor **rigorous experimental logic** (**Science**) have the role they should have.

III THE FUNDS NEEDED AND THE STATUS OF THE PLANETARY EMERGENCIES

Let me remind you that **the funds needed** to study the **Planetary Emergencies** compete with other projects like **man going to Mars**. But, while billions of Dollars are spent for going to Mars practically nothing has been devoted to the study of the Planetary Emergencies.

In order to allow decision-making leaders to focus their attention on Planetary Emergencies, we need to convince them that the new Science, the so-called **"Science of Complexity"**, is in our hands.

This is the reason why the World Federation of Scientists, through the National Scholarship Programme, has established a competition for all young fellows to think about TUD (see Addenda).

PLANETARY EMERGENCIES

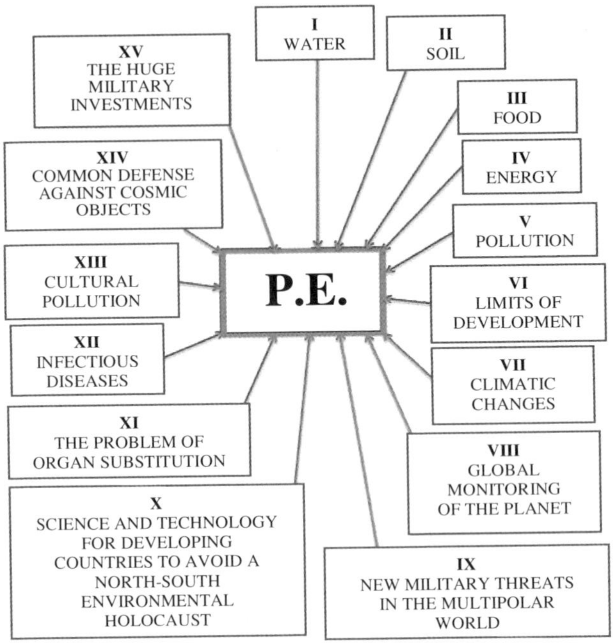

The Planetary Emergencies have been practically ignored by World Leaders, as can easily be deduced by the following telegraphic synthesis of worldwide affairs:

- Disarmament—22 years ago, we went from a bipolar world to a monopolar world and nothing seems to have changed. The arms race continues, local and regional wars are spreading out and the armament industry has not been regulated;
- Terrorism has thrived and terrorist cells have multiplied worldwide. Wars undertaken to contain and eliminate Terrorism were long and costly in terms of human casualties and economy, costing trillions of dollars, and have largely failed in achieving their aims. It seems to have instead exacerbated the problem and created well-trained and armed roving brigades of Jihadists, ready to lend support to every extremist party worldwide;
- Our Western economic systems are floundering, unable to restrain unbridled consumerism which, coupled to widely open markets, led to jobs being lost by the thousands and unemployment increasing in all levels of our societies. Banks have heavily contributed to the worsening of the economic situation by playing with markets in an uncontrolled fashion, generating losses by the trillions;

- Security at every level has become a real concern. Our civil societies have become more violent and their ethics have degraded. Unemployment plays a big role in some countries but, more generally, it is mostly due to moral degradation and lack of civic values. A new security concern has arisen in the last few years, which endangers what was shaping up as the most promising tool for our common future: the security of information. Hooliganism has thus entered the Golden informatics era;
- After a century of heavy industrialisation, pollution is now threatening our very existence. Our soils, our lakes, our rivers and oceans and the very air we breathe are polluted. Some of the chemicals used for agriculture and industry are causing mortal diseases and malformation in fetuses;
- Our earth is already now heavily overpopulated and the increase in population will accelerate. Natural non-renewable resources have become scarce and agriculture needs urgent action in order to be able to feed our population. There is a crying need for worldwide control to ensure a sustainable future;
- Despite the progress achieved in diagnosis and cure, new factors such as pollution, transport globalisation and overcrowding are creating new emergencies for health. For commercial reasons, some of the new diseases and drug-resistant bacteria are not being investigated as they should be;
- Climate has become a worldwide concern. Projections show a trend towards a global warming of the planet which needs two inputs: firstly to be confirmed and precisely evaluated and, secondly, to determine the percentage of the impact of human activities which is far from being quantitatively established. What is needed is an urgent worldwide preparation to mitigate consequences and adapt to possible changes.

The Planetary Emergencies have been at the centre of our attention since 1981 and here we are again, one hundred and eleven scientists from 40 countries, gathered in Erice to analyse a series of crucial multidisciplinary scientific issues. They are all part of the 71 Planetary Emergencies identified by the World Federation of Scientists 25 years ago.

Armenia
Australia
Austria
Azerbaijan
Belgium
Canada
Colombia
Czech Republic
Denmark
Egypt
France
Germany
Greece
India
Israel
Italy
Japan
Jordan
Kenya
Korea
Kosovo
Kuwait
Kyrgyzstan
Malaysia
Mexico
P.R. China
Philippines
Republic of
Moldova
Romania
Russia
Senegal
Slovenia
South Africa
Spain
Switzerland
Tajikistan
UK
Ukraine
USA
Vietnam

The main topics of this 45th Session are:

Why Science is Needed for the Culture of the Third Millennium (Xiii)
Mitigation of Terrorist Acts (Ix.3)
Climate & Climate Economics (Vii.3)
Global Nuclear Energy Issues (Iv.2)
Energy & Sustainability in Cities (V.5)
Global Food Production (Vi.6)
Forest Dynamics (Vii.3)
Water, Pollution (V.5) *& Terrorism* (Ix.3)
Brain Aging & Behaviour (Xii.6)
Frontiers in Fast Computing & Informatics (X.1)
Information Security (Xiii.5)

IV WHY THIS PROJECT

All I have said is a partial contribution to the reasons why WFS has taken the actions needed for the implementation of our Project. (Details in the Addenda).

Science, Technology and Culture in The Third Millennium: a Project for Mankind

Why this Project? Because we **care** about the consequences of the Planetary Emergencies on the future of the **human race**.

We want to overcome the danger of an **Environmental Holocaust**. **Environmental Holocaust** means spending enormous resources—billions of **Dollars/Euros**—for the solution of problems whose origin is believed to be known **but** it is not.

We are confronted with a very difficult task since, at the origin of the Planetary Emergencies, there is a **lack** of Knowledge, which in our days means a **lack** of Scientific Culture.

This is why we have to convince the great public that Science is needed in the Culture of the third millennium. We scientists cannot remain silent when the great public shows a vivid interest for topics such as those in focus at this 45th Session:

- Why Science is Needed for the Culture of the Third Millennium (XIII)
- Mitigation of Terrorist Acts (IX.3)
- Climate & Climate Economics (VII.3)
- Global Nuclear Energy Issues (IV.2)
- Energy & Sustainability in Cities (V.5)
- Global Food Production (VI.6)
- Forest Dynamics (VII.3)
- Water, Pollution (V.5) & Terrorism (IX.3)
- Brain Aging & Behaviour (XII.6)
- Frontiers in Fast Computing & Informatics (X.1)
- Information Security (XIII.5).

We have to convince the great public that the solution to all these **problems** require **clarity** and **rigour**. We have to convince people that the best way to study a problem, with clarity and rigour, is through **Science**.

V OUR CREDENTIALS

EMFCSC & WFS

Our community of Interdisciplinary Scientists has been able to contribute to overcome the danger of the Nuclear Holocaust, whose worldwide known symbol is the fall of the **Berlin Wall**.

Here comes a sequence of scientific leaders to whom we have dedicated our buildings. **Patrick M.S. Blackett** which is the **starting point** (my youth). **Isidor I. Rabi** a **decisive step** towards the creation of the Interdisciplinarity Scientific Community. **Eugene P. Wigner** a witness of the crucial steps towards the fall of the **Berlin Wall**. **Victor F. Weisskopf** whose support was decisive when this Institution was created at CERN-Geneva.

What have we been able to do in the past **decades are our credentials**.

ETTORE MAJORANA FOUNDATION AND CENTRE FOR SCIENTIFIC CULTURE

DATA ON ACTIVITIES SINCE 1963
124 INTERNATIONAL SCHOOLS, **1.594** COURSES,
113.684 PARTICIPANTS
(**125** OF WHICH NOBEL LAUREATES)
COMING FROM **932** UNIVERSITIES
AND LABORATORIES OF **140** NATIONS.

LECTURE HALLS AND CLOISTERS
Paul Adrien Maurice **DIRAC** - *LECTURE HALL*
John Von **NEUMANN** - *LECTURE HALL*
Robert **HOFSTADTER** - *LECTURE HALL*
Olof **PALME** - *DISCUSSION HALL*
Enrico **FERMI** - *LECTURE HALL*
JOHN PAUL II - *LECTURE HALLS*
Richard Phillips **FEYNMAN** - *LECTURE HALL*
Richard H. **DALITZ** - *LECTURE HALL*
Sandro **PERTINI** – *CLOISTER*
Dorothy C. **HODGKING** - *COURT*
John S. **BELL** - *CLOISTER*

9 October 1982 – The President of the Italian Republic, Sandro Pertini, a strong supporter of the Ettore Majorana Foundation and Centre for Scientific Culture, receiving the Erice Statement.

30 March 1979.

ERICE PRIZE

In order to promote the values of scientific culture worldwide and following a proposal by the WFS, a special law was voted unanimously by the Sicilian Parliament, in 1988, to establish the 'Ettore Majorana Prize—Erice—Science for Peace'. Every year, the Prize is awarded to distinguished scientists and world leaders who have contributed to the promotion of the values of scientific culture. So far sixty-two prizes have been awarded to: P.A.M. Dirac, P.L. Kapitza, A.D. Sakharov, E. Teller, V.F. Weisskopf, J.B.G. Dausset, S.D. Drell, M. Gell-Mann, H.W. Kendall, L.C. Pauling, A. Salam, C. Villi, R. Doll, J.C. Eccles, T.D. Lee, L. Montagnier, Qian Jiadong, J.S. Schwinger, U. Veronesi, G.M.C. Duby, R.L. Garwin, S.L. Glashow, D.C. Hodgkin, R.Z. Sagdeev, K.M.B. Siegbahn, Y.P. Velikhov, J. Karle, J.-M.P. Lehn, A. Magnéli, N.F. Ramsey, H. Rieben, J.J. van Rood, C.S. Wu, R.L. Mössbauer, A. Müller, H. Kohl, M.S. Gorbachev, H.H. John Paul II, R. Clark, M. Cosandey, A. Peterman, R. Wilson, Lord J. Alderdice, J.I. Friedman, M. Koshiba, S. Coleman, A.N. Chilingarov, P.C.W. Chu, L. Esaki, W.N. Lipscomb Jr., J. Szyszko, M.-K. Wu, H.A. Hauptman, D.H. Hubel, R. Huber, B.I. Samuelsson, H. Sun, A.E. Yonath, G. 't Hooft, Y.T. Lee, W. Arber and S.C.C. Ting.

SCHOLARSHIPS PROGRAMME

Afghanistan
Albania
Algeria
Angola
Argentina
Armenia
Aruba
Australia
Austria
Azerbaijan
Bangladesh
Belarus
Belgium
Bolivia
Bosnia
Botswana
Brazil
Bulgaria
Burkina Faso
Burundi
Cambodia
Cameroon
Canada
Central African Republic
Chile
China, Hong Kong
China, People's Republic
China, Taiwan
Colombia
Congo, Democratic Republic of
Cote d'Ivoire
Croatia
Cuba
Czech Republic
Denmark
Ecuador
Egypt
Eritrea
Estonia
Ethiopia
Finland
France
Gabon
Gambia, The
Georgia
Germany
Ghana
Greece
Guatemala
Guinea
Guinea-Bissau
Haiti
Honduras
Hungary
India
Indonesia
Iran
Iraq
Italy
Jamaica
Japan
Jordan
Kazakhstan
Kenya
Korea, Republic of
Kosovo
Cyprus
Kyrgyzstan
Laos
Lebanon
Liberia
Libya
Lithuania
Macedonia
Madagascar
Malaysia
Mali
Malta
Mexico
Moldova
Montenegro
Morocco
Mozambique
Nepal
Netherlands
Niger
Nigeria
Norway
Oman
Pakistan
Palestine
Panama
Paraguay
Peru
Philippines
Poland
Portugal
Romania
Russia
Rwanda
Senegal
Serbia
Slovakia
Slovenia
South Africa
Spain
Sri Lanka
Sudan
Sweden
Switzerland
Syria
Tajikistan
Tanzania
Thailand
Tunisia
Turkey
Uganda
Ukraine
United Kingdom
United States of America
Uruguay
Uzbekistan
Venezuela
Vietnam
Yemen
Zambia
Zimbabwe

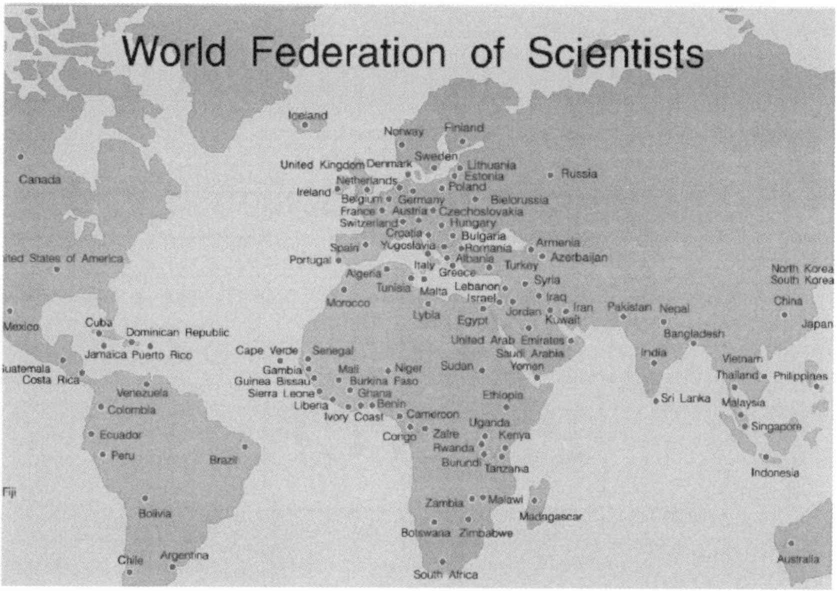

WFS

THE WORLD FEDERATION OF SCIENTISTS (WFS)

1966-2012

20 Agreements signed with Governments.

65 Scientific Collaboration Agreements signed with Governmental Institutions and Research Institutes worldwide.

43 Research Centres established in Developing Countries.

CONCLUSION

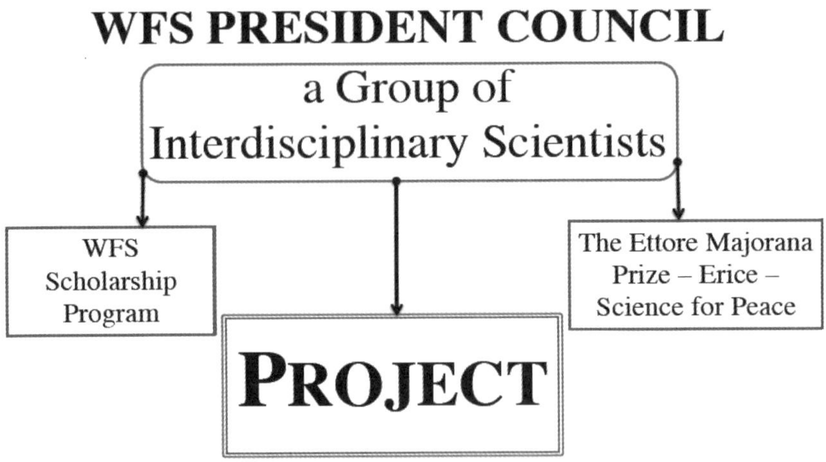

REFERENCES

1. Abdus Salam, the Nobel prize-winning physicist from Pakistan referred to by Robert Aymar, former Director General of CERN first published in Symmetry magazine, August 2006.
2. Announcement CERN: *On 31 July 2012 a Russian spacecraft was launched from the desert steppe of Kazakhstan. Its destination: the International Space Station (ISS). On board: five Timepix detectors developed by the Medipix2 Collaboration at CERN.*

CYBER RESILIENCE: THE ESSENCE OF CYBER PEACE

HAMADOUN I. TOURÉ
Secretary General, International Telecommunications Union
Geneva, Switzerland

Excellency Václav Klaus, President of the Czech Republic,
Professor Zichichi,
Lord John Alderdice,

Distinguished colleagues, Ladies and gentlemen,

It is a tremendous pleasure to be here with you in Erice, which really must count among the most beautiful historic locations—not just in Sicily, or even Italy, but in the whole world!

This morning I would like to take the opportunity to say a few words about cyber resilience, which I believe is the essence of cyber peace.

The World Federation of Scientists, Permanent Monitoring Panel played a key role in contributing to Nuclear Peace during the cold war era when the nuclear arms race was at its peak. Today, with the complexity of cyberspace and the fundamental role that ICTs play in social and economic development, I am glad that the WFS, PMP is again, enabling the world to benefit from the participation of science and scientist in contributing to the ever growing threats we face in cyber space.

Ladies and gentlemen,

We live in a world which now has more than six billion mobile cellular subscriptions, and where there will soon be two and a half billion people using the Internet.

This global hyperconnectivity allows us to leverage the power of technology—and especially mobile technologies—to make the world a better place.

Unfortunately, however, this indispensable new infrastructure also brings with it new challenges for preserving peace and stability.

Indeed, we have all seen the growing number of vulnerabilities and threats in cyberspace which have direct effects in the real world such as Stuxnet in 2010 which directly targeted the Iranian enrichment plant (a specific industrial system), Duqu in 2011 which, among others, was aimed at gathering information about industrial systems and Flame in 2012 which, among others, collected a wide range of intelligence to be used for other unlawful purposes.

Mobile is now the key driver of Internet growth, and by the end of last year there were almost 1.5 billion 3G subscriptions.

During 2011 alone, some 1.7 billion feature phones and smart phones were shipped worldwide.[1]

As the sales of smart phones in particular rise, mobiles offer new opportunities for potentially lucrative cybercrime and cybercriminals are adapting with extraordinary rapidity to this golden opportunity.

---

[1] Source: ABI.

Distinguished colleagues,

Last year was the first year when mobile malware became a tangible threat to enterprises and consumers, and we can only expect this trend to continue as malware authors explore new ways to attack mobile phones and tablets.

Indeed, in just the opening months of 2012, some 25,000 new threats were detected on one of the most common mobile platforms.[2]

One interesting new trend is that the spam phenomenon is changing in nature—with spam activity declining from 88.5% of all email traffic in 2010 to 75.1% of all email traffic last year.[3] It seems that spammers today are now focusing more on social networking, rather than purely on email.

Malicious code targeting Apple machines is also now on the rise, and this trend is expected to continue through 2012 as malicious code becomes more integrated within wider web-attack toolkits.[4]

In truth, none of this should come as any surprise to us, as cybercrime becomes increasingly profitable, with estimates no longer in the millions of dollars, but in billions of dollars annually.[5]

Cybercrime also has the potential to become much more dangerous than in the past, with ICTs increasingly used to control and monitor critical infrastructures, and nations becoming ever more dependent on them.

Last year we saw more targeted attacks; more politically and financially motivated attacks; and more data breaches and attacks on certificate authorities, than ever before. Government and public sector organizations found themselves specifically targeted by email attacks, with businesses of all sizes also increasingly at risk.[6]

I am pleased in this regard that ITU, jointly with Kaspersky Labs, and within the framework of the ITU-IMPACT collaboration, managed to discover Flame early on, and was able to immediately alert Member States and UN agencies.

And just about 10 days ago, we again partnered with Kaspersky in the discovery of yet another cyber weapon, called Gauss, which is specifically designed to monitor online banking accounts and related confidential financial information.

With malware and cyber weapons, we have seen how the power of ICT networks acts as a lure to terrorism and espionage, shaping a new concept of war—cyberwar.
Cyberwar is launched in cyberspace using ICTs—but it can quickly spread beyond the virtual world, affecting governments, businesses and individuals.

We need to start thinking seriously about the potential global negative impact that this may have on international security, and to put aside any political or other differences.

Ladies and gentlemen,

To ensure cyber peace, we need cyber resilience—and this means looking at the five strategic pillars of the Global Cybersecurity Agenda or GCA.

---

[2] Source: Trend Micro.

[3] Source: Symantec.

[4] Source: Symantec.

[5] Source: Trend Micro.

[6] Source: Symantec.

Cybersecurity is a multi-dimensional issue. Individual security includes the absence of cyber threats such as computer fraud, illegal electronic data processing, identity theft, child pornography, etc. Malicious cyber activities pose high-level threats to national security. Without secure individuals and secure countries, we cannot hope to achieve international security. To ensure global cybersecurity, an international cybersecurity culture must be created.

At ITU—the specialized agency of the United Nations responsible for ICTs, and a global, impartial organization—we are proud of the work we have done under the banner of our Global Cybersecurity Agenda, the GCA.

The GCA comprises five pillars, which together act as key measures in helping to achieve cyber resilience and foster cyber peace and cybersecurity.

The five pillars are:

- Legal measures—which are essential as a deterrent and in ensuring appropriate responses to cybercrime;
- Technical and procedural measures—which use technology itself to help increase cyber resilience;
- Organizational structures—which need to be in place to maximize cooperation and partnership between all stakeholders;
- Capacity building—which is of course essential in ensuring that people know and understand the technology they are using; and
- International cooperation—which I will come to in a moment.

As a result of work done within the framework of the GCA, it necessary to look into the following areas to have a comprehensive solution for cyber resilience:

- Access to the Internet;
- Protection of fundamental rights (including privacy and freedom of expression);
- State involvement; and
- International cooperation.

<u>Access to the Internet</u>

The Internet has become an indispensable tool for development—and universal **access to the Internet** should be a priority for all states.

Bridging the digital divide, however, requires more than simply having access to the Internet, as people must be able to make beneficial use of ICTs for their individual and collective development.

In their efforts to bridge the divide, states must attempt to provide universal access to information and knowledge, quality of education for all, respect for cultural and linguistic diversity, and protection of fundamental rights.

Furthermore, the development of the knowledge society demands that everyone be empowered with communication access, local infrastructure, affordable connectivity, and education on the use of ICTs.

States must effectively respond to online threats and vulnerabilities, and give special attention to the dangers imposed to children and young people, enabling digital citizens with effective mechanisms to protect themselves and their community.

Protection of Fundamental Rights (including privacy and freedom of expression)

Distinguished colleagues,

When taking measures to ensure the stability and security of cyberspace, and to fight cybercrime and counter online threats, states also need to respect **cyber freedom and the fundamental rights of users**.

This is a complex and sometimes controversial area, with a vigorous debate over the need to ensure security on the one hand, and to protect privacy and users' rights on the other. While some would want to oppose security with privacy, it is important to stress that these two goals are not mutually exclusive. Security is essential in guaranteeing rights such as privacy and freedom of expression. It is absolutely necessary that we achieve both goals.

When we discuss cyber resilience and cybersecurity it is sometimes too easy to forget the importance of fundamental rights—such as the right of freedom of opinion and expression; the right to information; the right to privacy and data protection; the right of assembly and freedom of association; and the right to non-discrimination.

It is also absolutely essential, I believe, to maximize free online public access to scientific research, which will help to drive innovation, improve welfare, and create the vital new employment opportunities we need in the 21st century—especially for women and for young people.

We must also do everything necessary to preserve culture and heritage online for current and future generations. We all know how important libraries, museums and archives are in the real world. And if anything, they are even more valuable resources online, where they are accessible to the entire world's people.

State involvement

Each action comprising **state involvement**—in the real world as well as in the virtual world—must take into account the existing provisions related to the respect of human rights, state territorial integrity and sovereignty.

Cyber attacks are no longer isolated events. States and other parties must be aware of the potential risk and dangers related to these types of actions.

Critical infrastructures are becoming common targets. Their destruction or damage could seriously compromise the security and safety of states, as well as human life.

With the willingness to preserve society, public services, etc., critical infrastructures should not be attacked. International rules for war must be taken into account concerning this type of incident.

International cooperation and commitment are required in this topic, in order to maintain a safer cyber environment/cyber space.

International cooperation

Ladies and gentlemen,

To achieve cyber resilience—and therefore ensure cyber peace—**international cooperation** will be required, and States will need to collaborate with each other and actively participate in joint international efforts to work together.

Cyberspace is global, and so building cyber resilience and ensuring cyber peace will require global efforts—ideally in the form of an international framework which takes into account the needs and wishes of all stakeholders.

Together, at the international level, we need to cooperate in the development and maintenance of secure technologies that protect users—particularly children and young people—and which together help us to build global cyber resilience.

States are also encouraged to share best practices and experiences and to transfer technologies which are able to strengthen confidence in cyberspace.

We also need to see strengthened international cooperation over financial transactions online, in order to prevent crimes such as money laundering, which can be facilitated in the virtual world via cyber payments and online gambling, for example.

States cannot do this alone. They need to work closely with other stakeholders. In this new world order, effective cooperation on cross cutting topics such as cyber resilience must involve civil society, the private sector, consumer groups, regional and international organizations and other relevant national and international players.

Finally, it is only through international cooperation that we can begin to discuss important issues such as limiting the proliferation of cyber weapons, and eventual cyber disarmament.

CONCLUSION

Distinguished colleagues,

Cybersecurity and cybercrime affect every country, every business, and each and every individual online. As we push forward the UN agenda for peace and safety, we must remember that cyber peace and cybersecurity are very much part of this in the 21st century.

As leader of the ITU, I am working to encourage the remaining nations which have not yet come on board to join the 144 countries which are already a part of the ITU-IMPACT initiative—the first truly global multi-stakeholder and public–private alliance against cyberthreats.

I am also working to encourage the private sector to come on board with ITU-IMPACT, along with intergovernmental agencies and non-governmental bodies.

We must work together to set international policies and standards, and to build cyber resilience through an international framework of norms and principles for cybersecurity and cyber peace.

Let me therefore invite you to continue doing what we all do best—which is to work together, listening to all the different stakeholders, and building a better future for all the world's people.

A world of cyber resilience—where cyber peace and cybersecurity replace cyber threats and cybercrime.

Before I conclude, let me ask a few pertinent questions:

How do we set aside our ideological and political differences and work towards developing globally acceptable norms and codes of conducts for individuals, governments and industry?

How do we continue to innovate in the development and use of ICTs and at the same time ensure that privacy and freedom of expression are preserved?

How do we put in place defensive, protective and reliable early warning systems that are verifiable to protect critical infrastructure?

Finally, are we mentally prepared to work together to achieve our common goal of ensuring cyber resilience?

Thank you.

RECENT DEVELOPMENTS IN TERRORISM

THE LORD ALDERDICE FRCPsych
House of Lords
London, UK

One of the differences between the work of colleagues in the physical sciences and those of us in the social sciences is that while in, for example, nuclear physics it is quite properly assumed that if I take the same observations now and in five years time, I will be able to replicate the outcome and so demonstrate the lasting significance of my hypothesis. In the social sciences it is highly possible that the psychosocial context will have changed so markedly in five years that my observations will be different, not only because the context has changed but because the phenomena that I am observing have themselves evolved. This, and the level of complexity of the phenomena we are observing, makes accurate prediction in this field impossible, though we can still develop worthwhile understandings.

In addition to this, while in the physical sciences we try to ensure that emotion is removed from scientific endeavour and replaced with rational argument, in the social sciences emotional responses are an essential element of what we observe, and it is not easy to isolate ourselves as investigators from the emotional material we are observing. For example in studying how people try to deal with the lasting emotional aftermath of the death of their children, or grievous physical injury as a result of terrorist attacks, even the scientific observer cannot but be humanly moved by what s/he is observing, without losing some of our essential humanity. This emotional component is especially important in dealing with terrorism, because, as the term suggests, abject fear and terror is not a mere unfortunate side-effect of terrorism, it is rather the key effective feature of this tactic of asymmetric warfare and must be appreciated and studied.

In the short time available I will have to pick and choose among the significant recent developments in the field of terrorism studies and of course others may make different choices, but with this audience I think it appropriate to start with the increasing significance of cyber-space.

Until recently there were four spaces in which we could function and relate as individuals and groups, and therefore engage in politically motivated-violence warfare and terrorism. Land is literally our home territory as human beings, and from ancient times we could also engage in warfare on water. More recently air and then space also became contexts where we could engage. However, in the last few decades a fifth space has opened up in addition to land, water, air and space—that of cyber-space. It is not just that we can use the cyber-world to assist in our prosecution of hostilities in the usual spaces—using internet drop boxes to arrange traditional terrorist incidents, or employing the internet to recruit young people to the cause, but because so much of our critical infrastructure is now mediated through information and communications technology, war can take place entirely through the medium of cyber-space. Before addressing terrorism let me point out that the overwhelming majority of hostile incidents that occur in cyber-space are not terrorist but are cyber-crime (credit card and other financial fraud, pornography, etc), cyber industrial espionage or cyber-warfare between states or state sponsored entities.

The most notable feature to me is that international law has not been able to keep up with these developments and so, for example, while we are generally very clear what constitutes a de facto declaration of war on land, sea, in the air or in space, this is not the case in cyber-space. If many of the massive hostile incidents recently seen in the cyber-world had been in any of the other four spaces they would undoubtedly have been regarded as declarations of war, however the position is not clear in the fifth space. Surely one of the key developments which must take place in the near future is a focus by lawyers, diplomats, scientists and politicians on developing international and indeed global understandings that address this issue. To do this they must be more attuned to recent technical developments than I have observed is the case with most domestic legislators who simply do not understand and cannot keep up with changes in ICT, even those on which they are legislating.

As far as terrorism is concerned my main concern was well described by Misha Glenny at the Second Worldwide Cyber-Security Summit at The Queen Elizabeth II Conference Centre in London, in June last year. Mr Glenny, a former BBC journalist whose most recent book, *Dark Market*, focuses on the world of cyber-crime, pointed out that although that conference, like others, was replete with a multitude of papers, presentations and discussions on the hardware and software of cyber-security, there was very little research being reported on the psychology of cyberspace. I regard this as a major gap in the field because the way people function, as individuals and as groups and networks in cyber-space has many features that are quite different from the other four spaces and this is therefore an area to which the WFS Permanent Monitoring Panel on Motivation for Terrorism is devoting some attention.

I turn now to developments in the study of more traditional terrorism. After many years of speculation and some observation about how and why terrorist campaigns commence and are sustained, more attention is now being paid to how terrorist campaigns end. As scientists we should not be satisfied with mere speculations about what might happen. We want to piece together the evidence about what actually happens, and one of the key works in this field is the study published in 2009 by Professor Audrey Kurth Cronin on 'How Terrorism Ends', in which she reviewed 457 terrorist organizations that had been active in running campaigns since 1968. There were a series of different outcomes, but only very rarely were the terrorists successful. Mostly they failed because of splits and in-fighting in the groups, a loss of momentum or relevance, massive state repression or decapitation of the organization through the killing or imprisonment of their leadership. In regard to this last outcome, imprisonment was generally more effective in destroying the myth of the leadership's power, while assassination could produce martyrs to the cause. In the minority of cases where there was some degree of success, it could come about through negotiations, however most negotiations with terrorist groups, while it could delay in the ultimate end of their operations, generally reduced the death and damage they caused and usually led to them re-orientating themselves in some way, for example into democratic politics. Other forms of re-orientation were more dangerous, for example when they transitioned into organized crime (such as the narco-terrorism of FARC in Colombia and the Shining Path in Peru), or morphed into an insurgency holding of territory, developing alternative state structures, and then sometimes sliding in into more traditional civil war or overt war scenarios.

This latter outcome, the destabilization of the state, is a major development for study.

In the late 19th and early 20th centuries terrorism was the tactic developed by weaker groups against a powerful state apparatus, and it saw some success in the national freedom movements of the 20th century. It is arguable how far the anarchist terrorists contributed to the end of empires that was the outcome of World War I, but what is much more clear is that after World War II, with the realization of the two new super-powers that a nuclear conflict could be catastrophic, terrorist groups became a modality though which super-power rivalries could be channelled without endangering world peace. The end of the Cold War brought this phase to a close and with it some long-standing terrorist campaigns were no longer sustainable, but a new set of problems have emerged.

If the tactic of terrorism was previously the strategy of the weak against powerful states, terror, chaos and massive violence as seen in many parts of Africa and increasingly in the Middle East is the result of the failure of many states to be able to sustain themselves against terrorist and other attacks and it is not clear how far this failure of states is a crisis of certain regions, or may herald a major shift towards instability and a crisis of democracy itself.

The paradoxical result of the globalisation that has been facilitated by the very positive scientific and resultant technological advances in travel, transport and communications and the collapse of the stable instability of the Cold War is that people have become more anxious, not more confident. Fear about the new uncertainties of globalisation and the return of earlier crises over commodities and profound economic uncertainty have seen many people searching out the false certitudes provided by nationalism or localism which go beyond patriotism and communal pride and become characterised by populism, racism and xenophobia. Religious ideas have also come to the fore in a way which must be a surprise to those who thought that the Enlightment would lead to the inevitable end of religious faith, as our colleague, Scott Atran has pointed out in a paper in paper this month the journal *Foreign Policy*. Let me quote the opening lines of Scott's paper, which are nothing if not dramatic in their implications:

> *"The era of world struggle between the great secular ideological -isms that began with the French Revolution and lasted through the Cold War (republicanism, anarchism, socialism, fascism, communism, liberalism) is passing on to a religious stage. Across the Middle East and North Africa, religious movements are gaining social and political ground, with election victories by avowedly Islamic parties in Turkey, Palestine, Egypt, Tunisia, and Morocco."*

This claim that we may be at a hinge of history is not related to observations in the Middle East alone. Despite years of communism in the old Soviet Union and in China, religious movements in those regions have proved profoundly resilient, and in the United States of America the place of religion in politics and government is much greater than the separation of church and state was assumed to permit.

Scott's paper goes on to point out that we should not be frightened about this because, contrary to what some believe, there is no evidence from an objective study of history that religious faith itself leads to war and internecine strife, although once conflicts start a religious tone entering in may make them more intractable. What Scott does insist, and I agree, is that instead of either ignoring or adopting religious beliefs, social scientists must try to understand them and the impact that they have in human affairs and in strategic decision-making at every level. This continues to be a key area of work for the PMP on Motivations for Terrorism with our work on 'sacred values', strategic trust and mutual influence.

As I have already pointed out however, the boundaries between terrorism and other tactics of warfare have become increasingly blurred with the use of terrorism and various tactics of terror by states. In this regard I must close with an expression of serious concern that goes beyond our work on terrorism, and may involve some of the rest of you.

I cannot think of a time in history when there was such profound economic difficulty and dislocation which did not result in violence of a major order. Sometimes the results were seen in revolutions, at other times civil wars and of course the economic depression of the 1930's is generally regarded as having played a role in creating the context for the terrible events of World War II. I fear that we will not escape this lesson of history. Already we have had the events of what was initially termed in the West, 'the Arab Spring', but which have turned out to be altogether more problematic and dangerous than was at first assumed. In addition in that region we have an on-going crisis between Iran and Israel. I am especially concerned to hear recent remarks emerging, apparently from semi-official sources, claiming that an Israeli attack on Iran's nuclear facilities would be over in 30 days with the loss of only some 500 Israeli lives. Previous generations, living in the aftermath of Hiroshima and Nagasaki, and educated by some of those in this very room and other colleagues in nuclear physics, were profoundly aware that the use of nuclear weapons power was a threat to our very survival as a species—a truly planetary emergency. Knowledge is not genetic and each generation must acquire knowledge for itself, albeit learning from the work of their predecessors. I am however concerned that this generation, which forgot, and is now learning by painful experience what a previous generation knew about economic uncertainty and instability, should not have to learn with the loss of many lives, about the dangers of nuclear conflict. On this front we in the PMP on Motivations for Terrorism look to our colleagues in nuclear physics, not just to assist us in addressing the nuclear dangers posed by small terrorist groups, but to take a lead in facing up to an emergent danger at the level of inter-state conflict involving Iran or another potential or actual nuclear power.

Whether in dealing with terrorism, an economic austerity whose extent is still unknown and which is likely to last a generation or more, the emerging crisis in democracy and democracies or whether in regard to the nuclear dangers which have not gone away, there are profound implications for the Erice commitment to "science in the service of humanity living at Peace" and an urgent mandate for us to find the passion, the resources and the creativity to continue our work.

Mr. Chairman and dear Professor Zichichi,

It is an honour and a pleasure to address such an exceptional assembly, and I will first address my congratulations to Professor Vaclav Klaus, President of the Czech Republic, for his seminal lecture on the status of climate issues. I would also like to convey my greetings to all the eminent scientists from 40 countries gathered in this Hall, to further the fight against the Planetary Emergencies.

Let me also express my admiration for what was achieved during the last 50 years, in this ancient city of Erice, by a young scientist who could have simply enjoyed his already brilliant scientific achievements and gone on to forge his exceptional career, inside his scientific "ivory tower". But he was among the first to realise that Modern Life was fast presenting a very complex challenge to humanity, which could only be met if the scientific community at large rallied together in an interdisciplinary effort, to avoid the looming catastrophes and provide a better future to Humanity.

A lot of what was analysed and claimed, in the last 30 years, in the Erice International Seminars on Nuclear War and Planetary Emergencies has been verified. Some very concrete results were also achieved, with the avoidance of the nuclear holocaust in the 80's and the successful implementation of 100 emergency-related scientific pilot projects in 50 developing countries. It continues with a very clever scheme for national scholarships in 30 countries, which is fast growing, while the Planetary Emergencies continue to be followed up by dedicated WFS Panels, year after year, during the Erice International Seminars.

The WFS was able to achieve some outstanding results in many areas, with very little funds, a lot of expertise and scientific volunteering, thus showing the way.

Over the last 30 years, the WFS has been identifying these planetary problems and many more. The WFS was able to analyse the problems and, in many cases issue recommendations to decision makers. The difficulty comes from the fact that each nation's governing structure is forced to take into consideration the immediate concerns of its industry, its commerce and its population. Taking the long view and imposing restrictions would be extremely unpopular. Any action in fighting planetary emergencies must come after educating the population and in a concerted effort by all nations. But before we get to this point what we need is to have a reliable diagnostic of what these emergencies entail and what solutions could be applied.

The WFS has consistently been proposing what it calls "A Project for Mankind", advocating the building up an international consensus on the necessity of establishing a dedicated Centre, to address, study and find solutions to the Planetary Emergencies. To be functional and efficient, this Centre would be endowed with sufficient funds, guaranteed year after year, and given the kind of structure and flexibility that would allow it to obtain fast results, free of administrative burdens and political pressures.

The time may have finally come when the nations of this world will realise that was has been advocated by WFS, for the last 30 years, is an absolute necessity. And Italy would be first among the nations to help establish the Centre for the Study of Planetary Emergencies.

Thank you for your attention.

Sen. Dr. Antonio D'Alì
President of the Senate Committee for Environmental Sciences
New York, 20 August 2012

SESSION 2

MITIGATION OF TERRORIST ACTS

DOSE THRESHOLDS FOR EVACUATION FOLLOWING EXPLOSION OF AN IMPROVISED NUCLEAR DEVICE OR RADIOLOGICAL DISPERSAL DEVICE

RICHARD L. GARWIN
Thomas J. Watson Research Center, IBM Research Division
Yorktown Heights, USA

ABSTRACT

A November 2011 report from the U.S. Department of Homeland Security provides substantial advance in understanding of damage from the explosion of an improvised nuclear device (IND) in the urban environment. Substantial reduction in fatalities and injuries would be achieved by the discipline of responding to the flash by immediately seeking cover under a desk or other protection against damage from wind-driven shattered windows. Furthermore, on the order of 100,000 lives might be saved by a rational response to taking shelter in existing buildings within minutes of the explosion, together with informed evacuation after some hours. For contamination and exposure to radioactivity from a radiological dispersal device (RDD) there is far less urgency to consider evacuation, since there is no early spike in dose rate. Rather than mandatory evacuation, an optional evacuation should be the primary response, with compensating damages paid to those who remain.

Far too little provision is made in implementing the preparation and dissemination of information to guide public response to an IND or an RDD, and that should continue to be a priority of our group. Over the ten years in which we have been considering these matters, the preferred option has evolved to be "push technology" that over the next months or years could result in highly specific information residing in the smart phones or other mobile IT devices of much of the populace.

PRESENTATION

Major effort worldwide is directed toward preventing terrorists from acquiring highly enriched uranium (HEU) or plutonium (Pu) for fabricating an improvised nuclear explosive device, but prudence demands also that attention be given to mitigating the consequences if such a device were actually exploded in an urban area. The more likely case of a radiological dispersal device, perhaps in the form of a "dirty bomb" in which conventional explosive is used to disperse an intense source of radioactive material normally used for cancer treatment in hospitals, industrial radiography, or irradiation of food or plastics, will also be addressed. It is quite different from a fission explosion because there is no early peak of dose rate, and far less radioactive material involved.

In addressing any question of public health, one must be concerned with different epochs. The most obvious is that which follows not only the analysis but consequent decisions, planning, and deployment; it might be characterized as more or less the "steady state." What would be the consequences for various attacks, and how can they be mitigated?

But long before the steady state, there is the question as to what should be done immediately after the publication of the paper or the delivery of this talk, before societies can be fully prepared to do "the right thing" following an attack. Although the probability of such an event in the first month or year may not be very high, the question should still be asked. And we will attempt to describe improvised capabilities for mitigation.

Concentrating first on the improvised nuclear explosion, we note that this has been addressed to some extent at the 2010 Planetary Emergencies seminar, in which I drew heavily upon an article that I published that spring, and a parallel piece by Brooke Buddemeier[1] of LLNL. In turn, this was supplemented by a publication[2] from the Department of Homeland Security (DHS) that, in particular, provided shielding factors for fallout radiation, for various types of buildings common in the United States.

We can now refer to a much more substantive 120-page DHS document[3] which refers to the National Capital Region (Washington, DC), but which provides methodology and data that could be used for almost any detonation point. Many of my illustrations are taken from that document.

As emphasized much earlier,[4] an IND is likely to be of much lower yield than the strategic nuclear weapons now common in the military inventories—which have yields of 100 kt or 500 or even 2000 kt. Instead, an IND is likely to aim for a yield of 10 kt but might achieve only 1 kt or even 0.1 kt. This last is still the equivalent of 100 tons of TNT (50 two-ton truck bombs detonated simultaneously at the same point). But with the smaller explosions, and a burst on the surface rather than at altitude, the various damage phenomena differ in relative importance from the strategic weapons.

Nuclear explosions cause damage by overpressure and high winds, from burns and fire from the large amount of the explosion energy that is radiated from the fireball as heat in a second or less from these low-yield explosions, and from ionizing radiation's impact on people. All of these effects could in principle but not in practice be mitigated by appropriate measures, even if not totally eliminated. Underground buildings are largely immune except in a region of very high blast overpressures, but no one is going to rebuild or build cities of totally different configuration to avoid damage from an IND. The flash burns from the low-yield explosion happen so quickly that little can be done to avoid or reduce them after the fact. Much of the damage from blast comes from broken glass, and tough plastic film on windows can reduce that hazard, as can tough blast curtains that allow light to filter through, but will retain broken window glass.

Where mitigation can really help is in reducing the death toll from the radioactivity instantly produced by the nuclear explosion.

---

[1] "A Nuclear Explosion in a City, or an Attack on a Nuclear Reactor," by R.L. Garwin, and "Reducing the Consequences of a Nuclear Detonation: Recent Research," by Brooke Buddemeier, *The Bridge*, Summer 2010, pp. 28-38. Both to be found at http://www.nae.edu/File.aspx?id=19815.

[2] "Planning Guidance for Response to a Nuclear Detonation", Second Edition, June 2010, at http://www.epa.gov/rpdweb00/docs/er/planning-guidance-for-response-to-nuclear-detonation-2-edition-final.pdf.

[3] "Key Planning Factors: Response to an IND in the National Capital Region" November 2011, http://www.fas.org/irp/agency/dhs/fema/ncr.pdf.

[4] "The Many Threats of Terror" and an Epilogue by R.L. Garwin in A New York Review Book *Striking Terror: America's New War*, edited by R.B. Silvers and B. Epstein, March 2002, pp. 235-256., http://www.fas.org/rlg/020300-strikingterror.pdf.

The explosion is accompanied by an instantaneous flash of neutrons and gamma rays from the fission reaction itself, much of it absorbed by materials of the bomb and attenuated at long distances by air itself. Beyond that, there is the local deposition of radioactive fission products from the bomb, mixed with soil and debris from the ground and buildings vaporized by the explosion. Much of this (perhaps half) falls to earth in the immediate neighborhood of the explosion, even in the region devastated by blast.

Beyond the immediate, local fallout, there is the portion that is lofted into the atmosphere, condensed of fine particles of debris rather than coarse, and that thus takes longer to deposit from the parcel of air in which the debris finds itself. It falls under the influence of gravity more slowly the finer the particle.

From the days of atmospheric testing of nuclear weapons, much is known about the distribution of fallout, and modern atmospheric modeling with fast computer complexes allows the real-time prediction of distribution of fallout and dose to exposed personnel. In similar fashion can be calculated the shielded dose that would be received by a person within a building of nominal construction of one kind or another. The figures below show nominal protective factors (PF) against fallout deposition on ground and roof.

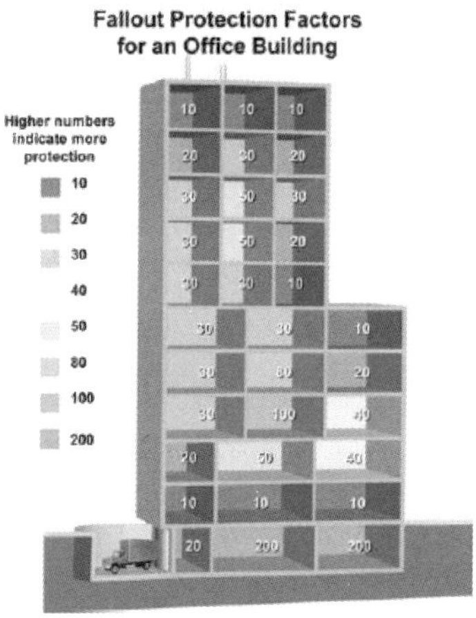

Fig. 1a: Building as shielding—numbers represent a dose reduction factor. A dose reduction factor of 10 indicates that a person in that area would receive 1/10th of the dose of a person in the open.

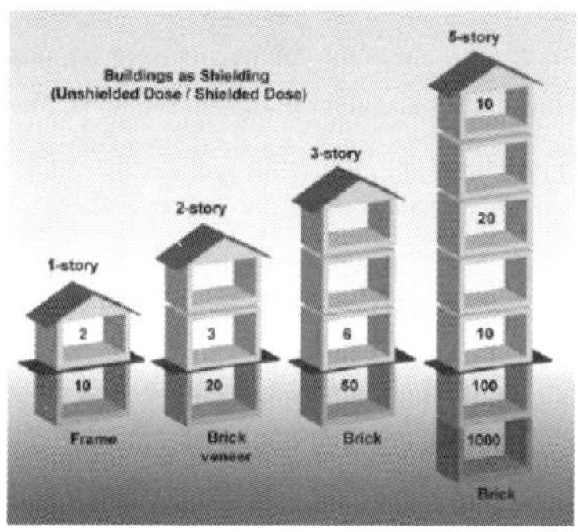

Fig. 1b: Building as shielding—numbers represent a dose reduction factor. A dose reduction factor of 200 indicates that a person in that area would receive 1/200th of the dose of a person in the open.

It has always been recognized that the "open field" values for blast and especially burns and prompt radiation are in reality affected and reduced by the presence of many buildings and structures, and with the increased computing power and sophistication of analysis, the most recent DHS paper reflects these modifications, as shown in its figure.

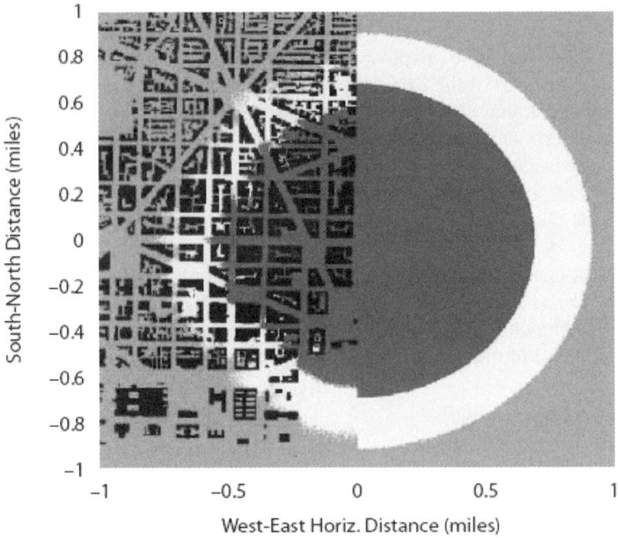

Fig. 2: Outdoor casualty areas for the illustrative scenario (left) and for an open field (right) from a 10-kt Hiroshima type device; red >800, yellow 100-800, green <100 rad.

As regards prompt radiation, a 10 kt explosion in a city might have the lethal radius of an 8 kt explosion in an open field.

The extent of structural damage to buildings is used to define the SDZ (Severe Damage Zone), the Moderate DZ, and the Light DZ. For a 10-kt ground burst at the intersection of K and 16th Streets in Washington, the SDZ extends to about 0.5 mile radius, the MDZ to one miles radius, and the LDZ to about three miles.

Although the explosion phenomena for a 10-kt burst happen within thousandths of a second, the blast propagates at the speed of sound and reaches the 2-3-mile radius in 10-15 seconds. This provides time for some people alerted by the flash from the explosion to take shelter under desks or behind cabinets so that they have less probability of being injured by windows that are likely to shatter under the blast load.

Fallout of bomb-produced radioactive material in the immediate area, and from the continued deposition as the radioactive material is moved out of the area by wind, leads to the definition of the Dangerous Fallout Zone (DFZ) and the Hot Zone (HZ). The DFZ is defined by radiation levels of 10 R[5]/hr or greater and is the region in which acute

---

[5] In this paper we use "R" for "roentgen," the unit of exposure to ionizing radiation. The health effect of radiation is better estimated from the Rem, but fortunately the R and the Rem are equivalent for beta and for gamma rays. In more modern units, 1 R = 0.01 gray (Gy); 1 Rem = 0.01 sievert (Sv). One gy is 1 joule/kg of energy deposition.

radiation injury or death is possible. In contrast, the HZ is defined by a dose rate of 0.1-10 R/hr and "could extend in numerous directions for hundreds of miles."

Little can be done to mitigate the radiation exposure from the prompt flash of gamma rays and neutrons from the explosion itself, which will give an effective dose to the body that decreases with distance, and in some azimuths is partially blocked by the mass of intervening buildings. But after a second or so, radiation will be dominated by that from bomb-produced fission products attached to coarse debris particles. People outside should get inside as soon as possible and crudely brush the debris from their clothing. Vehicles such as cars, trucks, and buses offer no protection against fallout,[6] and they will all be stopped in any case because of congestion and debris. Once inside, people continue to be exposed to radiation from the fallout on the ground or on ledges or roofs, with the protection factor offered by various types of building construction.

People should move to the regions of their building with higher PF, knowing that the dose rate will decline with time after the explosion. In fact, it has long been approximated that the fission product gamma-ray dose rate[7] decays as $T^{-1.2}$. The curve of Figure 3 illustrates the gamma-ray dose rate.

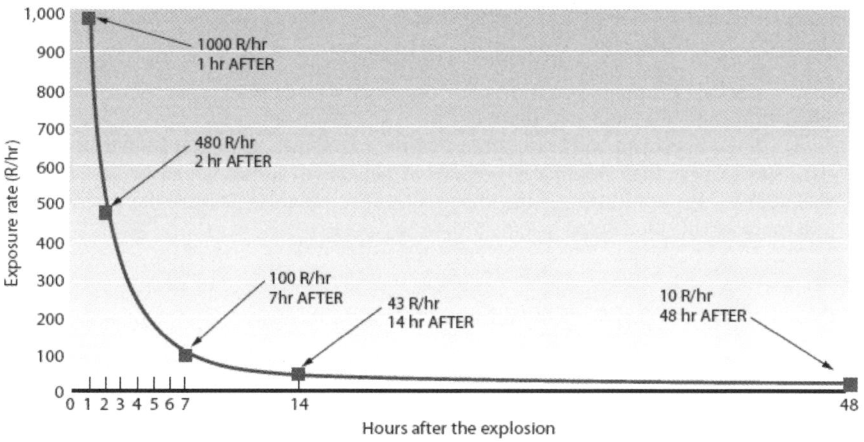

Fig. 3 (Fig 20 of ncr.pdf): "Radiation levels from fallout decrease rapidly over time, emitting more than half of their radiation in the first hour."

The whole-body dose for which about 50% of those exposed will die within a couple of weeks is about 400 R, and Table 1 shows the life-saving importance of proper shielding from the fallout for the substantial portion of people who without shelter would receive a lethal dose.

---

[6] Not quite true, since they do protect against beta-emitting fission products such a Sr-90 that can cause severe destruction of human skin.

[7] See "Radioactive Fallout From Terrorist Nuclear Detonations," by R.E. Marrs (2009), UCRL-TR-230908, at https://e-reports-ext.llnl.gov/pdf/347266.pdf.

Table 1. Modeled dose rates.

Time after detonation (hr:min)	Exposure rate (R/hr)
00:15	1,444
00:30	686
01:00	299
02:00	130
04:00	57
08:00	35
12:00	15
24:00	7
48:00	3
96:00	1

Evacuation after a couple of days, to reduce the additional dose to zero would be a good thing, and it is eminently feasible.

It is important, however, that evacuation routes do not inadvertently expose evacuees to very high dose rates because of pockets of initial fallout or accumulations due to rain. Figure 4 illustrates the importance of waiting to minimize the contribution to evacuation to the total dose.

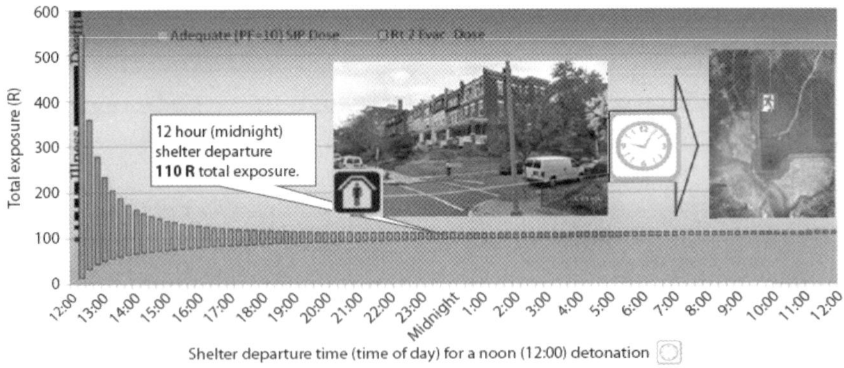

Fig. 4: Comparison of total exposure from various shelter departure times from Adequate Shelters (PF = 10) like the row homes using the default city evacuation route (green) is selected for evacuation.

Much can be done by moving to better shelters, while awaiting evacuation. A striking conclusion is that:

> *"The existing Washington, DC structures offer better than adequate protection. If all residents adopted a shelter-in-place strategy, it would reduce the number of potential acute radiation casualties by 98% (there would be ˜3,000 fallout casualties out of the ˜130,000 potential casualties of an unsheltered population)."*

As your PMP-MTA (Permanent Monitoring Panel on Mitigating the Effects of Terrorist Acts) has long emphasized, the DHS report recommends:

> *"Messages prepared and practiced in advance are fundamental to conveying clear, consistent information and instructions during an emergency incident. Planners should select individuals with the highest public trust and confidence to deliver messages. Such individuals should be prepared to deliver key information almost immediately to the public in affected areas about protection to maximize the number of lives saved."*

Thus far I have simply recounted the results of a major DHS study, but that has some value in itself, because few of you were probably aware of it. Now let's see what value we can add to that study.

First, to go back to basics, is there much sense on reducing exposure from 500 R to 200 R? How about the risk of cancer even at an exposure of 200 R? Indeed, there is great value in such a reduction. The probability of death at 500 R is on the order of 80%, whereas at 200 R, perhaps only 5% will succumb, although half will be sick. The probability of a lethal cancer is about 1/2000 R, so at the hypothecated 200-R exposure, it would be about 10%. Ten percent residual cancer mortality vs. 75% reduction in prompt lethality is a good payoff—65% of a human life saved—perhaps even more important if it is your own or the life of one of your family members. Furthermore, the death from cancer will not occur for years—perhaps 20 years on the average—a lot of life to be lived from the personal point of view, and invoking a discount rate from the economic or public planning approach.

HOW REALISTIC IS "INFORMED EVACUATION" AND SHELTERING?

First there is the question as to whether people can be made aware of the considerations of the 2011 "National Capital Region" paper and its summary here. That is involved with the "prepared communication" that gets only a single paragraph in the important DHS report but that has been treated at considerable length in the PMP-MTA papers. We propose not only the identification of people who deserve and indeed possess public trust (if any remain in the fractured American society) but also the dissemination of appropriate instructions and graphics that would reside on many, many home and office PCs and smart phones. Thus, even in the Moderate Damage Zone and heavy fallout areas a good fraction of the people could look up emergency summaries and graphics already resident on the smart phone to help guide their actions in the minutes following an IND.

How can we make this happen? In the United States, it will happen only if some contractor or consortium provides such a tool, either on speculation or in response to a solicitation and contract award from the federal government. This is not necessarily a long process, given fast-action elements of the U.S. government such as DARPA (Defense Advanced Research Projects Agency) and its younger sibling in the Department of Energy—ARPA-E.

But in the age of globalization, and in countries that have not yet been reached by the tsunami of privatization, this eminently governmental obligation and opportunity might be done within a government organization.

At this point one might have a short excursion as to why the contracting approach is faster than the in-house approach. Primarily it is because of the difficulty in diverting people from whatever they are doing to something that the management (occasionally leadership) of the organization decides they should do. It comes about because nobody is satisfied to appear to be doing nothing. It is bad for the soul and worse for the funding of the organization. You can write the rest of the story.

Even if there is a high probability of an urban IND somewhere, there is a low probability in any particular city. Therefore, whatever the merit of centrally-developed substance and communication, there is much less merit to its independent formulation in every potential target. Overall, if one assumes a 10% likelihood per year of an urban IND someplace, and considers that appropriate planning might save 100,000 people who would otherwise be killed by radiation from fallout, this would surely be an expenditure well justified on a world scale, even repeated several times in a competitive effort to produce a system that is considerably better than the uninformed response. Yes, I know that many millions of people are at risk from disease and poverty, and that a few million dollars in rational thought and allocation of resources could save them and that should be done, too. Here we are talking about people who have more or less functioning governments that are partially responsive to their demands and to rational thought.

Although the November 2011 DHS paper is a major advance over previous analyses, it does not do very much in the way of sensitivity analysis. That is, how accurate are the projections in the paper? And to what extent would knowledge of the local meteorology (winds and rain) reflect reality in the projections? It is a great achievement that the excellent center at LLNL, "NARAC" is able in real time to predict fallout, given only the magnitude and location of the IND detonation, but how accurate is that projection?

First, of course, it would be desirable to provide a proper input (explosive yield), and that is surely a government function that might be achieved from seismometers or from local barographs installed for the purpose. Location is readily and quickly determined seismically. Yield would then be determined by the low-frequency pressure pulse ("step") derived from the barograph. There is also a strong incentive to measure directly the fallout, in order to guide evacuation that might take place after a few hours or tens of hours. This could be done in a progressive manner, first with a low-resolution, less accurate fallout app or database, and later with increasing accuracy to guide evacuation routes and longer term planning.

As was recommended in my 2011 Erice paper,[8] it would be feasible and desirable to have a small system of 10-kg drone aircraft, self-navigating via GPS that would, via data link, provide a quantitative map of ground contamination of fission products. A single such vehicle flying at 50 km/hr could in 5 hours map a 15 km by 15 km square from an altitude of 0.5 km, with a resolution of about 0.3 km. Proportionally more quickly with more drones in operation. This would be enough to identify the broad patterns of ground contamination but not enough to provide a detailed map to help guide the sequence of evacuation.

That could be achieved by a lighter vehicle that would fly at an altitude of about 200 m, along potential evacuation routes, providing a ground resolution on the order of 50 m, to determine the likely exposure and the regions where it is important not to loiter.

EVACUATION IN RESPONSE TO AN RDD ATTACK?

Evacuation thresholds following use of a radiological dispersal device (RDD) as previous papers for and of the PMP-MTA have shown, the impact of contamination from an RDD can be large, even though the number of expected fatalities is small. That impact can be mitigated by appropriate analysis and communication strategy, including responsive monitoring of the environment following the RDD. Because industrial and commercial (including medical) sources of intense radiation all have lifetimes of years, there is far less urgency for evacuation or in fact any action following the dispersal of radioactivity than is the case with fallout from a nuclear explosion. For specificity, we take the illustration of Figure 5 in which 1000 Ci of Co-60 have been dispersed with 20% efficiency under certain atmospheric conditions. The contours shown are those corresponding to 5 (ck) years of residence in the region.

[8] "Learning More from Fukushima Dai-ichi," by R.L. Garwin, at www.fas.org/rlg/2011%20Erice%20Fukushima1a.pdf

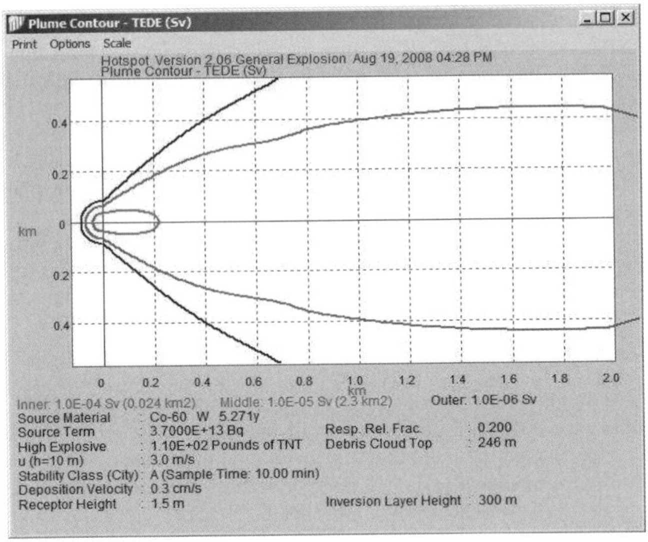

Fig. 5: Cobalt (Co-60)-based dirty dust—total effective dose equivalent (TEDE)-contour plot in an urban environment (Co-60 activity: 37 TBq or 1000 Ci; explosives: about 50 kg TNT) (F. Steinhauesler).

The middle contour of 10 microSv TEDE corresponds to about one additional death by cancer[1] for each two-million people exposed at that level, and is the limit sometimes recommended by the IAEA (for an unavoidable exposure deemed "negligible").[2] Many cities have peak population density regions of some 40,000 people per square kilometer[3], so that the 2.3 km^2 area of the middle contour would contain some 90,000 people. Without knowing the dose at each point within this 2.4 km^2 area, it is clear that it is less than the 100 microSv dose of the inner contour, so that the total collective dose cannot exceed 9 Sv; the expected cancer deaths without relocation are thus 0.05 x 9 = 0.45 total cancer deaths, compared with the 18,000 people expected to die of cancer of natural causes. Thus it is likely that there would be not a single cancer death outside the 0.024 m^2 contour, which itself might contain only 1000 people; it would be difficult to justify costly cleanup or restrictions on occupancy outside this 2.4 hectare (5.9 acre) boundary.

1 Using the ICRP coefficient of 0.05 cancer deaths per person-Sv.

2 "Health and Environmental Impacts of Electricity Generation Systems: Procedures for Comparative Assessment," IAEA Technical Reports Series No. 394 treats as "negligible additional radiation" exposures comparable with the natural background of some 3 milliSv per year. But one finds in IAEA-TECDOC-1484 Regulatory and management approaches for the control of environmental residues containing naturally occurring radioactive material (NORM), (www-pub.iaea.org/MTCD/publications/PDF/te_1484_web.pdf). P. 20:

"• The additional individual dose attributable to the exempted source should be of the order of 10 μSv per year or less; and
"• Either the collective dose to be committed by one year of performance of the practice should not be more than about 1 man-Sievert or exemption should be the optimum option."
3 http://www.demographia.com/db-citydenshist.htm *(Selected Current and Historic City, Ward & Neighborhood Densities, P. 4 of 24)*

Unlike the case with an IND, outside the tiny region in which people are likely to be killed by the ordinary explosives used in the dispersal (if, indeed, explosives are used rather than a vaporizer of a solution of the radioisotope), the question is solely one of reducing the human damage due to cancer, for which we take, again, a probability of lethal cancer of 1/2000 R (0.05 per Sv).

The United States Environmental Protection Agency guidelines mandate long-term evacuation if the dose expected within two years exceeds 2 R. This was, for instance, the guiding principle in the Fukushima disaster.[9] For the most part, the energy deposition from release of radioactive materials from an operating reactor is intermediate in nature between the contamination from an RDD and the fallout from a nuclear explosion. After the first seconds, the gamma dose rate from reactor fallout decreases as $T^{-0.2}$ rather than $T^{-1.2}$ from a nuclear explosion, as it should, being the integral over years of many fission events, each of which has the $T^{-1.2}$ behavior.

Unlike Fukushima, an RDD will take place in a society undamaged by earthquake or tsunami, in which many resources can be brought to bear on this rare event, including the resource of analysis that we are attempting to provide here, well in advance of any such event. At a committed whole-body dose of 2 R, the probability of lethal cancer is 0.1%, and with the planning factor in use in the U.S. government of $5 million per life saved (except in the Nuclear Regulatory Commission), the value of 0.1% of a lethal cancer avoided is $5000. In general, the abandonment of housing and the cost of displacing people is probably more in the range of $20,000—a cost that will surely be incurred, in contrast with the expectation of $5000 loss some 20 years in the future. So the EPA evacuation threshold is probably set too low by a factor four or five.

An additional element for the computation arises from the non-monetary costs of evacuation—even physical and psychological illness. On the other hand, many would worry about the validity of the estimate of harm if they remain in an area known to be contaminated with radioactive material at some level. For this, a personal dosimeter in the form of an easily read thermal luminescent detector (TLD) would be useful, together with a public system such that the TLD could be read by the individual at intervals of a few days or weeks, with the data going into a permanent database, and the TLD being reset.

However, there is a major public policy question as to whether evacuation should be mandatory or whether it should be optional. My own preference is for optional evacuation, with a payment by the government (that is, by the rest of society) of $5000 per person who chooses not to be evacuated from a potential 2 R exposure.

---

[9] In fact, in the confusion accompanying that disaster, some residents were evacuated from close-in regions to more distant locations that happened to have higher local fallout deposition.

Of course, for Cs-137, the exposure would go on well beyond two years, but probably for a period substantially less than the 30-year physical half-life of the isotope. We will have more information from Fukushima, but it appears from Chernobyl that the half-life in the environment is on the order of 5-15 years. I don't know the corresponding environmental half-life for Co-60 deposition.

Certainly a polluter who liberates to the environment radioisotopes that provide a dose to the public and thereby inflicts deaths from cancer at a cost of $2500 per person-R (on the assumption of $5 million per life lost) should pay a fine substantially higher than that, if the act is to be deterred and not simply assumed as a "cost of doing business." But once such dispersal occurs, whether intentional or not, mandatory evacuation is the wrong response.

SUMMARY

Rather than attempting to summarize this presentation before interaction with others in Erice, I provide highlighted points from the Nov. 2011 paper:

DUCK and COVER: After an unexplained dazzling flash of light, do not approach windows, and stay behind cover for at least a minute to prevent injuries from flying and falling debris, such as broken glass.
GO IN, TUNE IN: The best initial action immediately following a nuclear explosion is to take shelter in the nearest and most protective building or structure and listen for instructions from authorities. (EOP, 2010)
DON'T DRIVE: If in a car, try to find shelter immediately until given official information. A car does not offer protection.
STAY INDOORS: People should expect to remain sheltered for at least 12 to 24 hours. During that time, the intensity of fallout radiation will decrease greatly, allowing for less hazardous egress from dangerous fallout areas. (EOP, 2010)
GET CLEAN: Radioactive fallout particles can spread quickly and remain on the body and clothes until removed. Those in potentially fallout contaminated areas should take off the outer layer of clothing (including shoes) and wipe or wash exposed skin and hair upon leaving a contaminated area.
HELP OTHERS: Radiation injuries and fallout contamination on people do not represent a threat to others. People should allow others to enter their building, help decontaminate, render first aid, and share information.

This November 2011 report applies tools of large-scale computing and scientific analysis to the mitigation of the effects of the explosion of an improvised nuclear device, IND. But far more attention and investment are necessary to provide tools and motivation for the general public to be ready to act in their own interest and that of their society. These tools are likely to be highly prized in advertising and politics, but potential misuse should not impede their application in mitigating the consequences of disaster.

GAP ANALYSIS OF EU COUNTERTERRORISM RESEARCH INITIATIVES[1]

FRIEDRICH STEINHÄUSLER
Physics and Biophysics Division, University of Salzburg
Salzburg, Austria

ABSTRACT

Some of the 27 Member States in the European Union (EU) had gained extensive experience individually in counterterrorism over decades in their fight against well organized groups. However, it required the terror attacks in the U.S. in 2001, in Madrid 2004 and in London 2005 to shape the EU's reaction and to initiate coordinated research activities in the 27 Member States. The current 7th EU Framework Programme (FP7) has allocated € 1,350 million for *Security* for the period 2007 to 2013. The key FP7 activities relate to restoring safety and security in case of crisis. The paper provides an analysis of 17 major EU counterterrorism research projects and reviews the U.S. approach to security for comparison.

The following main gaps were identified in the implementation of the EU research initiatives: (1) diverse counterterrorism strategies between Member States; (2) insufficient consideration given to terrorism risk-assessment; (3) inadequate coordination/ cooperation between EU funded civil security research and similar research activities by the European Defence Agency (EDA) and European Space Agency (ESA); (4) dominance of EU defence industry over SMEs, universities and R&D organisations; (5) cumbersome bureaucracy in administrating research projects.

EU COUNTERTERRORISM RESEARCH STRATEGY

Up to 11 September 2001, Europe's approach to counterterrorism can be characterized as, *Terrorists are criminals who should be arrested by police and tried at court*. Considered mostly as a national security concern only, coordinated trans-boundary counterterrorism actions were rare. Some of the 27 Member States of the European Union (EU) had gained extensive experience individually in counterterrorism over decades in their fight against well organized, but relatively small groups, such as the Irish *Real IRA, Orange Volunteer Force* and *Red Hand Defenders*, Spanish *ETA* and *GRAPO*, Italian *Brigate Rosse* and *Nuclei Territoriali Antimperialisti*, French *AZF* group, and German *Baader-Meinhof Bande*.. However, it required the terror attacks in the U.S. in 2001, in Madrid 2004 and in London 2005 to shape the EU's counterterrorism reaction and to initiate coordinated research activities on this topic area in the Member States. The new threats, their pronounced trans-boundary character ignoring national borders and threatening the EU from within and also its outside interests, required a comprehensive EU strategy for security research. This process has a rather complex, partially event-driven history and

[1] This work was partially supported by the EU 7th Framework Programme (Project BRIDGE: "Bridging resources and agencies in large-scale emergency management", Grant agreement no: 261817).

continues to develop further, as shown schematically in Figure 1. In December 2003, the need for a coordinated action was recognized by the European Council and lead to the development of the comprehensive ***European Security Strategy*** (ESS). The ESS identified the following key terrorism-related threats for the EU:[2]

- *Pressure-induced terrorism.* The ESS identified as causes for the new threat environment as pressures of modernisation, cultural, social and political crises, and the alienation of young people living in foreign societies.[3]
- *Proliferation of weapons of mass destruction (WMD).* The ESS warned that (a) Advances in the biological sciences are likely to increase the potency of biological weapons; (b) Terrorist groups may acquire WMD.
- *Regional conflicts.* According to the ESS such conflicts, posing a threat to minorities, fundamental freedoms and human rights, can lead to extremism and terrorism, ultimately even provoke state failure and thereby adding to regional instability.
- *Organised crime.* The ESS linked the activities of organised crime (e.g., trafficking in drugs and weapons) with terrorism, in extreme cases potentially even dominating a State.

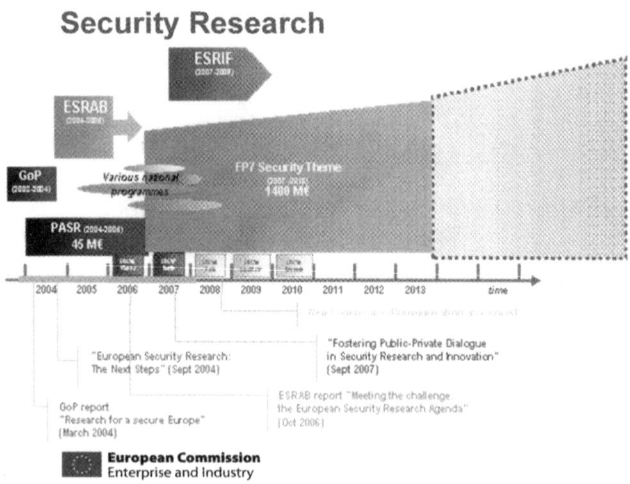

Fig. 1: Development of EU security research between 2004 and 2012 (ref: Paolo Salieri, EU security research related to restoring safety and security in case of crisis, *European Commission, Enterprise and Industry).*[4]

---

[2] European Security Strategy (ESS), *A secure Europe in a better world*, Brussels, 12 December 2003.
[3] This phenomenon is observed frequently among dissatisfied members of the 3rd generation of immigrants.
[4] http://www.civil-protection.com/Rueckblick/Reden-2010/binarywriterservlet?imgUid=bd210df6-ed56-2031-fe7b-ac907b988f2e&uBasVariant=11111111-1111-1111-1111-111111111111 (last visited 12 July 2012).

Subsequently, the ***Group of Personalities for Security Research*** (GoP) identified in 2004 the need to develop capabilities within the EU to protect its citizens at home and to deploy resources for peace keeping and humanitarian aid abroad.[5] The GoP emphasized in particular to use technology as a "force enabler", recommending a minimum of € 1 billion annually for security technology development.

In 2004, the first concrete R&D-actions in security were initiated within the ***Preparatory Action in Security Research*** (PASR), designed to bridge the gap between generic civil research and national and intergovernmental programmes oriented to defence procurement needs.[6] Starting with a rather modest budget of € 15 million, the final budget allocation was somewhat increased to € 65 million, distributed over the period of 2004 to 2006. This financial allocation was inadequate in view of the multiple fields of activity foreseen, such as:

- Improving situation awareness.
- Optimising security and protection of networked systems.
- Protecting against terrorism (including bio-terrorism and incidents with biological, chemical and other substances).
- Enhancing crisis management (including evacuation, search and rescue operations, active agents control and remediation).
- Achieving interoperability and integrated systems for information and communication.

Expectedly, the three annual calls for proposals were considerably over-subscribed and only 39 projects could be supported. It is noted that 60% of these projects were under the leadership of companies predominantly active in the defense sector, i.e., Thales (France), EADS (The Netherlands), Finmeccanica Companies (Italy), SAGEM Défense Sécurité (France), and AeroSpace and Defence Industries Association of Europe (Belgium); together with BAE Systems (UK), they participated in 67% of all 39 projects.[7]

Subsequently, the ***European Research Agenda*** was defined in the ESRAB Report in 2006.[8] In a combination of extensive experience and know-how altogether more than 300 experts from the end-users community jointly defined the required medium-term research development together with suppliers and scientists active in security research.[9] The report identifies three priority research paths for further investment:

---

[5] Report of the Group of Personalities in the Field of Security Research, *Research for a Secure Europe*, ISBN 92-894-66 11-1 (2004).

[6] European Commission Decision 2004/213/EC of 3 February 2004 on the implementation of the Preparatory Action on the Enhancement of the European industrial potential in the field of security research.

[7] Ben Hayes, *NeoConOpticon: The EU Security-Industrial Complex*, Transnational Institute (ISSN: 1756-851X), September 2009.

[8] European Security research Advisory Board, Meeting the challenge: the European Security Research Agenda, September 2006.

[9] The ESRAB Board was created by Commission Decision 2005/516/EC on 22 April 2005 and published its final report on 22 September 2006.

(1) Capability Development:
(1) Detection and identification capabilities (e.g., abnormal crowd behavior; small boats in blue borders; dangerous goods; biometrics); (2) Information management systems for first responders and decision makers (e.g., fusion of real-time sensor data; integration of modeling and situation analysis tools); protection of supervisory control and data acquisition (SCADA) systems; (3) Risk assessment, modeling and simulation tools (e.g., setting priorities for multiple threats; testing mitigation measures; improved training; development of intervention capabilities, such as decontamination).

(2) System Development:
Integration of different capabilities, technologies and disciplines into innovative combinations.

(3) System-of-Systems Demonstration:
Integrated and interoperable aftermath crisis management; European-wide integrated border control system; Logistic and supply chain security; Security of mass transportation; Integrated approach for CBRN threat assessment, consequence modeling, detection and identification of agents and devices, incident management tools, prevention, decontamination and medical care.

Taking advantage of EU´s technological strengths, this ultimately lead to the current ***7th EU Framework Programme*** (FP7) with an overall budget of € 50.5 billion, of which *Security* has been allocated € 1 350 million for the period 2007 to 2013. The key FP7 activities related to restoring safety and security in case of crisis are: (1) Security of citizens; (2) Security and decontamination of intentionally contaminated drinking water; (3) Security of infrastructure; (4) Border Security; (5) Crisis management; (6) Identifying the needs of medical first responders in disasters, and (7) Interconnected and interoperable systems.

In 2009, the ***European Security Research & Innovation Forum*** (ESRIF) released its Final Report.[10] This informal group of more than 600 experts and 65 distinguished personalities from Europe, set up in September 2007, provided a strategic innovation agenda for the EU. It defined several topic areas which should be addressed in formulation of the 8th EU Framework Programme (FP 8), looking at scenarios with a 2030 time horizon to frame. ESRIF emphasized that Europe's main objective is to preserve its values as an open society and respect for fundamental rights while addressing the increased security threat. However, it also acknowledged that the European society is highly dependent on technologies and critical infrastructures. In its vision of the future security needs the EU needs to embrace a wide range of risks, from natural to man-made incidents; capabilities and capacities need to be mobilized in order deliver equipment and services to deal with these risks.

[10] European Security Research & Innovation Forum (ESRIF), *ESRIF Final Report*, December 2009; http://ec.europa.eu/enterprise/policies/security/files/esrif_final_report_en.pdf (last visited 13 July 2012).

MAJOR EU COUNTERTERRORISM RESEARCH PROJECTS

The altogether 176 security research projects in EU FP 7 cover a wide range of subject areas related to counterterrorism, such as: CBRN attacks and countermeasures, first responder needs, critical infrastructure protection, increasing cyber security, enhancing transport security, toolboxes for on-scene deployment , virtual reality training methods, novel cross-border/multi-agency crisis management, optimized intelligence analysis, IED and EOD, UAV, underwater surveillance, real-time drinking water safety, impact on citizens rights, etc. Out of the 27 Member States 22 countries participate as Project Coordinators or members of a consortium.

This section provides an overview of selected projects already completed or currently ongoing, addressing counterterrorism-related topic areas.

DETECTER[11]—*Secure but with rights*

The EU is increasingly including requirements for human rights into the implementation of security projects. DETECTER investigated new intrusive surveillance technologies, which are of particular concern, ranging from "Smart CCTV" camera-surveillance of suspicious behaviour in public places to secret Internet monitoring and data-mining. In this project, high-tech developments in detection methods and their application by police were analyzed with regard to their compliance with ethical standards. The results of the project provide practically applicable advice on how to take counter-terrorism measures that cover both the security of European citizens and their human rights.

Contact: Tom Sorell, University of Birmingham, Birmingham, United Kingdom.

DECOTESSC1[12]—*CBRNE Deficits*

The CBRNE counterterrorism analysis carried out in DECOTESSC1 identified more than 150 gaps in CBRNE, of which the most important ones were:

> Too much focus on worst case scenarios and limited validation of scenarios and threat assessment; too little research/ identification of improvised CBRNE devices and production facilities; need for realistic training, possibly together with citizens and training with live agents; need for better deployment of equipment in terms of user friendliness (non-experts), mobility, risks etc.; limited national and international (structures for) cooperation/collaboration and sharing of (classified) information/knowledge/best practices; lack of uniform situation awareness and alert state, involving all parts of the society including the public; lack of stand-off detectors/Instant scanning detector: fast, sensitive, reliable, affordable, without disturbing business continuity; lack of possibility to tag chemicals, which are openly available but can be used as precursors; lack of minimum standards for security relevant infrastructures to ensure the use of adequate detection equipment

[11] Detection technologies, terrorism, ethics and human rights http://www.detecter.bham.ac.uk/index.html (last visited 14 July 2012).

[12] Demonstration of CounterTerrorism System-of-Systems against CBRNE phase 1 http://www.decotessc1.eu / index.html (last visited 14 July 2012).

and control barriers; lack of identification of CBRN-affected people; lack of detection systems that can detect multiple threats and detect degree of *hazard* (instead of *agents*); lack of capabilities for real time threat assessment; lack of adequate personal protection equipment (lighter and comfortable, allowing precise gestural acts); insufficient medical capability related to CBRNE-incidents (capacity, triage, treatment); insufficient equipment and techniques for decontamination of electronics, rough or porous surfaces (like concrete surfaces or wallpaper); how to perform safe sampling without putting First Responders at risk; gap in the leadership within the EU, such as the lack of an EU coordinating body, for keeping uniform rules for measures and for information.

Contact: Mark van den Brink, TNO, Delft, The Netherlands.

FORSEC[13]—*Future Insecurity*

The project focused on the obtaining a shared vision and coherent and holistic approach to current and future threats and challenges for European security and to identify future trends with relevance to European security. For this purpose it used a Delphi survey and the input from altogether 124 EU experts. Through this participatory foresight process the project facilitated the creation of a basis for the scenario building process, following the multiple threads of existing work on the future of European security. The results assisted in reducing security gaps by identifying knowledge gaps and uncertainties.

Contact: Kristina Rintakoski, Crisis Management Initiative, Helsinki, Finland.

PRISMATICA[14]—*Secure Transport*

The project involved important European transport operators (London, Paris, Brussels, Milan, Prague and Lisbon), research institutions, manufacturers and transport consultants and tasked them with analyzing personal security aspects, both real and perceived, understand what these factors are and to propose technical, operational and social solutions. The consortium developed a set of tools, enabling an instrumented detection/action environment. Thereby, control room operators obtain timely information to improve personal security (reported and perceived) in public transport sites (in particular in metropolitan railway systems). This includes the deployment of a range of sensors and transmission technology, combined with usable human machine interfaces. An essential part of the system is *MIPSA*, a supervisory computer providing a single point of contact with an operator and a means of controlling and communicating with intelligent devices over a scalable Local Area Network. A special audio surveillance system, consisting of a PC with dedicated DSP boards, has been developed to detect abnormal sound signatures as they would originate from a passengers shouting for assistance. The system allows for the automatic detection of potentially dangerous

[13] Europe´s evolving security: drivers, trends and scenarios http://www.foresec.eu/wp3_docs/Foresec_report.pdf (last visited 12 July 2012).

[14] Proactive Integrated systems for Security Management by Technological, Institutional and Communication Assistance http://www.ist-world.org/ProjectDetails.aspx?ProjectId=84b78bc677804ed0aa5fa3b29f999134&SourceDatabaseId=e4fcfde0182a45898e8741a1abae3984 (last visited 12 July 2012).

situations, such as violence, trespass, congestion, fires, suspect packages and suspect individuals, theft, vandalism, and ticket fraud.

Contact: Pierre Lagrange, Régie Autonome des Transports Parisiens (RATP), Paris, France.

NoAH[15]—*Cyber Security*

Cyber-attacks, deploying viruses, trojans, and other types of malicious software, have the capability to compromise tens of thousands of internet-connected computers within seconds. The project designed an infrastructure of affined, geographically dispersed *honey pots*, gathering and correlating data about attackers, their methods, and actions on the internet.[16] It developed techniques for the automatic identification of novel attacks and for the automated generation of corresponding signatures, enabling the effective containment of the spread of an attack. A pilot NoAH infrastructure demonstrated the effectiveness and provided comprehensive attack information, such as, where did the attacker come from, how did he enter the system, what did he try to do after entering, etc. In addition, NoAH delivered a distributed security analysis infrastructure for internet service providers, national research and education networks and security organizations.

Contact: Evangelos Markatos, Foundation for Research and Technology, Iraklion, Greece.

SAFEE[17]—*Automated Aviation Security*

SAFFEE provided methods to prevent acts of terrorism on board of an aircraft. For this purpose the consortium developed a spectrum of threat sensing systems, and the corresponding response actions against physical person(s) or electronic intruders. The consortium considered several threat scenarios, but excluded some it viewed as being outside of its project scope.[18] The onboard threat detection system, able to detect upcoming threats onboard an aircraft, is capable of detecting suspicious personal behavior and dangerous materials. The results obtained can be categorized as follows: (1) Onboard threat detection with an integrated threat detection system, based on multiple sensor information; (2) Threat assessment and response management system as an urgency decision making tool; (3) Flight protection, including an emergency avoidance system and an automatic control of the aircraft for a safe return; (4) Data protection system securing all the data exchanges (in and out the aircraft); (5) Security evaluation activities, including legal and regulatory issues about citizens' privacy and rights. The SAFEE systems have interfaces for the pilot in the cockpit, for the cabin crew and

---

[15] NoAH: a European Network of Affined Honeypots; http://www.fp6-noah.org/ (last visited 12 July 2012).

[16] *Honeypots* are computer systems that do not provide real production services, but are intentionally vulnerable, and at the same time closely monitored systems, that wait to be compromised by attackers.

[17] Security of Aircraft in the Future European Environment ; http://www.safee.reading.ac.uk/ (last visited 12 July 2012).

[18] These scenarios are: *biological attack*, *use another aircraft to crash into the 'target' aircraft* and *MANPADS*.

security staff in the cabin, on-board crew communication links, and air/ground (voice and data) communication links.

Contact: Daniel Gaultier, Sagem Defence Securite, Paris, France.

ISOTREX[19]—*Timely Detection of Explosives*

In view of the continuing threat from explosives carried on persons (suicide terrorist), in luggage or hidden in vehicles, fast screening sensors were developed, which are able to detect hidden explosives from their traces released in the environment (gaseous emissions, dispersed particles on packing surfaces or cloths).The consortium focused on a portable instrument for particle/vapour detection to be installed at large check points (e.g., airports, customs, main post-offices) and for special police teams. Due to the modular design of the system it can be separated into two sub-systems and placed in different checking points according to different scenarios. The principle for the detection of particle and liquid is the LIBS (Laser Induced Breakdown Spectroscopy) technique. Explosive vapour detection is based on IR (infrared) absorption methods, either cavity ring-down or laser photo-acoustic spectrometer. Software was developed for each instrument, allowing prompt explosive detection in accordance with a project-generated data base, thereby reducing false alarms.

Contact: Antonio Palucci, ENEA, Rome, Italy.

HAMLeT[20]—*Tracking HazMat*

The project assists security staff indoors in real-time to classify, track and localize potential threats. The system uses advanced sensors and multiple sensor fusion techniques. Person-tracking data (originating from a laser scanner) were fused with additional attribute information in order to identify threats coming from hazardous materials. Several metal oxide sensors for hydrocarbons (e.g., fuels, alcohols or solvents) were deployed, using the dependence of the surface resistance of a thin film of metal oxide particles on the amount of the analyte in ambient air. The sensor response yields mainly two parameters: (a) signal strength; (b) time delay; both are dependent from the distance between the person and the sensor system. Analyte diffusion, convection and transportation by a ventilation air stream reduce the gas concentration and influence both parameters. Many explosives or explosive related compounds (ERC) do not evaporate enough analyte molecules into air, except for TATP.[21] For this analyte a special multi sensor quartz microbalance sensors system was developed. However, the consortium recognized that up to now no technology for airborne ERC detection is sufficient for a fast and sensitive chemical multisensory system with stand-off properties.

Contact: Wolfgang Koch, Fachhochschule Bonn-Rhein-Sieg, 53757 Sankt Augustin.

---

[19] Integrated system for on-line trace explosives detection in solid and vapour state; ftp://ftp.cordis.europa.eu/pub/fp7/security/docs/isotrex_en.pdf (last visited 12 July 2012).

[20] Hazardous Material Localisation and Person Tracking; ftp://ftp.cordis.europa.eu/pub/fp7/security/docs/hamlet_en.pdf (last visited 12 July 2012).

[21] Triacetonetriperoxide , also referred to as "homemade explosive" ,often used by terrorists, has a relative high vapour pressure. With a sensor array the TATP can be unequivocally identified in a few seconds.

AEROBACTICS[22]—*Modeling Bio-Threats*

Biological crisis management, suffers from the current inaccuracy of predictive microbial dispersion models, since existing models are based on inadequate assumptions and parameters. The consortium developed and tested two models, one for background levels and long-range dispersion, and one for the aerial dispersion from an intentional release. Data were collected by sampling and analyzing representative sections of the atmosphere, as well as soils and plants. Both, natural events and intentional releases of micro-organisms were used as input for the development of improved predictive modeling tools for short and regional range aerial dispersion of micro-organisms from single sources. Micro-meteorological parameters relevant for the viability of microorganisms were accounted for.

Contact: Ulrich Karlson, National Environmental Research Institute, Roskilde, Denmark.

BEMOSA[23]—*Security through Airport Staff*

Airports continue to be apprehensive about the possibility of a terrorist attack. This project addressed the issues of the people operating the security systems and how their decisions impact on operations in the airport. The consortium observed the behaviour of security and non-security employees and carried out in-depth interviews with staff in four European airports. The results show that: (a) Airport employees don't tend to make individual decisions; (b) They are rarely alone; (c) Despite of the prevalence of group patterns, training for security and non-security professionals is still carried out on an individual basis; (d) When making an individual decision, employees tended not to follow the rules and regulations, whilst when in a group they tended to be more compliant; (e) Strong negative interaction exist with hierarchy; (f) Colleagues with an equal status tended to interact more effectively with each other. Therefore, actual security behaviour may deviate from rules and regulations to adapt to specific situations. An important finding was also that 2/3 of the security staff at airports had actually never faced a real security threat. Furthermore, security staff believe that in most cases airport top management does not intend to take their opinion into consideration. The results of the project were used for an innovative and cost-effective training programme that includes a predictive model of behaviour in real life crisis situations, blending current procedures with actual security behaviour.

Contact: Avi (Alan) Kirschenbaum, Technion—Israel Institute of Technology, Haifa, Israel.

[22] Assessment of the quantity, identity, viability, origin and dispersion of airborne microorganisms for application in crisis management tools; ftp://ftp.cordis.europa.eu/pub/fp7/security/docs/aerobactics_en.pdf (last visited 12 July 2012).

[23] Behavioural Modeling for Security in Airports; http://www.bemosa.eu/about-bemosa (last visited 12 July 2012).

CRISMA[24]—*Integrated Crisis Management*

At times of a large crisis frequently there is little coordination between the relevant parties' different systems. The CRISMA consortium develops a simulation-based decision support system (*Integrated Crisis Management System*—ICMS), which will provide crisis management with improved action and preparedness through modeling. By facilitating the simulation of realistic crisis scenarios, a crisis manager can envisage the following: (1) multi-sectored crisis scenarios and the consequences of an incident, (2) impacts resulting from alternative actions, (3) strategic decisions on capabilities, related investments, reserves and inventories, (4) deployment of resources dedicated to crisis response in-line with the evolvement of a crisis, and (5) action plans for preparedness and response phases of the crisis management. The provision of the IT-based support will be available for decision making and auditing of the decisions made during both real and simulated crisis events. Thereby, this improves the understanding how various accidents and crisis scenarios affect people, society, infrastructure, buildings, services and the economy. The project accounts inter alia for the impact of natural disasters on industrial activities and critical infrastructures. Also, the ICMS and associated simulation tools can be used to fuse information provided by different parties collaborating across borders or in vast areas.

Contact: Anna-mari HEIKKILÄ, VTT Technical Research Centre of Finland, Espoo, Finland.

SECURENV[25]—*Environmental Terrorism*

SECURENV contributes to the security of citizens from future threats caused by non-state actors carrying out deliberate attacks on the environment. The project analyses major industrial and environmental accidents from a security perspective. The methods deployed are: foresight methods, scenario-building techniques and expertise in the consortium (such as brainstorming, trend-spotting, mind-mapping and SWOT analyses). In addition, expertise outside the consortium was used through a Delphi survey, involving more than 600 experts in Europe and beyond, as well as scenario-building workshops with typically 15-20 experts. Furthermore, back casting was used at the end of the scenario workshops and during the final workshop to arrive at relevant policy recommendations. The threat analysis covered natural phenomena (fires, floods), industrial accidents (chemical, biological, etc.) among others. A key result is a database (MySql) on novel and emerging threats for the environment and means by which this can be deliberately influenced.. MySql database hosts over 330 records with qualitative and quantitative parameters used to categorize the entries. These data cover the wide spectrum of anecdotal evidence on deliberate destruction, to the actual incidents related to environmental terrorism in the recent past. Organised crime emerges in this domain with an increasing number of such incidents. MySql provides practically applicable information targeted for end-users.

---

[24] Modeling crisis management for improved action and preparedness; http://cordis.europa.eu/search/index.cfm?fuseaction=proj.document&PJ_LANG=EN&PJ_RCN=12636673&pid=20&q=FD8A9BBC079BD5FCECD584ADBD3CE6A7&type=adv (last visited 12 July 2012).

[25] Assessment of environmental accidents from a security perspective; http://www.securenv.eu (last visited 12 July 2012).

Contact: Balazs Bodo, Geonardo Environmental Technologies Ltd., Budapest, Hungary.

ODYSSEY[26]—*Sharing Gun-crime Information*

This project creates a secure platform for sharing information from multiple, heterogeneous sources on gun-crimes in all EU Member States. Data are combined in a high volume data repository. Using semantic knowledge extraction and data-mining, fast decision making and alert are enabled. A user interface permits automated generation of directed graphs through drag-and-drop of components, identifying crime incidents. The security architecture uses industry-standard encryption and applies this to inputs to and outputs from the message queue. An initial ontology and database structure demonstrates connectivity from the message queue to the back-end systems. The ontology has also been used to experiment with semantic querying. Results from querying both the ontological engine and the database are displayed in the same interface. Thereby, ODYSSEY creates European standards for ballistics data collection, storage and sharing of ballistics intelligence. Furthermore, it facilitates data mining and extraction of knowledge about gun crime across the EU, generating red flags through the analysis of complex data with multiple reference models. Ultimately, the consortium will extend the management of ballistics information across the wide range of IT platforms within the EU, including a secure ICT platform for the sharing of information about serious crime and terrorism where firearms are involved.

Contact: Simeon Yates, Sheffield Hallam University, Sheffield, United Kingdom.

SECUREAU[27]—*Safe Drinking Water*

Intentional contamination of drinking water distribution systems is likely to have major public health, economic and psychosocial consequences. The consortium develops methods to (a) limit the impact on the population due to contaminated networks, and (b) launch an appropriate response for rapidly restoring the use of the network after deliberate contamination. The project covers the complete cycle, i.e., from the rapid detection of unexpected changes in water quality (possibly indicating a deliberate contamination event), to adaptation of known analytical methods for the detection of specific CBRN contaminants in water and in bio-films, to localizing the point sources of contamination and subsequently the contaminated area. This includes the identification of efficient and realistic decontamination procedures, as well as controlling the efficacy of the corrective actions by analysing the water bulk and especially the pipe walls and the deposits.

Contact: Sylvain FASS, Universite de Lorraine, Nancy, France.

[26] Strategic pan-European ballistics intelligence platform for combating organised crime and terrorism; http://www.odyssey-project.eu (last visited 12 July 2012).

[27] Security and decontamination of drinking water distribution systems following a deliberate contamination; http://www.secureau.eu/ (last visited 12 July 2012).

CAST[28]—*Making It Fit*

Future security threats are likely to exceed the current standard training of first responders (FR)[29] in the EU member States. They must be prepared to face the deployment of weapons of mass disturbance, mass killing and mass destruction. Also, past experience has shown that EU first responders have significant difficulties dealing with extreme natural or technical catastrophes (*Flood of the Century*; explosions at chemical and nuclear facilities), in particular in cross-border events.

Training course curricula on disaster management for FR in EU member states have been comparatively assessed with a specially developed matrix-based software: (1) for all EU member states, (2) as derived from international best practices in the U.S., Russia, and Israel—all three countries with extensive experience in this field. The comparative assessment covered didactic areas (electronic and hardcopy teaching materials used, computer modeling, field exercises); subject areas (terror threats to FR; risk assessment and management; catastrophic terrorism; synchronization of response staff; comparative evaluation of training course curricula).

The results of the project were:

- Creation of an EU-security training curricula database;
- Identification of current gaps in the EU training curricula;
- Recommendation of an *Action Plan* for eliminating training deficiencies;
- Development of a standardized security-centered training curriculum for first responders on disaster management;
- Analysis of the application of computer modeling for enhancing situation awareness of first responders on scene;
- Development of a portable, *Virtual Reality Training Facility*, embedding the trainee in surround sounds, video film and smells of a scene simulating a terror attack. The system was field-tested with FR and Special Forces.

Contact: Friedrich Steinhäusler, University of Salzburg, Salzburg, Austria.

BRIDGE[30]—*Managing the Mega-Catastrophe*

Multiple events in the recent past have clearly shown that some crisis can exceed the capability of dedicated organizations due to the number of people and/or the size of the area affected, and/or the complexity of the crisis as it develops rapidly. The main goal of this project is the provision of technical support for multi-agency collaboration in large-scale emergency relief efforts, also in situations necessitating trans-boundary collaboration. The key to this is to ensure interoperability, harmonization and cooperation among stakeholders on the technical and organisational level. This is achieved by the development of the *BRIDGE Platform*, which will integrate existing emergency response systems and platforms; integrate laboratory tools into everyday working practice; represent and model organizational workflows and communication processes; comprise a

[28] Comparative assessment of security-centered training curricula for first responders on disaster management in the EU; http://castproject.eu (last visited 12 July 2012).
[29] Police, fire fighters, paramedics.
[30] Bridging resources and agencies in large-scale emergency management; http://www.bridgeproject.eu/en (last visited 17 July 2012).

middleware allowing data, system and network interoperability; advance human-computer interaction techniques. Several innovative tools are developed, ranging from 3D computer modeling of damaged buildings, victims´ injuries and contaminated areas for assisting in risk assessment, to smart phone apps allowing victims to establish ad hoc mobile networks, to a Master Table fusing all data describing the ongoing event and identifying the resources needed as well as those available. Field tests of the BRIDGE developments, simulating threat scenarios (e.g., explosions and vehicles on fire in a tunnel), are carried out under harsh environmental conditions in a dedicated underground facility in Flums (Switzerland). Yearly demonstrations of the different BRIDGE tools are planned across Europe, reviewed by a specifically established End User Advisory Board. Ultimately BRIDGE will provide innovative technical solutions in three different areas: Interoperability of data, systems and technology; Exploration of a common operational picture; Run-time inter-agency and inter-agent collaboration. In this manner the project will increase the security and safety of citizens by developing technical and organisational solutions that significantly improve management of large scale crisis and emergencies in the EU Member States involving multiple organisations.

Contact: Geir Horn, SINTEF, Oslo, Norway.

CATO[31]—*CBRN Tools for Practitioners*

CATO develops a comprehensive *Open Toolbox* for dealing with CBRN crises due to terrorist attacks using non-conventional weapons or attacks on facilities with CBRN material. For the development of this Tool Box CATO will combine CBRN knowledge, planning & response know-how across the EU in order to enable the development of a specific, holistic CBRN Decision Support System (DSS). The topic areas covered range from CBRN preparedness, detection, and response to recovery from a CNRB attack. In this manner CATO will support all parties involved, such as policy makers, decision makers, incident commanders, field teams and specialists alike. The in-depth analysis will address the complexities of CBRN preparedness and resilience, such as medical response, societal and psychological issues, organisational and operational approaches, and multiple-use equipment. Emphasizing practical application in its ultimate goal, CATO provides a dedicated simulation-centered *CATO Laboratory* (both virtual and hosted by some of the partners), enabling learning about CBRN attacks (e.g., new threats and responses) and the training of field teams. CATO's Open Toolbox will meet these challenges and will serve as an infrastructure for a dedicated and customized DSS that can be adapted to local, national, organizational, political and financial constraints as well as to different levels of exposure to CBRN threats.

Contact: Victor Remez, Tel Aviv, Israel.

[31] CBRN Architecture, Technologies and Operational Procedures; http://cordis.europa.eu/projects/index.cfm?fuseaction=app.details&TXT=&FRM=21&STP=10&SIC=&PGA=&CCY=ILCOUNTRY&PCY=&SRC=&LNG=en&REF=102095 (last visited 12 July 2012).

SOFIA[32]—*Flying out of Danger on Autopilot*

Civil aviation has been subject of multiple terror attacks, ranging from hijackers taking passengers as hostages to misuse of a hijacked plane as cruise missile attacking buildings. The objective of SOFIA was the strengthening of on-board systems in order to minimize this security threat. This project was a follow-up to the results obtained in the project SAFEE SP3, which had advanced a flight reconfiguration function (FRF) system that automatically returns the aircraft to the ground if it comes under a security threat. SOFIA assessed the suitability of the FRF concept for the intended purpose, made the necessary design refinements, developed the final configuration and validated the FRF system. It could be shown that the FRF concept - in the event of a security emergency such as a hijacking of the plane - would safely and automatically take over the piloting of the plane and land it safely at the nearest airport. Most importantly, safety and operability of these procedures were approved by ATCOs. Furthermore, the consortium initiated the assessment of regulatory and certification frameworks for SOFIA through discussions with ICAO, EASA, Eurocontrol, SESAR and several national CAAs. The discussions revealed that an authority and/or system (like a ground security decision system) at European level is needed in order to generate and track the flight plan for an FRF guided aircraft, and coordinate with national authorities, ANSP and airports.

Contact: Juan Alberto Herreria Garcia, Ingeniera de Sistema para la Defensa de Isdefe S.A., Madrid, Spain.

GAP ANALYSIS

The EU is facing the daunting task of creating a comprehensive, coordinated counterterrorism research programme to protect its approximately 400 hundred million citizens, subject to terror attacks for decades. In order to comprehend the magnitude of the issue this effort is compared with the response by the United States, having experienced a series of coordinated terror attacks on 11 September 2001.

U.S. Approach

In response to the terror attacks in New York City, Washington D.C. and Pennsylvania on 11 September 2001 the U.S. Government created the *Office of Homeland Security* (OHS) to coordinate "homeland security" efforts. Already 27 days after the attacks the Director of the OHS started his duties. On 25 November 2002 the OHS grew into the larger *Department of Homeland Security* (DHS) with more than 200 000 employees. It was allocated a budget of U.S. $98.8 billion in fiscal year 2011.[33] In addition to protecting the U.S. against terrorism DHS also provides grants to State and local governments for improvement of local homeland security. Since 2003 these governments received U.S. $31 billion to strengthen their counterterrorism capabilities.[34]

---

[32] 'Safe automatic flight back and landing of aircraft; http://cordis.europa.eu/search/index.cfm?fuseaction =result.document&RS_LANG=EN&RS_RCN=12365210&q= (last visited 12 July 2012).

[33] U.S. Department of Homeland Security Annual Financial Report FY 2010 (*vid.* pp.37,38,41). http://www.dhs.gov/xlibrary/assets/cfo-afrfy2010.pdf (last visited 16 July 2012).

[34] Priest, Dana and Arkin, William, Monitoring America, Washington Post, 20 December 2010; http://projects.washingtonpost.com/top-secret-america/articles/monitoring-america/?hpid=topnews

Since the terror attacks in September 2001, total U.S. Government expenditures on domestic homeland security have risen by $580 billion over those in place in 2001; if the private sector costs and opportunity costs of delays and inconveniences associated with enhanced security regulations are added, the increase in expenditures on domestic homeland security in the U.S. in the decade exceeds U.S. $1 trillion.[35]

The scope of responsibilities of DHS covers most aspects of homeland security; in some sectors DHS carries the responsibility for all elements of risk analysis. However, there is a prevalence to focus on the potential *consequences* without adequate regard to the *probability* for the event to happen. A typical example is the DHS focus on nuclear terrorism, i.e., deployment of a crude nuclear device by adversaries. In view of the proven difficulty of terrorists obtaining weapon-usable nuclear material the probability of terrorists building such a device is low.[36] An analysis by the U.S. Academy of Sciences concluded that DHS has *paid little effective attention* to *features of the risk problem that are fundamental.*[37] This becomes even more evident if the number of terror attacks against U.S. citizens abroad and at home are taken into account in order to judge the risk from terrorists on U.S. soil and in foreign countries (Table 1).

Table 1: Security risk due to terrorism for U.S. citizens in the homeland and abroad.

Type of Incident	No. of Incidents
Terror attacks against U.S. foiled since 11 September 2001[38]	50
U.S. citizens killed in terrorist incidents inside the United States between 2001 and 2011[39]	30
U.S. citizens worldwide killed as a result of incidents of terrorism in 2010[37]	15
U.S. citizens worldwide injured as a result of incidents of terrorism in 2010[37]	9
U.S. citizens worldwide kidnapped as a result of incidents of terrorism in 2010[37]	0

(last visited 16 July 2012).

[35] The costs of the terrorism-related wars in Iraq and Afghanistan are not included (ref.: Mark Stewart and John Mueller, Terror, Security, and Money: Balancing the Risks, Benefits, and Costs of Homeland Security, Oxford University Press (2011).

[36] Since 1991 altogether 27 cases of illicit trafficking involving weapon-usable/grade nuclear material were detected (Lyudmila Zaitseva and Friedrich Steinhäusler, Stanford-Salzburg Illicit Trafficking and Orphan Sources Database(DSTO).

[37] Committee to Review the Department of Homeland Security's Approach to Risk Analysis, Review of the Department of Homeland Security's Approach to RISK ANALYSIS, National Research Council of the National Academies, The National Academy Press, Washington, D.C., USA, ISBN—10: 0-309-15924-5 (2010).

[38] James Jay Carafano, Steven Bucci and Jessica Zuckerman, 50 terror plots against U.S. foiled since 9/11, Report Heritage Foundation (2 May 2012); http://www.securityinfowatch.com/news/10708919/50-terror-plots-against-us-foiled-since-9-11 (last visited 17 July 2012).

[39] Terrorist Attacks in the U.S. or Against Americans, http://www.infoplease.com/ipa/A0001454.html#ixzz20rw971jL; (last visited 12 July 2012); Global Terrorism Database, http://www.start.umd.edu/gtd/search/Results.aspx?chart=country&casualties_type=f&casualties_max=&start_yearonly=1970&end_yearonly=2010&dtp2=all&country=217 (last visited 17 July 2012).

For comparison, during the period 2002 to 2010 a total of 146 193 murders were committed in the U.S.[40] It is about 5,000 times more likely for one of the almost 314 million U.S. citizens to be murdered than becoming a victim of terrorism in the homeland. Using a conservative threat probability, the ratio of benefit to cost is only 0.03, i.e., the U.S. spending $1 on counterterrorism provides only 3¢ of benefits.[41]

EU Approach

EU security research focusing on counterterrorism is facing multiple obstacles. Most obvious is the fact that there is no EU Department of Homeland Security (EU DHS) at the cabinet level or similar like in the U.S. Managing the consequences of a terror attack is generally viewed by political decision makers worldwide as a national responsibility and opportunity for reinforcing their importance. As long as this attitude prevails there is little chance for the creation of an EU DHS.

Valuable input to counterterrorism research could come from the office of the EU *Counter-Terrorism Coordinator*[42] Gilles de Kerchove, who can approach directly all relevant commissioners, the High Representative, the President of the European Council and the pertinent ministries in the different Member States. However, he has to overcome an increasing *Counterterrorism Fatigue* in the EU Member States, since there had not been any major terror attacks in the EU since 2005.[43]

An important contributor could be *EUROPOL*, which had terrorism included in its mandate already in 1999. However, EUROPOL is dependent on the willingness of the Member States to contribute and share terrorism-sensitive information. Their preference is, however, to engage in bilateral cooperation and data sharing rather than sharing classified data with multiple Member States.[44] Whilst small EU Member States like Austria and Belgium promote the concept of an EU CIA-equivalent, traditionally large countries prefer classified information to remain under their sole control. Therefore, EUROPOL´s current contribution is the compilation of the annual data on terror incidents in the EU (TE-SAR Reports) and the provision of the terrorism threat assessment for the EU.[45]

EUROJUST is another candidate for an important input into EU counterterrorism research. However, so far its visibility in this field is largely limited to the provision of statistical data and information on terrorist convictions.

Optional input sources for such research could also be dedicated *working groups* assigned to terrorism issues, such as the Terrorism Working Group (TWG; internal threat assessments, coordination among EU organizations), the Working Party on Terrorism

---

[40] The Disaster Center, United States Crime Rates 1960–2010; http://www.disastercenter.com/crime/uscrime.htm (last visited 12 July 2012).
[41] Mark Stewart and John Mueller, Ten years and $1 trillion later, what has all our security spending achieved? ASK THIS, 2 June 2011; http://www.niemanwatchdog.org/index.cfm?fuseaction=ask_this.view&askthisid=00512 (last visited 17 July 2012).
[42] Currently the position is held by Mr. Gilles de Kerchove.
[43] Note by the EU Counter-Terrorism Coordinator to the Council 15359/09/REV 1, 26 Nov. 2009. http://www.europolitics.info/pdf/gratuit_en/261958-en.pdf (last visited 17 July 2012).
[44] Oldrich Bures,Europol's fledgling counterterrorism role, Terrorism and Political Violence, 20: 4, 2008.
[45] EUROPOL, European Police Office, TE-SAR 2011: EU Terrorism Situation and Trend Report 2011 ISBN:978-92-95018-86-0 (2011).

External Aspects (COTER; terrorism-related foreign policy issues) and the Working Group CP931 (terrorism lists).

Since the EU has recommended member states to establish dedicated *national security research programmes*, at least seven have done so by pursuing complementary research in support of national policies, e.g., Austria, Finland, Germany and Rumania.[46] But there is no risk-based coordination of the different national security research activities. Each Member State derives its national threat assessment and the identified specific national security needs to counter this specific threat from its own sources.

Strategic Deficits

The basis for the any counterterrorism strategy is the assessment of the terrorism risk. This implies determining the threat scenarios and the likelihood of them to occur. An important element in minimizing these risks—100% risk elimination is not feasible—is also the assessment of the resilience of the targeted society, itself depending inter alia on the training and equipment of the first responders. Table 2 shows the security threat to the 495 million EU citizens due to different forms of terrorism.

Table 2: Security risk due to terrorism for EU citizens in the homeland in 2010 (adapted from: EUROPOL TE-SAR 2011[44]).

Type of Incident	No. of Incidents
Terrorism in EU Member States:	
Failed, foiled, completed terrorist attacks	249
Arrested terrorist suspects	611
Threat statements against EU Member State	46
Islamist Terrorism:	
Islamist terror attacks in EU Member States	3
Arrests for Islamist terrorist offences	179
Arrests of Islamists for preparation of attack in EU	89
Separatist Terrorism:	
Separatist attack	160
Arrest for separatist terrorist related offences	349
Fatalities due to separatist terrorism	1
Left-wing and Anarchist Terrorism:	
Left-wing and anarchist terrorist attacks	45
Arrest for left-wing and anarchist terrorist activities	34
Fatalities due to left-wing and anarchist terrorist activities	6
Right-wing Terrorism:	
Right-wing terrorist attacks	0

In 2011 the number of terrorism-related incidents in Europe changed significantly in both directions:[47]

[46] Austria was the first EU Member State to create a national security research programme (KIRAS) in 2005.
[47] Rounded off figures, based on incidence data from EUROPOL, European Police Office, TE-SAR 2012: EU Terrorism Situation and Trend Report 2012 (2012).

- No. of terrorist attacks: minus 30%
- No. of terrorist arrests: minus 21%
- No. of fatalities due to terrorists: plus 1 000%.

The increased lethality is due to the killing by lone actors (2 victims in Germany, and 77 victims in the non-EU country Norway). In general, EUROPOL assesses the current threat from terrorism as remaining high in the EU, diversifying in scope and impact as it is reflected in the 316 individuals in concluded court proceedings for terrorism charges (an increase of 3% as compared to 2010). A realistic determination of the consequences, of the probability of the threats to occur, the most probable (not worst case) consequences and the terrorism resilience of society in EU Member States are largely missing.

By comparing Tables 1 and 2 it is noted that the higher impact of terrorism on EU citizens (number of incidents, arrested and legally prosecuted) and lower EU counterterrorism research budget are in stark contrast to the comparatively lower actual terrorism threat to U.S. citizens and significantly higher U.S. expenditure on counterterrorism-related activities (training, research, preparatory actions).

Typically, the national counterterrorism strategy of an EU Member State is guided by the national Ministry of Interior through a designated Federal Office for Counterterrorism or similar. The encompassing EU framework is defined by:[48]

- European Counter-Terrorism Strategy.[49]
- European Convention on Suppression of Terrorism and Amending Protocol (ETS 90).
- Council of Europe Convention on Prevention of terrorism (ETS 196).
- Council of Europe Convention on Laundering, Search, Seizure and Confiscation of the Proceeds from Crime and on the Financing of Terrorism (ETS 198).
- Convention on Cybercrime (ETS 185).

However, the counterterrorism policies actually implemented by the EU Member States are not uniform but cover rather a wide range, as shown in the different *Profiles on Counter-Terrorist Capacity*, published by the Council of Europe (CoE).[50]

EU counterterrorism research strategy is strongly influenced by the main players of the European defence industry, frequently seeking to readjust defence products for homeland security purposes. It is important to realize that EU counterterrorism research is embedded in the two main objectives of the FP7 programme: (a) Enhancing public safety through the development of security technologies; (b) Promoting the growth of a [still nascent and fragmented] EU homeland security community into a globally

[48] Additional, related legal requirements at the EU level are addressed in ETS 24, ETS 86, ETS 98, ETS 30, ETS 99, ETS 182, ETS 73, ETS 116 and ETS 141.
[49] Council of the European Union, 30 November 2005; http://europa.eu/legislation_summaries/justice_freedom_security/fight_against_terrorism/l33275_en.htm (last visited 17 July 2012).
[50] CoE country profiles; http://www.coe.int/t/dlapil/codexter/4_theme_files/country_profiles/default_en.asp (last visited 12 July 2012).

competitive European homeland security market. According to a *European Parliament study* "it is mostly large defence companies, the very same who have participated in the definition of EU-sponsored security research, which are the main beneficiaries of [European security research programme] funds".[51] Therefore, unlike other aspects of FP7, the European security research programme is managed by the European Commission's *DG for Enterprise and Industry* rather than *DG Research*. Also, significant additional funding—with aspects of security embedded—can be found in other themes of the FP 7 programme, such as in transport, energy, nanotechnology, etc.

Small and Medium-sized enterprises (SME)[52] as members of a project consortium face a funding scheme necessitating them to provide typically 50% of their total actual cost (as compared to 25% for universities) and high administrative burdens. Also, they appointed only infrequently as Project Coordinator. Not surprising their typical participation in EU funded security research projects is below 20%.[53]

An unusual feature of the EU security research programme is the significant involvement of non-EU countries in EU counterterrorism research, if they are categorized as *Associated Countries* due to former agreements signed with the Union, e.g., Switzerland, Norway and Israel.[54] Israel, for example, is currently one of the main foreign partners in the EU 7th Framework Programme (FP7) for scientific research. At present it plays a major role in eight EU security projects,[55] such as Project Coordinator for *CATO*.[56] Experts from Associated Countries are evaluating projects under FP 7. This can raise issues like, Israel not having signed the Non-Proliferation Treaty, but being involved in classified CBRN security research; or, representatives from non-EU countries deciding as evaluators on security research proposal submitted by consortia from EU Member States.

Implementation Deficits

It has become almost a routine that each Call for submission of security research proposals is significantly over-subscribed. Apart from the enormous waste of resources in

---

[51] Julien Jeandesboz and Francesco Ragazzi , European Parliament, DG for Internat Policies Policy Dept. C: Citizen´s Rights and Constitutional Affairs Civil Liberties, Justice and Hoe Affairs: Review of security measures in the Research Framework Programme, Bruxelles (2010); http://www.europarl.europa.eu/studies (last visited 17 July 2012).

[52] SME has to be an individual or organisation engaged in economic activity, employing less than 250 persons, having an annual turnover not exceeding € 50 million or a balance sheet total not exceeding € 43 million and also be 'autonomous', i.e., small commercial organisations cannot be regarded as an SME if they are owned by larger organisations.

[53] Centre for Strategy & Evaluation Services, *SMEs and their Participation in Security Research - Case Study* (January 2011); http://ec.europa.eu/enterprise/policies/security/files/doc/sme_case_study_cses_en.pdf (last visited 17 July 2012).

[54] Norway, an EEA EFTA state, funds its participation in EU programmes and agencies by an amount corresponding to the relative size of its gross domestic product (GDP) compared to the GDP of the whole EEA. The EEA EFTA participation is hence on an equal footing with EU member states. Israel has been an Associated Country since FP4 while it signed the Science and Technology Agreement for FP7 in 2007 (entered into force in February 2008).

[55] INFRA, ESS, HEMOLIA, TASS, FESTOS, SAFE-COMMS, NMFRDisaster.

[56] CBRN crisis management: Architecture, Technologies and Operational Procedures; http://cordis.europa.eu/search/index.cfm?fuseaction=proj.document&PJ_LANG=EN&PJ_RCN=12533818&pid=0&q=C26332F63230DFAA4A973946F1335A2A&type=adv (last visited 17 July 2012).

having to reject excellent proposals for the lack of funding,[57] the increasing number of proposals received each call is frequently associated with long Time of proposal approval-to-Grant approbation (TtG) times; TtG periods up to 18 months are not infrequent. Additional significant delays have been attributed to clarification of ethical questions and security clearance.

Frequently counterterrorism research results reflect science beyond-state-of-the art, but are not ready for commercialization yet. Further work is needed before these research results can be made operational and actually deployed by end-user organizations. Thereby, the sustainability of project outcomes becomes questionable without further follow-up research funding. Exploitation of research results, both at the project level and across projects, is often problematic due to the lack of cooperation during project implementation between different Project Coordinators. Some very innovative research results have suffered from the inadequate effort by the Commission to assist in the vital phase of disseminating these results to a wider audience in the mid-term, thereby assisting also in promoting their eventual use by the security community. Also, there is no effective mechanism in place for the involving the Commission in assisting in the topic areas *interoperability* and *standardization.* Project Coordinators should be obliged to include extensive research of the products and services available on the security market and benchmark their mostly pre-industrial prototypes resulting from their project against the best.

Often end users or large industrial companies are not sufficiently involved in a counterterrorism project because they hesitate to get involved in an R&D projects at an early stage.

Selecting the optimal size and membership of a project consortium addressing possibly classified information is difficult. Demonstrating good coverage between experts from "old" and "new" EU Member States has been found dominating over the question of adequate research capability fitting the research challenges.

With regard to cost efficiency of the counterterrorism research the topic of closer coordination and/or cooperation in areas of common interest should be openly discussed on defense research projects (EDA projects), space research (ESA projects) and EU funded civil security projects. A promising subject area would be, for example, CBRNE. Although the CBRN topic area is strongly supported by the Commission, inadequate testing under realistic conditions and certification in the CBRN domain are a known problem.

THE WAY FORWARD

Until now EU counterterrorism has been largely event driven and it took until 2007 when, for the first time, security-related research was given a prominent role in an EU FP, i.e., this type of research is still in its infancy in the EU.

Unless there is a major change in the EU counterterrorism strategy, in the midterm this research will continue to concentrate mostly on identifying ways to prevent acts of terrorism to happen in the first place through innovative means, such as: smart crowd behaviour analysis, enhanced signal analysis, aerial-, innovative land- and sea-based

---

[57] It typically takes several person-months for forming a suitable consortium and to prepare the proposal.

observation techniques, new stand-off detectors for hazardous materials, internet traps and data mining, etc. Despite the obvious deficiencies EU counterterrorism research has started on a promising path. It is likely to obtain broad acceptance by the EU citizen, if civil rights are neither perceived to be, or actually are threatened by the research results. This in turn will ensure their EU counter terrorism R&D is likely to obtain broad acceptance by the EU citizen, if civil rights are neither perceived to be, or actually are threatened by the research results

International competition outside the EU, particularly in the U.S., is significant already and will get stronger still in view of the larger U.S. financial resources dedicated to fighting terrorism. The current and potentially prolonged economic crisis in several EU Member States will add to the need for increased EU funding contributions in FP 8. This is likely to apply in particular to R&D facilities and SMEs in new Member States with limited financial resources already. It may be advantageous to aim for closer coordination and co-funding by Member States for EU projects which fund national counterterrorism research at present. If the pending issue of intellectual property rights for innovative solutions in counterterrorism research can be resolved, also large corporations operating in the EU may be willing to take on a larger financial share in a consortium consisting of experts from many EU Member States.

In summary, EU counterterrorism research is in its initial stage only and will require another ten years or so of high level funding and optimally coordinated support by the Commission before starting to reach its key objectives.

PREVENTING NUCLEAR-EXPLOSIVE TERRORISM BY ELIMINATING FISSILE MATERIAL: A PROGRESS REPORT

FRANK VON HIPPEL
Woodrow Wilson School, Princeton University
Princeton, USA

The mission of the International Panel on Fissile Materials is to develop the basis of initiatives to secure, to consolidate, and to reduce global stocks of nuclear weapon usable fissile materials. Fissile materials (primarily highly enriched uranium (HEU, uranium enriched to more than 20% U-235) and separated plutonium) are essential in all nuclear weapons, and controlling and reducing their stocks is critical to all efforts to reduce nuclear-weapon stockpiles, increase barriers to proliferation and prevent nuclear terrorism. This paper is a progress report on this effort.

GLOBAL STOCKS OF FISSILE MATERIAL

The fission of 1 kg of plutonium or HEU releases energy equivalent to the explosion of about 17,000 tons of chemical explosive—approximately the yield of the explosions that destroyed Hiroshima and Nagasaki. Figures 1a and 1b show national declarations and non-governmental estimates (indicated by asterisks) of stocks of HEU and plutonium of the weapon states, some key non-weapon states and the other non-weapon states collectively for 2011. The total is about 2,000 tons of fissile material, enough for over 100,000 nuclear explosives. In fact, in 1967, the U.S. had about 31,000 nuclear warheads and, in the mid 1980s, the Soviet Union is believed to have had about 40,000 warheads.[1]

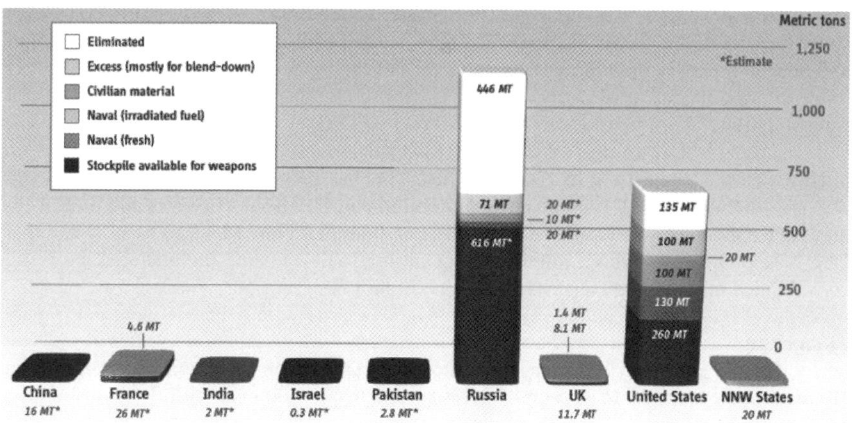

Fig. 1a: Global stocks of HEU currently total about 1500 tons—sufficient (at 25 kg per warhead) for 60,000 nuclear warheads (Ref. Global Fissile Material Report 2011, GFMR2011).

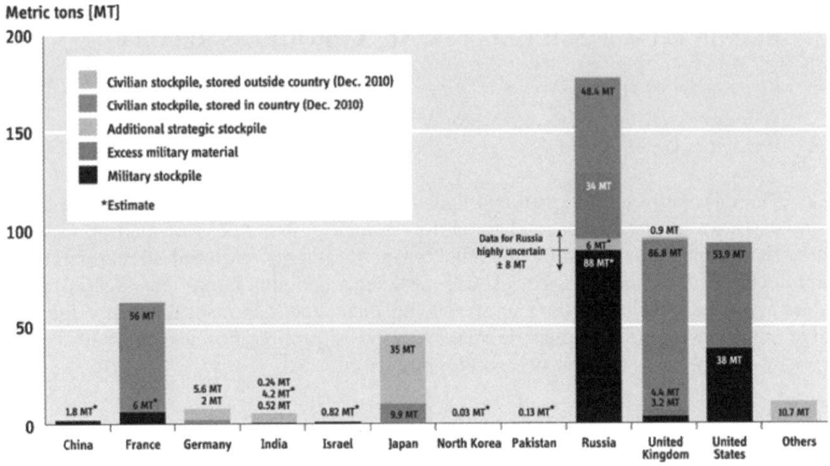

Fig. 1b: Global stocks of separated plutonium currently total about 500 tons—sufficient (at 8 or 4 kg per warhead for 60,000 or 120,000 warheads. (Ref. GFMR2011).

SECURING, CONSOLIDATING AND ELIMINATING NUCLEAR MATERIAL

Securing

Russia accounts for about half of the global HEU and one third of the separated plutonium stock. After the collapse of the Soviet Union, the U.S. was greatly concerned about the security of fissile materials and nuclear warheads in Russia and has spent about $10 billion dollars in upgrading security there. This effort continues today.

Consolidating

The U.S. Department of Energy (DoE) spends more than $1 billion per year on nuclear-material security at its own facilities. This has motivated a considerable amount of consolidation. The DoE is cleaning plutonium and HEU out of two of the three U.S. nuclear weapons labs (Livermore, California and Sandia, New Mexico) as well as other DoE sites. On sites where significant quantities will remain, the material is being consolidated into the high-security buildings (Figure 2).

Eliminating

Russia and the U.S. have eliminated a great deal of HEU and are planning on eliminating a considerable amount of excess weapons plutonium as well.

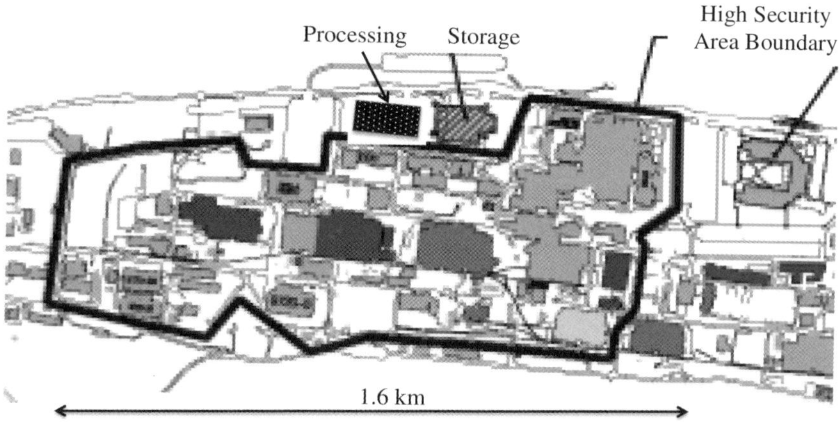

Fig 2: Consolidation at the Y12 HEU site in Oak Ridge, Tennessee. Historically, all U.S. HEU weapon component fabrication, dismantlement and HEU central storage has taken place in the complex of buildings within the heavy line. Two fortified buildings, will replace this complex: one for HEU storage (complete) and one for processing (under construction).

The white areas in Figure 1a show that Russia and the U.S. together have blended down about 600 tons of excess HEU to 4 to 5 percent low-enriched uranium (LEU). They are committed to blend down about 100 tons more. The LEU is being used mostly to fuel U.S. nuclear power plants (more than half of U.S. nuclear capacity). The blend-down of the 500 tons of HEU that Russia has committed to dispose has been at a level of 30 tons per year for about 15 years and will be completed in 2013. The blend-down rate of excess U.S. HEU has dropped from about 10 tons per year to about 3 to 4 tons a year, due to bottlenecks in the warhead and HEU-component dismantlement process and currently is expected to take decades or more to complete.

Much more HEU could be blended down. A global nuclear arsenal of about 10,000 nuclear warheads—roughly the situation today, not counting warheads in the dismantlement queue—would require only about 250 tons of weapon-grade (90+% U-235) HEU, vs. the approximately 1000 tons that are in the weapon-state military stockpiles.

Russia and the U.S. also each are committed to converting into fuel 34 tons of weapon-grade plutonium. Both are building "mixed-oxide" (MOX, uranium-plutonium) fuel-fabrication facilities. The U.S. plans to produce MOX for once-through use in light-water reactors while Russia plans to produce MOX for fast-neutron breeder reactors, which require plutonium recycle. Russia therefore will not be permanently disposing of its separated weapons plutonium but rather simply will be passing this plutonium through breeder reactors into its stockpile of separated civilian plutonium.

Much more weapons plutonium could be declared surplus. A global stockpile of about 10,000 warheads would require only about 40 tons of weapon-grade plutonium, vs. approximately 140 tons in the weapon stockpiles today.

REDUCING THE USE OF HEU AND SEPARATED PLUTONIUM FOR REACTOR FUEL

The principal non-weapon use of HEU and plutonium is in nuclear fuel.[2] HEU is used in research reactor fuel in about 30 countries, and in naval reactor fuel in the U.S., UK, Russia and India. Plutonium is recycled in MOX fuel in light-water reactors in France, and soon, in prototype breeder reactors in India and Russia. Japan has been planning to recycle plutonium in light water reactors. China has breeder reactor plans.

In all of these uses, there is a risk of diversion to weapons. After India's use of civilian plutonium for a "peaceful nuclear explosion" in 1974, the U.S. abandoned civilian reprocessing at home and actively tried to prevent the spread of plutonium separation to new countries and to minimize HEU fuel use in research reactors. Only in the case of HEU-fueled civilian research reactors, however, is there a real international consensus on minimization of the use of weapon-usable fissile materials in reactor fuel. Plutonium separation continues in Europe, Japan, Russia, India and China.

Research reactors

Since 9/11, the U.S. has led an intensified campaign to convert the world's HEU-fueled research reactors and medical fission isotope production targets to LEU (in practice uranium enriched to 19.75% U-235). This is an appropriate priority, since research reactor facilities are often not as secure as are facilities where nuclear weapons are assembled and disassembled and where naval reactor fuel is manufactured.

The technical approach that is being used is to develop fuels with roughly five times higher uranium density than the current fuels containing weapon-grade (> 90% U-235) uranium. This density increase makes it possible to add U-238 to reduce the enrichment of the uranium without reducing the concentration of the U-235. The HEU densities mostly are below 1.5 gms/cc. Thus far, fuels with densities up to 4.8 gms/cc have been developed and fuels with LEU densities up to 15 gm/cc are under development.

HEU has been "cleaned out" (reduced to less than 1 kg) in more than 20 non-weapon states (Figure 3). Most of the remaining HEU-fueled research reactors are in the weapon states—Russia, in particular. Also, as steady-power research reactors are converted to LEU (61 so far, with about 50 left to convert or retire), reactors with lifetime cores—especially HEU-fueled critical-facilities and pulsed reactors are becoming an increasingly large fraction of the HEU-fueled research reactor population (see Table 1).

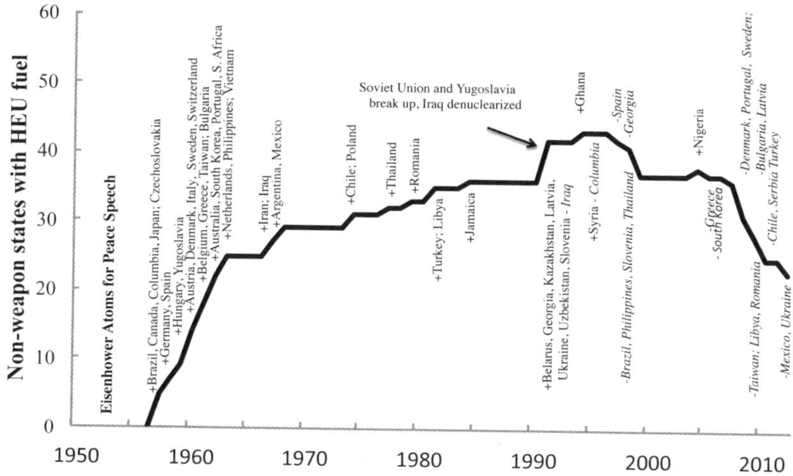

Fig. 3: Number of non-weapon states with plutonium (updated from GFMR2011*).*

There are at least three challenges to completing the task of phasing out HEU use in a research reactor fuel:

1. *Make it a priority for the Russian government.* Russia's Government has been happy to support the U.S. financed effort to convert Soviet-designed HEU-fueled reactors and retrieve Soviet and Russian-origin HEU fuel outside Russia—but has not given any priority to converting its own HEU-fueled reactors. In 2010, however, after repeated requests from the U.S. Government at the highest levels, Russia's Government agreed to allow the U.S. Global Threat Reduction Initiative (GTRI) to fund conversion feasibility studies for six of its civilian reactors. Also, since the March 2012 Nuclear Security Summit in Seoul, Rosatom has been showing more committed to the agenda of phasing out civilian HEU use.
2. *Higher-density uranium fuels.* To convert high-power reactors that require the highest uranium-fuel densities, the GTRI program is focusing on developing the most dense fuel possible: metallic uranium, alloyed with up to 10 weight-percent molybdenum to improve its mechanical properties. Such fuel would have a uranium density of about 15 gm/cc.

Table 1. The global population of HEU-fueled reactors includes much more than the steady-power research reactors that thus far have been the focus of the cleanout effort.[3]

	Critical Assemblies	Pulsed	Research		Isotope-Prod.	Breeders	Naval (including icebreakers)	Total
			≤0.25 MWt	0.25-250 (+targets)				
Russia &NIS	≈29	≈17	1	16	2	1	≈63	≈129
China	1	?	2	0	0	1	0	≈4
Europe	4	≈3	5	6 (+1)	0	0	10 (UK)	≈22
USA	≈5	2	1	6	0	0	100	≈113
Others	3	0	11	2 (+1)	0	0	1 (India)	≈17
Total	**≈42**	**≈22**	**20**	**30 (+2)**	**2**	**3**	**≈174**	**≈285**

3. *Critical assemblies and pulsed reactors.* Military research reactors include critical assemblies that mock up naval reactors and pulsed reactors to generate neutron bursts that can be used to test the effect on electronics of nearby nuclear explosions. Fortunately, computer simulations have made critical assemblies almost obsolete except for experiments need to benchmark the codes. Computer simulations probably also have taken over in Russia but RosAtom is retiring obsolete critical assemblies only slowly and still has them at about ten sites.

In the U.S., four of the five remaining HEU critical assemblies have been moved to the high-security Device Assembly Facility (DAF) on the Nevada Test Site from a location in Los Alamos that was not considered adequately secure (Figure 4).[4] Sandia National Laboratory also has shipped to the DAF the HEU fuel of its last pulsed reactor and is developing computer simulations of the effects of neutron bursts on the transistors in chips.

Most of Russia's dozen pulsed reactors are at its two weapons design laboratories. The U.S. had a similar number[5] but today, due to security concerns, reduced relevance of nuclear war-fighting scenarios and improved computational capabilities,[6] only has one pulsed reactor, the Army's Fast Burst Reactor at the White Sands Proving Ground in New Mexico. Recently, the Army put out a request for proposals for alternative ways to produce neutron pulses so that it can retire this last pulsed reactor.[7] Most military pulsed reactors have annular cores made up of massive pieces of solid uranium alloy.[8]

Fig. 4: The U.S. has consolidated all of its weapon-design-related critical assemblies in the high-security Device Assembly Facility at the Nevada Test Site.

More information is available about the designs of fast critical facilities than about military critical facilities. These designs are therefore useful for illustrating the security problem with critical facilities fueled with HEU and plutonium. The two BSF facilities at the Institute of Physics and Power Engineering, which are designed for modeling the cores of large fast breeder reactors, for example, are built up of about 100,000 different disks, including tens of thousands containing plutonium and HEU. These disks contain a total of 700 kg of HEU (90%), 500 kg of plutonium and 2800 kg of 36% HEU (36%) (Figure 5).

Fig 5: In the BSF-2 fast critical assembly at the Institute of Physics and Power Engineering in Obninsk Russia, disks of plutonium, HEU and depleted uranium can be stacked in tubes to test the criticality of breeder-reactor cores. The diagram inset shows the arrangement of tubes in the core into which the disks are stacked.[9]

Naval reactors

The U.S. and U.K. fuel their naval propulsion reactors with weapon-grade uranium. Russia and India are believed to use HEU that is less highly enriched. France and China are believed to use low-enriched uranium. Russia's Artic fleet of civilian nuclear-powered icebreakers and transports also uses HEU fuel.

The U.S. fabricates an average of about 2 tons of weapon-grade uranium into naval fuel per year and has established a reserve of 130 tons of weapon-grade uranium for future naval-reactor use. Although naval fuel cycle facilities are more secure than research reactors, thefts of HEU have occurred. It would be desirable to shift to LEU-fueled reactors. Also, if much lower levels of nuclear weaponry can be achieved, large stocks of HEU justified by the future needs of naval reactors would raise concerns about breakout.

In 1994, the U.S. Congress required the U.S. Navy to study the possibility of moving to LEU-fuel. The report from the Director of Naval Nuclear Propulsion argued that the costs would be high and the benefits insignificant.[10] In 2008, the Senate Armed Services Committee, in its report on the Defense Authorization Act for Fiscal 2009, stated that:[11]

"The committee directs the Office of Naval Reactors to review carefully options for using low enriched uranium fuel in new or modified reactor plants for surface ships and submarines."

The Office has ignored this language. As of 2012, the design of a new aircraft-carrier propulsion reactor was about 96 percent complete and the design of the reactor for a new ballistic-missile submarine was launched in fiscal year 2012.[12] Both designs are HEU fueled. The HEU-fueled *Virginia*-class attack submarine will be in production for at least another decade.

Nevertheless, it would be desirable to continue the effort to establish the use of low-enriched uranium as a norm for naval as well as other reactors. If, in the future, the production of HEU for weapons is banned but production for naval fuel is not by a Fissile Material Cutoff Treaty, one of the most difficult verification challenges will be difficult to detect governmental diversion of HEU from naval fuel cycles.

What are the objections of the U.S. nuclear navy to shifting to LEU? The objections in the 1995 related primarily to cost. It was assumed that conversion to LEU would require either:

1. Retreat from lifetime cores to refueling (3 times in a submarine lifetime and two times in an aircraft carrier lifetime). The cost of purchasing the extra cores was estimated at about $100 million each for submarines and $600 million for twin aircraft carrier cores. It also was estimated that refueling would extend the duration of major overhauls in port by 10 months for submarines and 22 months for the aircraft carriers and the cost of each overhaul by about $60 million for the submarines and $300 million for aircraft carriers. Because of the extra time in port, it was assumed that the Navy would have to buy an additional four attack submarines, one ballistic-missile submarine and one aircraft carrier to maintain the same at-sea strength.
2. Build reactors with bigger cores. It was estimated that this would cost about $300 million extra for the submarine reactors and $1.2 billion extra per aircraft carrier. In addition, it was estimated that the size of attack submarines would have to be increased to accommodate the larger cores and reactors at a cost of $150 million each. (The estimated cost increases for the larger ballistic missile submarines and aircraft carriers were relatively minor.)

Thus the lifetime core option was found to be less costly at about an extra $1 billion/year vs. about $2.5 billion/year for the return to refueling (1995$). The report concluded, however, that, "[n]either option for using LEU in place of HEU offers the Navy a technical, military or economic advantage. Either option would be extremely costly."

The UK, which obtains HEU and its fuel design from the U.S., would change to LEU only if the U.S. did. France and China, however, have made different decisions.
In the case of France, there appear to be three differences in the cost-benefit calculation:

1. France, unlike the U.S. does not have a huge stockpile of excess Cold War HEU and did not want to replace its costly HEU-production plant, which was

based on gas-diffusion technology and was expected to expire around 2000.[13] In its newest nuclear submarine, it will move to fuel using the same enrichment used in power reactors, about 5% enriched uranium;[14]

2. French submarines have large hatches that allow refueling without an extension of refit time,[15] whereas refueling U.S. submarines requires cutting a hole in the hull; and
3. The fabrication cost for U.S. naval-reactor fuel is very high.[16]

China too may not have a large stockpile of HEU in excess to its weapons needs. Also, its first submarine reactor design was based on published descriptions of the propulsion reactors of Russia's first nuclear-powered icebreaker, whose fuel was 5 percent enriched.[17]

Plutonium use in power reactors

Separation of civilian plutonium was launched by the major industrialized countries in the 1960s and 1970s because of twin beliefs: i) Global nuclear power capacity would quickly grow to thousands of gigawatts, and ii) The world does not have enough low-cost uranium to support on that scale nuclear power based primarily on a once-through fuel cycle not involving plutonium recycling (Figure 6).

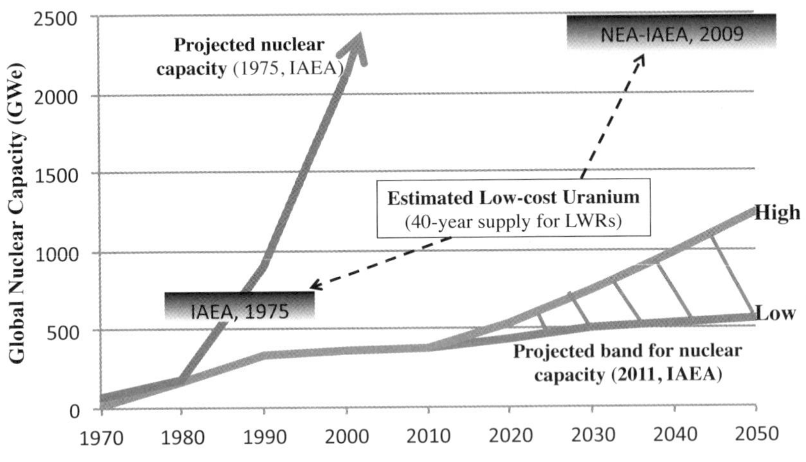

Fig 6: IAEA projections in 1975 and in 2011 of global nuclear power growth and of global resources of low-cost (< $130/kgU) uranium.

On that basis, several countries decided to launch major efforts to develop plutonium breeder reactors whose ultimate fuel would be the U-238 that constitutes 99.3% of natural uranium. Despite the expenditure of about $100 billion on attempts to commercialize fast-neutron sodium-cooled breeder reactors, however, the effort failed. And, because of slow growth of global nuclear generating capacity and growing estimates of global uranium resources, most countries concluded that breeder reactors would not in any case be required for more than a century.

One legacy of this effort is about 250 tons of separated civilian plutonium from reprocessing programs that were originally launched to obtain plutonium for initial breeder reactor cores. In India and Russia plutonium separation continues on a small scale (1 to 2 tons per year) with this rationale and China has built a pilot reprocessing plant (1 ton/year if it is operated at full capacity) and is considering building a larger reprocessing plant. All three countries are still concerned that available uranium supplies may not be able to keep up with their ambitious plans for nuclear power expansion.

The large reprocessing plants are in France, Japan and the UK, however–although the two U.K. reprocessing plants are only operating at partial capacity and Japan's reprocessing plant is not operating at all. France recycles its separated plutonium in "mixed-oxide" (MOX, uranium-plutonium) fuel into the light water reactors from whose spent uranium fuel the plutonium was separated. Spent MOX fuel is stored pending decisions on whether to dispose of it in an underground repository or to reprocess it and use the recovered plutonium for either startup fuel for future breeder reactors or perhaps repeated recycle in light-water reactor fuel. In the wake of the Fukushima accident, Japan is currently debating whether or not to continue with its reprocessing plans. The UK has decided to end reprocessing when its foreign and domestic contracts have been completed—currently estimated in 2018.

Japan is starting to build a MOX plant, despite the fact that its efforts to recycle MOX fuel coming back from France had been delayed by a decade by public opposition even before the Fukushima accident. The U.S. is mid-way in building a MOX plant for its excess plutonium but has encountered huge cost overruns and schedule delays. The UK has accumulated approximately 100 tons of separated civilian plutonium that it now has to dispose of. It built one MOX fuel fabrication plant that only operated at about one percent of its design capacity for ten years before being abandoned in 2011. It is currently planning to build a second MOX plant to produce fuel that would be used in light-water reactors that it hopes will be built in the UK (the UK currently as only one LWR). Some of us have suggested direct disposal of plutonium as an alternative to MOX.[18]

CONCLUSIONS

1. Reduction of the Cold War stockpiles of HEU has begun on a substantial scale and reductions of the weapons stockpiles of plutonium are planned.
2. There is an emerging consensus to phase out the civilian use of HEU.
3. There has been an intense debate over plutonium separation and recycle for almost 40 years. Most countries have abandoned reprocessing because of its large costs and few benefits, but a few major countries remain committed to reprocessing.
4. A policy debate over whether to move away from the use of HEU fuel in naval propulsion reactors has not really begun.

REFERENCES

1 United States, U.S. Department of Defense, "Increasing Transparency in the U.S. Nuclear Weapons Stockpile," Fact Sheet, 3 May 2010; Soviet Union/Russia, Natural Resources Defense Council estimate, http://www.nrdc.org/nuclear/nudb/datab19.asp.

2 About 50 kg of weapon-grade uranium annually has been used in neutron "targets" to produce fission radioisotopes for medical purposes. This use is being phased out.

3 Update of table in Ole Reistad and Styrkaar Hustveit, "HEU Fuel Cycle Inventories and Progress on Global Minimization," *Nonproliferation Review*, Vol. 15, # 2, 2008, p. 265.

4 The fifth U.S. critical assembly is associated with the Advanced Test Reactor at the Idaho National Laboratory. It is also possible that the U.S. has a naval critical facility.

5 C. Paxton, *A History of Critical Experiments at Pajarito Site* (Los Alamos National Laboratory, LA-9685-H, 1983 H.), p. 13.

6 In 2004, U.S. Secretary of Energy Spencer Abraham announced that "the Sandia Pulsed Reactor [at the Department of Energy's Sandia National Laboratory, where nuclear-weapons work is done] will no longer be needed because computer simulations will be able to assume its mission…When its mission is complete, this reactor's fuel will be removed…allowing us to reduce security costs at Sandia and further consolidate our nuclear materials", "Remarks prepared for Energy Secretary Spencer Abraham for the Security Police Officer Training Competition," May 7, 2004, reprinted in *U.S. Nuclear Weapons Complex: Homeland Security Opportunities,* Project on Government Oversight, May 2005, Appendix B. Sandia has a program on Qualification Alternatives to the Sandia Pulsed Reactor (QASPR), which focuses on computer modeling of neutron effects on transistors. Predictions are compared to experimental results obtained by the Ion Beam Laboratory.

7 U.S. Army, "Alternative Source for Neutron Generation," 2012), http://www.dodsbir.net/sitis/display_topic.asp?Bookmark=42661.

8 The Idaho National Laboratory has not used the last U.S. civilian pulsed reactor, the Transient Reactor Test (TREAT) reactor, for 20 years. This reactor is a lesser security concern, however, because its HEU is in solution-impregnated graphite—one atom per ten thousand carbon atoms.

9 BFS-2 image captured from the website of the Institute of Physics and Power Engineering, Obninsk, Russia. Inset diagram from Alessandro Marinoni *et al,* "Analysis of the BN-600 fast-spectrum core mock-up at BFS-2 zero-power facility using MCNPX," *Annals of Nuclear Energy,* Vol. 44 (2012) p. 26.

10 *Report on Use of Low Enriched Uranium in Naval Nuclear Propulsion,* Director, Naval Nuclear Propulsion, 1995.

11 U.S. *Senate Armed Services Committee fiscal year 2009 Authorization Report* (p. 515).

12 *National Nuclear Security Administration, FY 2013 Congressional Budget Request,* p. 484.

13 Y. Girard, Techmcatome of France, presentation, Conference on The Implication of the Aquisition of Nuclear-Powered Submarines (SSN) by Non-Nuclear Weapons States, MIT, 27-28 March 1989;

14 "France's Future SSNs: The Barracuda Class," *Defense Industry Daily,* 21 Dec. 2011.

15 Y. Girard, *op. cit.*

16 A U.S. *Trident* submarine contains about 1 ton of HEU and a *Virginia*-class submarine about 0.7 tons, IPFM, *Global Fissile Material Report 2010,* endnote 131. On this basis, the figures quoted above from the 1995 report by the Director, Naval Nuclear Propulsion suggest that the fabrication costs for these cores were about $100 million or about $100,000/kgU.

17 John Lewis and Zue Litai, *China's Strategic Seapower* (Stanford University Press, 1994), p. 31; *Global Fissile Material Report 2010, op. cit.,* endnote 282.

18 Frank von Hippel, Rodney Ewing, Richard Garwin and Allison Macfarlane, "Time to Bury Plutonium," *Nature,* 10 May 2012, p. 167.

PUBLIC PERCEPTION OF NUCLEAR RADIATION: ITS IMPORTANCE FOR NUCLEAR ENERGY, RDD, TERRORISM AND NUCLEAR DISARMAMENT

RAMAMURTI RAJARAMAN
School of Physical Sciences, Jawaharlal Nehru University
New Delhi, India

Before discussing why and how the public needs to be better educated about the dangers of nuclear radiation, we must recall how their existing perceptions came to be.

Ever since the holocaust inflicted over Hiroshima and Nagasaki, the specter of the damage that nuclear radiation can do to human beings has been deeply imbedded in the public psyche. While the exact count of fatalities still varies from estimate to estimate, roughly between 100,000 and 150,000 people died within days of those two terrible bombings. Most were killed by a combined onslaught of intense heat, blast wave, and nuclear radiation, any one of which hazard was dangerous enough to be potentially fatal.

The bomb blasts and the fires initiated within seconds by each of two nuclear weapons were much more intense than what a score of conventional bombs could have produced. Nevertheless, images of death and destruction caused by blasts and fires were not qualitatively new. Four years of exposure to the ongoing World War II had by then quite inured people to seeing towns being razed and buildings on fire. What was new and different was the effect of nuclear radiation.

Even after the damage due to the blast waves, heat waves and fires had been brought under control in Hiroshima and Nagasaki, and fatalities caused by them began to taper down, the effects of radioactive damage continued to persist. They lingered for weeks, months and in terms of initiating cancer of various forms, for years. Gut-wrenching photographs of victims suffering from radiation sickness and of genetic deformation of children born to them were soon seen by millions of people around the world, along with accounts of the suffering from the victims and from commentators and experts.

Radiation sickness was a new form of horror, completely unfamiliar to people until then. There was also a scary mysteriousness to nuclear radiation. It can penetrate the body, destroy internal organs, cause cancer and induce grotesque birth defects. Yet it is invisible, odorless and colorless. There are few immediate external signs apart from some skin burns, especially at low but still eventually lethal dosage, of the terrible damage done to the interior of the body. The inability of normal human senses to detect nuclear radiation and the possibility of dreadful after-effects developing years later, has created an almost irrational dread in people's minds about radioactivity.

Use of the phrase "irrational dread' in connection with radioactive hazards may be offensive to some people. The phrase does carry the implication that perhaps the fear is sometimes out of proportion to the real and actual danger. Understandably this may be offensive to environmentalists, peoples' groups and individuals who have been valiantly struggling for years to raise people's consciousness about such hazards, particularly in the context of nuclear energy in the aftermath of Chernobyl and Fukushima.

But much as we do not wish to cause offense on this very sensitive and touchy area, we feel compelled to argue in this article that in some ways and for some situations the fear of radioactive hazards has been exaggerated. Furthermore this may have some harmful consequences for society.

The concern I will address is based on new developments in this millennium. For decades before that there was no real benefit to be had by ensuring that the public got a balanced education on the damage that radioactivity could or could not do. With the threat of a nuclear war looming real in the early years of the Cold War, the emphasis was on raising consciousness about radioactivity as much as possible. The governments of nuclear weapon states, particularly the UK and the U.S. made a major effort towards developing civil defense measures, in the process educating the public about nuclear dangers through various do's and don'ts in the event of a nuclear attack. The philosophy was that the more people were concerned about radioactivity, the better they were likely to take immediate precautions and avoid being exposed to it, should radioactive fallout be released into the atmosphere by a nuclear blast even far away.

In addition, public fear of nuclear hazards has also played an extremely valuable role for mankind as a whole in keeping nuclear war at bay, until today. Although this fear could not generate enough support for anti nuclear movements to stop the Cold Warriors from building up absurdly large arsenals, it has contributed to maintaining the "nuclear taboo". There have been no nuclear bombings since 1945. We believe that as things stand today, no state actor would initiate a nuclear first attack, whether they have a stated policy of No-First-Use or not—even countries that are judged to be rogue states by the West. No responsible country seriously considers initiating a deliberate nuclear attack, notwithstanding all the U.S. rhetoric of "keeping all options open". Not even battlefield artillery nukes have been used so far, even though conventional bombs of similar explosive power are used without censure. Things may of course change for the worse if some state were to be taken over by suicidal groups—a possibility that cannot be ruled out.

Fear of radioactivity has also played a tacit role in pushing for nuclear arms reduction in countries like the U.S., Russia and the UK.

Therefore there is no question that the fear of nuclear radiation deeply inculcated in the public has been a good thing, in the interests of public safety and also as a driving force behind continued efforts to reduce nuclear arsenals and fissile material stocks around the world.

Having said that however, a time has come when the message to be conveyed to the public needs to be more nuanced because of two separate quite unrelated developments. One is the prospect of terrorists using an RDD Device (the "Dirty Bomb"), which is viewed by security experts as real possibility. The other development was what promised to be a "nuclear renaissance" thanks to increasing concerns about fossil fuel emissions and global warming—a process that soon got stalled by the Fukushima accidents. The main cause of the setback has been the revival of the public fear of radioactive hazards, fortified and made more worldwide by the Fukushima explosions

Although these two developments are on quite different fronts each of them, for its own reasons, demands that the public carry a balanced impression of what radiation can and cannot do in different situations. They require us to re-think the strategy of public education about the hazards of radiation. While as a general rule, nuclear radiation is harmful to people, all forms of radiation in all situations cannot be covered by a single blanket condemnation. People have to understand the distinction in scale between a true fission-generated nuclear explosion and a reactor explosion—even a large one like those in Chernobyl or Fukushima. And similarly the difference between those two and a Dirty Bomb exploded by a terrorist.

Let us take up both these topics, starting first with the "Dirty Bomb", formally called a Radiation Dispersal Device (RDD). It refers to a terrorist blast in which a container full of hazardous nuclear material is exploded with ***conventional*** explosives. The explosion will release into the air the assorted radioactive material that the terrorist has gathered for the purpose. It is not a nuclear weapon, in the sense that there will be no nuclear fission and therefore will not cause even a fraction of the physical damage or radiation release that a nuclear weapon will. The scale of the blast and fire will be set by the conventional explosives in the bomb and may have a range of no more than a hundred feet. The casualties due to the radiation will also be far fewer than in a Chernobyl-type reactor explosion. Some of damage, in the form of latent cancer, will not be immediately visible. The zone of strongest radioactive dose will overlap with the zone of physical damage. So the number of deaths due to the radiation, over and above what the dynamite blast and fires would cause, can be contained if people are evacuated from the area soon enough. All in all, the RDD device is not a Weapon of Mass Destruction (WMD).

Nevertheless, it can be a very valuable part of the terrorist's plans. The value addition obtained by lacing a conventional RDX bomb with radioactive junk is immense, and requires a far smaller effort than stealing fissile materials or a fully assembled nuclear weapon. The nuclear material used an RDD device needs not be fissile at all. It could be any radioactive material—hospital wastes, university research samples and so on—material that is are far less seriously guarded than fissile materials like plutonium or enriched uranium and far easier to accumulate, in bits and pieces, Although the scale of physical damage will not be very different from that of conventional RDX blasts, there will be the additional radiation induced injuries. Depending on the quantity and type of radioactive isotopes used, dozens or even hundreds of cancer deaths can occur over the years due to the radiation released from the radioactive junk dispersed into the air. In addition, the entire area of several city blocks would remain contaminated by the radioactivity for a long time. So, apart from injury to humans, whatever activity was going on in that area would come to a grinding halt for months if not years. One can imagine the economic or political damage done if this dirty bomb were dropped on some financial or key administrative district. Therefore even if the Dirty Bomb were not a weapon of mass destruction it will certainly be a "weapon of mass disruption".

DIRTY BOMB AND PANIC AS THE WEAPON

A less discussed but perhaps far more serious consequence of a Dirty Bomb explosion will be the panic that would result at the site of the explosion, if the public became aware that radiation has been released. This panic will be strongest element of the damage that the dirty bomb can do, given the largely uneducated and irrational public fear of radiation that we mentioned at the beginning. Even though the police, doctors and first responders may carry radiation measurement counters and may be well educated about the actual extent and nature of radiation hazards, the damage due to panic would have been done even before they get there, if rumors of the bomb being radioactive were to spread on the spot. People will run helter-skelter, even in places well removed from the blast center. There can be massive stampede, traffic jams and riots.

Worse still, the terrorist need not even actually detonate a Dirty Bomb. It will suffice if he spreads the rumor that a particular conventional explosion is in fact a Dirty Bomb. There is no way the people present on the spot can tell the difference. Not even the police can, unless they routinely carry radiation detectors. This is what makes the Dirty bomb such a potent weapon. You don't even need one, but just enough misleading propaganda apparatus—agents planted near the site of an explosion mingling with the public and claiming that it is radioactive or that some future one will be.

A public over-anxious about radiation hazards will harm its own interests even beyond the immediate site of the Dirty Bomb explosion. People will desert the entire city even if the effects of radiation may, in objective terms, render only a few square kilometers significantly dangerous, except along the wind direction. Of course the civic authorities should as quickly as possible announce the prevailing speed and direction of the winds and accordingly inform them where the radiation dangers continue to exist.

Even if such clear instructions were given in time by the authorities, the public may or may not believe them. Over the years, some distrust has been built up between the public and the officialdom on matters of environmental hazards. People don't trust official assurances. Environmental disaster movies portraying governmental and industrial leadership as callously indifferent to the public welfare, with convincing portrayals by popular actors, although possibly motivated by a few genuine real life examples, has ended up painting essentially all industry and government agencies as untrustworthy villains in this regard. No assurances from them will persuade people from making up their own minds and playing it extra safe. Upon hearing of an RDD explosion in some part of their city, their idea of playing it safe would be dash out of town as quickly as possible by any mode of transport available—a recipe for mayhem on the roads, airports and the railways. This could result in far more casualties that what the original terrorist weapon could have caused directly.

Similarly, even long after the explosion, when the attempts are made to decontaminate the affected area and return it to the activities it used to host, it will be difficult to persuade the public to return to that area. In today's atmosphere such reluctance will be shown by all segments of the public—whether they be blue collar workers, shopkeepers, business managers or educated bureaucrats. If a financial hub like Wall Street or a transportation hub like Grand Central station were to be in the neighborhood of ground zero of the RDD attack they would remain unusable long after they have been cleaned up declared safe. This would be very detrimental to the public good.

It is clear that an important way to contain the damage in such cases is to educate the public about the actual, relatively smaller extent of damage that a dirty bomb will cause, as compared to a nuclear weapon or Chernobyl type of reactor explosion. It is vital to psychologically immunize the public, through a credible public education campaign, about the ***comparatively*** limited consequences of a dirty bomb, and its differences from a nuclear weapon. This must be done a in a balanced fashion so that their legitimate fear of radioactivity—so vital for their protection and survival, remains intact. They must be provided with accurate information on the likely radiological effects presented by such an event, and a clear and unambiguous set of emergency response instructions to deal with its consequences. As the U.S. National Academy of Sciences has suggested, "To optimize the overall health and well being of the population, and to improve the overall response to terrorism events, it is necessary that (psychological reactions) be addressed preventively as well as throughout the phases of an event."

Most countries, including India are actively training first responders and the police on detecting and handling radiation disasters. This of course is a must. But it should be emphasized again that the people on the spot during an explosion will be members of the general public, not the trained first responders. Unless the public is somehow trained to react without panic, the worst damage due to panic would have happened before the trained personnel even get to the spot.

How to train the general public of the dozens of cities and tourist resorts which could be potential Dirty Bomb targets in a large country like India or in the West is a huge problem that disaster managers will have to contend with. Nevertheless this has to be done, not only to mitigate the effects of a dirty bomb attack but equally, to ***reduce the motivation*** for the terrorist in resorting to one, by removing the element of panic from his arsenal.

NUCLEAR ENERGY PROGRAMS AND REACTOR ACCIDENTS

The other domain, where exaggerated fears of radioactivity can end up doing more harm than good to society as a whole, is nuclear power. As you all know, the development of civilian nuclear energy worldwide has had a roller coaster history. When the nuclear era dawned Atomic Energy seemed the ultimate solution to mankind's energy needs. The U.S. started building nuclear reactors rapidly. Europe, the Soviet Union followed suit. Thanks to Eisenhower's Atoms for Peace program, developing countries were also encouraged to go nuclear.

But a few decades' later skepticism and opposition began to gather against nuclear energy in many countries. The contributing factors to this downturn were:

1. Major reactor accidents like the 3-mile Island in the U.S. and Chernobyl in the USSR,
2. Greater safety and environmental consciousness, and the emergence of Green Parties in Europe, and
3. Awareness of complexity and expense of Reactor-decommissioning and Spent fuel disposal.

Although some countries like France, Russia, China and India continued with their programs the biggest producer, the U.S., stopped licensing new reactors. After building 100 reactors by the 1970s, the U.S. has not licensed any additional reactors since then. Germany also stopped increasing its nuclear capacity (We will return to Germany's post-Fukushima nuclear policy later). Some advanced nations like Norway never went for nuclear power.

But with the start of the new Millennium with its own new concerns, there seemed to be a revival in nuclear energy's fortunes, motivated by:

1. Growing concern over climate change due the high Carbon emission by coal and hydrocarbon fuels.
2. Improved reactor safety features and passage of two more decades of reactor use with no major accidents, making them 'statistically' safer (until Fukushima happened later).

The U.S. government through its Nuclear Power 2010 Program and Energy Policy Act has initiated building of new nuclear power plants, and was expecting over 20 applications for 34 new plants. Other countries showed signs of following suit, especially smaller nations with no nuclear program as of then.

THEN CAME FUKUSHIMA

The nuclear reactor explosions at Fukushima are well known to all of you. It was caused by the twin blows of a magnitude 9 earthquake on March 11, 2011, which led to the collapse of the electric grid in the region, followed within an hour by a Tsunami wave of height variously estimated as between 9 meters and 14 meters, which drowned out the emergency secondary power supply for pumping in the coolants. I need not recount here the emergency venting and other details.

It is the response of the public and the world media to those events that concerns us here. It illustrates our point that the public and media response was out of proportion to the danger, real and serious though the latter was.

First and foremost we must clarify that our comments are not intended to apply to fears about radioactive fallout by people living in Japan especially within a few hundred miles of Fukushima, including Tokyo. All the daily bulletins in Tokyo about contamination of drinking water or milk were fully justified and not paranoid. So were the decisions of the Japanese government, after perhaps some initial reluctance, to get to the bottom of possible lapses and failures of policy by the nuclear utilities. It has revealed many worrisome acts of commission and omission from which not only Japan but all other countries can learn. Indeed, anyone who saw the images of how the Japanese responded—without panic, with great discipline, dignity and stoicism—to the massive death, damage and dislocation they suffered due to the earthquake and tsunami could not but admire them.

What we are concerned about was the level of anxiety in the rest of the world in the aftermath of the reactor explosion, which was quite disproportionate to the actual threat to other countries. Partly what happened was caused by the hand of fate. The "double whammy" of the earthquake and tsunami had already drawn hundreds of reporters and TV channels from all over the globe to the area to cover the massive natural disaster. Meanwhile everyone knew that there was a reactor complex there and quite justifiably the media began to raise the possibility of the reactors being affected. Soon the world's TV stations, already represented in full strength there because of the earthquake began focusing on the reactors. When the first reactor's roof exploded it was seen live on TV screens in hundreds of millions of homes around the world. These were unprecedented images in the history of nuclear energy

Panic spread all over. In New Delhi, SMS messages were circulated warning people not to go out in the rain lest they be showered by radioactivity coming from Japan.

In China, people started consuming iodized salt in large and potentially harmful quantities. Assurances from the Chinese officials that fallout radiation in Beijing was well below danger levels did not make sufficient impact. Eventually, the WHO was compelled to call for calm on its Twitter page, which too did not work. Much farther away, drug stores in Russia's Far East and British Columbia reported shortages of iodine pills. California, always fertile soil for health paranoia, also joined in stocking up with iodine pills.

The biggest culprits responsible for this panic were the electronic and tabloid media around the globe. By the time the roof of Reactor no.1 exploded, the nuclear hazard occupied as much space-time in the media as the suffering and rehabilitation of the Japanese who had lost over 10,000 people as dead with many more injured and homeless from the tsunami/earthquake. This was despite the fact that not even one member of the public had, at that stage, died because of the reactor explosion. Without discounting in any way the importance or news value of a reactor building exploding, this was not balanced reporting.

Apart from this, there were daily reports of finding traces of fallout at various places in distant continents. I have no particular reason to doubt that these reports were correct. Radiation detecting techniques have come a long way since the early Geiger counters and minute traces of nuclear emissions ***can*** be detected far away.

Unfortunately, however, these reports typically failed to mention how small those doses were. In the first two weeks after Fukushima this writer scanned nearly a hundred such news reports. Almost none of those reports gave quantitative details—either on the amount of radioactivity emitted by those traces, or, for purposes of comparison, some information on how that dose matched against, say, ambient natural radioactivity [Average annual cumulative dose of 2.4 milliSieverts (mSv)] or standard medical procedures like CT Scans (about 10 mSv) or just X rays (less than 0.5 mSv). Knowing the prowess of even cub reporters in gathering information fast from their contacts and over the Net, I was not persuaded that this omission was due the urgency of filing reports. Rather, it may have been because such clarification could have diminished the "newsworthiness" of the item.

In fact the fallout in, say, California from Fukushima would have been minuscule, as could have been predicted in advance. The Chernobyl fallout has been studied extensively and what fell on the U.K. at that time was less dangerous, cancer-wise, than smoking a cigarette a year.[1]

I think it would be fair to say that largely as a result of such one-sided and partial (although technically not incorrect) news reports day after day following the Fukushima explosions, there descended a different kind of fallout around the globe—one of gloom and doom regarding all things nuclear—not just the bomb but also peaceful nuclear energy.

If this had resulted only in a few days of relatively harmless panic over the Fukushima explosions and a run on iodine pills, it would not matter so much. But in fact it made a big impact on one of the world's most serious problems—of a looming energy scarcity which in turn has to be met without worsening global warming. It galvanized anti

---

[1] The Daily Telegraph (London), Editorial, March 16, 2011.

nuclear activists and NGOs into renewed opposition to nuclear energy, but now with far greater public support than ever before.

A groundswell of opinion was generated in many countries to demand that in the light of the Fukushima disaster their nuclear programs should be stopped or reversed. This trend was strengthened by the decision of the German government to completely close down its entire nuclear program within a decade. When a technologically advanced, and sophisticated country like Germany decides in its wisdom that nuclear energy is too dangerous, should other countries, especially developing nations, continue with their nuclear plans and programs? There was pressure on other countries to follow suit. With that also grew concern among utilities and reactor manufacturers in supplier nations that further investment in that area may be far too risky. For a while it looked that the nuclear renaissance had died stillborn.

That, we believe would be an over reaction and a retrograde development. We are not blindly advocating nuclear energy here, as against wind, solar and other renewable sources, nor claiming that it will substantially mitigate global warming. Admittedly, these are all debatable. Public discussion on them has been going on for years. And the pros and cons were well known, long before Fukushima. Nor are we arguing that there are no lessons to be learnt from Fukushima. Indeed there are several. Criteria for locating reactors in areas prone to natural disasters like earthquakes, tsunamis and hurricanes should be made even more stringent. In designing backup systems, as for instance for cooling the reactor, one should try to ensure that the different layers of back-up are not vulnerable to the same external disaster. Their utility as backups then fails, as happened in Fukushima. The problem of over-pressurization in containment vessels and the resultant need to vent radioactive gases has to be addressed more stringently. There is a long list which post-Fukushima safety reviews in many countries are addressing.

Furthermore, individuals and groups with anti-nuclear energy positions include some highly respected experts, who no doubt have good reasons for their opposition to nuclear energy, based on any number of plausible arguments—not just its radiation hazards, but its cost, the problems of waste disposal, etc. If governments are not paying sufficient attention to their arguments, it is fair to demand that they do.

But the overwhelming factor in shaping public opinion against nuclear energy was not these or other objective arguments. Each of them, whether it be the economics of nuclear power or even the actual extent of hazard that different low levels of radiation represent carries a vast literature with differing conclusions. To make one's own educated judgment on any of them would require plowing through that literature and absorbing its essence—something that even most scientists are not in a position to do, let alone the lay public. Large protest movements have to be driven by emotions rather than logic. In their meetings and pamphlets the main strategy used to persuade the public to oppose nuclear energy was to stoke their fear of the radiation dangers. Gruesome images from Hiroshima and Nagasaki were freely mixed in with those of Fukushima, even though the two are vastly different in their levels of damage. Anecdotal accounts rather than comprehensive statistics of radiation induced cancers were offered. The dangers of a reactor accident were mixed up with those of a normally functioning plant. The credibility of nuclear utility spokesmen and government officials in the respective countries was ridiculed so that any attempt by them to rebut the arguments of the activists fell on pre-set deaf ears.

In places like India, the genuine public protest was a local phenomenon—largely in areas like Jaitapur and Koodangulam where new nuclear plants are being built. It was primarily caused, not so much by fear of radiation hazards but by the displacement of the village people in those areas from their homes and established modes of livelihood, without adequate compensation or sensitivity. When the Fukushima disaster happened, anti-nuclear groups provided an alternative and more potent cover to the villagers for raising their otherwise legitimate frustrations. They sat in protest demanding the closure of the reactors being built there. This began steamrolling from "not in my backyard" to "not in my country", with demands that the Indian government pullback from the ambitious nuclear energy expansion it was trying to put in motion, to meet its massive energy deficit.

The story was similar in some other countries. The most prominent example was Germany's decision to completely phase out of its nuclear energy program within a decade, even though it will presumably continue to get nuclear-tainted power from neighboring France. What is more, some of the French reactors are located right across the German border posing almost as much of a threat to Germany as its own reactors would have. Nevertheless that was the sovereign decision of the German government, and I personally **welcome it and admire them** for it. The decision will not only fulfill their people's preferences but also benefit the rest of the world. If there is any country in the world whose economy and technology can handle shelving all its nuclear energy and eventually replacing it by developing more benign sources like solar and wind or by reducing transmission losses, it is Germany. One hopes that the new technologies they introduce will become available to the rest of the world too.

What distressed me was not the German decision *per se*, but the manner in which it was taken—within a few days of the Fukushima explosions, and based not on a technical analysis of why Fukushima called for such closure, but only to appease of public fears of nuclear energy. That too, after having had a full examination of the nuclear program earlier and finding it safe enough to announce extension of life for some of its old reactors. It was not clear which aspect of the Fukushima explosion demanded the complete reversal of that policy, rather than just tightening safety features further. I believe that the compulsions of a regional election at that time also contributed to that speedy reversal.

Worldwide, the abandonment or significant roll back of nuclear energy expansion would have been a retrograde step, especially for developing nations whose per capita availability of electrical energy is a tenth of what, say, Germany enjoys. The world needs every source of energy including nuclear. If there is some danger associated with nuclear energy, there are other hazards associated with lack of electricity for the hundreds of millions of people in developing countries. In India over 30% of rural households and a hundred thousand villages do not have electricity. Unlit roads are unsafe for women, unlit homes are bad for children's eyesight and education, rural hospitals with constant power breakdowns can be deadly and cooking with wooden sticks or cow dung is bad not just for global warming but also for the respiratory health of the women and girl children hunched over those stoves. The actual hazards of nuclear power have to be in balanced some meaningful sense against no power or alternate power sources, if available.

True, using alternate sources may be safer, but neither solar not wind energy, both of which are beginning to be used worldwide has matured into a universally available, inexpensive, grid friendly form. People who advocate going back to coal instead of nuclear do not offer data on how much cancer is caused by coal plants.

Fortunately many other countries, although initially taken aback by the Fukushima explosions, reacted more moderately. Variously, they ordered a review of all existing reactor safety procedures and of future expansion plans. That was perfectly in order. Eventually many of them have decided to go ahead cautiously with their reactor programs. India, China and Russia are major examples. So, I believe, has the U.S., although a recent ruling by their regulatory Board has suspended final decisions on licensing and relicensing applications for new and existing nuclear plants, pending a better plan for waste disposal. But the processing of the applications will continue.[2]

So the prospect of nuclear energy being globally abandoned seems to have been avoided thus far. But the process of implementing nuclear plans has become much more difficult in the face of the continuing and largely irrational fear of the public about radiation. Certainly nuclear energy carries its hazards. But this hazard has to be quantified in a manner which allows the pubic to compare it with other risks they have decided to live with and make some sort of a rational decision. For instance, India suffered one of world's worst industrial disasters when a chemical factory in Bhopal leaked poisonous gases into the surroundings. Thousands were killed and many more seriously injured for life. One of the longest dedicated protest movements has been on for 25 years to demand help and compensation for the victims. Yet no one has called for the abandonment of all chemical industry involving poisonous material. In a modern industrial society that just cannot be done.

Is this undue fear of radioactivity the fault of the public? Certainly not. The public is not expected to have expertise in radiation hazards, and cannot be blamed if it fears radiation dangers excessively. In fact even among the best experts on radiation hazards, there is no full unanimity on the extent of damage that radiation can do, especially low level radiation persisting for years from fallout far from the explosion.

But governments and others who shape public opinion can be blamed for public ignorance. Unfortunately, sometimes they too pander to populist fears. The scientific community must also share the blame for not having stayed in communication with the general public over the decades to give them a balanced education on the benefits and hazards of nuclear energy. That would have won them the confidence of the people. School textbooks rarely touch upon radiation safety. This must change.

Governments as well as social activists must be pro-active in educating the public about nuclear hazards, but in a responsible and balanced manner. Neither bland assurances that there is no danger, nor the stoking of hysterical fears will serve the public good. Raising people's consciousness about legitimate nuclear hazards, but doing this in such a way as not to contribute to panic will be an intellectual and pedagogical challenge, quite apart from the problems of implementation. But it must be done before some future Dirty Bomb attack or a nuclear accident results in additional deaths because of public ignorance about radiation hazards.

Finally, this is a good opportunity to demand greater transparency from nuclear establishments all over the world. That is overdue. The Fukushima disaster demonstrates

---

[2] NRC ruling credit neutral for U.P. public power Fri Aug 10, 2012 http://in.reuters.com/article/2012/08/10/idINWNA324320120810.

the lack of credibility of nuclear agencies in the public's mind, as evidenced by the latter's lack of faith in official assurances. This is not an easy problem to solve. We live in an age where people want the benefits of extremely complex technology, whether it is cell phones or nuclear energy or genetically modified food, but are also suspicious of it, in some cases rightly. Yet they do not have the technical background for making up their own minds about them. They have to take someone's word for it. Who will give them required assurances when the people best qualified to do so are increasingly disbelieved?

SESSION 3

CLIMATE & CLIMATE ECONOMICS

THE IMPORTANCE OF ECONOMICS IN THE FRAMEWORK OF CLIMATE ISSUES

CHRISTOPHER ESSEX
Department of Applied Mathematics, University of Western Ontario
London, Canada

Good afternoon, it's a pleasure to once again speak to you in the plenary sessions at Erice. I am co-chairing this session on climate, with Professor Zichichi and Lord Monckton. This session follows directly from the presentation made this morning by President Klaus in which climate and economics were central topics.

Whenever I co-chair or organize such a session, I like to spend a few minutes briefly framing it with explanations as to its broader significance. In this case, because this session, and President Klaus's session this morning, are both facilitated through the Climate PMP, the question arises as to why we are discussing economics and not just climate. Economics can seem quite remote from the physics, chemistry, or mathematics of climate, and yet there is an entire field known as climate economics. Indeed, when I was a student I would never have imagined that I would spend as much time with economists as I have in recent years. Although I am not one of them, for the most part my time among them has turned out surprisingly well. So I will attempt to explain why economics matters on the fundamentally physical issue of climate.

Central to popular and policy interest in the science of climate is concern over projected deleterious changes in climate, particularly any potential deleterious changes caused by humans that are preventable by humans. Of course there are academics who would be interested in climate even if there were no risk or danger, but the broader society cares more about the risk and danger and less about the natural wonder. The obvious human response to such fears is to alter actions, in hopes of sparing us all of the potential negative outcomes. But, alas, it is not that simple. Not only are the negative outcomes difficult to nail down, but so are climate and climate change highly problematic and elusive in terms of rigorous physical thinking. I have found over the years that if you present this unambiguous reality in detail to people involved with this topic, many will respond by raising the "precautionary principle."

In case you don't know what that is, I will attempt to explain. The precautionary principle is a policy doctrine for facing the unknown, and what better fits into the category of the "unknown" than topics of scientific research, of which climate is surely one. If you have two paths, where one is safe and the other potentially dangerous, the precautionary principle calls on people to choose the safe path, unless it can be proven that the novel one is safe too. Pretty straightforward, isn't it? Uncertainty, particularly scientific uncertainty, then is faced by practicing risk avoidance.

Of course simple risk avoidance is not a realistic, let alone effective, policy. To illustrate how the real world can intrude, I will provide a couple of examples, not from grand policy issues, but from real individual lives:

1. Near the end of the cold war, a Canadian living on the west coast of Canada was concerned about the risk of nuclear war. He applied the precautionary principle to the problem (without using the term, I imagine). He concluded

that western countries were potential targets, and so resolved to personally avoid the risk by moving as far from them as he could. The southern hemisphere was about as far away as he could go. He would further insulate himself from the risk by finding an island far from potential land targets. So in accordance with his application of the precautionary principle, he set himself up as a sheep farmer in the Falkland Islands only to find himself in the middle of a real war and not just a hypothetical one.

2. When I was with the Canadian Climate Centre's general circulation modeling group, one of my colleagues revealed that he came to Canada to avoid military service that might require him to fight in a war. After arriving in Canada as a young man, he wanted to become a scientist. He liked physics but he was afraid of how physics had become embroiled with politics. So as a precautionary measure he deliberately chose to study a field that he could not imagine would ever be of political interest: meteorology and climate.

While it may or may not have useful legal applications, the precautionary principle simply cannot stand as the basis of coherent policy toward the unknown of real events if for no other reason that it presumes to know of a pristine road through the unknown that is risk free. That is it copes with the unknown by presuming to know the unknown.

It is easy to decide which way to go if one path is known to be dark, dangerous, and foreboding, while the other has only butterflies, rainbows, sunshine, and unicorns to offer. But in the real world things can go wrong on either path, while there may also be unique advantages to either. The only rational way to decide which road to take is to have a way to compare the risks and advantages.

An engineer's classical way of discussing risk is in terms of probability. What are the odds, say, that a bridge will fail, or a building will collapse? The relative risk, in this picture, between seemingly unrelated things can be compared on a number line of the relative likelihood of desirable or undesirable events.

It is true that probability in the real world is not quite like the clean probability you may have learned about it school, where you select, say, black or white marbles from a bowl. In the real world probabilities are hard to know formally, and they are conditional in complicated and devilishly subtle ways, leading to many famous paradoxes like the prisoner's paradox or the Monty Hall problem. I have seen knowledgeable people get into huge disputes over such things. I would love to tell you about the look on the face of a classically trained probabilist meeting a Bayesian probabilist for the first time. I have seen this. There are also some great jokes about probability that arise from such things. But in the interest of time, we will not pursue these things today. I will of course be willing to share probability jokes with anyone who is interested at another time.

So then, assuming that probabilities can be well-defined and determined, we can compare the risk of getting struck by lighting to the risk of being bitten by a dog, to a dam breaking or even to the likelihood of winning a prize. But what this picture lacks is a sense of magnitude. A dam breaking is in a different league of calamity than being bitten by a dog, although you may be less affected personally in a particular realization of the former than the latter.

We say these things have different "magnitudes" but different magnitudes of what? These events are not one-dimensional in a common physical quantity. They are qualitatively different from each other, and yet at the human level we have an intuitive sense of rank ordering. Fortunately humans long ago devised an index to compare the human significance of things that may be physically qualitatively quite different from each other. And that index is of course known as money.

At the human level for every event we may assess a value in terms of money. That means that every event can be assigned a positive or negative monetary value and a probability, placing all events in this context into a two-dimensional space rather than a one-dimensional number line. This is depicted in Figure 1.

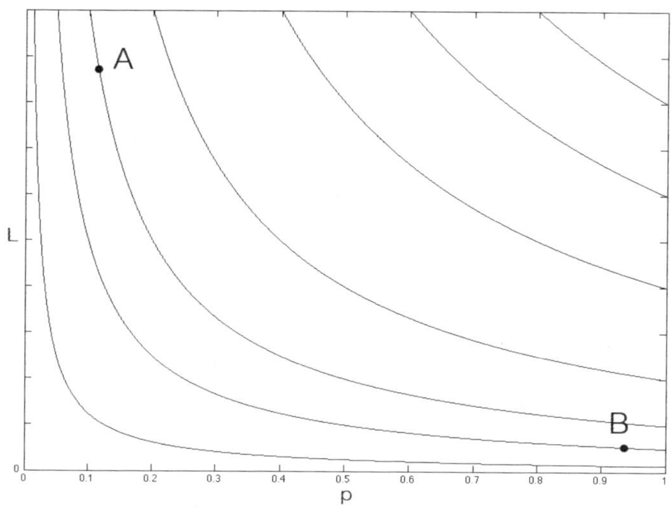

Fig. 1.

In this case the probability, p, of a deleterious event is along the horizontal axis while the cost represented as L, for losses, is along the vertical axis. Lines of constant risk in the classical engineering picture would be vertical lines. But if we bring the issue of cost into the picture, the lines of constant risk look qualitatively quite different.

Risk in this picture is measured in terms of total dollars paid out, or received. Thus it's clear that a rare but costly event, A, can actually be riskier in terms of losses than a common, but less expensive, event B. Risk becomes the product of the probability and the cost. Lines of constant risk then become hyperbolas rather than vertical lines. It makes an event in the upper left of the diagram equal, in terms of risk, to events in the lower right. It also means that risk increases upward and to the right rather than horizontally. I think it's fascinating how introducing the valuation of an event completely changes the shape of lines of constant risk. That shows us that how one determines the meaning of the various quantities involved has a significant effect on how we will make decisions.

And that is exactly what economics concerns itself with: not just the costs of doing something but also the costs of not doing something. You may feel safe hiding in a closet, but you lose out on all of the benefits that you might gain by going out into the world from your closet. You may also find unanticipated risks from staying in the closet. Making a full assessment of these, such as may be possible, is the kind of thing that economists do. In this framework the precautionary principle is completely superseded for the purposes of sound policymaking. So while economics is not primarily concerned with climate, the importance of climate for humanity must ultimately involve economics. And that is why we find ourselves talking about climate economics today.

We may worry about the consequences of producing carbon dioxide in large enough quantities to affect climate change. But it is not correct to presume that there are no consequences to *not* producing carbon dioxide. Rational decisions about what we should do ultimately then have to weigh the economic consequences against the climatological consequences that we may imagine. Among economists there is more than one school of thought on this as one might naturally expect. We have attempted to represent this to a limited extent. We did invite Lord Stern, who is a well-known protagonist for one pole of the spectrum of views, but as you can see he is not present today. I am grateful to our own Dr. Bruce Stram who has kindly agreed to represent that side of the picture.

What I hope you'll pay close attention to in all of the talks is that we are *not* dealing with the case of action versus doing nothing, as contemporary demagoguery continues to offer us, but the relative costs and benefits of different scenarios. I would also like you to bear in mind that extreme (implausible) economic scenarios are not trivial. They can be every bit as apocalyptic as extreme (implausible) scenarios for climate change. In either case, extreme scenarios call for narrow focused efforts which tend not to be robust against the unknown.

THE ROLE OF SCIENCE IN THE THIRD MILLENNIUM

LORD NIGEL LAWSON
House of Lords
London, UK

This is an economic session within a conference organized by the World Federation of Scientists.

So it is appropriate, perhaps, to begin by asking who, so far as climate change is concerned, have let us down more—the scientists or the economists.

In my judgment it is the scientists, not least because we expect so much more of them.

Science is so much greater than economics, and—as a result—the achievements of scientists have, over the years, benefited mankind so much more than the achievements of economists.

Those achievements have been based on one overarching principle: the application of reason, rather than authority, to our understanding of the natural world.

The long march from the middle ages to the enlightenment was the transition from relying on religion and the authority of the church to explain the natural world to relying on the application of reason and the open-minded study of empirical evidence.

That is what science is, or should be, about.

But in the case of climate science that is no longer the case.

There is now a new religion—the AGW religion, of which scientists are the new priesthood, preaching their dogma with precisely the same claim to authority as the mediaeval catholic church.

The truth is, as the best scientists recognise, that the greenhouse effect is a highly complex phenomenon, and the scale of the climate sensitivity of carbon is hugely uncertain.

Moreover, the greenhouse effect is only part of an even more complex and only partially understood overall climate system, and the notion that this can be reliably captured in a computer model is arrogant folly.

But if we have been let down badly by the scientists, the economists are not free from blame, either.

The most culpable economist of all is surely Nicholas Stern, now Lord Stern, whose 700-page review of the economics of climate change purported to show that the cost of a crash programme to reduce carbon emissions would be far outweighed by the benefits of doing so.

Although the Stern review is now almost six years old, it remains the explicit basis on which my own country's damaging climate and energy policies are officially justified, and Stern is taken seriously as an authority worldwide.

And although a number of economists—unlike the politicians, the media, and the environmentalists, few of whom had even read it—found fault with the Stern Review when it was first published, few have stepped up to the plate since.

Next month, the Global Warming Policy Foundation, the think-tank I founded some three years ago, will be publishing the most comprehensive examination of the Stern Review to have appeared anywhere.

It shows the economic analysis of the Stern Review to be so shoddy as to border on the fraudulent, and wholly unfit to be used as a basis for important policy decisions.

I have time today to mention only a few of the defects the GWPF study exposes:

- It calculates the long-term benefits of decarbonisation by using an abnormal and ludicrously low discount rate, while assessing the costs of this policy by using a normal discount rate.
- It compares the cost of limiting the extent of global warming with the benefit of eliminating warming altogether.
- It assumes that there will be no behavioural adaptation that would reduce the adverse effects of a warmer planet, even using existing technological resources, let alone as a result of future technological development.
- It also assumes that future technological development will greatly reduce the cost of non-CO_2-emitting energy.
- In particular, it assumes the adoption of large-scale carbon capture and storage, a technology that does not at present exist and, which, on present evidence, if it ever were to exist, would be prohibitively costly.
- It ramps up the alleged cost of warming on the basis of wholly unreliable non-peer-reviewed alarmist literature. A subsequent World Bank study, for example, has shown that Stern's forecasts of damage to infrastructure from more powerful storms are up to 100 times too large.
- It suggests that the losses from global warming (it wholly ignores, incidentally, any offsetting benefits from global warming) will be at least 5% of global GDP—and I quote— "each year now and forever"; concealing the fact that *even on Stern's own phoney figuring* the cumulative costs of the crash programme he advocates would exceed the cumulative benefits for the next 100 years and more.

Uneasily aware—although not publicly admitting—some of these cumulatively fatal flaws, Stern and his team at the hugely well-funded Grantham Institute he now heads are increasingly falling back on the Weitzman thesis, the so-called 'dismal theorem' or 'fat tail' argument.

Harvard's Martin Weitzman is a more distinguished economist than Stern.

He claims that Stern's conclusion is right: it is merely his reasons for that conclusion that are wrong.

In sharp contrast to Stern, Weitzman maintains that cost-benefit analysis is inapplicable: the uncertainty is far too great.

What matters in his eyes is that we cannot rule out the possibility that, however small the risk, global warming will increase to the point where it threatens the very survival of humanity on this planet.

That being so, we must do whatever it takes, now, to prevent it. The cost of doing so is, essentially, irrelevant.

But it is immediately clear that this is absurd, for two compelling reasons.

First, as Peter Lilley, the author of the GWPF study, points out, this climate Armageddon can occur only if the climate sensitivity of carbon is very high indeed.

The fact that there has been no recorded global warming trend for the past 15 years or so, during which CO_2 emissions have grown very rapidly indeed, thanks in large part to remarkable economic growth in China, suggests that the climate sensitivity of carbon, so far from being as high as Weitzman fears, is in fact less than is incorporated in the models on which, for example, the IPCC relies.

But we cannot entirely rule out the possibility that, over the past 15 years, purely temporary and short-term climate factors have offset and obscured the true underlying warming trend.

If so, however, these will by their very nature before long come to an end, and we shall very soon discover whether the climate sensitivity of carbon is in fact as high as Weitzman fears it might be, and will be able to act accordingly.

The second reason why the Weitzman thesis is absurd, is that there are a host of possible planetary emergencies, the principal subject matter of this conference.

We simply cannot spend unlimited resources on seeking to eliminate them all.

And common sense suggests that the menace of manmade global warming, so far from being, as Weitzman claims, in a league of its own, is probably the least of them.

But there is another and even more fundamental reason why Stern's decarbonisation prescription (whether based on his own phoney cost-benefit analysis or on Weitzman's 'dismal theorem') is—or should be—a non-starter.

Strictly speaking, and unlike physics, economics—or political economy, as it used more accurately to be known—is not a science; and the attempts by economists to pretend that it is renders them both ridiculous and of diminishing use to policy makers.

Economics is an important aspect of the study of human behaviour. And, unlike the case with science, which is concerned with the natural world, in considering human behaviour we have always to take also into account two non-economic dimensions: ethics and politics.

If I had to name one reason, more than any other, why I object to the policy prescription of the Stern review, which at the end of the day is only slightly more extreme than that of the IPCC and indeed the UN, of which the IPCC is an offshoot, is that it is profoundly immoral.

For what it calls for is *global* decarbonisation.

And the reason why the world uses hydrocarbons for its energy needs is simply that they are far and away the cheapest source of energy available, and are likely to remain so, not for ever, but for a very long time to come.

Even if (and I rather doubt it) we in the western world feel that we are rich enough to move from relatively cheap energy to much more expensive energy, the notion that the developing world should unnecessarily impoverish its people by doing so is, I believe, profoundly immoral.

Particularly when, in their case, impoverishment means, for hundreds of millions of their people, destitution, malnutrition, preventable disease, and premature death.

All this, moreover, in the name of arguably benefiting distant future generations who, thanks to economic growth, will be many times better off than those who are asked to sacrifice their living standard, and that of their children, today.

Happily, of course, this won't happen.

The Stern/IPCC/UN prescription is not merely ethically unacceptable: it is also, partly for that very reason, politically unattainable.

Copenhagen and its successors have demonstrated that no global decarbonisation agreement will ever be delivered—and quite right, too.

The position of China is particularly salient in this context.

According to the most recent projections made by the International Energy Agency, China alone will account for getting on for half the total growth in global energy demand by 2020.

And with its overwhelmingly coal-based economy, China has no intention whatever of taking the decarbonisation route, or of signing up to a global treaty requiring it to do so.

There are some credulous souls who affect to believe that China is in the process of switching over to renewable energy.

It is true that—rightly or wrongly—it sees great export opportunities in selling wind turbines and solar panels to the West. And it is indeed striking that Chinese solar panel manufacturers, most of whom did not even exist 10 years ago, now satisfy a third of the world's demand for solar panels.

Admittedly most of these Chinese manufacturers are now on the brink of bankruptcy, as Western demand has failed to match their expectations.

But what matters in this context is not Chinese exports to foolish foreigners, but how China plans to meet its own rapidly expanding energy needs.

And being an authoritarian regime it does indeed have a published plan.

This shows that it firmly intends to remain a carbon-based economy—which of course is why it is spending so much economic and political capital on securing coal, oil and gas resources in sub-Saharan Africa, where it has become the new imperial power, and oil resources in Latin America.

As for renewables, the plan indicates that, by 2020, wind-power will be supplying something under 1 1/2% of China's energy consumption.

Even this may be an overestimate. At the present time, almost half of China's wind turbines are there largely to impress credulous foreigners. Unconnected to the grid (this would cost too much) they are incapable of providing any electricity at all.

And solar? According to the plan, by 2020 this will meet—wait for it—less than 0.1 per cent of China's total energy consumption.

In short, climate economics—present company excepted—has been hijacked by a mixture of pseudo-scientific quasi-religion and political fantasy.

It is time to return to reason.

Rational climate economics means, quite simply, that, should there be global warming, for whatever reason, and of course there might well be, we will do, in the most cost-effective way that developing technology permits, what humankind has always done, throughout the world and throughout the ages.

We will adapt.

In that way we will minimise the disbenefits of any warming that occurs, and pocket the benefits.

In the words of the poet, "That is all ye know on earth, and all ye need to know."

A SIMPLE STATE-CONTINGENT PRICING POLICY FOR COMPLEX INTERTEMPORAL EXTERNALITIES

ROSS MCKITRICK
Department of Economics, University of Guelph
Toronto, Canada

ABSTRACT

Uncertainties over the future path of global warming and the underlying severity of the problem make it difficult to specify the intertemporally-optimal emissions price on carbon dioxide emissions, and create practical problems for securing an agreement. The dominant methods for dealing with uncertainty involve trying to derive an emissions price or insurance premium to which agents are expected to make a long term commitment. This chapter explores an alternative approach based the concept of state-contingent pricing, in which agents commit to a pricing *rule* rather than a *path*. The rule ties current values of the emissions price to currently-observed temperatures, since these integrate all known information about the effect of past emissions on the climate as well as constraining the future movements of the tax to future changes in the climate. In essence, if the climate warms, the tax goes up, and vice versa. A derivation is provided showing how such a rule yields an approximation to the unknown optimal dynamic externality tax, yet can be computed using currently-observable data. A recently-proposed extension coupling the state-contingent tax with a tradable futures market in emission allowances would yield not only a feasible mechanism for guiding long term investment, but an objective prediction market for climate change, in essence creating the world's best climate model. The advantage of the state-contingent approach for facilitating coalition-formation is also discussed, as are directions for research.

INTRODUCTION

Suppose we have a time machine that allows us to visit the year 2040 just long enough to collect some climate data. Figure 1 shows the post-1979 globally-averaged lower tropospheric air temperature anomaly averaged over the two satellite series developed by, respectively, Spencer and Christy (1990) and Mears and Wentz (2005). This is only one of many data series people use to try and represent the global climate as a univariate time series, but it will do for the current illustration. Figure 1 shows the observed data from 1979 up to the end of 2010 (shown by the vertical line), and then runs the series forward using assumed trends and random numbers to conjecture two quite different futures. In the gray dots the next three decades exhibit continued variability but no upward trend, and even a slight downward trend. The black dots show variability and a strong upward trend. Now suppose that, given an identical future greenhouse gas emissions trajectory, the data we collect in 2040 will look like one of those two paths. If we could find out which one would be observed, would it affect today's policy choices?

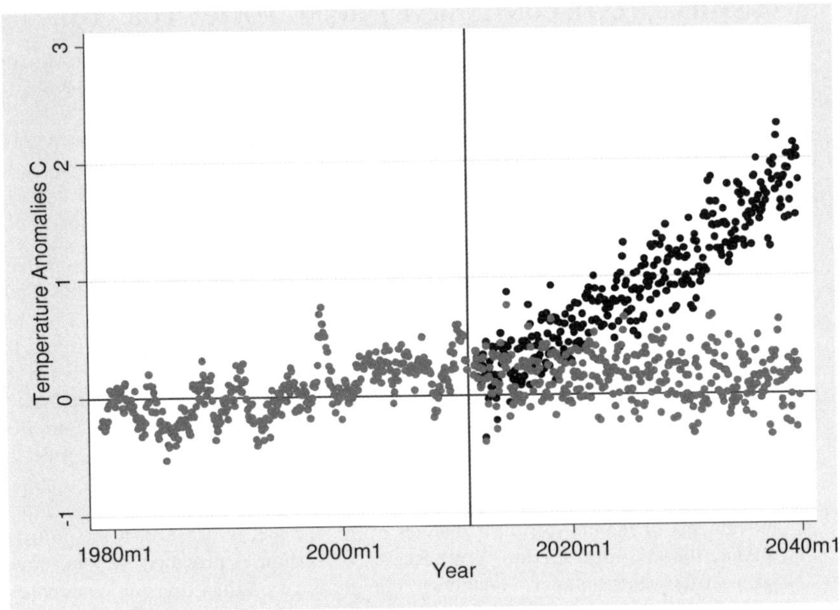

Fig 1: Two conjectured atmospheric warming paths 1979-2040. Left of the vertical line are observations from weather satellites of the global average lower troposphere temperature anomaly. Right are conjectured using trends and random numbers.

Obviously the answer is yes. The fact that we do not know what the graph will look like has led to longstanding and well-known political difficulties in devising policy strategies. In this paper I will critically review the main current approaches to dealing with the uncertainty, and then propose an alternative that I believe is more likely to lead to the right policy outcome than any others currently being examined. Briefly stated, my argument is as follows.

- Forecast-based proposals (such as from Integrated Assessment Models, or IAMs) for making optimal climate policy decisions effectively assume we can agree what the data to the right of the line will probably look like, and we only need to resolve the time-sequence of emission pricing. While the optimal time-sequence is a significant puzzle to be solved, framing the issue in this way assumes away all the real uncertainty that makes the problem difficult in the first place. If we make a commitment to a long-term policy based on IAM analysis, when we get to, say, 2040, there is a high probability we will realize that we followed the wrong emission pricing path.

- Bayesian updating and other learning strategies involve placing bets on the unknown future, then observing the effects of the policy decisions and revising our strategy when we have learned enough to figure out if the bet was right or wrong. The main lesson from studies of these approaches is that this kind of learning is too slow to yield the necessary information. Consequently, when we get to 2040, we will likely not know if we were on the correct path or not.
- Each of the futures in Figure 1 implies a corresponding optimal emissions price path, which, for instance, might look something like those in Figure 2. If we knew the future temperatures with certainty, we would, in principle, be able to work out the optimal emission tax path.
- The state-contingent approach involves starting an emissions tax at the current best guess as to its optimal level, then specifying a rule that updates it each year based on the observed climate state. As of the present we do not know what the path will look like, but if we choose the rule correctly, *we can know today that as of 2040 we will have followed the closest feasible approximation to the optimal price path.* Furthermore, the greatest economic gains will accrue to agents that make the most accurate forecasts about the climate state, and hence, the emissions price.
- Under a state-contingent pricing rule, the need for accurate forecasts of the future tax for the purpose of guiding investment decisions will create market incentives to make optimal use of available information and construct the most objective climate forecasts to guide optimal behavioural responses to the policy. Consequently, while the state-contingent price path will only track the actual optimum within error bounds, there is no information currently available that could identify a better price path than the one generated by information markets induced by a state-contingent pricing rule. Likewise, expectations of the future tax path will provide an objective indication of the most likely future path of the climate itself.

The rest of this chapter explains these ideas in more detail.

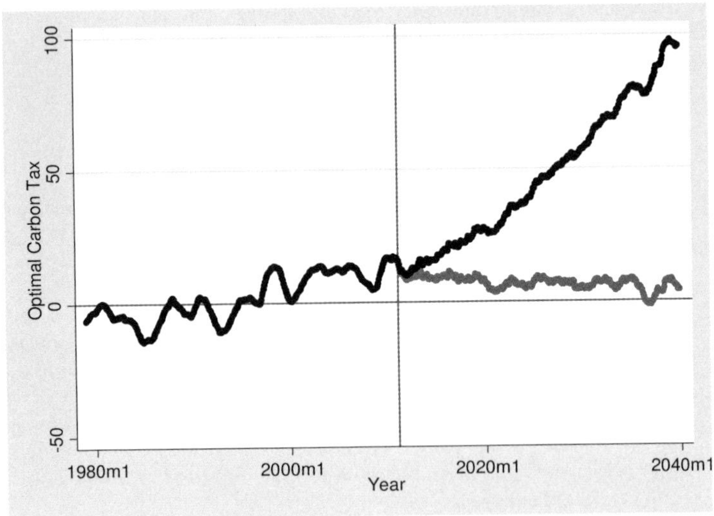

Fig. 2: Possible optimal emission price paths corresponding to future warming scenarios in Figure 1.

UNCERTAINTY AND INERTIA PROBLEMS IN CARBON DIOXIDE EMISSION PRICING

Sources of uncertainty

Analysis of carbon dioxide (CO_2) emission pricing is simple compared to other air emissions such as sulphur dioxide (SO_2) or particulate matter (PM) since there is no CO_2 scrubber technology. The volume of fuel consumed yields a close estimate of the volume of CO_2 released, whereas fuel consumption can be quite uncorrelated with other emissions depending on the pollution controls and combustion technology in place. However, the time element that connects CO_2 to its external costs is considerably more complex than for other emissions.

a) The atmospheric residency of CO_2 is measured in decades, so emissions today could potentially have effects many years into the future, and each year's emissions have marginal effects that accumulate with those of other years;
b) The response of the climate system to changes in the atmospheric stock of CO_2 may be slow, especially if the ocean acts as a flywheel, delaying effects for decades or centuries;
c) Since there are no scrubbers for CO_2, emission reductions must take the form either of changes in combustion efficiency, fuel-switching or reductions in the scale of output, all of which take time to plan and implement;
d) Economically-viable technology for generating electricity and converting fossil fuels into energy is subject to innovation over time, and while it is

reasonable to assume that some innovations and efficiency gains will be realized, the effects and timing of changes can only be conjectured;

e) Since the stock of CO_2 mixes globally, actions of individual emitters are negligible. The only policies that would affect the atmospheric stock must be coordinated among all major emitting nations, and such processes are slow and subject to uncertain success.

Parson and Karwat (2011) frame these issues under the headings of *inertia* and *uncertainty*. If we faced only inertia, we could sit down today and devise an optimal intertemporal policy plan that would yield the right sequence of interventions at the right time through the future. On the other hand, if we faced only uncertainty, we could make short-term decisions on the expectation that new decisions would be made at each point in the future as circumstances change. Faced with both uncertainty and inertia, Parson and Karwat conclude that sequential decision-making is necessary, though they do not spell out how such a process would work in practice. The state-contingent approach, it will be shown, attempts to create a formal structure for sequential decision-making in light of both uncertainty and inertia.

With regard to climate change there are two very large sources of anxiety that have fueled decades of intense controversy. On one side are those who believe the threat from CO_2 and other greenhouse gas emissions are substantial, and who fear that inadequate policy actions are being taken, so that future generations will experience serious welfare losses due to global warming. On the other side are those who believe the threat from CO_2 and other greenhouse gas emissions are small, and who fear that implementation of policies sufficiently stringent to achieve large emission reductions will impose costs on current and future generations far larger than any benefits they yield. For the first group, the fear is that by the time enough information is obtained to resolve uncertainty about the environmental effects of CO_2 it will be too late to avert intolerable environmental damages. For the second group, the fear is that if we act now to try and prevent such damages we will have incurred intolerable economic costs by the time they are shown to have been unnecessary.

The so-called "no regrets" policies that are sometimes invoked to try and make this wrenching dilemma disappear are not relevant. Some argue that, in light of the threat of catastrophic (or even somewhat harmful) global warming, we must act, and by saving energy and reducing air pollution we'd be better off anyway, so on balance it is better to implement them. This argument fails once the details are examined. Energy efficiency measures would not yield the scale of emission reductions necessary to substantially affect the future stock of global atmospheric CO_2. Improvements in energy efficiency that actually make consumers and firms better off are automatically adopted by rational economic decision-makers anyway. Yet CO_2 emissions continue to rise globally as population and income rise. And air pollution is already subject to regulation throughout the developed (and much of the underdeveloped) world. If we assume that households, firms and governments have already made reasonably efficient decisions as to energy efficiency and pollution reduction, further large-scale reductions in CO_2 emissions must be, on net, costly. In other words, policies that might have a trivial cost will only have trivial effects. The policies that actually have an effect on the climate must entail a large economic cost. The dilemma is real.

Integrated assessment models and pseudo-optimal solutions

The IAM approach of Nordhaus (2007) and coauthors yields a solution that can be described as "pseudo-optimal." It assumes the modeler knows the key parameters that govern the economy and the climate, and solution of the model yields a smooth policy "ramp" in the form of an escalating tax on CO_2 emissions over time. This solution can only be considered optimal if we assume the model parameters are correct. But strong assumptions about key functional forms and parameter values are not put to the test by implementation of the policy. If decision-makers were to commit to a policy path based on the IAM analysis, it would amount to acknowledging the inertia but not the uncertainty in the policy problem. The lack of recognition the extent of uncertainty in the IAM approach is one of the bases of the criticism of Weitzman (2009).

Bayesian learning models

Kelly and Kolstand (1999) and Leach (2007) introduced learning into the IAM framework by supposing that we observe the response of the climate to policy innovations, and then we use such information in a Bayesian updating routine. The goal is to accumulate enough information that the policymaker can decide, at 5 percent statistical significance, whether or not to reject the hypothesis that the correct policy is being implemented. Uncertainty and inertia interact in an interesting way: uncertainty about even one or two key inertia (lag) parameters is sufficient to delay for hundreds of years the identification of an expected-optimal policy rule. With only two model parameters subject to uncertainty, Leach (2007) showed the learning time ranges from several hundred to several thousand years, depending on the base case emissions growth rate. An expanded version of the model, incorporating simple production and an intertemporal capital investment structure, not only yields a time-to-learn measured in centuries, even when most model parameters are assumed known, but depending on which of several climate data sets are used to form the priors, the policy path may never converge on the correct target.

It is an illusion to suppose that the IAM, or pseudo-optimal approach, is better, because we apparently follow an optimal path from the outset. The difference between them is that in the Bayesian approach we eventually learn if we are on the wrong path and in the IAM approach we never do.

Insurance and fat tails

Weitzman (2009) looked at the global warming problem as one of trying to price an insurance contract when there is a nontrivial probability of extreme damages. Geweke (2001) had shown that a basic insurance problem can become degenerate if a few features of the set-up are chosen in a particular way. If the risk is distributed normally and utility is of the constant relative risk aversion form, and the change in consumption over the insured interval is expressed as e^C where C is future consumption relative to current consumption, then the expected cost of insuring future consumption under some general conditions can be shown to take the form of a moment generating function for the t distribution, which does not exist, or is infinitely large, making it impossible to place a finite value on a full insurance contract. Weitzman's adaptation of this model to the climate case depends on some specific assumptions, some of which are conventional and some of which are not. One unusual assumption is that there is a possibility of infinite (+

or -) climate sensitivity, or in other words, that while the possibility of an extreme change in the climate (twenty degrees or more) may be small, it cannot be ruled out, no matter how large. To perform conventional cost-benefit analysis it is necessary either to truncate the range of climate sensitivities or assume that the distribution has "thin tails." But, as Weitzman points out, this implies that the optimal insurance policy depends on assumptions about the distribution of possible climatic changes in regions where there are too few observations to know for sure. Hence cost-benefit analysis using IAMs assumes away extreme risks, and cannot therefore provide an economic case for ignoring them. Nordhaus (2009), Pindyck (2011) and others have critiqued the Weitzman model, especially for its assumption of infinite marginal utility as consumption gets very low.

The state-contingent approach

McKitrick (2010) proposed an alternative approach to the pricing of complex intertemporal externalities which focuses on developing an adaptive pricing rule, rather than a long term emissions path. In the standard economic model of pollution pricing, current damages are a direct function of current emissions:

In simple problems of this sort, the solution is to impose an emissions price

$$\tau = D'(e_t) \qquad (1)$$

where D is the damage function and e_t is the optimal emissions level. In the case of CO_2, for any individual country the damage function is linear (since no country on its own can substantially affect the concentration of carbon in the atmosphere) so D' is constant and the optimal value will equal that corresponding with the current emissions level, which is observable. In the presence of inertia, emissions may have lagged effects, and the length of the delay may itself be unknown. If so, then we are currently experiencing the effects not only of present-day emissions, but also of emissions that have occurred at some point in the past. We therefore consider the case in which, instead of directly causing damages, emissions affect some aspect of the environment (such as the average air temperature, possibly with lags due to geophysical processes), and it is those changes that cause damages. In that case the above diagram would be redrawn as follows.

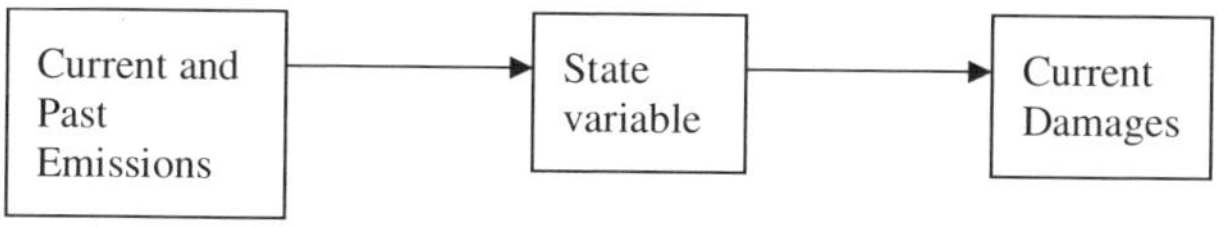

Emissions affect an observable state variable $s(t)$ and current-period damages are a function of the current state variable $D(s(t))$. The state variable, in turn, is determined as

$$s(t) = s(e_t, e_{t-1}, \ldots, e_{t-k}) \tag{2}$$

where the lag length k is unknown. Equation (2) is sufficiently general to handle stock-flow problems. If the stock of emissions is denoted M_t, then the state may be described as the function $s(M_t)$, where $M_t = \delta M_{t-1} + e_t$ and δ is the stock decay rate. But since the emissions flow is the policy target it will be necessary to include e_t in the specification of (2). A form like $s(M_t)$ would not lead to a policy-relevant conclusion since the policy maker cannot control M_t (or past values of e_t).

Note also that $s(t)$ is not, in most cases, the same as the stock of the pollution (i.e. the atmospheric concentration). It is therefore necessary to identify the correct state variable. Hsu (2011) discusses the state-contingent pricing option for carbon dioxide emissions and proposes a list of possible candidates for the state variable. The list will be critically evaluated below, but for the moment our point is simply that for a complex issue like global warming it can be difficult to agree on what to measure, and not all proposed values of the state variable make sense for the purpose of configuring an optimal policy mechanism.

Setting that aside for now, given a definition of s the Discounted Present Value (DPV) of damages is

$$V(t) = \sum_{j=0}^{\infty} \beta^j D(s(t+j)) \tag{3}$$

where β^j is the discount factor j periods ahead. The optimal emissions price is:

$$\tau(t) = \frac{\partial V(t)}{\partial e_t} \tag{4}$$

The influence of current and past emissions on the state variable is complex and uncertain. While this adds to the difficulty of determining how current emissions ought to be priced, it also implies that the state variable contains information about the effect of emissions over time, and this information can be used to reduce uncertainty. The next section explains how to use observations on s to approximate $\tau(t)$.

DERIVATION OF THE STATE-CONTINGENT PRICING RULE

Required Assumptions

Following McKitrick (2010) the derivation requires a number of assumptions.

(A1) ASSUMPTION 1: The state function s is homogeneous of degree c.

This implies

$$s(\lambda e_t,...,\lambda e_{t-k}) = \lambda^c s(e_t,...,e_{t-k}) \tag{5}$$

We do not necessarily assume linear homogeneity.

(A2) ASSUMPTION 2: Over the interval $(t-k,...,t+k)$, ∇s is locally autonomous, that is,

$$\frac{\partial s(t+i)}{\partial e_t} = \frac{\partial s(t)}{\partial e_{t-i}} \text{ for all } i = 0,...,k\,.$$

This imposes slightly more structure on *s*, as it implies the marginal effects over a lag of length *i* are independent of *t*. For example, suppose *s* depends on emissions out to three lags:

$$s(t) = s(e_t, e_{t-1}, e_{t-2}, e_{t-3})\,.$$

At time *t*, the partial derivative with respect to the second argument is $\frac{\partial s(t)}{\partial e_{t-1}}$. At time *t*+1, the partial derivative with respect to the second argument is $\frac{\partial s(t+1)}{\partial e_t}$. (A2) requires that these partial derivatives be equal. This will be true, for instance, if *s* is a function of a weighted sum or moving average of the *e*'s. It will not be true if *s* is nonlinear in the individual *e*'s, in which case it will be true only approximately, where the approximation will depend on how much "curl" *s* has over time.

(A3) ASSUMPTION 3: The function $s(t)$ is locally linear in e_t, and current period emissions must have a non-zero effect on $s(t)$.

This assumption states that in the direction e_t, $\frac{\partial s}{\partial e_t} = v$ in the neighbourhood of e_t, where v is a positive constant and may be arbitrarily small.

(A4) ASSUMPTION 4: At each time *t*, the damage function $D(t)$ can be approximated by a step-wise quadratic (in emissions) function, such that at time *t*, $\frac{\partial D(t+j)}{\partial e_{t+j}} \approx \delta e_t$ for $j = 0,...,k$.

This is different from, and slightly more restrictive than, assuming *D* is quadratic. It states that the slope can be extrapolated forward at time *t* over *k*-1 subsequent periods, and that the extrapolation will be re-set each period. (A4) can be pictured as an approximation for D by a porcupine-like pattern of tangent lines, as shown in Figure 3. The approximation will get worse the larger is k. However, in cases where k is large, effects mix slowly across long time spans and we would not expect total damages to be strongly nonlinear (convex) in current emissions, so the extrapolation is automatically used across a shorter interval in circumstances where it is less accurate over long intervals. In other words, the larger is k (implying greater potential inaccuracy in the slope extrapolation), the less is the likely curvature in D (implying less inaccuracy in the slope extrapolation). If D has a strong curvature then current emissions must have a rather strong immediate effect, so we expect the tangent lines in Figure 3 to be shorter, mitigating the extrapolation error. (A4) also implies that marginal damages are zero when emissions are zero, which is a common and reasonable assumption.

(A4) combined with (A3) implies

$$\frac{\partial D(t+j)}{\partial s_{t+j}} = \theta e_t \text{ for } j = 0,\ldots,k \tag{6}$$

where $\theta = \delta / v$, a positive constant. Neither c nor θ are typically encountered in ordinary environmental policy models. θ measures the change in damages due to a change in the state variable, per unit of emissions. c is the degree of homogeneity of the state variable, and thus equals the sum of the partials of $s(t)$ with respect to k emission lags, with each term multiplied by $e_{t-k} / s(t)$. If we assume emissions are constant for k periods, the product $c\theta$ can be shown to equal the sum of the partials of the damage function with respect to k lags of e, all divided by $s(t)$. In other words, $c\theta$ can be thought of as, approximately, the marginal damage rate, or marginal damages proportional to the value of the state variable.

The next assumption eliminates the role of discounting when summing marginal damages over the lag interval k. While it introduces a form of imprecision in the tax instrument, it is conservative in the sense that it will tend to overstate rather than understate the value of the sequence of damages.

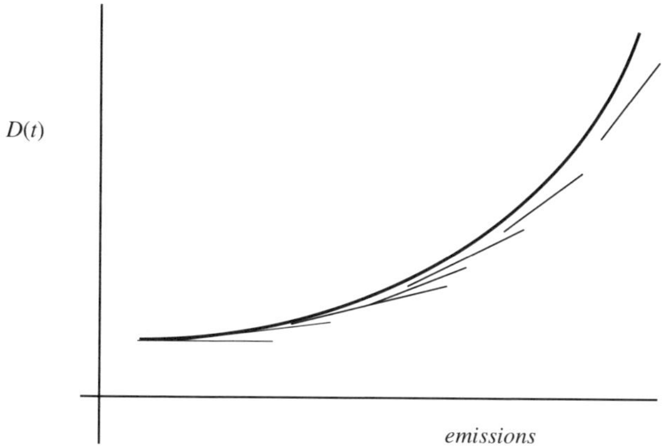

Fig. 3: Approximation of damage function embodied in Assumption A4.

(A5) ASSUMPTION 5: In the evaluation of τ using equation (3), β^j is set equal to unity for $j = 1,\ldots,k$.

The discounted present value of damages is defined by Equation (3). Since emissions only have an effect for up to *k* lags we can write this as

$$\frac{\partial V(t)}{\partial e(t)} = \sum_{j=0}^{k} \beta^j \frac{\partial D(s(t+j))}{\partial s(t+j)} \frac{\partial s(t+j)}{\partial e_t}$$

$$= \theta e_t \sum_{j=0}^{k} \beta^j \frac{\partial s(t+j)}{\partial e_t} \qquad \text{by (6)}$$

$$= \theta e_t \sum_{j=0}^{k} \frac{\partial s(t)}{\partial e_{t-j}} \qquad \text{by (A2) and (A5)} \qquad (7)$$

Now denote, respectively, moving sums, moving averages and weights of e_t as $E_t \equiv \sum_{i=0}^{k} e_{t-i}$, $\bar{e}_t \equiv E_t/(k+1)$ and $\varepsilon_i(t) \equiv e_{t-i}/E_t$. Note that $\sum \varepsilon_i = 1$.

(A1) implies (by Euler's theorem)

$$c \cdot s(t) = \sum_{j=0}^{k} e_{t-j} \frac{\partial s(t)}{\partial e_{t-j}}$$

Hence

$$c \cdot s(t) = E_t \sum_{j=0}^{k} \varepsilon_{t-j} \frac{\partial s(t)}{\partial e_{t-j}} \tag{8}$$

Equation (8) states that $cs(t)/E_t$ equals a weighted average of the partial derivatives of s. In situations where the emissions do not vary too much in percentage terms, over the k-period interval an unweighted mean will be a reasonable approximation to the weighted mean, hence:

(A6) ASSUMPTION 6: $c \cdot s(t)/E_t \approx \frac{1}{k+1} \sum_{j=0}^{k} \frac{\partial s(t)}{\partial e_{t-j}}$.

Combining (A6) and Equation (7) yields

$$\frac{\partial V(t)}{\partial e_t} \approx \theta e_t (k+1) c \cdot s(t)/E_t$$

Hence

$$\tilde{\tau}_t = \gamma \frac{e_t}{\bar{e}_t} s(t) \tag{9}$$

where $\gamma = c\theta$ and ~ denotes the approximation to the optimum.

Equation (9) is the state-contingent pricing rule. It is an easily-calculated approximation to the marginal damages of the complex intertemporal externality shown in equation (3). Comparing it to equation (4), its usefulness has to do with how knowledge of the future is represented. In a system with inertia, emissions today will have an effect on the state variable over an uncertain span into the future. But this means that the current value of the state variable must also reflect the influence of past emissions. If the set of lag relationships extending over the past is structurally similar to the set of lag relationships that will extend into the future, then the current observation of s contains information about how past emissions affected today's climate, and hence how today's emissions will affect the future climate. Equation (9) uses that information to guide the emission price path.

Note that although (9) only uses information dated at time t, the underlying form of the tax is determined by equation (4). That means that, in principle, equation (9) charges firms today for the future value of their emission damages as well as the current value.

There remains one assumption to invoke: $\bar{e}_t$ is not known exactly since it depends on the unknown lag length k, so a lag length must be selected. However, trailing averages are smoothing devices, so unless the emissions series is extremely volatile, $\bar{e}_t$ will be relatively stable across a range of choices of lag length.

Figure 4, taken from McKitrick (2010), shows the implied value of the state-contingent emissions tax over the 1979-2010 interval, using the mean temperature of the tropical troposphere (see next section) and calibration of the free parameter to yield a value of $15 per tonne in 2002. The thick line is a 3-year moving average.

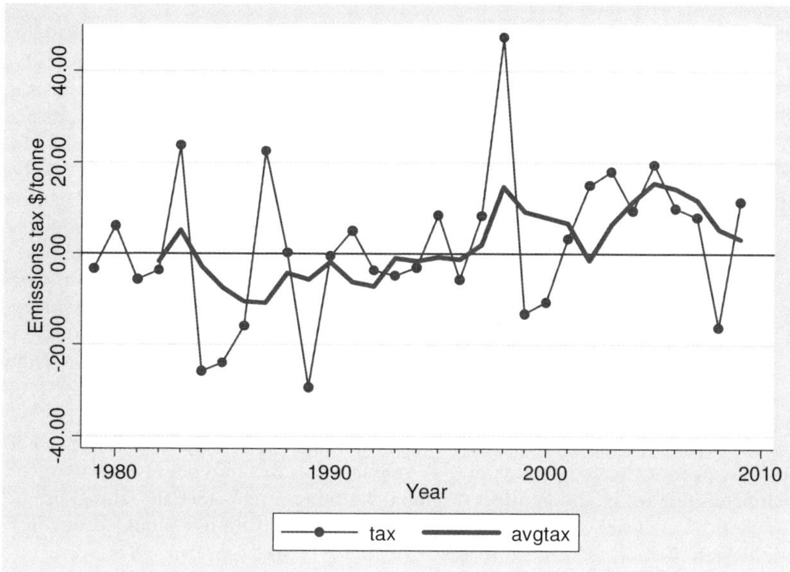

Fig. 4: Value of state-contingent tax on greenhouse gas emissions since 1979. 'avgtax' denotes 3-year moving average.

McKitrick (2010) presented synthetic examples calibrated to the stylized facts of global warming, in which a series of complex and varied simulations into the future are shown, and in each case the state-contingent mechanism $\tilde{\tau}_t$ closely followed the unobservable optimum, even when known errors in approximation were introduced, with correlations typically in the >95% range.

Choice of State Variable

Assumption (1) requires that the state variable be chosen carefully, so that while it represents the factor that influences damages, as much as possible it represents the influence of emissions rather than exogenous factors. This would certainly hold if we simply used total emissions (or the atmospheric stock) as the state variable, but that would not make sense unless it were known that the sensitivity term $ds(t)/de(t) = 1$ for all t, which is not the case, or at least it is not known to be the case. Imposing it by assumption means we are assuming away all the uncertainties over the magnitude of $ds(t)/de(t)$. In the extreme, it would imply that e_t should still be priced the same regardless of whether $ds(t)/de(t) = 0$ or $ds(t)/de(t) \rightarrow \infty$, which is clearly not credible. CO_2 has not historically been regulated as an air pollutant because it is naturally occurring and harmless for humans except at extremely high concentrations. It is only a candidate for policy intervention if it turns out to have a strong effect on the climate by changing atmospheric temperatures. The state variable should be chosen to capture that contingency.

Hsu (2011) correctly points out that the state variable must be non-manipulable and reliably, regularly and uncontroversially measurable. He then proposes a basket of measures arising from the tort principle of focusing on the harms done by climate change, which would include (a) the global mean temperature, (b) counts of days of "unusually" high or low temperatures, (c) counts of extreme rainfall or drought events, (d) rises in sea level, (e) ocean acidity and (f) numbers of hurricanes above a certain intensity level. The weighting scheme is not specified.

However the tort concept is not the appropriate criterion for picking a state variable, since it presupposes since our concern is the extent to which greenhouse gas emissions affect the state variable, not the extent to which changes in the state variable harm potential victims. The latter becomes relevant only when the former is known, but that is precisely what we are uncertain about. Items (b), (c) and (f) are sub-grid scale weather phenomena with poorly-understood and controversial connections to greenhouse gases. The history of global weather shows that extended intervals of elevated drought conditions or excessive moisture can and do arise purely due to natural variability, so these measures do not satisfy Assumption (1). In other words, such index terms could go up (or down) for extended periods even if greenhouse gases turn out to have no effect on the climate. Item (e) is not an effect of global warming, it is potentially a measure of the atmospheric CO_2 level since acidification (or reduced alkalinity) would occur if large enough quantities of CO_2 were to dissolve in the world's oceans. While use of this measure would make sense if the state variable were ocean alkalinity (and the damage function were defined accordingly), if the damages are connected to atmospheric warming the use of ocean alkalinity would, at best, be equivalent to making emissions the state variable, which is a flawed concept for the reason noted above. Item (d) has a somewhat controversial connection to global warming, as Hsu notes, with sharply varying projections as to the rate of rise over the coming century. The main deficiency as a state variable is that sea levels have apparently followed a steady upward trend for centuries, and what is at issue is not whether this will continue but whether global warming will cause it to accelerate. So the more relevant concept would be the acceleration of sea level rise. But the oceanic system is so slow to respond that any acceleration may not be apparent for a long time, due to the inertia involved. Finally, item (a) suffers from controversy and data quality problems (see, for instance, de Laat and Maurellis 2006, McKitrick and Nierenberg 2010, Fall et al. 2011), especially over the ocean (Thompson et al. 2008, Christy et al. 2001), as well as being subject to many influences other than greenhouse gases, which is why a satellite-derived temperature would be a better state variable in this context.

McKitrick (2010), by contrast, proposes just one measure, the mean temperature of the tropical troposphere. This is based on the argument that, as well as being non-manipulable and reliably measurable, the state variable must represent an index of the unique effect of emissions with as rapid a response time as possible. The climate modeling work surveyed in the 2007 IPCC Report (IPCC 2007) and Karl et al. (2006) clearly points to the tropical troposphere as a place in the climate system where an unusually strong and rapid signal of the effects of greenhouse gases should be observable. Thorne et al. (2011) and Fu et al. (2011) provide updated analyses emphasizing the importance of the response to greenhouse forcing in the tropical troposphere as a metric of climate sensitivity, since this is where climate model behaviour is strongly constrained. The Arctic surface temperature is also believed to be strongly influenced by greenhouse gases, but is also strongly influenced by natural phenomena and, being an oceanic region, is subject to poor spatial sampling.

A PERMITS-BASED PREDICTION MARKET TO GENERATE FUTURE PRICES

Use of a state-contingent, myopic pricing rule does not mean that we only put a price on emissions after the damage is done, since the approximation is to the discounted present value of current and future emissions. Also, businesses are forward-looking and investment plans are based not only on today's prices, but on expectations of future prices. While those forecasts may be wrong, there is no incentive for firms to make systematically biased forecasts, since in the future the actual prices will be revealed and mistakes will be costly. However, the basic proposal in McKitrick (2010) does not reveal the expectations in the market, except as they are implicit in firms' investment decisions.

In order to reveal the future price expectations, Hsu (2011) proposed extending the tax rule of McKitrick (2010) to include a tradable exemption permits market for future periods. The permits market would allow firms in year t to purchase a per-tonne exemption from paying the tax in years $t+1$, $t+2$, etc., out to the limit of the forecast horizon. Hence it would establish, in effect, a prediction market for the state-contingent tax, and by implication, the state variable. As with any prediction market, the maximum financial rewards would accrue to those who make the best forecasts, so the published list of price futures would then imply an optimal and objective set of climate forecasts. If agents believe the permit price for year $t+10$ (say) is below the likely value of the tax rate, or in other words if agents believe people are underestimating the amount of likely warming between now and then, they will have an incentive to buy permits, and will thereby drive up the price. And likewise, agents who believe warming is being overestimated will have an incentive to sell permits, or short the market.

Hsu (2011, pp. 33-35) captures well the incentives this would create for basing forward investment behaviour on the most accurate possible scientific information about climate:

> The price bid by emitters for say, permits to emit in 2020, would speak volumes about private expectations of the consequences of climate change, free from, as climate skeptics claim, conspiracies by climate scientists to shore up their research grant fiefdoms, or desires by radical environmentalists who really wish to use climate change as an excuse for imposing environmental restrictions. ...Without an obvious ideological horse in the race, emitters like [American Electric Power, AEP] will brutally and honestly evaluate the credibility of climate science, and spend its climate investigation money carefully. It is the participation of large emitters in a cap-and-trade program for emissions futures that is likely to make or break the credibility of climate science. In essence, this proposal uses markets to turn the evaluation of climate science over to those emitters that will

> potentially rely on those permits for their emitting operations. And such a liability could be very significant: in 2005 AEP emitted approximately 161 million tons of CO_2; if one assumed a very modest carbon tax that was set to five dollars per ton at current climate outcomes, AEP's annual carbon tax liability would be about $805 million. If climate outcomes increased by say, twenty-five percent, its annual carbon tax liability would top one billion dollars. All 101 electricity generators in the EPA's Egrid database would have a combined current carbon tax liability (assuming a rate of five dollars per ton of CO_2) of $8.75 billion. Environmental advocates may chafe at the notion that the greatest greenhouse gas emitters will have such a large say in evaluating the quality of climate science, but $8.75 billion is a lot of impetus for honestly evaluating climate science.

Some of the concern expressed by authors like Stern (2006) and Weitzman (2009) is that the climate may be subject to nonlinear effects of greenhouse gases (Hsu 2010 also raises this concern). Future greenhouse warming may be subject to sudden, rapid acceleration into catastrophic levels of environmental damage. But when we consider how to integrate this possibility into current decision-making, we confront the same dilemma as before, namely between the fears of highly costly potential environmental damages and highly costly potential policy mistakes. The prospect of sudden catastrophic change only amplifies the magnitude of each of these two apparently awful options. But now imagine a state-contingent tax is implemented along with a thirty-year sequence of prediction markets. Suppose initially the prediction markets show a slow, smooth ramp in CO_2 emission prices. This would indicate the market discounts the possibility of a serious bifurcation or (so-called) "tipping point" into catastrophic change. But to the extent that anyone can construct a *credible* scientific argument that a bifurcation is approaching, they would know that those emission futures are underpriced and they be able to invest profitably in them, knowing that once their scientific arguments were digested by the prediction market, there would be a predictable ramp in prices. The prediction market would be the most reliable instrument available for generating a rational signal of a coming bifurcation. Again, the reason is that firms would have a strong financial incentive to get the climate science right, to whatever extent possible, and not to make a systematic forecast error or adopt a disingenuous view of the underlying problem. If an objective information-processing system like a prediction market existed, research warning of a climate nonlinearity would be "brutally and honestly" evaluated, and could potentially be the basis for a spike in emission price futures, thereby providing an instantaneous signal of the gravity of the threat. For this reason, while the state-contingent price mechanism is not guaranteed to resolve the uncertainty at the core of Weitzman's analysis, it provides incentives for the maximum possible resolution; in other words, no other policy path could provide a more objective basis for forming expectations about possible future catastrophes.

The tradable futures market would not allow firms to evade paying the tax on future emissions, since they would have to buy the exemption permits. Instead it would allow firms to trade on changes in expectations about the future path of the state variable and the emissions price. This illustrates another distinction between the state-contingent approach and the IAM approach. In the latter, it is assumed that we possess correct, unbiased scientific information regarding climate, and all we require is a mechanism to implement the optimal price. No allowance is made for the possibility that key parameter estimates (such as climate sensitivity) might be biased due to distorted incentives for scientists who prepare such forecasts. In the IAM approach, implementation of the policy does not induce an improvement in the scientific basis of policy. But the approach in this chapter does not assume we have correct scientific information. It works even if the information we have today is incorrect, either through technical inadequacy or researcher bias. The policy mechanism rewards those agents that eliminate bias and inaccuracy in their forecasting work, by allowing them to trade in futures markets on the difference between the current market price and their expectations of how it will evolve.

Some implementation issues that need to be addressed for other types of policy would also need to be addressed in the present analysis. For instance, ideal implementation would be at the global level with each country imposing the tax based on damage valuations over the whole world, rather than at the national level. But incentives favour free-riding, making it difficult to ensure global participation. On the other hand, the tax approach has the advantage that revenues can be retained domestically to reduce the excess burden of the tax system, reducing the macroeconomic cost of implementation, and the state-contingent nature of the policy provides reassurance that the stringency will be increased only if the underlying problem is shown objectively to merit such tightening, and both these features may increase incentives for multilateral cooperation compared to alternative policies.

Another potential issue is that as new information becomes available, policymakers may decide the initial calibrated value of the tax was incorrect and must be revised. This too would be a problem for any policy mechanism. In the state-contingent case an argument can be made that incentives favour efficient acquisition of the information required to optimally calibrate the tax. In this regard the feedback between policy and science would be particularly fruitful, since emitters subject to the tax will have an incentive to pay for the best, most objective information they can get, and there is no assumption that the forecasts made prior to implementing the policy are correct, or will be validated in the future.

EXTENSIONS

Endogenous emissions response

Since no information about abatement costs are used in deriving the tax t, it may seem that it cannot be a complete policy prescription. The tax paths derived in integrated assessment models are solutions to a two-sided optimization problem, with intertemporal damages netted against intertemporal abatement costs. However, it is important to bear in mind that the formula above does not prescribe a policy *path*, it yields a *rule* that ties the tax rate to the environmental state. The actual path of taxes over time will be determined by the evolution of the state variable, and the ensuing level of abatement will be determined by emitters who respond to the current and expected future tax rates according to their current and future marginal abatement costs. If the capital stock is highly variable then firms will respond to current emission tax rates as they would to any variable input costs. If capital is fixed and time-to-build lags are long, firms will need to form forecasts of the future values of the tax rate, which in turn will depend on future values of the temperature variable, and the usual structure of optimal investment under uncertainty will ensue.

The simulations in McKitrick (2010) assume that the emissions tax $\tau(t)$ is a function of the future state $s(t)$, but not vice-versa. The coherence between the approximate tax, given by equation (9), and the actual optimum is demonstrated with this assumption in force. However, it is likely to be the case that $s(t)$ is a function of $\tau(t)$ as well. This will certainly be true if emission tax increases reduce future temperatures, but it will also be true as long as they reduce future emissions since $e(t)$ enters equation (9) directly, as well as through $s(t)$, and the emissions path in McKitrick's simulations are exogenous. The rationale is that emissions are added up globally whereas the tax is imposed by single governments, and no one country can do much to reduce global total emissions through unilateral action. However, if all countries (or a substantial majority) were to enact the tax, total emissions would be affected by the path of $\tau(t)$. In that situation it has not been shown that the state-contingent pricing rule would yield a stable approximation to the true optimum. This is something that needs to be addressed in subsequent research.

Coalition-formation

An interesting feature of the state-contingent tax is its potential ability to appeal to a broad coalition of interests.[1] People with conflicting expectations about the future evolution of the state variable will nevertheless each expect to observe his or her preferred policy path. Those who think emissions have no effect on climate will expect low emission taxes to prevail in the future, and those who think they have strong effects will expect the tax to increase rapidly. Since each agent expects to get his or her preferred outcome, it may be easier to get agreement for implementation. One of the challenges of climate policy is the need to get agreement at the global level. Different regions have different views on the urgency of the problem and how it compares to their domestic economic priorities, which makes it all but impossible to get agreement on emission targets, or to ensure compliance with earlier agreements. Asking policymakers around the world to agree on a state-contingent tax might be easier. The tax revenue would stay within each country, reducing the burden of inequality across different nations. And during the negotiations, there would be no reason for countries that took opposing views on the likely future path of temperatures to take opposing views on whether the tax is desirable, since each party will expect to get what they consider to be the "correct" outcome.

Suppose the stringency, and hence costliness, of a policy can be summarized as a parameter z, where a higher value of z corresponds to a more stringent policy. A potential voter (person i) has a private view of the optimal value of z given his or her beliefs about the marginal effect of emissions on $s(t)$, which we denote by s^i. Their preferred policy is thus $z_i(s^i)$. If $z < z_i(s^i)$ then the proposed policy is deemed too lax, and vice versa.

Typical median voter models only require $z \geq z_i(s^i)$ to ensure voter i's support, namely people are satisfied as long as z equals or exceeds their preferred policy. But

[1] As an anecdotal illustration, Hsu (2011) and McKitrick (2010) hold very different views on the underlying threat of global warming, yet both advocate the same policy mechanism, albeit with different recommendations as to the appropriate state variable.

suppose the voter's support for a policy z declines based on the distance ($z > z_i(s^i)$), i.e. z can be too strict even for someone who prefers a relatively high value. In this case, obtaining majority support can be difficult since it faces two-sided opposition. For example, a moderate emissions price might be opposed by those who prefer it to be much higher, as well as much lower. But proposals to adjust z up or down may alienate as many supporters as they would attract, making it impossible to get a majority.

Suppose a potential voter uses a quadratic loss function like $L_i = [z_i(s^i) - z]^2$ to determine his or her degree of opposition to the policy. Then the greater the variance of beliefs about s^i, the smaller the coalition of support for any policy. The state-contingent approach can potentially alleviate this problem, however. The policymaker no longer proposes a fixed value of z, but instead proposes a function of the observed state $z(s(t))$ over time. Each agent will then expect future values of s to be correlated with s^i, hence the sum of the expected loss terms L_i will be smaller than before, even if the variance of beliefs about s^i remains large. Intuitively, by proposing a policy target that is dependent on the actual future state, each agent "expects" to get his or her preferred outcome. The one who expects the emissions to have a large effect expects the policy to end up being stringent, while the one who expects emissions to have little effect expects the policy to end up being lax. Consequently both types of agents expect small losses from the policy, and have an equally strong incentive to support it, even though they have conflicting views about the form it will actually take.

CONCLUSIONS

Uncertainties over the future path of global warming and the underlying severity of the problem make derivation of an intertemporally-optimal emissions price on carbon dioxide both theoretically and politically very difficult. IAM-based approaches to the problem assume knowledge of a lot of key parameters, while learning models suggest it will take too long to resolve those uncertainties to be of much use in the current debate. If all conceptually-possible climate risks are considered, it may be impossible to place a finite value on full insurance against those risks without arbitrarily truncating the range of extreme outcomes being considered.

A fundamental problem with the existing analyses of carbon dioxide pricing is that agents and policymakers cannot commit to a long term emissions price. The issue is polarized such that fear of two very large potential mistakes seems to have paralyzed the decision-making process, namely fear of climate catastrophe due to failure to act, and economic catastrophe from inept action. The nature of the climate issue makes these fears justified.

This chapter explores an alternative approach based the concept of state-contingent pricing, in which agents commit to a pricing *rule* rather than a *path*. The rule connects current values of the emissions price to observed temperatures at each point in time. In essence, if the climate warms, the tax goes up, and vice versa. A derivation is provided showing how such a rule yields an approximation to the unknown optimal dynamic externality tax, yet can be computed using currently-observable data. A recently-proposed extension coupling the state-contingent tax with a tradable futures market in emission allowances would then yield not only a feasible mechanism for guiding long term investment, but an objective prediction market for climate change.

There are many potential advantages of the state-contingent approach. For one thing, people with divergent views on the nature of the climate issue can still potentially commit to the same instrument, since each one would expect to get his preferred outcome. The rule is structured such that, however the future unfolds, in retrospect we will know that we followed a reasonably good approximation to the optimum, and the incentives along the way favour the use of unbiased forecasts of the pricing path to guide investment decisions. Consequently there are informational, as well as theoretical and practical, advantages to the state-contingent approach, which make it worth exploring in more depth as a potentially viable tool for implementing sound climate policy.

REFERENCES

1. Christy, John R., David E. Parker, Simon J. Brown, et al. (2001) "Differential trends in tropical sea surface and atmospheric temperatures since 1979." *Geophysical Research Letters* 28, no. 1: PAGES 183-186.
2. de Laat, A.T.J., and A.N. Maurellis (2006) "Evidence for influence of anthropogenic surface processes on lower tropospheric and surface temperature trends." *International Journal of Climatology* 26 (June 1): 897-913.
3. Fall, S., A. Watts, J. Nielsen-Gammon, E. Jones, D. Niyogi, J. Christy, and R.A. Pielke Sr., (2011) "Analysis of the impacts of station exposure on the U.S. Historical Climatology Network temperatures and temperature trends." *J. Geophys. Res.*, in press.
4. Fu, Qiang, Syukuro Manabe and Celeste M. Johanson (2011) "On the warming in the tropical upper troposphere: Models versus observations" *Geophysical Research Letters* VOL. 38, L15704, doi:10.1029/2011GL048101, 2011.
5. Geweke, John (2001) "A note on some limitations of CRRA utility," *Economic Letters, 71, 2001, 341–345.*
6. Hsu, Shi-Ling (2011) A Prediction Market for Climate Outcomes (February 26, 2011). Available at SSRN: http://ssrn.com/abstract=1770882
7. IPCC (2007) *Climate Change 2007: The Physical Science Basis.* Contribution of Working Group I to the Fourth Assessment Report of the Intergovernmental Panel on Climate Change. Solomon, S., D. Qin, M. Manning, Z. Chen, M. Marquis, K.B. Averyt, M. Tignor and H.L. Miller (eds.). Cambridge University Press, Cambridge, United Kingdom and New York, NY, USA.
8. Karl, T.R., Susan J. Hassol, Christopher D. Miller, and William L. Murray (2006) *Temperature Trends in the Lower Atmosphere: Steps for Understanding and Reconciling Differences.* Synthesis and Assessment Product. Climate Change Science Program and the Subcommittee on Global Change Research. http://www.climatescience.gov/Library/sap/sap1-1/finalreport/sap1-1-final-all.pdf. Accessed August 3 2010.
9. Kelly, D.L., and C.D. Kolstad (1999) "Bayesian learning, growth, and pollution." *Journal of Economic Dynamics and Control* 23, no. 4: 491-518.
10. Leach, A.J. (2007) "The climate change learning curve." *Journal of Economic Dynamics and Control* 31, no. 5: 1728-1752.
11. McKitrick, Ross R. (2010) A Simple State-Contingent Pricing Rule for Complex Intertemporal Externalities. Energy Economics doi:10.1016/j.eneco.2010.06.013

12. McKitrick, Ross R. and Nicolas Nierenberg (2010) "Socioeconomic Patterns in Climate Data." *Journal of Economic and Social Measurement*, 35(3,4) pp. 149-175. DOI 10.3233/JEM-2010-0336.
13. Mears, C.A., and F.J. Wentz (2005) "The effect of diurnal correction on satellite-derived lower tropospheric temperature." *Science* 309: 1548-1551.
14. Nordhaus, William D. (2009) "An Analysis of the Dismal Theorem" Cowles Foundation Discussion Paper No. 1686.
15. Nordhaus, William D. (2007) "To Tax or Not to Tax: Alternative Approaches to Slowing Global Warming." *Rev Environ Econ Policy* 1, no. 1 (January 1): 26-44. doi:10.1093/reep/rem008.
16. Parson, Edward A. and Darshan Karwat (2011) "Sequential Climate Change Policy" *WIREs Climate Change* DOI: 10.1002/wcc.128.
17. Pindyck, Robert (2011) "Fat Tails, Thin Tails and Climate Change Policy" *Review of Environmental Economics and Policy* 2011, forthcoming.
18. Spencer, R.W., and J.R. Christy (1990) "Precise monitoring of global temperature trends from satellites." *Science* 247, no. 4950: 1558-1562.
19. Stern, Nicholas (2008) *Stern Review on the Economics of Climate Change* London, UK: Her Majesty's Treasury.
20. Thompson, David W.J., John J. Kennedy, John M. Wallace, and Phil D. Jones (2008) "A large discontinuity in the mid-twentieth century in observed global-mean surface temperature." *Nature* 453, no. 7195 (May 29): 646-649. doi:10.1038/nature06982.
21. Thorne, Peter W. , Philip Brohan, Holly A. Titchner, Mark P. McCarthy, Steve C. Sherwood, Thomas C. Peterson, Leopold Haimberger, David E. Parker, Simon F. B. Tett, Benjamin D. Santer, David R. Fereday, and John J. Kennedy (2011) "A quantification of uncertainties in historical tropical tropospheric temperature trends from radiosondes." *Journal of Geophysical Research-Atmospheres* VOL. 116, D12116, doi:10.1029/2010JD015487, 2011.
22. Weitzman, Martin L. (2009) "On Modeling and Interpreting the Economics of Catastrophic Climate Change." *Review of Economics and Statistics* 91(1) 1-19.

'CLIMATE ECONOMICS': THE LITERATURE AND ITS UTILITY

MICHAEL JEFFERSON
International Business and Sustainability Center, London Metropolitan Business School, London, UK

ABSTRACT

Despite the uncertainties surrounding the precise causes of climatic change, few 'climate economists' have focused on what truly sound precautionary policies, measures and relevant technologies may best serve to avoid possible severe disruption of future lifestyles and even lives. Instead, over the past twenty years—beginning with the works of William Cline and William Nordhaus—economists claiming to specialise in the field of climate economics have seen "the economist as saviour", assumed global climatic change between 1970 and 2000 (at least) must have had anthropogenic causation, and come forward with over-ambitious ideas and techniques.

Ignoring the implications of "the tragedy of the commons", which in principle excludes key issues from resolution in the market place and places them uncertainly in the political and negotiating arena, they have all too often proceeded to try to value costs and benefits, make an integrated assessment of them all, pluck long-term discount rates out of the air, place scientific and economic uncertainty aside, and make challengeable assumptions about distribution between generations and nations. Economists would be better advised to be more modest in what they seek to achieve, and recall the long tradition within economics of recognising the "unknowable". Their techniques may be better applied in assessing the respective costs and benefits of taxes or regulation, and in exposing the internal contradictions of many claims for renewable energy and the assumption that technological innovation will continue to meet human needs and aspirations. They should recall the remark of England's King Charles II: "Things impossible do excuse themselves in not being done." With world population continuing to grow, many resources under pressure, and widespread evidence of sub-optimal decision-making, climate economics should return to "the art of the possible".

INTRODUCTION

Some of us have taken a close interest in weather conditions and climatic change for a very long time: in the case of this writer, for over fifty-five years. We were brought up to acknowledge that there had been considerable increases in global near surface temperature, of varying duration, long before the human species inhabited the Earth. This was certainly in the Carboniferous and Permian periods, and some others. It was widely recognised that solar variation, water vapour, clouds and albedo all played their part in shifting the climate, as could volcanic eruptions; and that climatic change could be, in part, ascribed to human activities. The impact of human actions on local climate has been noticed for over 2,500 years. Scientists and would-be scientists have been speculating on the causes of warming for over 300 years, ever since Edmé Mariotte first focused on the issue of "chaleur de feu". The names of Fourier, Tyndall, and Arrhenius have become familiar in recent years.

By 1962, at a UNESCO Symposium in Rome, stress continued to be placed on solar variation, clouds and albedo, and volcanic activity, but the speaker's conclusion was that forecasts based solely upon such evidence could not be sufficiently substantiated. The speaker then referred to the carbon dioxide content of the atmosphere having risen from 290 ppm at the end of the 19th century to 330 ppm by the late 1950s. Yet, quoting Revelle and Suess (1957), he commented that fossil fuel combustion could only account for 2-4 % of this 12% increase. He stated that "there are serious objections against any over-emphasis of the CO_2 theory" given that: "it deals with one single term of the global heat balance", and the historical record. He concluded from the historical record: "Such examples seem to demonstrate that the CO_2 effect cannot be considered as the single (nor even as the main) cause of climatic variations, but certainly as an essential contributing factor." (Flohn, 1963, page 341) The idea that an increase of 104 parts per million in the carbon dioxide content of the Earth's atmosphere over the past 250 years (or an increase of 140 parts per million in carbon dioxide equivalent) has been enough to explain the increases, and other variations, in mean near surface global temperature continues to elude many serious observers of weather conditions and climatic change.

The year before the Symposium was held, President Kennedy had appealed at the UN for "future cooperative efforts between all nations in weather prediction and eventually weather control". Stimulated by this remark, the Global Atmospheric Research Programme was set up in 1967 with "the main aim to observe the global atmosphere in sufficient detail so as to investigate the way in which different scales of atmospheric motion are organised and how they interact, hence to determine the extent to which the larger-scale motions can be predicted by numerical models." (Houghton, J., 1984, page 3). The three decades preceding 1984 had not appeared to provide much evidence of near surface global warming but, in retrospect, in a period when global cooling concerns were being widely voiced—in the early 1970s (somewhat earlier in the Southern Hemisphere) —a renewed period of warming appears to have begun, and clearly continued for some thirty years.

The U.S. National Research Council had been limbering up to make a contribution to the debate on global climatic change as early as 1978 (with a 1979 publication on carbon dioxide and climate), and in 1983 published: "Changing climate". There, names of economists who have become closely associated with the climate change debate first appeared before a wider audience, including William Nordhaus, Thomas Schelling, and Gary Yohe. Nordhaus and Yohe wrote on future carbon dioxide emissions from fossil fuels (Nordhaus, W., 1983), and Schelling on the implications for welfare and policy of climate change. (Schelling, T., 1983)

Businesses, especially in the energy field, were not far behind. The Report which emerged from the 1985 World Meteorological Conference was taken to heart in some of Shell's scenario planning work in 1986, for example, despite this being a period of recovery from the 'oil price collapse' which had accelerated in December, 1985. It was pointed out that:

> "Much improved energy-efficiency, or a major move towards nuclear power and renewable energy, are alternative ways of reducing carbon dioxide emissions. Obviously, all of these have major problems associated with them, not least the

need for concerted global action from both industrialised and newly industrialising countries alike." (Shell, 1986, page 2)

Shell's report concluded this section with the words: "the pro-nuclear, pro-renewable energy and pro-conservation lobbies can all be expected to cite the potential enormity of the Greenhouse Effect as a strong argument in their favour." (Shell, 1986, page 3) Prescient words, one might have thought, twenty-five and more years ago. Yet, apart from some expansion of renewable energy and exaggerated claims thereon, what has really been achieved?

Three years later Shell raised the question what value people would really place on environmental goods, the difficulties of value measurement, and the temptation for current rather than postponed material gratification—or, to quote: "the social time preference rate of discount". The list of environmental concerns began with "global warming", and there was discussion of "marketable permits", tradable or otherwise, in order to gain "credits" from reduction of emissions. One scenario was entitled "Sustainable World", with tighter targets for carbon dioxide emissions and the introduction of carbon taxes. (Shell, 1989)

Within a year the IPCC's first Scientific Assessment had appeared, in one chapter of which it was concluded that: "The latest atmospheric models, while by no means perfect, are thus sufficiently close to reality to inspire some confidence in their ability to predict the broad features of a doubled CO_2 climate at equilibrium, provided the changes in sea-surface temperature and sea-ice are correct." (Houghton, J. et al. 1990, page 126) However, Working Group III (Formulation of Response Strategies) of the 1990 IPCC Assessment found that, in general, "the most effective response strategies, especially in the short-term, are those which are beneficial for reasons other than climate change and justifiable in their own right, for example increased energy efficiency and lower greenhouse gas emission technologies, better management of forests and other natural resources...economically efficient and cost effective (responses) in particular those that use market mechanisms." (IPCC, 1990, page 125)

In January, 1991, the IPCC's Energy and Industry Subgroup reported—warning that the costs of limiting 'greenhouse gas' emissions were likely to be substantial although the Subgroup was "not yet able to provide informed advice on the costs associated with the measures and response strategies discussed in this report." (IPCC Energy and Industry Subgroup Report, January, 1991, page 137) In his Foreword to this Report co-Chair Keiichi Yokobori had written:

> "The report also acknowledges many uncertainties and ambiguities which prevented the production of a clear-cut emissions scenario. They include uncertainties over future economic development paths and the consequential energy requirements, the lack of data and information on developing countries and centrally planned economies and the lack of agreement on cost-assessment methods and estimates."

<u>What is new, one may ask.</u>

Meanwhile, in 1990 the World Energy Council had appointed a Commission to look into the energy and energy-related issues confronting the world, and issued its

Report (written by the present writer) in 1993 as: "Energy for Tomorrow's World—the Realities, the Real Options and the Agenda for Achievement". Many issues of high priority were discussed in addition to climatic change, but the Report did call for: "precautionary measures to reduce the emissions of greenhouse gases since scientific evidence does not so far justify any other policy"; abatement policies based on the principle of expenditure optimisation across the globe, not just on a national basis to secure national targets irrespective of global impact; and adaptation policies "in the event that enhanced global warming and climate change due to anthropogenic greenhouse gas emissions becoming confirmed, and realistic prospects of fossil fuel demand indicate that if there is confirmation then adaptation policies are already unavoidable." (WEC, 1993, page 239) Among the other issues covered in the WEC Report were the importance of economic growth and the scope for improving energy efficiency (the latter being the writer's contribution—"Energy Efficiency and Sustainability"—to the 44th Session of these International Seminars).

The IPCC's 1990 Report opened the floodgates to a surge of books and papers on 'climate economics'. From the outset the lines of battle were drawn. William Cline considered that "social benefit-cost ratios are favourable for an aggressive program of international abatement. The difference stems in part from a longer-term perspective that takes account of much greater warming and damage." (Cline, W., June, 1992, page 4) He considered that: "In sum, under risk aversion it appears sensible on economic grounds to undertake aggressive action to curtail the greenhouse effect". He concluded that: "If by, say, the year 2015 progress in slowing and reducing carbon emissions seemed inadequate, it could become necessary to shift from a carbon tax regime (internationally coordinated, nationally levied) to a carbon quota regime with tradable permits..." (Cline, W., May, 1992, page 94) He contrasted his position with that of William Nordhaus who, he mentioned, had found "that only modest action is warranted."

William Nordhaus himself considered: "The need to address the potential issues raised by future climate change is one of the most challenging economic problems of today, and it is daunting for those who take policy analysis seriously." He did suggest at this time (1994) that "a massive effort to slow climate change today would be premature given current understanding of the damages imposed by greenhouse warming", but went on (given the risks of catastrophic and irreversible changes):

> "Economics does not rule out these outcomes. If scientific evidence indicates that calamitous consequences are likely to accompany global warming, then our economic models will not only signal that a strenuous effort to slow or prevent future climate change is necessary but help devise the time and scope of policy responses. Our future lies not in the stars, but in our models." (Nordhaus, W., 1994, page 6)

It is time to examine this hubris more closely.

<u>The Burgeoning of Climate Economics Literature</u>

From the early 1990s there was a rapid expansion of books and peer-reviewed papers on climate economics. Such was the rate of expansion and the tone of many of the contributions that it reminds one of the sub-title of Robert Skidelsky's second volume of his biography of Keynes: "The Economist as Saviour". One is further reminded by the following words:

> "Everyone agrees that Keynes was the most intuitive of men...He used to say that his best ideas came to him from 'messing around with figures and *seeing* what they must mean.' From his earliest years Keynes was fascinated by numbers, which were perhaps more 'real' to him than people or situations. Yet he was famously sceptical about econometrics—the application of mathematical and statistical methods to the analysis of economic data, chiefly for the purpose of explanation and forecasting. The truth seems to be that numbers were for him simply clues, triggers of the imagination, rather like anecdotes are for the non-mathematically minded." (Skidelsky, R., 1992, page xix)

Climate economists—in general—appear to have forgotten the lessons which may be learned from this revelation. They also appear to be unaware of a lengthy and important strand of economic thinking: the relevance of subjectivism in economics and the persistence of the unknowable, which should require us to recognise "the bounds of unknowledge", or that "things impossible do excuse themselves in not being done." (Wiseman, J., (Ed.), 1983) Yet also: "The assignment of a low probability to a potential disaster might tempt a decision maker to dismiss that contingency. That temptation must be eliminated." (Shackle, G., 1982) They do not recognise, some would claim, the existence of fundamentally different approaches and value systems, even the concept of market prices "do not matter much, if at all, in the contemporary theory of environmental economics or in its way of doing cost-benefit analysis"; willingness to pay is the only value recognised by the economic analyst; and moral, aesthetic and theological judgments are set to one side. (Sagoff, M., 2008, pages 34-35, 40, and 206-207)

Others have claimed, as Simon Dresner has, that: "The starting point for sustainable development was the aim to integrate environmental considerations into economic policy. More profoundly, it was conceived as an attempt to bring environmentalist ideas into the central area of policy, which in the modern world is economics. It was to be the ground on which the mainstream was to consider the environmentalist case." (Dresner, S., 2nd ed., 2008, page 69)

But some have increasingly considered that there has been a strong bias in the mainstream towards mitigation of climate change, rather than adaptation to it. Although the four IPCC Assessments to date have considered both mitigation and adaptation, the contributions of 'climate economists' almost entirely appear in the Working Group III Reports—on Mitigation for the Third and Fourth Assessments, on Economic and Social Dimensions for the Second Assessment. In seeking to rebalance efforts for mitigation *and* adaptation, one recent set of essays has examined the ways in which economies have adjusted to past climatic change, based principally upon US experience. The motivation for the book of essays was: "The types of adaptation to climate that are described in this volume have received less attention in the economics literature than have policies for mitigation of emissions." (Libecap, G. and R. Steckel (eds.), 2011, pages 2-3) Motivation, fashion, vested interests, and political influences have all probably played a part in this bias.

William Nordhaus, in introducing the subject: "Assessing the Economics of Climate Change" (in Nordhaus, ed., 1998) claimed: "Understanding the science, economics, and policy aspects of global warming has proven one of the most exciting and challenging tasks facing the natural and social sciences over the last decade." He went on:

> "Nowhere has research been so intense as in the field of climate change. The intellectual and policy challenges here have mobilized a small army of researchers to investigate every conceivable aspect of the problem. Although economics was a late entrant into the research process, today there are dozens of individual researchers or teams in all major industrialized countries looking at the different questions raised by the threat of climate change." (Nordhaus, 1998, page 1)

In his Introduction, as many others have written before and particularly since, Nordhaus acknowledges that "there are vast uncertainties in the field, but these should not lead people to conclude that the issue can be ignored." (*supra*, page 5) The present writer agrees, but the central issue here is whether 'climate economics' has been usefully focused and has produced useful results. And on that issue the writer has grave doubts. Instead of economists focussing on more specific issues such as the scope for and costs/benefits of raising energy efficiency, promoting energy conservation, discouraging deforestation, harnessing land and water resources for optimal societal outcomes, and promoting truly effective harnessing of renewable energy sources (fully recognising their low power densities) and technologies, they have strayed into 'climate economics' where much of the work done has been of little practical benefit.

William Nordhaus had explained in his 1994 publication that his book was the result of almost two decades of research into climate economics—supported by the U.S. National Science Foundation, Yale University, and the U.S. Environmental Protection Agency. Here we find reference to his DICE Model—the Dynamic Integrated model of Climate and the Economy—and his recognition that "the earlier studies had a number of shortcomings, but one of the most significant from an analytical point of view was the inadequate treatment of the dynamics of the economy and the climate". (Nordhaus, W., 1994, page 4)

In William Nordhaus's latest book, "A Question of Balance" (2008), he concludes:

> "We begin with some reservations that should be kept in mind in weighing the results of this book. The first reservation is that the structure, equations, data, and parameters of the model all have major uncertain elements. Virtually none of the major components is completely understood. Moreover, because the model embodies long-term projections of poorly understood phenomena, the results should be viewed as having growing error bounds the further the projections move into the future." (Nordhaus, 2008, pages 192-193)

Nordhaus then refers to possible impacts of uncertainties on policies, the fact that his DICE model is just one approach to understanding the economic and policy issues involved in global warming, and that the model itself "contains highly simplified representations of the major relationships among emissions, concentrations, climate change, the costs of emissions reductions, and the impacts of climate change." (*supra,* page 194). His position is that the use of highly aggregated relationships is primarily motivated by the extremely complicated nature of the linkages, particularly as they involve long time dynamics. He asserts, therefore, that it is useful to work with a model that is as simple and as transparent as possible.

Nordhaus's openness is to be commended but, after nearly forty years of working in this field, it raises the question whether his (and others') efforts have been worthwhile in terms of value of output. Perhaps the goal has been out-of-reach because too complex, too uncertain, and frankly not possible. In that event Nordhaus's work exposes a fundamental flaw in climate economics as its mainstream exponents have understood it. The fact that Nordhaus concludes that "an ideal and efficient climate-change policy would be relatively inexpensive and would have a substantial impact on long-run climate change" needs to be seen within the cautionary comments which precede it, and may be considered as little more than a "best-guess".

One of the most balanced examinations of the costs and benefits of limiting carbon dioxide emissions came from Alan Manne and Richard Richels in their "Buying Greenhouse Insurance: The Economic Costs of CO_2 Emissions Limits" (1992). Already, as they pointed out, "the greenhouse debate is short on facts and long on rhetoric." (page 1) Their work was based upon a computer model, Global 2100, but their analysis covered a careful examination of energy supply prospects, not least the scope for greater efficiency in the supply and use of energy reflecting in part rising prices as conventional oil and natural gas resources become depleted, and the limitations placed upon coal usage as carbon limits are imposed. (pages 15-19) For these authors "over the long term, the CO_2 problem is primarily a coal problem, and nearly 90 per cent of the world's coal resources are contained in the United States, Soviet Union, and China." (page 27) Their work attracted favourable comment from a number of leading climate economists, and the U.S. Electric Power Research Institute was widely commended for its financial support. Interestingly, it is one of numerous books covered here now regarded as surplus to the requirements of some UK universities.

Another book published by the MIT Press shortly before that of Manne and Richels has also not stood the test of time on the shelves of some UK universities (the writer has no evidence for the situation in other countries): "Global Warming: Economic Policy Responses", edited by Rudiger Dornbusch and James Poterba. (Dornbusch, R. and J. Poterba, 1991) This involved some twenty-six 'climate economists' writing or commenting upon different aspects of the theme. They had the original Working Paper by Manne and Richels (1990) already to hand. They also had a balanced overview by Andrew Solow on the historical background of global warming to draw upon. The then usual crop of climate economists were present: Alan Manne, William Nordhaus, Lars Bergman, David Pearce, John Martin, Thomas Schelling, and William Cline. Contributions were fairly balanced, although the contribution on tax policy to combat global warming was dismissed by the Commentator as premature and requiring a careful cost/benefit analysis which had not been carried out (page 100); the Commentator on the contribution on technological options stressed the scepticism surrounding the many estimates of costs, especially as the authors had plunged ahead to cite estimates without devoting some attention to the key issue of the reliability of such estimates (page 161);

the Commentator on "the European perspective" expressed concern that the chapter failed to address the strategic conflict of interests between producers and exporters of petroleum on the one hand and (large) importers on the other (page 190); and David Newbery made robust comments on the weaknesses of the chapter on the international incidence of carbon taxes and efforts to model this. The most telling comments were made in Thomas Shelling's contribution. He pointed out a number of challenges—from world population growth to his view that the U.S. economy would not be among the first to be troubled by enhanced global warming—but perhaps his most prescient comment was: "I expect that the current momentum will not lead to serious greenhouse abatement commitments but will become absorbed in institutional arrangements." Then would start "the long, slow process of examining the hard facts that underlie the greenhouse dilemma, both in the international organizations set up for such a purpose and, more importantly, within national governments. And then there will begin the arduous process of negotiating commitments to certain goals, criteria, or quantitative schedules and targets." (page 206) It is worth emphasising that these words were written before the UN Framework Convention on Climate Change, never mind Kyoto, Copenhagen, Durban or Rio+20.

Probably the most powerful critique of efforts by economists to model climatic change has come from Stephen DeCanio. In "Economic Models of Climate Change: A Critique" DeCanio finds such models inadequate for explaining consumers' preferences and market demand, for the treatment of time and overlapping generations, for the representation of production, and for long-term prediction. Matters of distribution, multiple equilibria, and dynamics are ignored or downplayed. He points out that: "Energy modelling in economics has a long enough history that the match between forecasts and actual outcomes can be compared. The results of such comparisons are not encouraging for those who would invest economics with an aura of scientific precision. This does not mean that economists should abandon the effort to gain insight into the forces shaping the long run; it only means that they must be conscious of the limits to their ability to do so." (DeCanio, S., 2003, page 152) DeCanio was particularly scathing about energy/economic modelling, its poor predictive performance, and the failure of its practitioners to realise that the value of their work lies principally in the possibility of providing insights and testing assumptions. He considered that more recent economic models of climate change owe too much to the flawed energy/economic modelling of the 1970s (he spends some time commenting on Nordhaus's work in that period). The contrasting historical performance of sound scenario development, consciously preferred over global energy/economic modelling, reinforces this point. (Jefferson, M., 2012)

DeCanio in fact believes that traditional modelling has concentrated on efficiency, and has over-estimated costs while exhibiting a bias against bold and timely action. This may be a matter for contention. However, DeCanio makes two more fundamental points about many economists and the state of their discipline (if discipline is what it can fairly be termed):

> "Despite the limitations of their models, economists have been willing to play a central role in the policy discussion, and there is no doubt that they have had a great deal of influence on the debate. Power is seductive; susceptibility to its attractions is not evidence that economists as a group are more self-promoting or

venal than natural scientists or other social scientists. Instead, the failure of economics derives from a kind of *hubris* on the part of the profession. The technical achievements of economics, particularly of mathematical general equilibrium theory, have engendered an attitude of superiority vis-à-vis other disciplines. More seriously, the limitations and unanswered questions that lie at the heart of general equilibrium theory have been ignored in the drive to provide answers that can carry political weight." (DeCanio, S, 2003, page 158)

In addition, DeCanio proceeds:

"It is perhaps understandable that economists desire to emulate the natural scientists by seeking sparse and elegant mathematical descriptions of the social system. Unfortunately, this quest has so far proven to be futile. It has been known for some time within economics that the most parsimonious representations of rationality do not provide sufficient restrictions on the behaviour of aggregates of diverse individuals to determine a unique equilibrium. There are limits on what economics can say, given only the requirements of formal rationality. By disregarding these limits in carrying out climate policy analysis based on neoclassical general equilibrium theory, economists have constructed an imposing structure on unsound foundations. While the edifice has the trappings and appearance of scientific rigour, its 'results' are in fact derived in large part from unverified or untrue assumptions." (DeCanio, S, 2003, page 158)

These are neither original (Jefferson, M, 1971) nor old-fashioned (Hall, C. & K. Klitgaard, 2012) criticisms. They remain well-directed.

But DeCanio also has pertinent things to say about discount rates, a key element in the application of climate economics. The 'paradox of discounting' is that a discount rate which appears to mirror expected or commercial rates of return on investment precludes the possibility of anything happening in the distant future mattering for the decisions taken today. DeCanio takes the example of a discount rate of 7% per annum, and GDP growing at 2% per year, to consider the present value of the world in 300 years' time. The amount is so trivial in per capita terms that the loss of all economic output at that time would be worth very little to people today. On the other hand, if the discount rate were zero then even the tiniest increase in future consumption would theoretically have an infinite present value. In the latter case, any sacrifice of current consumption would be justified. DeCanio (pages 58-59, and 93) quotes Shane Frederick et al.:

"The discounted utility (DU) model, which continues to be widely used by economists, has little empirical support." (Frederick, S. et al. 2002, page 393)

Given this background, it appeared somewhat surprising that in 2006 there appeared "The Stern Review" aka "The Economics of Climate Change". In the version published by Cambridge University Press the following year, covering 692 pages, it was claimed: "*The Review* considers all aspects of the issue." Sir Nicholas Stern, already a Knight of the Realm although at that time rather uneasily positioned in UK Treasury, headed the Review and was subsequently elevated to become a Peer of the Realm. In its

Introduction the Review stated that: "Economics has much to say about assessing and managing the risks of climate change, and about how to design national and international responses for both the reduction of emissions and adaptation to the impacts we can no longer avoid. If economics is used to design cost-effective policies, then taking action to tackle climate change will enable societies' potential for well-being to increase much faster in the long run than without action." (Stern, N., 2007, page xiii)

The Stern Review contained much material of interest. It recognised "the limitations inherent in formal integrated models" (page 188) but attracted less support for its view that "within the case of marginal perturbations, the key concept is the discount factor" (page 59), a view which had already been foreshadowed in the second chapter of the Review: "Economics, Ethics and Climate Change" (at pages 36-37). In the appendix to this chapter (Chapter IIA: "Ethical Frameworks and Intertemporal Equity") there occurs a discussion of welfare economics and the use of discounting when addressing intertemporal equity and possible actions. Thus the Review supported the use of a 0.1% discount rate, largely reflecting its assessment of pure time preference based upon an assumed future human lifetime of close to 100 years rather than any wealth-based component. A key result was that largely based upon this premise the Review concluded that unless early and effective action is taken to counter climate change then the costs and risks "will be equivalent to losing at least 5% of global GDP each year—now and forever." (page xv)

The Review provoked a massive response. William Nordhaus reviewed the report as a whole and concluded:

> "The Review's radical revision of the economics of climate change does not arise from any new economics, science, or modelling. Rather, it depends decisively on the assumption of a near-zero time discount rate combined with a specific utility function. The Review's unambiguous conclusions about the need for extreme immediate action will not survive the substitution of assumptions that are more consistent with today's marketplace real interest rates and savings rates. Hence, the central questions about global-warming—how much, how fast, and how costly —remain open. The Review informs but does not answer these fundamental questions." (Nordhaus, N., 2007, page 701)

In the October–December, 2006, issue of "World Economics" an even more devastating and wider ranging critique was provided. Drawing on a comment made by Nicholas Stern himself, Part I of the critique pointed out that Stern had little idea about what the greenhouse effect was until July or August, 2005; failed to acknowledge the uncertainties in climate science and model-based explanations; and arguably grossly overstated climatic change impacts. (Carter, R., et al. 2006, pages 168, and 193)

Part II of the critique, on Economic Aspects, concluded that—assisted by its selective use of sources—the Review understated uncertainties in the economic field as well, mishandled data, systematically overstated projected costs of climate change and understated likely costs of mitigation, and proposed a specially low discount rate "on the basis of inadequate analysis and without regard for the problems and risks that would result." (Byatt, I., et al. 2006, page 224)

In the same issue of "World Economics" appeared a paper by two authors better known in the field of climate economics (with no disrespect intended to the distinguished authors of Part II above), and who clearly stated: "We think—and this is supported by a vast collection of scientific studies—that it is in the self-interest of the vast majority of people to support climate policy." (Tol, R. and G. Yohe, 2006, page 245) [The present writer also believes that, in principle, precautionary measures should be taken to tackle possible global climatic change—many of which can be justified on grounds other than curbing so-called greenhouse gas emissions—but given the many sub-optimal policies, measures, investments and subsidies now around, would emphasise the need to support *sound* climate policy.] Richard Tol and Gary Yohe have independently produced a number of papers commenting on the Stern Review. Here they criticised it on six counts:

- Failure to provide new estimates of either the impacts of climate change or of costs of emissions' reduction, yet produced results "that are so far outside the range of previous published literature." (page 233);
- The high valuation of climate change reflected the double-counting of risk, and the assumption that vulnerability is constant for two or more centuries (both points the authors considered the "products of substandard analysis"), and use of a very low discount rate;
- The likely economic repercussions of dearer energy and capital investment in the energy sector are simply ignored, and the time horizon is unduly truncated to 2050;
- The costs and benefits estimates provided do not match the policy conclusions since, if the former are so low and the latter so high, an atmospheric concentration target should be much more stringent than 550 ppmv;
- The case for emissions reductions, even in the near term, need not be grounded on suspect valuations and inappropriate summing across the multiple sources of climate risk; and
- "Alarmism supported by dubious economics born of the *Stern Review* may further polarize the climate policy debate." (page 234).

In the following issue of "World Economics" Wilfred Beckerman and Cameron Hepburn took a different path to criticising the discount rate adopted in the Stern Review. They focused on ethical issues—risk aversion, aversion to inequality, aversion to inequality through time, and pure time preference. They noted at the outset of their critique that John Rawls considered that the issues relating to inter-generational justice are ones that subject ethical theory to "severe if not impossible tests." (Rawls, J., 1972, page 284) They concluded that the 'revealed ethics' of the market place had limited application to climate policy, and that 'climate ethics' were the province of elites and philosopher-kings who have down the ages sought to promote social goals that have been a common cause of misery to millions of people down the ages—as Isaiah Berlin had concluded in "The Proper Study of Mankind" (1997). They called for a middle path between the dictates of elites and the 'revealed ethics' of the market place, discovering what social preferences actually exist.

Martin Weitzman reflected many of the above criticisms of the Stern Review, together with a clear explanation of the complexities of integrated assessment models, before expressing the opinion that instead of "trying to go through the back door" with dubious assumptions and claims:

> "it is much better to go directly through the front door with the legitimate concern that there is a chance, whose subjective probability is small but diffuse (thereby resulting in a dangerously thickened left tail of comprehensive consumption growth rates), that global warming may eventually cause disastrous temperature and environmental catastrophes." (Weitzman, M., 2007, page 721)
>
> Weitzman lists many questions which need answers before concluding:
>
> "In my opinion, public policy on greenhouse warming needs desperately to steer a middle course, which is not yet there, for dealing with possible climate-change disasters." (page 722)

Weitzman is dismissive of the Stern Review for not "more openly disclosing the full extent to which (its) radical policy recommendations depend upon controversial extreme assumptions and unconventional discount rates that most mainstream economists would consider much too low." (page 724)

Other economists have criticised the Stern Review on the grounds that it took insufficient account of the wealth-related component of the discount rate (thereby paying too little attention to the poor), and adjusting for this a discount rate of around 1.4% could be argued. (Dasgupta, P., 2007; Ackerman, F., 2009) There has been little discussion since the Stern Review appeared of the distinction that can be made between an ethical or prescriptive approach to discounting (which suggests either an elitist approach or one particularly motivated by concern for the poor, rather than one which economists may feel naturally inclined to) and a descriptive approach (more reflective of the expectations of savers and investors). Paul Baer has criticised the Review on the grounds that it probably understated the risks of catastrophic impacts of climatic change possible within its target for climate stabilisation. (Baer, P., 2007) Frank Ackerman and colleagues at the Stockholm Environment Institute have concluded that, on the basis of their PAGE model, the costs of inaction of climate change had been greatly understated in the Stern Review, especially for the USA and 'Other OECD'. (Ackerman, F. et al. 2008). Writing in *World Economics,* Anderson, D.; Arnell, N. et al.; Dietz, S. et al.; and Mitchell, J. et al. all 2007, considered the Stern Review had been unfairly criticised. All eight of the authors of the first three papers were closely associated with the Stern Review.

In the wake of the Stern Review some considered it had provided further evidence that early and strong action is required to tackle climatic change and its adverse consequences. Geoffrey Heal was among those who considered that we have learned a lot from the outpouring of literature as a result of the Stern Review. He concluded: "I think this should change the presumption that economists hold about the need for strong action on climate change, which prior to Stern was negative, to positive. We can see many ways in which we can make a case for strong action now, and few in which we can deny it." (Heal, G., 2008, page 22) Admittedly, Heal remained critical of most models, believing the way forward for modelling is to make them more disaggregated. Nevertheless, for many economists, this writer suspects little has changed as a result of the Stern Review –

and in some respects (as Tol and Yohe feared might happen)—the climate of debate has deteriorated.

Unless and until economists can successfully get to grips with more modest decision-making arenas their pretensions and hubris will give grounds for scorn. Instead they should be pointing out the absurdities of simply burning palm oil in European electricity generators, placing wind turbines where there is little wind, subsidising solar PV where solar insolation levels are low, and exploiting first-generation biomass and biofuel sources and technologies with scant regard for impacts on food availability and prices. There is scope for advising on optimal pricing and taxation schemes to inhibit carbon dioxide and other 'greenhouse' gas emissions, especially those which encourage greater efficiency in our use of resources and enhance the prospects for conservation of resources. As was long ago pointed out, there is plenty of scope for taking steps on grounds other than simply concerns about enhanced global warming. Whether persistence in the attempts to estimate the total economic effects of global warming will produce useful results remains a very open question despite determined efforts to do so. (Tol, R., 2011) On the other hand, approaches which call for global benefits to exceed global costs, for every country to gain individually from agreement, for broad participation, and for agreements to be enforced, seem banal however well intended and obvious in a perfect world. (Barrett, S. and M. Toman, 2010)

More encouraging is the symposium put together by Dieter Helm and Cameron Hepburn as "The Economics and Politics of Climate Change" (2009). The paperback edition of 2011 notes the events at Fukushima and considers that shale gas "promises to be a game changer from the climate-change perspective" by both slowing down the use of coal and showing that fossil fuel availability overall (taking into account shale oil and coal bed methane) is greater than many had imagined. The present writer finds it strange that there is a widespread impression that no-one had considered recovering shale gas resources before 2011, whereas most energy economists would have had some estimated ultimately recoverable figure in their back pocket. Dieter Helm recasts the economics of climate change, away from "an idealized time-preference rate which dictates the case for urgent action" towards a simpler line of argument reliant not on idealized technical supply functions but upon cost assumptions based on empirically observed data and the consequences of seeking to substitute the environment for man-made capital, including the implications of a lower conventional economic growth rate and rising mitigation costs. (page 35) The same author later critiques EU climate change policy, concluding that its "early promise of a market-based approach has been gradually emaciated by the politics." (page 244) The EU Emissions Trading Scheme also comes in for technical criticism in what is one of the most useful contributions to the economics of climate change mitigation, reviewing a wide range of policy instruments (OECD, 2009).

Finally, here, where has the IPCC been in all this? The Working Group III Report of the Second Assessment (1995) contained chapters on discounting and intertemporal equity, cost/benefit analysis, mitigation costs (the word adaptation appeared, but was not followed through), integrated assessment models (most of them produced since 1992, and thus scarcely off the press), and an economic assessment of policy instruments. The present writer was a member of the Synthesis Report drafting team for this Assessment and recalls an interesting discussion on discount rates where one of the co-authors of

Chapter 4 (an economist of world-renown) debunked the very idea of trying to fix on a discount rate over the length of time relevant to the issues reviewed.

Working Group III's contribution to the Third IPCC Assessment (2001) had a different structure to its predecessors, but got little further in its review of discounting. Some considered that the review of modelling was over-generous to macro-economic modelling and integrated assessment modelling, and the claim that since the Second Assessment was published: "much progress has been achieved in the development of consistent and transparent approaches to assess climate change mitigation costs" was much exaggerated. (IPCC, 2001, page 494) Similar optimism was in evidence by the statement: "Since (the Second Assessment) the most important advance is the treatment of new topics related to linkages between national policies and the international framework of these policies in the context of the pre-Kyoto and post-Kyoto negotiation process." (page 503) This was merely the first, of three, "evolutions" in this chapter which exhibited a considerable degree of unreality. Of greater concern to this writer, who was an Editorial Reviewer in Working Group III for this Assessment, was one of process. The IPCC had set up a system of governmental and expert review, and the role of Editorial Reviewers was to ensure review comments were fully considered and taken on board unless there were solid reasons for not doing so. The system in theory was above criticism, but not all Editorial Reviewers were firm enough—and certainly not all authors willing enough – to implement the process put in place. These are matters of individual personality, and strength of character, but there was evidence that Editorial Reviewers from many developing countries felt disinclined for cultural reasons to have a set-to with recalcitrant authors—some of whom occupied senior academic positions in major industrialised countries.

The IPCC's Fourth Assessment (2007) gave less ground for such criticisms in the contribution by Working Group III, although there were concerns expressed that once again some interests had been given more coverage than others (a criticism more stringently expressed with the later IPCC Special Report on Renewables).

Perhaps the greatest challenge arising from the IPCC's work is not so much the sense, that many have, that its past Scientific Assessments have been led by those who, though natural scientists, pursued their belief in anthropogenic causation with almost religious zeal, as the fat upper tail of temperature changes emerging from climate modelling dynamics and emissions scenarios. (IPCC, 2000) This is largely the result of combining three forms of uncertainty: the attempted formulation of damages, choice of pure time preference discounting, and covering possible future temperature dynamics. These combine to dominate the standard economic analysis of climate change in recent times, a particular application of cost/benefit analysis which some believe is "more inherently prone to being dependent on subjective judgments about structural uncertainties than most other, more ordinary applications." (Martin Weitzman in Libecap, R. and R. Steckel, 2011, page 45)

CONCLUSIONS

The debate on enhanced global warming, its causes and possible future trajectories, has become one of the most polarised of modern times in international affairs. Claims are made of intolerance on both sides of the divide—those who consider anthropogenic causation obvious on the one hand; 'naysayers' on the other. The writer has been attacked by some fossil fuel interests on the one hand; and by wind energy interests on the other. Both sets of polarised interest have sought to silence him for seeking a balanced view and reliance on evidence. He remains convinced that, given the scientific and economic uncertainties, sound precautionary measures which are consistent with curbing so-called 'greenhouse' gas emissions should be taken. Regrettably, many of the policies, measures, investments and subsidies now in place or proposed are not sound. Just as regrettable is the fact that, as the run-up to Rio+20 has shown, little progress or agreement on what should be done has been achieved over the past twenty years. This should not cause surprise.

This paper has largely side-stepped the debate on whether there is a need to panic about global warming, and whether there is ingrained bias in the IPCC process. Instead, it has focused on 'climate economics' and the large volume of literature which has emerged on this subject, especially over the past twenty years. Unlike Paul Krugman, it is not at all confident that when the support for political action on climate change occurs "the economic analysis will be ready". (Krugman, P., 2010)

Even where full recognition is given to the scientific uncertainties surrounding enhanced global warming and its prospects, and a clear exposition is given of the uncertainties surrounding work in the field of climate economics, there has been a tendency to exaggerate (or fail to circumscribe) what economists can really offer in the field. Thus even the clearest statement of the economics of climate change known to this writer is not free of this charge. (Congressional Budget Office, 2003) The efforts to value costs and benefits of so many, so complex and differentiated, factors; to do so when the background is one where markets are absent in some crucial areas and the time horizon 50 to 500 years forward; to make an integrated assessment of all these factors and over many generations; to pluck discount rates almost out of the air to reflect "pure time preference"; and to make assumptions about economic growth and its distribution between generations and nations, is surely to attempt the impossible at this point of human evolution. Thus even worthy efforts to present the basic elements of climate economics without a robust critique of their limitations have modest value. (Owen, A. and N. Hanley (eds.), 2004; Goodstein, E., 2008)

Yet climate economists appear to suffer from hubris: they seem to consider they have a crucial role to play in taking over where a coterie of climate scientists leave off, and to guide policy makers and negotiators on the way forward whether societies in general like it or not. An examination of the literature of climate economics suggests that their ambitions should be regarded as farfetched.

Note on author:
Michael Jefferson is Professor of International Business and Sustainability at the Centre for International Business and Sustainability, London Metropolitan Business School. [E-mail address: m.jefferson@londonmet.ac.uk]. He is also Visiting Professor in the Department of Economics and International Studies, University of Buckingham. [michael.jefferson@buckingham.ac.uk]. He is a consultant on energy and environmental matters [jeffers@dircon.co.uk] He is a former: Group Chief Economist of Royal Dutch/Shell, Head of Planning and of Oil Supply Appraisal for Shell International in Europe; Deputy Secretary-General, World Energy Council; Lead Author, Contributing

Author, Editorial Reviewer, and Expert Reviewer for the Intergovernmental Panel on Climate Change (who granted him their Certificate for his contributions to their award of the Nobel Peace Prize in 2007); a Convening Lead Author for the UNDP's "World Energy Assessment" (2000); and for many years chaired the Policies Committee of the World Renewable Energy Network/Congresses.

REFERENCES

1. Ackerman, F. (2009) *Can We Afford the Future? The Economics of a Warming World.* Zed Books, London, England.
2. Ackerman, F. et al. (2008) "Did the Stern Review underestimate U.S. and global climate damages?" (October), *Stockholm Environment Institute Working Paper WP-U.S.-0802*, Stockholm, Sweden.
3. Anderson, D. (2007) "The Stern Review and the Costs of Climate Change Mitigation: A response to the 'Dual Critique' and the misrepresentation of Tol and Yohe" *World Economics*, Vol. 8, No. 1, 211-219. Also Dietz, S., Anderson, D., Stern, N. et al. (2007) "Right for the Right Reasons" *World Economics*, Vol. 8, No. 2, 229-258.
4. Arnell, N. et al. (2007) "Response to the Stern Review: A Duel Critique" *World Economics,* Vol. 8, No. 229-231.
5. Barrett, S. and M. Toman (2010) "Contrasting Future Paths for an Evolving Global Climate Regime, Policy Research Working Paper 5164" (January), Environment and Energy Team, Development Research Group, The World Bank, Washington, DC, USA.
6. Baer, P. (2007) "The Worth of an Ice-Sheet: A Critique of the Treatment of Catastrophic Impacts" in *the Stern Review* www.ecoequity.org/docs/WorthOfAnIceSheet.pdf
7. Beckerman, W. and C. Hepburn (2007) "Ethics of the Discount Rate in the Stern Review on the Economics of Climate Change" *World Economics*, Vol. 8, No.1, (January-March), 187-210.
8. Byatt, I. et al. (2006) "The Stern Review: A Dual Critique—Part II: Economic Aspects" *World Economics*, Vol. 7, No. 4, (October-December), 199-228. See also: Robinson, C. (ed.) (2008) *Climate Change Policy: Challenging the Activists*, Institute of Economic Affairs, London, England.
9. Carter, R. et al. (2006) "The Stern Review: A Dual Critique—Part I: The Science" *World Economics*, Vol. 7, No. 4, (October-December), 167-198.
10. Cline, W. (1992) *Global Warming: The Economic Stakes*, May, Institute for International Economics, Washington, DC, USA.
11. Cline, W. (1992) *The Economics of Global Warming*, June, Institute for International Economics, Washington, DC, USA.
12. Congressional Budget Office (2003) *The Economics of Climate Change: A Primer*, University of the Pacific, Honolulu, Hawaii.
13. Dasgupta, P. (2007) "Comments on the Stern Review's Economics of Climate Change", *National Institute Economic Review*, 199 (1), 4-7.
14. DeCanio, S. (2003) *Economic Models of Climate Change: A Critique*, Palgrave Macmillan, Basingstoke, England.

15. Dietz, S. (2007) "Right for the Right Reasons", *World Economics*, Vol. 8, No. 2, 229-258.
16. Dornbusch, R. and J. Poterba (eds.) (1991) *Global Warming: Economic Policy Responses*, MIT Press, Cambridge, Massachusetts, USA.
17. Dresner, S. (2008) *The Principles of Sustainability, 2nd ed.*, Earthscan, London, England.
18. Flohn, H. (1963) "Theories of Climatic Change from the viewpoint of the Global Energy Budget", in *Changes of Climate: Proceedings of the Rome Symposium organized by Unesco and the World Meteorological Organisation,* Liège, Belgium.
19. Frederick, S. et al. (2002) "Time Discounting and Time Preference: A Critical Review", *Journal of Economic Literature*, Vol. 40, No. 2 (June), 351-401.
20. Goodstein, E.(2008) *Economics and the Environment*, John Wiley, New Jersey, USA.
21. Hall, C. and K. Klitgaard (2012) *Energy and the Wealth of Nations—Understanding the Biophysical Economy*, Springer, New York, USA. Also review by Jefferson, M. in *Energy Policy*, 2012, Vol. 42, 720-721.
22. Heal, G. (2008) "Climate Economics: A Meta-Review and Some Suggestions, National Bureau of Economic Research Working Paper 13927", April, *NBER*, Cambridge, Massachusetts, USA. http://www.nber.org/papers/w13927.
23. Helm, D. and C. Hepburn (2009) *The Economics and Politics of Climate Change*, Oxford University Press, Oxford, England. Paperback edition with new Preface, 2011.
24. Houghton, J. (Ed.) (1984) *The Global Climate*, Cambridge University Press, Cambridge, England.
25. Houghton, J. et al. (1990) Climate Change: The IPCC Scientific Assessment, Report prepared for IPCC by Working Group I, 1990, Cambridge University Press, Cambridge, England.
26. Intergovernmental Panel on Climate Change (IPCC) (1990) Policymaker Summary of Working Group III (Formulation of Response Strategies), (reprinted 1992), WMO/UNEP, IPCC, Canada.
27. Intergovernmental Panel on Climate Change (1991) Energy and Industry Subgroup Report, January, U.S. EPA, Washington DC, USA.
28. IPCC: Special Report on Emissions Scenarios (2000) Cambridge University Press, Cambridge, England.
29. Jefferson, M. (1971) Home of Lost Economics, *Twentieth Century* Vol. CLXXIX, No. 1047, 28-29.
30. Jefferson, M. (2012) "Shell scenarios: What really happened in the 1970s and what may be learned for current world prospects", *Technological Forecasting & Social Change*, Vol. 79, 186-197.
31. Krugman, P. (2010) "Building a Green Economy", April 7, *The New York Times*, NY, USA. http://www.nytimes.com/2010/04/11/magazine/11Economy-t.html
32. Libecap, G. and R. Steckel (eds.) (2011) *The Economics of Climate Change: Adaptations Past and Present*, University of Chicago Press, Chicago, USA.

33. Manne, A. and R. Richels (1992) *Buying Greenhouse Gases: The Economic Costs of CO_2 Emission Limits*, Massachusetts Institute of Technology, Cambridge, Massachusetts, USA.
34. Mitchell, J. et al. (2007) Response to Carter et al. 2007, *World Economics*, Vol. 8, No. 1, 221-228.
35. Nordhaus, W. and G. Yohe (1983) "Future carbon dioxide emissions from fossil fuel emissions", in *Changing Climate*, National Research Council, National Academy Press, Washington, DC, USA.
36. Nordhaus, W. (1994) *Managing the Global Commons: The Economics of Climate Change*, MIT Press, Cambridge, Massachusetts, USA.
37. Nordhaus, W. (2007) A Review of the *Stern Review on the Economics of Climate Change, Journal of Economic Literature,* 2007 (September), Vol. XLV, 686-702.
38. Nordhaus, W.: A Question of Balance—Weighing the Options on Global Warming Policies, 2008, Yale University Press, New Haven, USA.
39. OECD (2009) *The Economics of Climate Change Mitigation: Policies and Options for Global Action Beyond 2012*, OECD, Paris, France.
40. Owen, A. and N. Hanley (eds.) (2004) *The Economics of Climate Change*, Routledge, Abingdon, Oxford, England.
41. Rawls, J. (1972) *A Theory of Justice*, Oxford University Press, Oxford, England.
42. Revelle, R. and H. Suess (1957) Carbon Dioxide Exchange Between Atmosphere and Ocean and the Question of an Increase of Atmospheric CO_2 during the Past Decades, *Tellus,* Vol. IX, 1, 18-27.
43. Sagoff, M. (2008) *The Economy of the Earth—Philosophy, Law, and the Environment*, 2nd Edition, Cambridge University Press, Cambridge, England.
44. Schelling, T. (1983) "Climate change: Implications for welfare and policy", in *Changing Climate,* National Research Council, National Academy Press, Washington, DC, USA.
45. Shackle, G. (1982) "To Cope With Time", unpublished paper, 15 February, 1982. See also: Loasby, B.: Uncertainty and imagination, illusion and order: Shackleian connections, 2011, *Cambridge Journal of Economics*, **35**, 771-783.
46. Shell UK (1986) Towards *2010: Scenarios for the long-term business outlook*, October 1986, Shell UK Planning, London, UK.
47. Shell UK (1989) *UK Scenarios: Has the Biggest Challenge Begun?* November, 1989, Shell UK Planning, London, UK.
48. Skidelsky, R. (1992) *John Maynard Keynes: Volume Two—The Economist as Saviour, 1920-1937*, Macmillan, London, UK.
49. Stern, N. (2007) *The Economics of Climate Change: The Stern Review*, Cambridge University Press, Cambridge, England.
50. Tol, R. and G. Yohe (2006) A Review of the *Stern Review*, *World Economics*, 2006 (October-December), Vol. 7, No. 4, 233-250.
51. Tol, R. (2011) "The Uncertainty about the Total Economic Impact of Climate Change, Economic and Social Research Institute", (April), Working Paper No. 382, *ESRI*, Dublin, Ireland.
52. Weitzman, M. (2007) A Review of *The Stern Review on the Economics of Climate Change*, *Journal of Economic Literature,* (September), Vol. XLV, 703 724.

53. Wiseman, J. (1983) *Beyond Positive Economics?* Macmillan, London, UK.
54. World Energy Council (1993) *Energy for Tomorrow's World—the Realities, the Real Options and the Agenda for Achievement*, Kogan Page/St. Martin's Press, London, UK & New York, USA.

THOUGHTS ON THE SOCIAL RATE OF DISCOUNT

BRUCE STRAM
Element Markets
Houston, USA

INTRODUCTION

Two recent evaluations of potential global warming policy are starkly at odds. The "Stern Review" (Nickolas Stern for the British Treasury) and "A Question of Balance" (William Nordhaus) are nearly an order of magnitude apart on what level of current resources should be directed toward reducing carbon emissions. This difference is all the more striking because careful examination of the two documents reveals that both use very similar estimates of future damages from warming. The difference almost entirely boils down to one number: the rate of discount applied to those future damages. Nordhaus uses a discount rate comparable to that used for most benefit cost analyses which are based on market rates of return. Stern uses a very low rate, a choice driven by the fact that this analysis is intergenerational in nature, which per Stern requires a very low discount rate to avoid generational bias.

Social rate of discount has been a theoretical issue for economists for many years, beginning at least with Ramsey (1928). Its application circumscribes many important policy issues associated with economic evaluation of long term consequences versus actions that might be taken today. Unquestionably, however, the dialogue regarding global warming has given substantial new impetus to the discussion.

Much of the dispute about the "social rate of discount" centers on the fear that the current generation will be selfish with regard to preventing future damage to the environment, consuming now and leaving great costs for future generations. And this despite the fact that a market-like rate of discount applied to both private and public cost benefit analysis encourages deferral of current consumption in favor of investment and provision for the future.

Models optimizing long term economic and environmental consequences of human activity seem to yield, at least for some observers, acceptance of too much environmental harm. Since market-based rates of discount seem to reduce the present value of dire environmental consequences to very low values over very long time periods, such observers logically question whether this discounting applied to time periods extending well beyond current generations has inherent flaws. The hypothesis here is that such flaws, if they exist, are occasioned by more mundane shortcomings in the valuations placed on environmental damage in the future.

DISCOUNTING COSTS AND BENEFITS

Cost benefit analysis (CBA, public sector) or discounted cash flow (private sector) follows the same general methodology. Values which occur in the future are factored by a compounded discount rate which resembles a rate of interest (i.e., 1.06). These discounted values are then compared to current values.

The basic rationale for such discounting is that economic resources which are withheld from current consumption can be diverted to projects or activities which on average provide for greater future benefit. Such "diversions" include investments in physical capital, research and development of technology, and education.

Economists point to the markets for savings and lending to provide guidelines for establishing criteria for selecting desirable investments. These markets in effect aggregate willingness to defer consumption versus receiving future reward. Investments which do not yield returns at least equal to these rates diminish overall welfare (for public investments) or reduce the value of a private sector business or activity. Both governments and businesses borrow funds which are repaid with interest (i.e., a discount rate). Businesses also raise equity capital for projects with the expectation that there will be a positive return, also an implicit discount rate.

Earning a positive return on invested capital is part of an amazing and highly distinct period of economic development in relatively recent history.

Figure 2.1: Population and Income, England: 0-1999

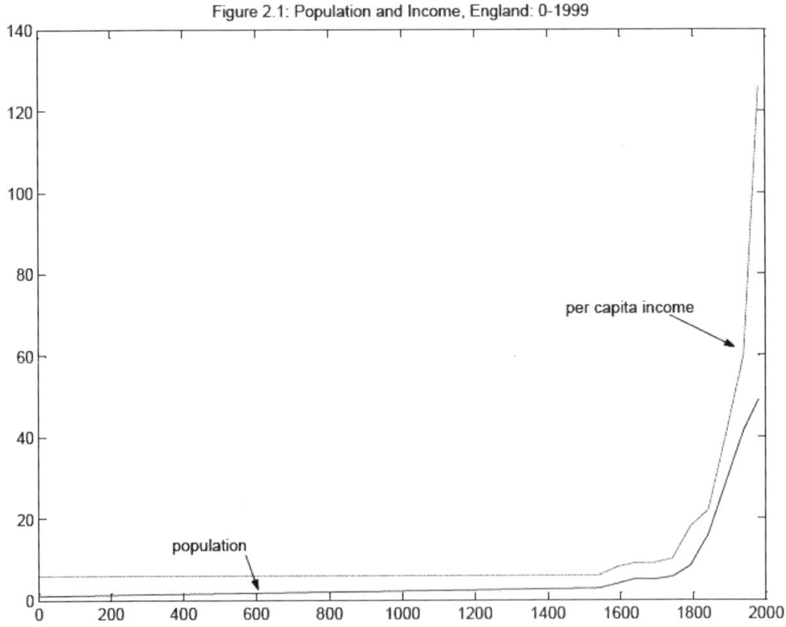

"Was Malthus Right?" Economic Growth and Population Dynamics, Jesus Fernadez-Villaverde, University of Pennsylvania, Dec. 1, 2001.

Economists strongly believe that the notion of a focus on capital formation and the concomitant formalization of a market for capital is an integral part of the foundation of this economic growth. This development, along with entrepreneurship, vision, risk taking, and "animal spirits" has created an engine of economic growth. CBA and discounted cash flow is a methodology that is a codification of this process in order to provide criterion for efficient investment. This methodology is used almost universally for evaluating governmental and business investments.

Objections have arisen, however, regarding very long term evaluations. It begins simply as a matter of arithmetic. Typically a discount rate is estimated in the range of 4% to 6% in real terms. Thus a project "paying off" in 10 years must return 1.8 times its cost. Alternately, an asset valued at 1.8 10 years from now would be valued at 1 today.

Longer term compounding has a much more dramatic effect. Fifty years from now, something valued at 18 would be deemed to be worth 1 today. For 100 years the ratio to current value would be 340:1, and for 200 years, an astonishing 115,000:1.

Needless to say, applying such discounts to the value of hypothesized adverse effects from global warming (or other environmental damages) is viewed with some substantial concern. That has led to a vigorous renewal of the social rate of discount analysis.

A BRIEF EXAMINATION OF SOCIAL RATE OF DISCOUNT THEORY

Much thought has been devoted to this fundamental issue of weighing the value of the future to value in the present. These include the following: The Stern approach; various rationales for declining discount rates (over time, DDR), and a "Fat Tails" approach. These will be outlined in turn, followed by an alternative approach.

Stern

The Stern Report justifies the use of a very low [relative to a typical 4.5% to 6% real often used for cost benefit analysis (CBA)], numerically 1.4%. The rationale is based on moral grounds. It devolves from a well known mathematical and highly theoretical economic model depicting economic growth based on capital formation and population growth and optimization of welfare over time. An end result is a simple formula:

$$r = \rho + \theta g = \delta$$

the letter r represents the marginal rate of return to capital, ρ the pure rate of time preference, i.e., human's preference for gratification now as opposed to the future, θ is the elasticity of utility (or welfare) with respect to income, g per capita income growth, and δ the social rate of discount.

This may be interpreted as follow: societies' future welfare should be discounted according to the marginal rate of return to capital (r and δ being equal), and in addition by decline in the value of future income increase fostered by growth in per capita income. [Given the abstract nature of the analysis, many real world conditions (taxes, externalities, etc.) that would disturb the equality are ignored.]

From this foundation, the Stern Report makes the following assertion: the pure rate of time preference observed for the current generation should not be visited on future generations to their disadvantage. Therefore that component of the formulation should be set to zero. Since it is take to be 3%, the social discount rate is appropriately 1.4% rather than say 4.5% which would represent the marginal return to capital.

In this, the Stern Report follows a number of eminent British (and other) economists—Ramsey, Harrod, etc.

However, this formulation would have the current generation defer consumption to achieve much greater benefit for future generations than would be the case applying the standard methods. Such an approach seems to ignore the fact that a high discount rate implies continued economic growth such as has occurred over the last few hundred years. Thus those future generations for whom greater sacrifice is deemed appropriate will be much wealthier than we currently. Equivalently, this is akin to asking that our ancestors alive in 1800 and much poorer than we today, be compelled to reduce consumption in order to additionally benefit us. By this formulation, future generations are to be treated better than we have been by our ancestors, and better that we treat ourselves.

UNCERTAINTY AS TO FUTURE GROWTH RATES

It is certainly the case that for at least the last two hundred years, "Western" economies have enjoyed historically very high growth rates driven in significant part by a high marginal productivity of capital. Much of the rest of the world (weighted by population) seems finally to be catching on and achieving this by similar means. But this growth is not axiomatic. There is no certainty this effect will continue in the future, and the less so the farther off the future. Such uncertainty has given rise to several analyses which seem to imply that use of DDR is appropriate.

Financial markets are often looked to find measures that reflect a rate of return to capital and hence a CBA discount rate. But forward markets, which are not perfect predictors of the future, do not extend to the future being contemplated in very long term analyses and therefore provide very limited guidance.

Martin Weitzman tries to capture this effect roughly as follows: he defines future states of the world which embody different returns to capital and hence appropriate discount rates. He then proposes to evaluate the future in total by summing these weighted by their probability and discounted by their own return to capital. As time goes to infinity, the future world with the "smallest possible" discount rate dominates the weighted average.

This hoisting of high discount rates on their own petard certainly deserves a smile. However, some objections come immediately to mind. What is the smallest possible discount rate, zero? Certainly this and even negative discount rates are contemplated in the literature on this subject. An infinite time horizon makes no sense. If nothing else, the earth and the universe have a limited life. And finally there is a certain circularity to the result: if the conclusion of the analysis is that the "lowest possible" discount rate is the one appropriate, shouldn't it be applied to all the future states, obviating the conclusion?

But these objections don't apply to a more mundane point. Given uncertainty as to potential future growth rates and hence appropriate discount rates, a normal probability distribution of such would suggest that an expected value is a reasonable measure to use for determining expected future income. However, the marginal value of that income will be higher in lower growth situations then higher. That is, even if uncertain future income is normally distributed, future welfare is not. Further, one can demonstrate that there is a DDR which captures the welfare effect of this asymmetry. (Whether using a DDR to reflect this is the best practice is another matter and will be considered below.)

Weitzman also raises a "Fat Tails" argument. The supposition is that the probability of a catastrophic outcome, i.e., representing near infinite loss from global warming remains high enough so as to more than offset cumulative exponential discounting. This really amounts to a strong suggestion that cost benefit analysis cannot be reasonably applied to global warming concerns given potential catastrophic impacts. But a resource allocation problem remains: there are many unknowable probability catastrophic events (meteor, hadron collider creating a black hole on earth, nano technology gone wild, etc.), But only finite resources are available now to be allocated toward avoiding possible infinite outcomes in the future.

Perhaps a more useful approach to such concerns is to these dangers in terms of "backstops". That is, instead of evaluating the consequences of the hypothesized events, examine the actions that might be taken to avert them. For global warming these might include the cost of chemical carbon capture and storage, genetic engineering of carbon eating and sequesting plants (Dyson) and/or geoengineering. CBA is clearly appropriate for evaluating such alternatives.

SUBSTITUTABILITY

CBA or discounting makes sense because the flow of reproducible or manufactured goods can be increased by deferral of consumption and concomitant capital formation. But environmental assets are not generally reproducible. That natural endowment can be and demonstrably is being reduced by human activity. If it is reasonable to suppose that future generations will benefit from high economic growth rates (with appropriate high discount rates), the availability of environmental assets will diminish both in a relative sense and absolutely. Therefore, the value of such assets will logically increase relative to reproducible goods. Further, one may readily observe that richer societies spend a greater proportion of income on environmental protection and the enjoyment of such amenities. One can posit that as environmental assets diminish and reproducible goods increase, that an increase in value of the former could outweigh the growth rate of the latter, and even be greater than the discount rate. Such circumstances potentially provide for an additional offset for the discount of potential environmental damages in the distant future. (Krutilla and Fisher)

CONCLUSION

No one believes that economic analysis of the far off future necessary to evaluate the potential dangers of global warming (or any other very long term situations) is anything but problematic. Straightforward use of a "standard" discount rate yields a devaluation of future environmental damages which is disturbing to many. That has led to much reevaluation of standard CBA methods and in particular, analysis of the appropriate social rate of discount. Various methods have been proposed to revise this rate to reflect various concerns. These have been briefly considered in this paper and are certainly in whole, worthy of consideration. However, there is no inherent need to incorporate such adjustments in a singular discount rate number. At best, doing so is simply an analytical convenience. Each of the adjustments considered here could be incorporated into an overall analysis as a separate item in an overall analysis.

Such an approach would be eminently more functional than relying on a single discount rate factor. The logical need for such adjustments remains open to question. Further in application, the elements of these adjustments are themselves highly uncertain. For example the long term decline of the marginal utility of income at much higher per capita income levels is highly speculative. So too is any uncertainty in long term rate of economic growth and the return to capital. Including any such adjustments explicitly rather than subsuming them into a single discount rate provides for much more clarity and transparency.

This analysis very strongly implies that collaborative action is needed between environmental scientists and economists to get more correct answers resulting from more careful examination of potential future income levels, their impact on environmental amenities and systematic estimation of how the valuation of environmental benefits changes with income level.

REFERENCES

1. Gollier, Koundouri and Pantelidis (2008) "Declining discount rates: Economic justifications and implications for long-run policy," *Economic Policy, CEPR & CES & MSH*, vol. 23, pages 757-795, October.
2. Krutilla, J. and A.C. Fisher (1975) *The Economics of Natural Environments*. Baltimore: Johns Hopkins University Press.
3. Nordhaus, William (2008) A *Question of Balance: Weighing the Options on Global Warming Policies*, Yale University Press, New Haven, CT.
4. Ramsey, F.P. (1928) New Haven, CT, "2008s on Global Warming Policies,long-run policy." (Ramsey F.P. (1928), "A Mathematical Theory of Saving," *Economic Journal*, Vol. 38, No 152, pp. 543 559).
5. Stern, N.H. (2006) "The Economics of Climate Change" available at: http://www.hm-treasury.gov.uk/independent_reviews/stern_review_economics_climate_change/stern_review_report.cfm.
6. Weitzman, M. (1994) "On the 'Environmental' Discount Rate." *Journal of Environmental Economics and Management* 26(2), pp. 200-209.
7. Weitzman, M. (1998) "Why the Far-Distant Future Should Be Discounted at its Lowest Possible Rate." *Journal of Environmental Economics and Management* 36(3), pp. 201-208.
8. Weitzman, M. (1999) *Just Keep Discounting, But...* in P. Portney and J. Weyant (eds), Discounting and Intergenerational Equity, pp. 23.
9. Weitzman, M. (2001) "Gamma Discounting." American Economic Review 91(1), pp. 260-271.
10. Weitzman, M.L. (2004) "Statistical Discounting of an Uncertain Distant Future." mimeo, Harvard University.

IS CO_2 MITIGATION COST-EFFECTIVE?

CHRISTOPHER MONCKTON OF BRENCHLEY
Monckton Enterprises Ltd.
Carie, Rannoch, UK

ABSTRACT

A much-simplified inter-temporal appraisal method using results from the Intergovernmental Panel on Climate Change (IPCC) permits non-specialist policymakers rapidly to obtain a first approximation of how much global warming a given CO_2-mitigation policy may abate; the cash, per-capita, and global-GDP cost of abating all global warming to a target year by measures as cost-effective as the policy; the policy's mitigation cost-effectiveness in dollars per Kelvin abated; and the cost/benefit ratio. Case studies illustrating the method's utility in rapidly comparing competing policy options indicate that government projections tend to understate the cost of climate action and overstate the welfare loss from inaction. At a market discount rate the global abatement cost of the Australian carbon tax is 48 times the welfare loss from inaction. Focused adaptation to any adverse consequences of global warming will prove more cost-effective than attempted mitigation. No policy will be cost-effective solely on grounds of the welfare benefit foreseeable from CO_2 mitigation alone. Mitigation policies inexpensive enough to be affordable will be ineffective: policies costly enough to be effective will be unaffordable.

INTRODUCTION

Hitherto, economists have chiefly addressed the cost-effectiveness of climate mitigation globally. Here a much-simplified method, based on results of the Intergovernmental Panel on Climate Change (IPCC), taken as normative *ad argumentum,* is intended to enable even non-specialist policy-makers rapidly to estimate not only how much global warming any proposed CO_2-reduction policy may be expected to abate but also, on the assumption that the cost-effectiveness of all mitigation strategies worldwide is equivalent to that of the policy, its global abatement cost (which is the cost of abating the projected warming over the term of the policy by measures of equivalent cost-effectiveness) and its mitigation cost-effectiveness in U.S. dollars per Kelvin of warming abated. Brief case studies compare the costs of competing CO_2-reduction policies with one another and with published estimates of the welfare loss arising from unmitigated climate change.

As benchmarks, Stern (2006, p. vi), adopting an inter-temporal discount rate not exceeding 1.4%, estimates that the cost of abating the 3 K 21st century global warming the IPCC expects will be 0-3% of GDP (mean 1.5%) and that 1% of 21st century global GDP would suffice to abate 5 K global warming to 2100, against a global inaction cost of 5-20%, while Garnaut (2008, p. 253, Figure 11.2, & p. 270, Table 11.3), projecting 5.1 K unmitigated global warming to 2100 and using a 1.35-2.65% discount rate, puts Australia's 21st century mitigation and inaction costs at 3.2-4% and 6% respectively. The economic literature puts the inaction cost at 1-4% of global 21st century GDP and regards an inter-temporal discount rate of 5% as normative (e.g., Nordhaus, 2008; Murphy, 2008).

Warming beyond 2100 is not considered here. Since equilibrium temperature will not be reached for 1000-3000 years (Solomon et al. 2009), it is centennial-scale transient warming that is policy-relevant today: subsequent warming will occur too slowly over the millennia to do unavoidable harm. Costs external to the policy and benefits external to mitigation of CO_2 forcing are beyond the focus of this paper, but the method may readily be adapted to encompass them. The IPCC's central estimate of 3 K warming by 2100 and 3% uniform GDP growth in the 21st century are assumed: little error arises from assuming uniformity, and other rates may be chosen. A 5% discount rate is assumed on account of centennial-scale uncertainties, but lesser rates are also considered.

PROJECTED 21ST CENTURY CO_2-DRIVEN WARMING

Stern (2006) concluded that, though previous projections had indicated 2-3 K anthropogenic warming by 2100, causing a permanent global output loss estimated at 0-3% (mean 1.5%), more recent evidence suggested 5-6 K warming by the end of this century, and possibly 10-11 K warming by 2200.

However, IPCC (2007) presented estimates of radiative forcing and warming this century under six emissions scenarios (*op. cit.*, p. 18), to each of which it accorded equal weight. Taking their mean, a central estimate ΔT_{C21} of 21st century warming is 2.8 K (Table 1), of which 0.6 K is already committed, so that the implicit central estimate of mean warming from 2000-2100 consequent upon all greenhouse-gas emissions since 2000 is 2.2 K, of which 70%, or 1.5 K, is CO_2-driven.

Table 1. Projected 21st century anthropogenic warming $\Delta TC21$ (IPCC, 2007, p. 13, Table SPM.3) and warming $\Delta T200$ and total radiative forcings $\Delta F200$ from all greenhouse gases for 1900-2100 on all emissions scenarios, and CO_2 concentration $C2100$ in 2100 (IPCC, 2007, p. 803, Fig. 10.26); and, derived from these, the 200-year transient-sensitivity parameter $\lambda 200 = \Delta T200/\Delta F200$; the CO_2 radiative forcing $\Delta F200,CO_2 = 5.35 \ln(C2100/C1900)$ from 1900-2100, taking $C1900$ as 300 ppmv; and the ratio $q = \Delta F200,CO_2/\Delta F200$ of CO_2 forcing to total greenhouse-gas forcing.

Scenario	ΔT_{C21}	ΔT_{200}	ΔF_{200}	C_{2100}	Λ_{200}	$\Delta F_{200,\,CO2}$	q
A1B	2.8 K	3.0 K	6.2 W m^{-2}	700 ppmv	0.5 K W^{-1} m^2	4.5 W m^{-2}	0.7
A1FI	4.0 K	4.5 K	9.1 W m^{-2}	960 ppmv	0.5 K W^{-1} m^2	6.2 W m^{-2}	0.7
A1T	2.4 K	2.5 K	5.1W m^{-2}	570 ppmv	0.5 K W^{-1} m^2	3.4 W m^{-2}	0.7
A2	3.4 K	3.8 K	8.0 W m^{-2}	840 ppmv	0.5 K W^{-1} m^2	5.5 W m^{-2}	0.7
B1	1.8 K	2.0 K	4.1 W m^{-2}	520 ppmv	0.5 K W^{-1} m^2	2.9 W m^{-2}	0.7
B2	2.4 K	2.7 K	5.6 W m^{-2}	610 ppmv	0.5 K W^{-1} m^2	3.8 W m^{-2}	0.7
Mean	2.8 K	3.1 K	6.3 W m^{-2}	700 ppmv	0.5 K W^{-1} m^2	4.4 W m^{-2}	0.7

To reflect the IPCC's wide error-intervals, where λc is a central bicentennial-scale climate-sensitivity estimate $\lambda 200$ will fall on the 1 σ interval $[0.8\lambda c, 1.2\lambda c]$ (from the ±0.69 K 1σ error-bar in IPCC, 2007, p. 798, box 10.2), or on the >66%-probability interval $[0.6\lambda c, 1.4\lambda c]$ (derived *ibid.*, p. 12). So, where $\lambda c = 0.5$ K W^{-1} m^2, λ_{200} will fall on [0.4, 0.6] to 1σ, and on [0.3, 0.7] with probability >0.66.

Recall that IPCC (2001, p. 358, Table 6.2), following Myhre et al. (1998), takes the CO_2 forcing in W m^{-2} as 5.35 times the logarithm of a given proportionate change Cb/Ca in CO_2 concentration, where Ca is the unperturbed value. Note that there was no statistically-significant warming from 1997-2012 (RSS, 2012; UAH, 2012). Then projected CO_2 concentration $C2100$ in 2100, as the six-scenario mean, is 700 ppmv against 391 ppmv in 2011 (Conway & Tans, 2011), implying CO_2-driven warming from 2011-2100 of $\lambda 100[5.35 \ln(C2100/C2011)] = 0.4[5.35 \ln(700/391)] = 1.25$ K, similar to the 1.5 K established earlier from IPCC (2007, Table SPM.3).

Observed warming since 1950 has occurred at a rate equivalent to 1.2 K/century (HadCruT3, 2012). Of this, 70%, or 0.8 K/century, is attributable to CO_2; and, since IPCC (2007) finds that up to half of the warming since 1950 might be natural, the centennial rate of CO_2-driven warming could have been as low as 0.4 K/century for the 21st century may accordingly be best seen as an upper bound rather than as a central estimate.

Table 2. Summarizes official projections and observations of 21st century warming.

Projected anthropogenic warming, 2000-2100	Source	$\Delta TC21$
High-end projection	Stern (2006)	10-11 K
Central projection	Stern (2006)	5-6 K
Low-end projection	Stern (2006)	2-3 K
Mean projection: all anthropogenic warming	IPCC (2007)	2.8 K
...of which, warming not already committed	IPCC (2007)	2.2 K
...of which, fraction caused by CO_2 emissions	IPCC Tbl. SPM.3	1.5 K
...of which, fraction caused by calculation	(*v.* text *supra*)	1.25 K
Observed warming rate/century, 1950-2011	$HadCRUt_3$	1.2 K
Observed warming rate/century, (from CO_2 only)	$0.7\,\Delta T$	0.8 K
(from CO_2 if half of warming was manmade)	$0.7\,\Delta T/2$	0.4 K

METHOD

In this deliberately very simple method, only two case-specific inputs are required: Cy, the projected business-as-usual CO_2 concentration in the target final year y of the policy, and p, the proportion of projected global business-as-usual CO_2 emissions till year y that the policy is intended to abate.

Eq. (1) determines Cpol, the CO_2 concentration in parts per million by volume (somewhat below Cy) in year y that may be achievable by following a given policy to mitigate the radiative forcing from atmospheric CO_2 enrichment from 2010 (when $C2010$ = 390 ppmv) till year y. Eq. (1) also determines ΔTnix, the quantum (in K) of transient global warming that the policy will abate if pursued until year y. Eq. (1) may be tuned to represent any forcing (see Table 4 below): but only warming attributable to the CO_2 forcing is demonstrated here.

$$\Delta T_{\text{nix}} = \lambda_{\text{tra}}\,\Delta F_{\text{nix}} = 0.4\left[5.35\ln\left(\frac{C_y}{C_{\text{pol}}}\right)\right]$$

$$= 2.14\ln\left(\frac{C_y}{C_y - p(C_y - 390)}\right). \qquad (1)$$

Mitigation cost-effectiveness is here defined as the cost of abating 1 K CO_2-driven global warming on the assumption that all measures to mitigate all CO_2-driven warming to year y are as cost-effective as the policy under consideration. On the same assumption, the policy's *global abatement cost* is defined as the total cost from 2010 to year y (as a global cash cost, or a per-capita cost, or a percentage of real global GDP) of abating all anthropogenic warming that the IPCC projects will occur by year y (i.e., over the policy term) without mitigation.

Eq. (2) gives the mitigation cost-effectiveness M in U.S. dollars per Kelvin of global warming abated. The lesser the value of M, the more cost-effective is the policy, enabling policymakers rapidly reliably to compare the estimated mitigation cost-effectiveness of competing mitigation proposals.

$$M = \frac{x}{\Delta T_{\text{nix}}}. \qquad (2)$$

Global abatement cost: Where w = 7 x 109 is world population, q is the fraction of total anthropogenic forcing attributable to CO_2, and, from (1), $\Delta Ty = 2.14 \ln(C_y / C_{2010}) / q$ is the projected anthropogenic global warming to year y, Eqs. (3-5) give the policy's global abatement cost over the term to year y in cash; per head of global population; and as a percentage of real global 21st century GDP r over the term:

Cash	Per capita	As % global GDP
$G = M\,\Delta T_y$	$H = G / w$	$J = 100G / r$
(3)	(4)	(5)

Derivation of Equation (1)

Where λ is a climate-sensitivity parameter in K W^{-1} m^2, consequent global warming in K may be expressed generally by Eq. (6):

$$\Delta T = \lambda\,\Delta F = \lambda\,[5.35 \ln(C_b/C_a)]. \qquad (6)$$

As a check, taking λ_{tra} = 0.4 K W^{-1} m^2 for 1900-2100, at CO_2 doubling Eq. (6) gives 0.4[5.35 ln(2)] ≈ 1.5 K, falling on the model-derived transient-climate-response interval [1, 3] K (IPCC, 2007, p. 749).

Where p, on [0, 1], is the fraction of future global emissions that a given CO_2-reduction policy is projected to abate by a target calendar year y, and Cy is the IPCC's projected unmitigated CO_2 concentration in year y, Eq. (7) gives C_{pol}, the somewhat lesser concentration in ppmv that may be expected to obtain in year y if the policy is followed:

$$C_{\text{pol}} = C_y - p(C_y - C_{2010}). \qquad (7)$$

Table 3 gives central estimates of projected decadal values of Cy for 2010-2100.

Table 3. Business-as-usual CO_2 concentrations, 2010-2100, as the mean of the central projections for all emissions scenarios (IPCC, 2007, p. 803, Fig. 10.26).

y	2010	2020	2030	2040	2050	2060	2070	2080	2090	2100
C_y	390	410	440	480	510	550	590	630	660	700

Eq. (8), of similar form to Eq. (6), determines how much warning ΔTnix, in Kelvin, a specific policy intended to cut CO_2 emissions will abate in the 21st century:

$$\Delta T_{\text{nix}} = \lambda_{\text{tra}} \left[5.35 \ln(C_y/C_{2010}) - 5.35 \ln(C_{\text{pol}}/C_{2010})\right]$$
$$= \lambda_{\text{tra}} \left[5.35 \ln (C_y/C_{\text{pol}})\right]. \qquad (8)$$

The second expressions of Eq. (8) and Eq. (1) are equivalent.

OTHER GREENHOUSE GASES

The method sketched here may be adapted to determine the mitigation cost-effectiveness and global abatement costs of greenhouse gases other than CO_2. Table 4 summarizes some climate-relevant radiative forcing functions.

Table 4. Forcing functions for methane (CH_4), nitrous oxide (N_2O), the chloro-fluorocarbons CFC11, CFC12, sulfur hexafluoride (SF6), and sulfur dioxide (SO_2). Global concentrations for SO_2 are unknown because its concentration is highly variable both spatially and temporally. Source: IPCC (2007).

Trace	**Radiative forcing ΔF**	**Concentration**
Gas	g_y= business-as-usual; g_{pol} = lesser conc. after policy	in 2010
CH_4	$0.036(\mu y^{0.5} - \mu_{\text{pol}}{}^{0.5}) + f(\mu_{\text{pol}}, \nu_{\text{pol}})^\dagger - f(\mu y, \nu_{\text{pol}})^\dagger$	1816 ppbv
N_2O	$0.12(\nu_y{}^{0.5} - \nu_{\text{pol}}{}^{0.5}) + f(\mu_{\text{pol}}, \nu_{\text{pol}} - f(\mu_{\text{pol}}, \nu_y)^\dagger$	324 ppbv
CFC_{11}	$0.00025(\beta y - \beta_{\text{pol}})$	238 pptv
CFC_{12}	$0.00033(\gamma_y - \gamma_{\text{pol}})$	532 pptv
SF6	$5.2\text{-}4(\varphi_y - \varphi_{\text{po}}\text{l})$	7.31 pptv
SO_2	$-[0.03\ \psi_y/\psi_{\text{pol}}+0.08 \ln(1+\psi_y/34.4)/(1+\psi_{\text{pol}}/34.4)]$	Unknown

$^\dagger f(\sigma,\tau) = 0.47 \ln[1 + 2.01 \text{ x } 10^{-5}(\sigma\tau)^{0.75} + 5.31 \text{ x } 10^{-15}\ \sigma(\sigma\tau)^{1.52}]$

THE INTER-TEMPORAL DISCOUNT RATE

By how much should future costs and benefits be discounted to net present value to take account of the uncertainties inherent in any long-term investment appraisal such as that of a given policy's effect in reducing global warming? Stern (2006) adopts a discount rate not exceeding 1.4% (it may in practice have been as low as 0.1%: Stern Review team, pers. comm., 2006), well below HM Treasury's standard 3.5% "Green Book" rate, which is in turn somewhat below the 5% rate typical in the literature (e.g., Nordhaus, 2008; Murphy, 2008).

Stern justifies his rate as follows: "The most straightforward and defensible interpretation (as argued in the Review) of [the utility discount factor] δ is the probability of existence of the world. In the Review, we took as our base case δ = 0.1%/year, which gives roughly a one-in-ten chance of the planet not seeing out this century. … [Per capita consumption growth g is on average ~1.3% in a world without climate change, giving an average consumption or social discount rate across the entire period of 1.4% (being lower where the impacts of climate change depress consumption growth)" (Dietz et al. 2007).

HM Treasury has moved in Stern's direction by adopting two reduced "climate-change" discount rates that are initially commercial and are reduced after year 30 and again after year 75 to allow for "very-long-term, substantial, and for practical purposes irreversible wealth transfers between generations" (Grice, 2011; Lowe, 2008).

Over a century these variable rates are equivalent to uniform rates of 3.22% and 2.75% respectively—closer to the Treasury's standard 3.5% "Green Book" discount rate than to Stern's 1.4% or the 1.35-2.65% in Garnaut (2008).

However, Nordhaus (2008) and Murphy (2008) argue that the discount rate should be around the market rate of 5%, and that it should be close to the real rate of return on capital, for that is the rate that represents the opportunity cost of climate mitigation. This rate will be adopted in the illustrative case studies that follow, though other rates will be considered later.

Since warming is not occurring at the rate the IPCC predicts, the extreme warming rates of 5-6 K or even 10-11 K in Stern (2006) are not modeled here, though the method allows for them if required.

It will be assumed that the IPCC's central estimate of 2.8 K anthropgenic warming will occur by 2100, and that the welfare loss arising from climate inaction will be 1.5% of GSP, on Stern's interval [0, 3] % of GDP.

WELFARE LOSS FROM INACTION

Table 5 adjusts the projected 21st century 1.5%, 3%, 5% and 20% global costs of inaction given in Stern (2006), in line with the published discount rates shown:

Table 5. Estimates of the welfare loss over 100 years owing to climate inaction at the discount rates shown are determined by multiplying Stern's 1.5%-, 3%-, 5%- and 20%-of-GDP estimated 21st century inaction costs by the ratio of 21st century GDP discounted at Stern's 1.4% rate to 21st century GDP discounted at the mean rates shown. Annual GDP growth in the 21st century is assumed uniform at 3% throughout the table.

Stern (2006) inaction costs adjusted for discount rate	Dsct. Rate	Yr 0 -30	Yr 31 -75	Yr 76 -100	Cost: % of	Cost: % of	Cost: % of	Cost: % of
Units	%	%	%	%	**GDP**	**GDP**	**GDP**	**GDP**
Stern (2006)	1.40				150%	3.0%	5.0%	20%
Garnaut (2011) high	2.65				0.72%	1.5%	2.4%	9.7%
HM Tsy low (Lowe, 2008)	2.75	3.00	2.57	2.14	0.69%	1.4%	2.3%	9.2%
HM Tsy high (Grice, 2011)	3.22				0.54%	1.1%	1.8%	7.3%
HM Tsy Green Book rate	3.50				0.48%	1.0%	1.6%	6.4%
Market rate (see text)	5.00				0.26%	0.5%	0.9%	3.5%

Adjustments between a given inaction cost Zs mentioned by Stern on the basis of a 1.4% discount rate ds and the equivalent inaction cost Zm on the basis of a 5% market rate dm may be conveniently made using (9), where t is the term of the policy in years; a is each successive year from 1 to t; g is an assumed uniform annual percentage growth rate (here, 3%); dm is the preferred discount rate (here, the 5% minimum market rate); ds

is the discount rate at which the inaction cost was determined (here, Stern's 1.4%); $|n|$ is the absolute value of n; and $\mathrm{sgn}(n)$ is the signum function.

$$Z_m = Z_s \frac{\sum_{a=1}^{t}\left(1 + \frac{|g - d_m|}{100}\right)^{a\,\mathrm{sgn}(g-d_m)}}{\sum_{a=1}^{t}\left(1 + \frac{|g - d_s|}{100}\right)^{a\,\mathrm{sgn}(g-d_s)}} \quad (9)$$

Published estimates of the welfare loss from failing to take action to mitigate warming by reducing future emissions of CO2 follow. For comparison, the reviewed literature, summarized by Tol (2009ab), suggests that the global inaction cost will be 1-5% of GDP: 1.0 K warming will cost 2.5% of GDP (Tol, 2002); 2.5 K will cost 0.9% (Nordhaus, 2006), 1.4% (Fankhauser, 1995), 1.5% (Nordhaus & Boyer, 2000), 1.7% (Nordhaus & Yang, 1996), 1.9% (Tol, 1995), 2.5% (Plamberk & Hope, 1996), or 0.0-0.1% of GDP (from market impacts only, Mendelsohn et al. 2000); 3 K will cost 1.3-4.8% of GDP (Nordhaus, 1994ab).

THE COST/BENEFIT RATIO

Since this much-simplified model excludes all costs and benefits external to those of CO_2 mitigation, the ratio of the global abatement cost J (Eq. 5) of climate action, expressed as a percentage of GDP over the term of the policy, to the market GDP cost Zm of the climate-related damage that is projected to arise from inaction is, for present purposes, the policy's cost/benefit ratio.

Nordhaus (2012) argues that the absolute difference between cost and benefit, rather than the ratio, should determine CO_2 mitigation policies:

"Suppose we were thinking about two policies. Policy A has a small investment in abatement of CO_2 emissions. It costs relatively little (say \$1 billion) but has substantial benefits (say \$10 billion), for a net benefit of \$9 billion. Now compare this with a very effective and larger investment, Policy B. This second investment costs more (say \$10 billion) but has substantial benefits (say \$50 billion), for a net benefit of \$40 billion. B is preferable because it has higher net benefits (\$40 billion for B as compared with \$9 for A), but A has a higher benefit-cost ratio (a ratio of 10 for A as compared with 5 for B). This example shows why we should, in designing the most effective policies, look at benefits minus costs, not benefits divided by costs."

However, the true choice is not between a small but cost-effective investment and a larger though less cost-effective investment that will yield a greater absolute benefit. A rational choice surely depends upon appraising the cost of acting to prevent global warming by way of a given CO_2 mitigation policy compared with the cost of the climate-related damage that might arise from inaction over the term of the policy. In the latter context, the estimated cost of inaction is fixed, but various policy options for action to mitigate CO_2 emissions are available, wherefore the cost/benefit ratio is of no less relevance than the absolute cost or benefit of action against inaction.

ILLUSTRATIVE CASE STUDIES

In the brief illustrative case studies that follow, uniform real GDP growth of 3%/year from \$60 tr/year in 2010 (World Bank, 2011) is assumed in all cases, with a further 2% cost escalator for the Australian emissions trading scheme. Since the 5% discount rate prevalent in the literature rather than Stern's 1.4% is adopted here, Stern's estimated maximum 8 welfare loss of 20% of GDP arising in the absence of any mitigation falls to 3.5% of GDP (Table 5). Since there is no sign of warming above the IPCC's central estimate, an inaction cost of 1.5% of GDP discounted over the term [on Stern's interval [0, 3]% of GDP, is the basis for comparison with the discounted costs of action to establish the cost/benefit ratio in each case study.

Case study 1: U.S. carbon-trading Bill

At \$180 bn/year for 40 years, total \$7.2 tr, discounted to \$5 tr at p.v., the climate Bill (HR 2454, 2009, s. 311) would have abated 83% of U.S. CO_2 emissions by 2050. The U.S. emits 17% of global CO_2 (derived from Olivier & Peters, 2010, table A1). Thus $p = 0.1411$. From Table 1, business-as-usual CO_2 concentration in 2050 would be 510 ppmv, falling to 493.1 ppmv (from Eq. 7) via the Bill. From Eq. (1), forcing abated is 0.2 W m^{-2} and warming abated is 0.07 K; from Eq. (2), mitigation cost-effectiveness is \$69 tr/K; from Eq. (3), the global abatement cost of all projected warming to 2050 is \$56 tr, or (from Eq. 4), \$8,000 per capita of global population, or (from Eq. 5), 3.4% of global GDP to 2050. The mitigation cost exceeds fivefold the benefit in preventing climate damage.

Case study 2: UK Climate Change Act

At an officially-estimated cost of \$1.2 tr by 2050, discounted to \$835 bn, the Climate Change Act [2008, s. 1(1)], aims to cut 80% of UK emissions, which are 1.5% of world emissions (derived from Olivier & Peters, 2010, Table A1). Thus $p = 0.012$. Business-as-usual CO_2 concentration of 510 ppmv in 2050 would fall to 508.6 ppmv via the Climate Change Act. Forcing abated is 0.015 W m^{-2}; warming abated is 0.006 K; mitigation cost-effectiveness is \$138 tr/K; and global abatement cost to 2020 is \$113 tr, or \$16,000/head, or 6.8% of global GDP to 2050. Cost exceeds benefit 9-fold.

Case study 3: EU carbon trading

EU carbon trading costs \$92 bn/year (World Bank, 2009, p. 1), here multiplied by 2.5 (implicit in Lomborg, 2007) to allow for non-trading mitigation measures. Total cost is \$2 tr at p.v. by 2020. The EU aims to halt 20% of its emissions, which are 13% of global emissions (from Boden et al. 2010ab). Thus $p = 0.026$. Business-as-usual CO_2 concentration of 410 ppmv in 2020 would fall to 409.5 ppmv via the policy. Forcing abated is 0.007 W m^{-2}; warming abated is 0.003 K; mitigation cost-effectiveness is \$763 tr/K; and the global abatement cost of \$117 tr is \$17,000 per capita, or 21.5% of GDP to 2020. Mitigation costs 17.5 times the cost of climate-related damage in the absence of mitigation.

Case study 4: Californian cap and trade

Under the cap and trade Act (AB 32 of 2006), which took full effect in August 2012, some $182 billion per year (Varshney & Tootelian, 2009) will be spent for a decade on cap and trade and related measures. The report has been criticized for overstating costs: accordingly, one-quarter of this value will be taken over a ten-year term, giving a discounted cost of $410 billion, to abate 25% of current emissions, which represent 8% of U.S. emissions, which represent 18.7% of global emissions (derived from Olivier & Peters, 2010, table A1). Thus $p = 0.0033$. CO_2 concentration would fall from a business-as-usual 410 to 409.93 ppmv by 2020. Forcing abated is 0.001 W m^{-2}; warming abated is 0.00034 K; mitigation cost-effectiveness is $1200 tr/K; global abatement cost is $183 tr, or $26,000/head, or 34% of global GDP to 2020. Action costs 28 times inaction.

Case study 5: Thanet Wind Array

Subsidy to the world's largest wind-farm, off the English coast, guaranteed at $100 mn annually for its 20-year life, is $1.6 bn at p.v. Rated output of the 100 turbines is 300 MW, but wind-farms yield only 24% of rated capacity (Young, 2011, p. 1), so total output, at 72 MW, is 1/600 of mean 43.2 GW UK electricity demand (Department for Energy and Climate Change, 2011). Electricity is 33% of UK CO_2 emissions, which are 1.5% of global emissions, so $p = 8.333 \times 10^{-6}$. Business-as-usual CO_2 concentration in 2030 would be 440 ppmv, falling to 439.9996 ppmv as a result of the subsidy. Forcing abated is 0.000005 W m^{-2}; warming abated is 0.000002 K; mitigation cost-effectiveness is $800 tr/K; and the global abatement cost of almost $300 tr is $42,000/head, or 30% of GDP to 2030. Action costs 29 times inaction.

Case study 6: Australia cuts emissions 5% in 10 years

Carbon trading in Australia, as enacted the Clean Energy legislation (Parliament of the Commonwealth of Australia, 2011), costs $10.1 bn/year, plus $1.6 bn/year for administration (Wong, 2010, p. 5), plus $1.2 bn/year for renewables and other costs, a total of $13 bn/year, rising at 5%/year, or $130 bn by 2020 at n.p.v., to abate 5% of current emissions, which represent 1.2% of world emissions (derived from Boden et al. 2010ab). Thus $p = 0.0006$. CO2 concentration would fall from a business-as-usual 412 to 411.987 ppmv after ten years. Forcing abated is 0.0002 W m^{-2}; warming abated is 0.00006 K; mitigation cost-effectiveness is $2,000 tr/K; global abatement cost of projected warming to 2020 is $300 trillion, or $45,000/head, or 59% of global GDP to 2020. Action costs 48 times inaction.

Case study 7: Oldbury Primary School wind turbine

On 31 March 2010 Sandwell Council, England, answered a freedom-of-information request, disclosing that it had spent $9694 (£5875) on buying and installing a small wind-turbine like one at a primary school in Oldbury which had in a year generated 209 KWh—enough to power a single 100 W reading-lamp for <3 months. Assuming no maintenance costs, and discounting revenues of $0.18 (11p)/KWh for 20 years to p.v. of $623, net project cost is $9070. $p = 209$ KWh / 365 days / 24 hrs /43.2 GW x 0.33 x $0.015 = 2.76 \times 10^{-12}$. CO_2 concentration of 440 ppmv will fall to 439.9999999999 ppmv.

Forcing abated is 0.000000000002 W m^{-2}; warming abated is 0.000000000001 K; mitigation cost-effectiveness is $13,500 tr/K; and the global abatement cost, at close to $5,000 tr, is $700,000/head, or 500% of global GDP to 2030. Action costs almost 500 times inaction.

Case study 8: London bicycle-hire scheme

In 2010, the Mayor of London set up what he called a "Rolls Royce" scheme at U.S.$ 130 mn for 5000 bicycles (>$26,000 per bicycle). Transport represents 15.2% of UK emissions (from Office for National Statistics, 2010, Table C). Cycling represents 3.1 bn of the 316.3 bn vehicle miles traveled on UK roads annually (Department for Transport, 2011). There are 23 mn bicycles in use in Britain (Cyclists' Touring Club, 2011). Global emissions will be cut by 1.5% of 15.2% of 3.1/316.3 times 5000/23 mn. Thus p = 4.886 x 10^{-9}. If the lifetime of bicycles and docking stations is 20 years, business-as-usual CO_2 concentration of 440 ppmv will fall to 439.9999998 ppmv through the scheme. Forcing abated is 0.000000003 W m^{-2}; warming abated is 0.000000001 K; mitigation cost-effectiveness exceeds $110,000 tr/K; and the global abatement cost of $40,000 tr is $5.8 mn per capita, or 4000% of global GDP to 2030. Action costs almost 4000 times inaction.

RESULTS

Government estimates of abatement cost (cases 1-2) are of the same order as those in Stern (2006), Garnaut (2008) and the reviewed literature. However, the costs of specific measures prove considerably higher than all such estimates, which have proven optimistic. Gesture policies (cases 7-8) are particularly cost-ineffective.

This analysis is strictly confined to comparing the costs of taking climate action now with those of climate-related damage that might arise if no action were taken, excluding all other costs and benefits. In particular, benefits from investment in alternative or renewable energy are excluded, since they are likely to be comfortably exceeded by the opportunity losses arising from the diversion of substantial resources from the productive sector in the form of mitigation costs. Opportunity losses are also excluded from the accounting.

This analysis is not a complete study of all the costs and benefits of attempted climate mitigation. Its focus is on enabling policy-makers to understand the relationship between the IPCC's implicit central estimates of sub-centennial-scale climate sensitivity and of the forcing and warming likely to be forestalled by a given mitigation policy, allowing a first approximation of how much (or how little) global warming that policy may forestall. The equations derived here may readily be adapted for detailed cost-benefit analyses and comparisons beyond the scope of the present paper.

Table 6 summarizes the results of the case studies:

Table 6. Summary of case-study results, assuming a 5% intertemporal discount rate, uniform 3% annual GDP growth, and a 1.5%-of-GDP inaction cost.

Case Study [#]	Warming abated (K by year *y*)	Mitigation cost-effect. ($ tr/K)	Abate. Cost (%GDP)	Action/inaction ratio
[1] U.S. cap & trade	0.07 K by 2050	$69 tr/K	3.4%	5 x
[2] UK Climate Act	0.006 K by 2050	$138 tr/K	6.8%	9 x
[3] EU carbon trading	0.003 K by 2020	$763 tr/K	22%	18 x
[4] California AB 32	0.0 K by 2020	1,155 tr/K	33%	26 x
[5] Thanet Wind Array	0.0 K by 2030	$803 tr/K	30%	29 x
[6] Australia 5% cut	0.0 K by 2020	$2082 tr/K	59%	48 x
[7] School windmill	0.0 K by 2030	$13,500 tr/K	504%	488 x
[8] London cycle hire	0.0 K by 2030	$109,000 tr/K	4090%	3956 x

In Table 7, the effect of various inter-temporal discount rates is illustrated by comparing the mitigation and inaction costs of the Australian Government's carbon dioxide tax policy (case study 5) after applying Stern's and Garnaut's rates, as well as the Treasury's standard 3.5% flat rate and the 5% minimum market rate. Over longer periods than a decade, differing discount rates have a greater impact.

Table 7. The policy cost *x* of case study 6; its mitigation cost-effectiveness *M;* the per-capita global abatement cost *J*; the cash global abatement cost *H* in cash and as a percentage of GDP; the upper and lower bounds of the global welfare loss interval *I* arising from inaction, expressed as percentages of GDP; and the mitigation cost/climate benefit ratio of *H* (expressed as a percentage of GDP) to the bounds of *I*. N.B.: mn = million; bn = billion; tr = trillion; qd = quadrillion.

	Stern	**Garnaut 2**	**Green Book**	**Market**
Discount rate	**1.4%**	**2.65%**	**3.5%**	**5.0%**
Case study 5: cost *x*	$159 bn	$148 bn	$141 bn	$130 bn
Mitigat. Cost-effect	$2.8 qd/K	$2.6 qd/K	$2.5 qd/K	$2.2 qd/K
Global abatement cost	$378 tr	$352 tr	$336 tr	$310 tr
Global abatement per capita	$54,000	$50,000	$48,000	$44,000
Global abatement as % GDP	59%	59%	59%	59%
Global inaction cost	1.5-20%	1.4-18.6%	1.3-17.8%	1.2-16.4%
Cost/benefit ratio	2.9-38	3.1-41	3.2-42	3.5-48

Even the use of Stern's minimalist discount rate shows that the global abatement cost of the carbon tax—i.e., the cost of abating all global warming over the decade to 2020 if all measures to mitigate global warming from all anthropogenic causes were as cost-effective as the Australian Government's proposal—will greatly exceed even Stern's maximum 20% cost of climate-related damage arising from worldwide inaction.

DISCUSSION

For the sake of simplicity and accessibility, the focus of the method is deliberately narrow. Potential benefits external to CO_2 mitigation, changes in global-warming potentials, variability in the global-GDP growth rate, or relatively higher mitigation costs in regions with lower emission intensities are ignored, for little error arises. GDP growth rates and climate-inaction costs are assumed as uniform, though in practice little climate-related damage would arise unless global temperature rose by at least 2°C temperatures.

Given the small quanta of warming abated by CO_2-reduction policies, as well as the breadth of the intervals of published estimates of inaction and mitigation costs, the greater complexity of adopting non-uniform GDP growth rates and climate-inaction costs may in any event be otiose.

The case studies suggest official projections may be optimistic against the cost-effectiveness of specific policies. Based on the U.S. and UK Governments' estimates, the global abatement cost of their policies would be 5 and 9 times the cost of inaction respectively: however, the global abatement costs of EU's carbon trading scheme and the taxpayer subsidy to the world's largest wind farm would be 18 and 29 times inaction respectively, with smaller schemes proving considerably less cost-effective still. In general, smaller projects seem less cost-effective than larger projects: but projects of any scale are cost-ineffective.

A substantial reduction in global CO_2 emissions, maintained over centuries, might offset some of the warming caused by the pre-existing increase in atmospheric CO_2 concentration from 278 ppmv in 1750 to 390 ppmv in 2010.

After a sufficiently long period of global emissions reduction ($y >> 2100$), it may become justifiable to reduce the value 390 in the denominator of Eq. (1) stepwise towards the pre-industrial CO_2 concentration 278 ppmv, increasing ΔTnix and consequently improving cost-effectiveness by reducing M. However, within the 21st century even the immediate and total elimination of CO_2 emissions would only abate ~1.5 K global warming.

For numerous reasons, Eq. (1) and the case studies tend to overstate the warming that any CO_2-reduction policy may abate, and also to overstate cost-effectiveness. The IPCC takes CO_2's mean atmospheric residence time as 50-200 years: if so, little mitigation will occur within the 21st century. It is here assumed that any policy-driven reduction in CO_2 concentration occurs at once, when it would be likely to occur stepwise to year y, halving the warming otherwise abated by that year and doubling the cost-ineffectiveness.

In some cases, it is assumed that the policy will meet the emissions-reduction target on its own, ignoring the often heavy cost of all other mitigation measures intended to contribute to achievement of the target. In most case studies capital costs only are counted and running costs excluded. Capital costs external but essential to a project, such as provision of spinning-reserve generation for wind turbines on windless days, are excluded. Emissions from project construction and installation, such as concrete bases for wind turbines, are ignored, as are costs and CO_2 emissions arising from necessary external operating expenditures such as spinning-reserve for wind turbines.

If the IPCC's central projections exaggerate the warming that may arise from a given increase in atmospheric CO_2 concentration, the warming abated may be less than shown. Though emissions are rising in accordance with the IPCC's A2 emissions scenario, concentration growth has been near-linear for a decade, so that outturn by 2100 may be considerably below the IPCC's mean estimate of 700 ppmv.

The climate-sensitivity parameter λ_{tra} used in the case studies is centennial-scale: accordingly, over the shorter periods covered by the studies a somewhat lesser coefficient (allowing for the fact that longer-term temperature feedbacks may not yet have acted) and consequently less warming abated would reduce mitigation cost-effectiveness.

Finally, all opportunity losses from diverting resources to global-warming mitigation are ignored.

CONCLUSIONS

The case studies indicate that government's initial abatement-cost estimates have proven optimistic. It is unlikely that any policy to abate global warming by taxing, trading, regulating, reducing, or replacing greenhouse gas emissions will prove cost-effective solely on grounds of the welfare benefit foreseeable from global-warming mitigation alone.

High abatement costs, and the negligible returns in warming abated, imply that focused adaptation to the consequences of such future warming as may occur may be considerably more cost-effective than any attempted mitigation. Mitigation policies inexpensive enough to be affordable are likely to prove ineffective, while policies costly enough to be effective will be unaffordable. Since the opportunity cost of mitigation is heavy, the question arises whether mitigation should be attempted at all.

ACKNOWLEDGEMENTS

This paper is based on lectures to the 43rd (2010) and 45th (2012) Seminars on Planetary Emergencies of the World Federation of Scientists at Erice, Sicily, to the Prague School of Economics in May 2011, and to the Third Los Alamos Conference on Global and Regional Climate Change, Santa Fe, in November 2011. I am most grateful for comments from Dr. Petr Chylek of the Los Alamos National Laboratory; Professor Tim Congdon, formerly of the Monetary Policy Committee of the Bank of England; Dr. Christopher Essex, Professor and Departmental Chair of Applied Mathematics in the University of Western Ontario; Dr. David Evans, formerly of the Australian Government's Carbon Accounting Office; Dr. Laurence Gould, Professor of Physics in the University of Hartford; Dr. Vaclav Klaus, President of the Czech Republic; and Dr. Fred Singer, Professor Emeritus of Environmental Sciences in the University of Virginia.

REFERENCES

1. AB 32 (2006) *Global Warming Solutions Act*, California State Legislature.
2. Boden and Marland (2010a) *Global CO_2 Emissions from Fossil-Fuel Burning, Cement Manufacture, and Gas Flaring, 1751-2007*. Carbon Dioxide Information and Analysis Center, Oak Ridge, Tennessee, USA.
3. Boden et al. (2010b) *Ranking of the world's countries by 2007 total CO_2 emissions from fossil-fuel burning, cement production, and gas flaring*. Carbon Dioxide Information and Analysis Center, Oak Ridge, Tennessee, USA.
4. Climate Change Act (2008) http://www.legislation.gov.uk/ukpga/2008/27/section /1.
5. Conway, T., and P. Tans (2011) Recent trends in globally-averaged CO_2 concentration, NOAA/ESRL, http://www.esrl.noaa.gov/gmd/ccgg/trends/global. html #global.

6. Cyclists' Touring Club (2011) Cyclists' Touring Club Facts and Figures. http://www.ctc.org.uk/resources/Campaigns/CTC-Facts+figs_rpt.pdf.
7. Department for Energy and Climate Change (2011) UK Energy Statistics: Electricity.www.decc.gov.uk/en/content/cms/statistics/source/electricity/ electricity.aspx.
8. Department for Transport (2011) Table TRA0101, Road traffic by vehicle type, Great Britain, 1950-2009. http://www.dft.gov.uk/pgr/statistics/datatables publications/ roads/traffic/#tables.
9. Fankhauser (1995) Valuing Climate Change—The Economics of the Greenhouse, 1st edn. EarthScan, London, UK.
10. Garnaut (2008) The Garnaut Climate Change Review: Final Report. Cambridge University Press, Port Melbourne, Australia, 680 pp, ISBN 9780521744447.
11. Grice (2011) The Green Book: Appraisal and Evaluation in Central Government: Treasury Guidance. The Stationery Office, London, UK. http://www.hm-treasury.gov.uk/d/green_book_complete.pdf.
12. HadCRUt3gl (2011) Monthly global mean surface temperature anomalies, 1850-2011. http://www.cru.uea.ac.uk/cru/data/temperature/hadcrut3gl.txt.
13. HR 2454 (2009) American Clean Energy & Security Bill. 111th Congress, Washington DC.
14. IPCC (2001) Climate Change 2001: The Scientific Basis: Contribution of Working Group I to the Third Assessment Report of the Intergovernmental Panel on Climate Change [Houghton, J.T., Y. Ding, D.J. Griggs, M. Noguer, P.J. van der Linden, X. Dai, K. Maskell and C.A. Johnson (eds.)]. Cambridge University Press, Cambridge, United Kingdom, and New York, NY, USA.
15. IPCC (2007) Climate Change 2007: the Physical Science Basis. Contribution of Working Group I to the Fourth Assessment Report of the Intergovernmental Panel on Climate Change, 2007 [Solomon, S., D. Qin, M. Manning, Z. Chen, M. Marquis, K.B. Avery, M. Tignor and H.L. Miller (eds.)]. Cambridge University Press, Cambridge, United Kingdom, and New York, NY, USA.
16. Klaus (2011). The Global Warming Doctrine Is Not a Science. Conference on The Science and Economics of Climate Change, Downing College, Cambridge, UK, May 10.
17. Lomborg (2007) Perspective on Climate Change, Testimony before the Committee on Science and Technology of the U.S. House of Representatives, March 21.
18. Lowe (2008) Intergenerational wealth transfers and social discounting: supplementary Green Book guidance. HM Treasury, London, ISBN 978-1-84532-419-3.
19. McCauley (2011) Letter of response to a request under the Freedom of Information Act received from Peter Day, Sandwell Borough Council, Oldbury, England, 31 Mar, www.whatdotheyknow.com/request/wind_turbine#incoming-163689.
20. Mendelsohn et al. (2000) "Country-specific market impacts of climate change." *Climatic Change* 45:3-4, 553-569.
21. Murphy (2008) "Some Simple Economics of Climate Changes." Paper presented to the MPS General Meeting,Tokyo, September 8.

22. Myhre et al. (1998) "New estimates of radiative forcing due to well mixed greenhouse gases." *Geophysical Research Letters* 25:14, 2715-2718, doi:10.1029/98GL01908.
23. Nordhaus, W.D. (1994a.) "Expert opinion on climate change." *American Scientist* 82:1, 45-51.
24. Nordhaus, W.D. (1994b) *Managing the global commons: the economics of climate change*. MIT Press, Cambridge, Mass., and London, England.
25. Nordhaus, W.D. "Geography and macroeconomics: new data and new findings," *Proceedings of the National Academy of Sciences* 103:10, 3510-3517.
26. Nordhaus, W.D. *A Question of Balance: Weighing the Options on Global Warming Policies*, Yale University Press.
27. Nordhaus, W.D. (2012) "Why the global warming skeptics are wrong," *New York Review of Books,* March 22.
28. Nordhaus and Boyer (2000) *Warming the world: economic models of global warming*. MIT Press, Cambridge, Mass., and London, England.
29. Nordhaus and Yang (1996) "RICE: a regional dynamic general-equilibrium model of optimal climate-change policy." *American Economic Review* 86:4, 741-765.
30. Office for National Statistics (2010) "Statistical Bulletin, Greenhouse gas emissions intensity falls in 2008," www.statistics.gov.uk/pdfdir/ea0610.pdf.
31. Olivier and Peters (2010) Mondiale emissies koolstofdioxide door gebruik fossiele brandstoffen en cementproductie, 1990-2009. PBL Netherlands Environmental Assessment Agency, Den Haag, Netherlands.
32. Parliament of the Commonwealth of Australia (2011) Exposure Draft of the Clean Energy Bill. http://www.climatechange.gov.au/government/submissions/clean-energy-legislative-package/~/media/publications/clean-energy-legislation/exposure-draft-clean-energy-bill-2011-pdf.pdf.
33. Plamberk and Hope (1996) "PAGE95—an updated valuation of the impacts of global warming." *Energy Policy* 24:9, 783-793.
34. Ramanathan, et al. (1985) "Trace gas trends and their potential role in climate change." *Journal of Geophysical Research* 90, 5547-5566.
35. RSS (2012) Monthly MSU global mean land and sea lower-troposphere temperature anomalies, http://www.remss.com/data/msu/monthly_time_series/RSS_Monthly_MSU_AMSU_Channel_TLT_Anomalies_Land_and_Ocean_v03_3.txt.
36. Solomon et al. (2009) "Irreversible climate change due to carbon dioxide emissions." *Proceedings of the National Academy of Sciences* 106:6, 74-79, doi:10.1073/pnas:0812721106.
37. Stern, N. (2006) *The Economics of Climate Change: The Stern Review*. Cambridge University Press, Cambridge, United Kingdom, and New York, NY, USA.
38. Tans, P. (2011) NOAA global monthly mean CO_2 concentration trends dataset, ftp://ftp.cmdl.noaa.gov/ccg/co2/trends/co2_mm_gl.txt.
39. Tol (1995) "The damage costs of climate change: toward more comprehensive calculations." *Environmental and Resource Economics* 5:4, 353-374.

40. Tol (2002) "Benchmark and dynamic estimates of the damage costs of climate change." *Environmental and Resource Economics* 21:1, 47-73, and 21:2, 135-160.
41. Tol (2009a) "The Economic Effects of Climate Change." *Journal of Economic Perspectives* 23:2, 29-51.
42. Tol (2009b) *An analysis of mitigation as a response to climate change.* Copenhagen Consensus Center, Copenhagen Business School, Frederiksberg, Denmark.
43. UAH (2012) Monthly global mean land and sea lower-troposphere temperature anomalies, http://vortex.nsstc.uah.edu/data/msu/t2lt/tltglhmam_5.4.
44. Varshney, S.B., and Tootelian, D.H. (2009) *Cost of AB 32 on California small businesses*, California Small Business Round Table.
45. Wong (2010) Portfolio Budget Statements 2010-11: Budget-Related Paper No. 1.4. Climate Change and Energy Efficiency Portfolio, Commonwealth of Australia, Canberra, Australia.
46. World Bank (2009) State and Trends of the Carbon Market, Washington DC. http://siteresources.worldbank.org/INTCARBONFINANCE/Resources/State_Trends_of_the_Carbon_Market_2009-FINAL_26_May09.pdf.
47. World Bank (2011) Gross Domestic Product 2009, World Development Indicators. http://siteresources.worldbank.org/DATASTATISTICS/Resources/GDP pdf.
48. Young, S (2011) *Analysis of UK Wind Power Generation, November 2008 to December 2010,* John Muir Trust, Edinburgh, 25 pp: http://www.jmt.org/assets/pdf/wind-report.pdf.

SESSION 4

GLOBAL NUCLEAR ENERGY ISSUES

WORLD OUTLOOK FOR NUCLEAR POWER AFTER FUKUSHIMA

HANS-HOLGER ROGNER
Planning and Economics Studies Section, Department of Nuclear Energy
International Atomic Energy Agency, Vienna, Austria

THE IMMEDIATE IMPACT OF THE FUKUSHIMA DAIICHI ACCIDENT

The Fukushima Daiichi accident of 11 March 2011 that was caused by a devastating earthquake and subsequent tsunami has re-ignited the debate about the role of nuclear power in the current and future global energy mix. Initial government policy responses, in large part fuelled by sensationalist media reports and public pressure, pointed towards an even more uncertain future of the technology than before.

Germany, within weeks after the accident, renounced its September 2010 policy decision to extend the lifetime of its reactor fleet by an average of twelve years and closed for good eight nuclear reactors that were off-line for refuelling, maintenance or repairs. Moreover, an accelerated phase-out of the remaining nine reactors by 2022 was decided upon which corresponds to an average plant-lifetime of approximately 30 years. In order to mitigate the impact of the lower total generating capacity in the country, the government announced a fundamental energy transformation ("Energiewende") based on renewables and energy efficiency improvements throughout the economy. The German phase-out-path is considered to be the fastest possible way of shutting down the remaining nuclear power capacities without running into critical system instabilities.

The Japanese government ordered the shut-down or blocked the restart of most reactors in the country for the conduction of mandatory two-phase stress tests.[1] They further ceased all construction activities of new builds and decided to decommission the Fukushima Daiichi Units 1 to 4. By May 2011 (two months after the accident), only 26 of the remaining 50 reactors were in operation and this number successively declined to zero on 5 May 2012 when Tomari-3, the country's only reactor left in operation, was taken off the grid for maintenance and refuelling. Several units that had completed planned maintenance and refuelling outages remained shut down because of public opposition or lack of local authorities' endorsement of a restart.

On 16 June 2012, Prime Minister Noda announced the government's approval, with agreement of the local authorities, to restart Units 3 and 4 at Kansai Electric Power Company's (Kepco's) Ohi plant in the Fukui prefecture. Both plants were shut down shortly after the Fukushima Daiichi accident for scheduled outages and have since successfully passed phase 1 of the mandatory stress test. After two months without

---

[1] Phase 1 requires utilities while units are shut down for inspections to examine the safety margins of important reactor components with regard to their ability to withstand beyond design-based earthquakes and tsunamis, and to put in place measure to cope with extended station black outs, etc. Based on the results of these tests, the government will decide whether a reactor can or cannot resume operation. In addition, any restart requires a go ahead from Japan's regulatory organizations and approval from local prefectural governments. Phase 2 of the stress tests will involve a comprehensive safety assessment of all reactors with the objective to enhance the reliability and effectiveness of safety checks. This overall assessment resembles the stress tests carried out in the European Union and elsewhere.

nuclear power in Japan, Ohi-3 started generating electricity again on 5 July 2012 and was followed by Ohi-4 two weeks later.

In late May 2011, the government of Switzerland allowed the country's five existing reactors to continue operating, but without replacement at the end of their service lives of approximately 50 years, with the last reactor going offline in 2034. In September, the Swiss parliament endorsed this long-term phase-out plan. In addition to operating safety concerns and lack of resistance to natural and man-made disasters, the rationale was that new-build nuclear power would lose its competitive advantage due to the added costs associated with new (post-Fukushima) safety and security standards.

In Italy, a referendum, planned before 11 March 2011, on the reintroduction of nuclear power which was abandoned in the late 1980s by a similar referendum, was held in June 2011 and rejected the reestablishment of a national nuclear power programme by a large majority.

In November 2011, Taiwan, China, announced a new nuclear energy policy of phasing out nuclear power, however, no specific time frame has been outlined other than: "once their licenses expire" (Taipei Times, 2012). The plants currently under construction will be completed and become operational contingent upon passing all safety inspections.

In Belgium in October 2011, the 2003 decision to shut down the country's oldest nuclear power reactors in 2015, which had been reconsidered in 2009, was reinforced. Moreover, the government proposed a doubling of the special tax on nuclear power paid annually by the nuclear industry. The decision effectively ruled out the building of new nuclear plants in the country. On 23 July 2012, the Belgian government announced the schedule for the total phase out of nuclear power by 2025 complementing the closures for three units announced on 4 July 2012. No particular justification was provided. All Belgian units are to close between 2015 and 2025 which corresponds roughly to a service life of 40 years per unit. Only one unit, Tihange 1, is permitted to operate to 50 years of age; an exception made specifically to avoid blackouts.

Other countries pursued a less drastic approach. Clearly, a disaster such as the Fukushima Daiichi accident calls for reflection. In most countries with nuclear programmes, the immediate response to the accident was to conduct comprehensive safety reviews including reassessing the safety margins of nuclear plants of all nuclear installations. In addition, many countries introduced improvements to their frameworks for dealing with nuclear accidents and associated emergency preparedness and communication activities. In addition, many countries revisited their long-term energy development plans and policies and analysed their national energy demand and supply options. In order to enhance nuclear safety, many countries ordered the implementation of stress tests of nuclear power plants to evaluate their capacity to withstand extreme natural and often also man-made events. Measures have been initiated to: improve preparedness for prolonged power blackouts; protect sources of backup electrical power; and to assure the availability of water for cooling, even under severe accident conditions. Emergency preparedness and response capabilities came under critical review, and plans to strengthen these are currently under development. Most importantly perhaps is the radical revision of worst-case assumptions for safety planning.

In addition, some countries with explicit intentions to explore the nuclear option or introduce nuclear power to their national energy mixes have announced that they will no longer do so (e.g., Kuwait, Senegal, Venezuela) or that plans will be delayed (e.g., Indonesia, Thailand) while governments continue to reassess their plans for their future use of nuclear power.

THE IMPACT ON THE NUCLEAR INDUSTRY

The Fukushima Daiichi accident definitely undermined public confidence in nuclear power. Past experience from the Three Mile Island and Chernobyl accidents have shown that in many countries it will take decades of successful operating experience of the global fleet of nuclear power plants to restore a certain level of public confidence. In the wake of these accidents, the industry improved its overall safety culture, particularly with regard to all operational aspects of nuclear power installations.

The memories of the Three Mile Island and Chernobyl accidents eventually faded, in large part due to the demonstrated safe and economic operation of reactors around the world. In addition, during the early years of the 21st century, several issues began to dominate public energy policy debates, namely: mounting concerns about energy security; volatile fossil fuel prices after a long period of stability at low levels; climate change considerations; and continuously rising electricity demand. As nuclear power can play a pivotal role in mitigating all these issues, political and public opinion progressively tilted towards a higher level of tolerance of nuclear technology in many countries.

As of 2004, construction starts per year began to rise and reached 16 new builds by 2010—a level of construction starts not witnessed since 1985 (see

Fig. *1*). All starts since 2000 have occurred in countries with already existing nuclear power plants, with Asian countries taking the global lead (see Fig. 2).

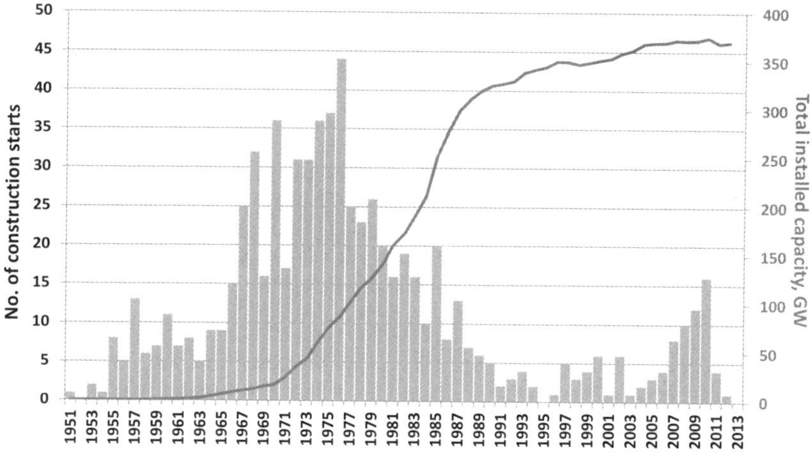

Fig. 1: Number of construction starts globally and total installed generating capacity, 1951-2012 (30 June 2012). Source: IAEA, 2012a.

Before the Fukushima Daiichi accident many countries without nuclear power programmes expressed interest in adding nuclear to their national electricity generation portfolios. This interest varied from country to country. A dozen countries had decided to implement national nuclear power programmes and started preparing the necessary infrastructure. Another seven countries started preparations, however without a final decision. More than 40 countries either considered nuclear power programmes or were interested in obtaining relevant information on the technology and its prerequisite infrastructure requirements. Together with the surge in construction starts in countries with already operating nuclear power plants, this rise in interest in, and popularity of, nuclear power after 20 years of quasi stagnation (except for Asia) was popularly labelled as a "nuclear renaissance".

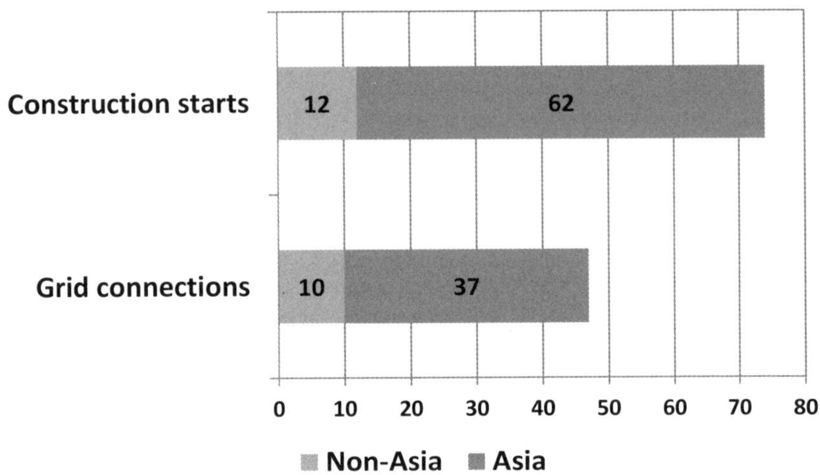

Fig. 2: Construction starts and grid connections since 01 January 2000. Source: IAEA, 2012a.

Figure 1 shows the global impact of the Fukushima Daiichi accidents on construction starts. Construction starts in 2011 plummeted to four reactors (although all starts occurred after 11 March 2011) and to one new build so far in 2012. This decline is the result of several factors as several countries suspended new construction permits pending: site re-evaluations and seismic re-qualifications; reactor design reviews especially with respect to withstanding multiple, concurrent external initiating events; and updated safety and regulatory guidelines based on the "lessons-learned" from the accident. However, the pre-Fukushima trend of uprates through reactor component refurbishments and renewed or extended licences for many operating reactors continued in 2011 and the first half of 2012.

<u>Nuclear Operating Safety</u>

"One of the statutory functions of the IAEA is to establish or adopt standards of safety for the protection of health, life and property in the development and application of nuclear energy for peaceful purposes, and to provide for the application of these standards to its own operations as well as to assisted operations and, at the request of the parties, to operations under any bilateral or multilateral arrangement, or, at the request of a State, to any of that State's activities in the field of nuclear energy" (IAEA, 2003).

The IAEA safety standards establish fundamental safety principles (IAEA, 2006), requirements and measures aimed at the protection of people and the environment against radiation risks by achieving the highest practicable safety levels in nuclear installations. It also involves the control of radiation exposure of people and the release of radioactive material to the environment, restriction of the likelihood of events that might lead to a loss of control over a nuclear reactor core or a radioactive source and the mitigation of the consequences of such events, if they were to occur.

A fundamental safety principle is that the prime responsibility for safety must rest with the person or organisation responsible for facilities and activities that give rise to radiation risks, i.e., nuclear facility operators are ultimately responsible for the safety of their facility. Another principle is that an effective legal and governmental framework for safety, including an independent regulatory body, must be established and sustained. National governments, therefore, are responsible for regulations that govern how safety at nuclear facilities is maintained, as well as for reducing radiation risks, including emergency response and recovery actions, monitoring releases of radioactive substances to the environment, and regulating the safe decommissioning of facilities, and the disposal of radioactive waste.

Nuclear accidents do not respect borders. Therefore, the primary responsibility of operators and Member States, enshrined in the IAEA Fundamental Safety Principles, must be backed by an international approach to safety. The IAEA, through the Department of Nuclear Safety and Security, works to provide a strong, sustainable and visible global nuclear safety and security framework for the protection of people, society and the environment. This framework provides for the harmonized development and application of safety and security standards, guidelines and requirements. But the IAEA does not have the mandate to enforce the application of safety standards within a country. Safety standards are only effective, if they are properly applied in practice. National regulators, therefore, have a pivotal role in ensuring continued nuclear safety and are encouraged to adopt the IAEA's safety standards for use in their national regulations.

<u>IAEA response to the Fukushima Daiichi accident</u>

The IAEA responded to the accident by activating the Incident and Emergency Centre in Vienna and organizing specialized expert missions to Japan to gain an understanding of the accident and to provide assistance and expert advice. On 30 March 2011, the IAEA Director General called for a Ministerial Conference on Nuclear Safety to be held in June 2011 to consider the implications of the accident. The objectives of the Conference were to:

- Undertake a preliminary assessment of the accident;

- Identify actions needed to strengthen the national and international emergency preparedness and response capabilities;
- Identify those areas of the global nuclear safety framework that may require strengthening.

The Conference adopted a Ministerial Declaration which requested the Director General, inter alia, to prepare an IAEA Action Plan on Nuclear Safety.

The 12 point IAEA Action Plan on Nuclear Safety was developed following intensive consultations with Member States (IAEA, 2011). The Plan built on the Declaration by the Ministerial Conference, the conclusions and recommendations of the three conference working sessions as well as other relevant sources. It was adopted by the Board of Governors in September 2011 and unanimously endorsed by all Member States at the 2011 IAEA General Conference. The ultimate goal of the Action Plan is to strengthen all aspects of nuclear safety worldwide.

The 12 points of the Action Plan with short annotations are listed in the Annex. Initial calls for a comprehensive and globally binding convention of nuclear operating safety, e.g., mandating all countries with nuclear facilities to adopt and fully implement current IAEA Safety Standards with verification of adherence, were not supported by Member States. Nuclear operating safety remains a national responsibility. Instead, the plan emphasizes voluntary peer reviews of the operating safety of all nuclear power plants as well of the effectiveness of national regulatory bodies.

The Action Plan, therefore, is not a blue print for the employees of the IAEA Secretariat. Its success will hinge on the level of ownership adopted by Member States, regulators, nuclear operators, vendors, international and intergovernmental organizations, i.e., it requires inclusive participation and full cooperation of all involved in nuclear matters. Actions, therefore, are explicitly addressed either to Member States, or to the IAEA Secretariat or to other stakeholders, or to all.

It should be noted that, operationally, the level of nuclear power plant safety around the world remains high. Safety indicators, such as those published by the World Association of Nuclear Operators (WANO) and reproduced in

Fig. 3, improved dramatically in the 1990s along with plant performance measured in annual plant availabilities. To a large extent, this attributable to the efforts of regulatory bodies over the years as well as the standards that they have set in design and operation. In recent years, in most areas the situation has stabilized. However, the gap between the best and worst performers is still large, providing substantial room for further improvement.

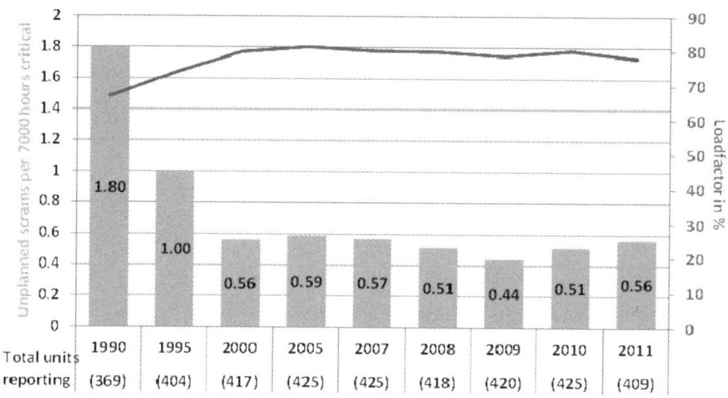

Fig. 3: Unplanned scrams per 7000 hours critical (left axis) and load factors (right axis). Source: WANO (2012).

<u>Nuclear power today</u>

As of 25 of July 2012, the global fleet of nuclear power plants consisted of 435 reactors with a combined installed nuclear generating capacity of 370 GWe (375.5 GWe on 10 March 2011). Note: the total includes plants that currently are off-grid such as the remaining 48 reactors in Japan but not declared as permanently shut down. In 2011 nuclear power accounted for 12.3% of global electricity supply down from 13.5% the year before. Fig. 4 (left) depicts the regional distribution of nuclear generating capacities.

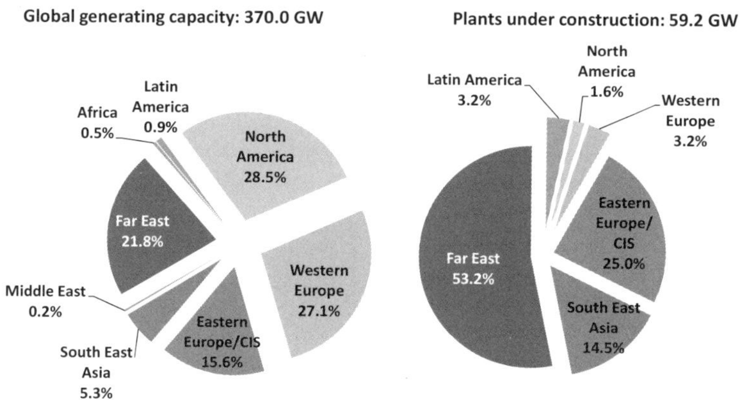

Fig. 4: Global generating capacity (left) and plants under construction (right) - as of 25 July 2012. Source: IAEA, 2012a.

Despite the Fukushima Daiichi accident, the trend of uprates and renewed or extended licences for many operating reactors continued in many countries. Uprates of some 400 MWe and the grid connection of 8 new build reactors (totalling 5.7 GWe since the accident helped mitigate the lost capacities of the 15 reactors (11.4 GWe) declared shut-down for good.

The right panel of Fig. 4 shows the regional distribution of the 62 reactors currently under construction with a combined generating capacity of 59.2 GW. It highlights the shift in expansion from the traditional nuclear countries in North America and Europe which dominate the current regional capacity distribution to Asia where the long term growth prospects remain centered.

Outlook

Every year the IAEA develops projections of global nuclear power development derived from a country-by-country 'bottom-up' approach. They are established by a group of experts and based upon a review of nuclear power projects and programmes in Member States. The projections cover the period to 2030 with an outlook to 2050. The IAEA estimates should be viewed as very general trends whose validity must be constantly subjected to critical review as energy and electricity demand and the role of nuclear power are subject to economic growth and structural economic change, demographic developments, technology performance and costs, energy resource availability and future fuel prices, and energy and environmental policy. The projections provide a plausible range of nuclear capacity growth by region and worldwide. They are not intended to be predictive nor to reflect the whole range of possible futures from the lowest to the highest feasible.

The LOW projection accounts for new builds already firmly in the pipe line including scheduled retirements, planned uprates and license renewals, plus projects in Member States that can be reasonably assumed to materialize. Naturally, the Fukushima accident was on the minds of experts when the 2011 and 2012 projections were compiled. Likewise, the continued financial and economic crises and associated uncertainties affected the projections, especially in the short run. However, the past drivers behind the use of nuclear energy have not changed (population and economic development, environmental protection and climate change, energy security considerations and fossil fuel price volatility) and will continue to fuel nuclear power development. The 2012 projection shows a 90 GWe reduction compared with 2010—the last projection before the Fukushima Daiichi accident (or 456 GWe instead of 546 GWe for 2030).

The longer term outlook to 2050 with 469 GWe is rather modest. The projections reflect a future where replacement investments of retired capacities rather than expansion characterize the nuclear sector after 2030. The absence of growth, however, is consistent with the philosophy of the LOW projection as only few projects can be considered 'firm' beyond 2030.

Additionally, the LOW projection does not automatically assume that ambitious targets for nuclear power growth in a particular country would necessarily be achieved. These assumptions are relaxed in the HIGH projection which is much more optimistic, but still plausible and technically feasible. The high case assumes that the current financial and economic crises will be overcome in the not so distant future and past rates of economic growth and electricity demand, especially in the Far East, would essentially resume. In addition, the high case assumes the implementation of stringent policies globally targeted at mitigating climate change. The Fukushima impact was a reduction of 63 GWe (down from 803 GWe in the 2010 projection) by 2030.

The longer term outlook to 2050 (HIGH) is characterized by a temporal shift of about one decade to the future—a continuation of the delay effect and 'wait and see' behaviour regarding the technology in many, especially newcomer countries. In 2010 global nuclear generating capacities for 2050 were estimated at 1415 GWe and about 1110 GWe for 2040. In 2012 the value for 2050 was lowered to 1137 GWe or about the 2010 estimate for the year 2040.

Fig. 5 displays the historical expansion of nuclear generating capacities and the annual projections for the period 2005 to 2012. For the LOW projections the financial and economic crises since 2008 did not affect the outlook notably and projections still pointed to continued nuclear expansion. In contrast, the HIGH projection was affected by the overall, rather dire global economic, outlook and the successively higher projections since 2005 came to a halt in 2010.

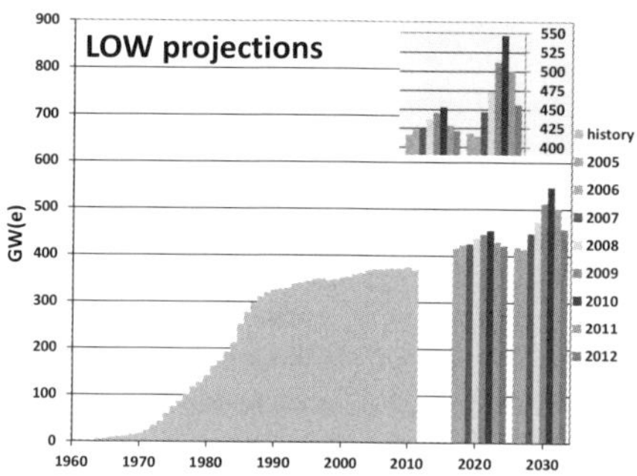

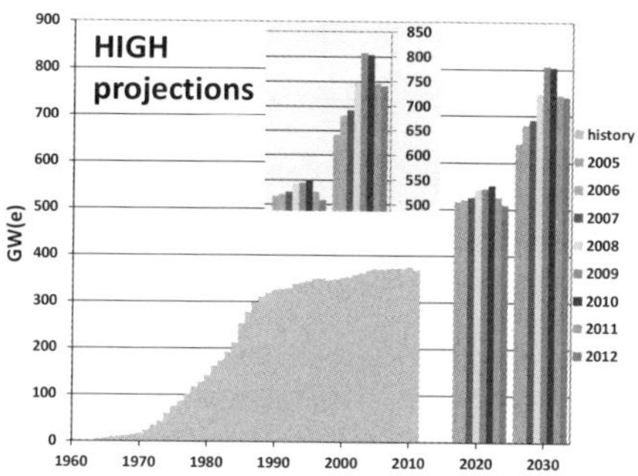

Fig. 5: Development of successive annual projections of total nuclear generating capacities, IAEA LOW (left) and HIGH (right) projections, 2005 - 2012. Source: IAEA, 2012b.

After 2010 the Fukushima impact is clearly evident and, in the LOW projection, suggests a delay in nuclear new builds of about a decade and the 2030 projection of 2012 now falls in the range of pre-Fukushima projections for the year 2020. The sharp drop of construction starts in 2011 and 2012 (see Figure 1) is the consequence of the suspension of construction permits in several countries, the renewable directives in many jurisdictions, the emergence of potentially large amounts of relatively inexpensive shale gas and the decline in public acceptance in many countries. The HIGH projections for 2030 show a lower reduction of nuclear generating capacities than the LOW projections in both absolute and relative terms. This is consistent with the HIGH projections' overall more optimistic outlook on economic development, a new and effective binding international agreement on mitigating climate change and the restoration of public confidence in the technology.

Fig. 6 shows the projected nuclear capacity developments for major world regions. In both projections, future nuclear development will be centered in Asia. This shift is already well in progress as demonstrated by the location of the reactors currently under construction (see Fig. 2). As the momentum is rising to Asia, especially non-OECD Asia, Europe shows the biggest negative difference between the 2011 installed capacity and the LOW projection—a contraction of 41 GWe from 125.8 GWe at the end of 2011 to 85 GWe in 2030. North America (-3.5 GWe) and Pacific OECD (-.4 GWe) are further regions with a contraction of nuclear generating capacities by 2030. The expansion leaders are non-OECD Asia and the Commonwealth of Independent States (CIS) with 97 GWe and 28 GWe additional nuclear generating capacity. Modest nuclear growth occurs in the Middle East, Africa and Latin America.

In the HIGH projection nuclear power capacities expand in all regions, most prominently in Non-OECD Asia which account for 55% (or net 205 GWe) of the expansion alone. Europe and Pacific OECD reverse the negative trends portrayed in the LOW projection and increase their respective current generating capacities by about 30 GWe each. CIS and North America add 42 GWe and 33 GWe while Africa, Middle East and Latin America experience the beginning of a bullish nuclear sector.

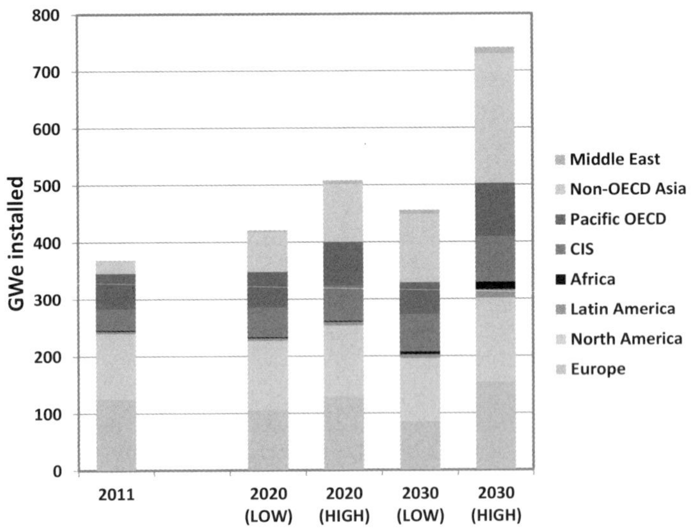

Fig. 6: Regional nuclear capacity developments in the 2012 IAEA LOW and HIGH projections. Source: IAEA, 2012b.

CONCLUSION

The Fukushima Daiichi accident and subsequent safety reviews and stress tests have significantly affected the global prospects for new nuclear build. The 2012 projections of total nuclear power generating capacity in 2030 is 8-16% lower than before the accident.

However, globally, the accident is expected to slow down or delay growth in nuclear power but not to reverse it. On a nation-scale, the impact is more diverse and measures taken by countries have been varied: a number of countries announced reviews of their programmes; some took steps to phase out nuclear power entirely while others re-emphasized their expansion plans. In fact, most countries with nuclear power have reaffirmed their commitment to expanding nuclear power while incorporating the lessons-learned from the accident. In countries considering the introduction of nuclear power, interest also remains high. Of the countries without nuclear power that before the accident had strongly indicated their intentions to proceed with nuclear power programmes, a few have cancelled or revised their plans, while others have taken a 'wait-and-see' approach. Most have continued in their pursuit of introducing nuclear power into their energy mixes. The factors that had contributed to the renaissance of interest in nuclear power before the Fukushima accident largely remain the same.

Questions about the future of the technology remain. In order to regain public confidence in nuclear power, the lessons learned and those yet to be learned from the Fukushima accident have to be implemented into national nuclear programmes in a timely, transparent and inclusive manner - a process that will extend well into the future.

ANNEX

The Action Plan

1. *Undertake assessments of the safety vulnerabilities of nuclear power plants in the light of lessons learned to date from the accident:* Member States assess the design of their nuclear installations against site specific extreme natural hazards and implement the necessary corrective actions.
2. *Strengthen IAEA peer reviews in order to maximize the benefits to Member States:* Member States incorporate the accident's lessons into IAEA peer reviews, apply these more broadly to address regulatory effectiveness, operational safety, design safety, and emergency preparedness and response. The results should be made transparent. Member States are also requested to provide experts for peer review missions.
3. *Review and strengthen emergency preparedness and response arrangements and capabilities:* Member States conduct national reviews of their emergency preparedness and response arrangements and capabilities, with IAEA providing assistance through Emergency Preparedness Review (EPREV) missions. The IAEA, Member States and relevant international organizations (IOs) review and strengthen the international emergency preparedness and response framework encouraging greater involvement of the relevant IOs in the Joint Radiation Emergency Management Plan JREMP). Bolster the assistance mechanisms to ensure that necessary assistance is made available promptly, and fully utilize the IAEA Response and Assistance Network (RANET), including expanding its rapid response capabilities.
4. *Strengthen the effectiveness of national regulatory bodies:* Regularly review (e.g., through IAEA Integrated Regulatory Review Service—IRRS—missions) of national regulatory bodies, particularly their independence and resources, and strengthen them as needed. The IAEA Secretariat to enhance

the IRRS for peer review of regulatory effectiveness through a more comprehensive assessment of national regulations against IAEA Safety Standards.

5. *Strengthen the effectiveness of operating organizations with respect to nuclear safety:* Regularly review (e.g., through IAEA Operational Safety Review Team—OSART—missions), and strengthen the management systems, safety culture, human resources management, scientific and technical capacities in operating organizations. Each Member State with nuclear power plants is requested to voluntarily host at least one OSART mission during the coming three years, with the initial focus on older nuclear power plants. Thereafter, OSART missions are to be voluntarily hosted on a regular basis.
6. *Review and strengthen IAEA Safety Standards and improve their implementation:* Improve the effectiveness of the international legal framework and work towards a global nuclear liability regime that addresses the concerns of all States that might be affected by a nuclear accident. Member States implement the IAEA Safety Standards as broadly and effectively as possible, in an open, timely and transparent manner.
7. *Improve the effectiveness of the international legal framework (ILF):* Improve the effectiveness of the ILF and work towards a global nuclear liability regime that addresses the concerns of all States that might be affected by a nuclear accident. Member States explore mechanisms to enhance the implementation of the Convention on Nuclear Safety, the Joint Convention on the Safety of Spent Fuel Management and the Safety of Radioactive Waste Management, the Convention on the Early Notification of a Nuclear Accident and the Convention on Assistance in the Case of a Nuclear Accident or Radiological Emergency and consider proposals made to amend the Convention on Nuclear Safety and the Convention on the Early Notification of a Nuclear Accident. Member States work towards establishing a global nuclear liability regime that addresses the concerns of all States that might be affected by a nuclear accident with a view to providing appropriate compensation for nuclear damage. Member States give due consideration to the possibility of joining the international nuclear liability instruments as a step toward achieving such a global regime.
8. *Facilitate the development of the infrastructure necessary for Member States embarking on a nuclear power programme:* Help countries planning to start a nuclear power programme to create an appropriate nuclear infrastructure based on the IAEA Safety Standards. Member States are encouraged to voluntarily host Integrated Nuclear Infrastructure Reviews (INIR) and relevant peer review missions, including site and design safety reviews, prior to commissioning the first nuclear power plant.
9. *Strengthen and maintain capacity building:* Member States with nuclear power programmes and those planning to embark on such a programme strengthen, develop, maintain and implement their capacity building programs, including education, training and exercises at the national, regional and international levels; to continuously ensure sufficient and competent human resources necessary to assume their responsibility for safe, responsible

and sustainable use of nuclear technologies; and to incorporate lessons learned from the accident into their nuclear power programme infrastructure.

10. *Ensure the on-going protection of people and the environment from ionizing radiation following a nuclear emergency:* Member States, the IAEA Secretariat and other relevant stakeholders facilitate the use of available information, expertise and techniques for monitoring, decontamination and remediation of both on- and off- nuclear sites and the IAEA to consider strategies and programmes to improve knowledge and strengthen capabilities in these areas; to facilitate the use of available information, expertise and techniques regarding the removal of damaged nuclear fuel and the management and disposal of radioactive waste resulting from a nuclear emergency; and to share information regarding the assessment of radiation doses and any associated impacts on people and the environment.
11. *Enhance transparency and effectiveness of communication and improve dissemination of Information:* Member States, with the assistance of the IAEA Secretariat, strengthen the emergency notification system, and reporting and information sharing arrangements and capabilities; to enhance the transparency and effectiveness of communication among operators, regulators and various international organizations, and strengthen the IAEA's coordinating role in this regard, underlining that the freest possible flow and wide dissemination of safety related technical and technological information enhances nuclear safety. The IAEA Secretariat facilitates and continues sharing with Member States a fully transparent assessment of the accident at TEPCO's Fukushima Daiichi Nuclear Power Station, in cooperation with Japan. The IAEA Secretariat and Member States, in consultation with the OECD/NEA and the IAEA International Nuclear and Radiological Event Scale (INES) Advisory Committee review the application of the INES scale as a communication tool.
12. *Effectively utilize research and development:* To conduct research and development in areas highlighted by the accident, such as extreme natural hazards, management of severe accidents, station blackout, loss of heat sink, spent fuel accidents, and post-accident monitoring systems in extreme environments.

REFERENCES

1. IAEA (International Atomic Energy Agency), 2003. Fundamental Safety Principles: Safety Standards Series No. NS-R-3, Vienna, Austria.
2. IAEA (International Atomic Energy Agency), 2006. Fundamental Safety Principles: Safety Fundamentals, IAEA Safety Standards Series No. SF-1, Vienna, Austria.
3. IAEA (International Atomic Energy Agency), 2011. IAEA Action Plan on Nuclear Safety. Vienna, Austria.
4. IAEA (International Atomic Energy Agency), 2012a. Power Reactor Information System. http://www.iaea.org/pris. Vienna, Austria.

5. IAEA (International Atomic Energy Agency), 2012b. Energy, Electricity and Nuclear Power Estimates for the Period up to 2050, 2012 Edition. Reference Data Series No. 1 (RDS-1/32). Vienna, Austria.
6. Taipei Times, 2012. Nuclear power phase-out plan confirmed. 15 May 2012. http://www.taipeitimes.com/News/taiwan/archives/2012/05/15/2003532872
7. WANO (World Association of Nuclear Operators), 2012. Performance Indicators 2011. London, United Kingdom.

OUTLOOK FOR NUCLEAR POWER IN JAPAN

TATSUO MASUDA
The Graduate School, Nagoya University Commerce and Business
Tokyo, Japan

HISTORICAL BACKGROUND

History matters in discussing the nuclear future of Japan as the Fukushima nuclear disaster of March 2011 and the resulting collapse of credibility in nuclear safety has its root causes in the early phase of nuclear power introduction into Japan. When the news ran through the world that Otto Hahn and Fritz Strassmann conducted experiments to discover nuclear fission at the end of 1938, the world military organizations started to develop nuclear weapon almost at once as it was on the eve of World War II. During this period, Japan also started to make research on nuclear weapons like other countries. "Rikagaku Kenkyujo" (The Institute of Physical and Chemical Research) played a key role in the effort to develop nuclear bombs. The Imperial Army strongly backed up this project. However, the research in Japan was no comparison to the Manhattan Project in its size and speed, with no tangible outcomes contributing to actual military application any time soon.

After the unconditional surrender of Japan in August 1945, the General Head Quarters, the Supreme Commander for the Allied Powers, completely forbid any research related with nuclear in Japan. The moratorium was lifted when Japan recovered its independence in April 1952. The complete absence of nuclear research for this prolonged period might have left invisible but significant impact on the course of nuclear development. Looking back, there were seven groups, among others, influencing the nuclear future of Japan particularly in the early years: scientists, politicians, the government, equipment and engineering industry, utility industry, the general public and the USA.

Scientists

In the early days after the lifting of the moratorium, scientists were largely divided on the way to do with nuclear. This was typically the case when the Science Council of Japan discussed whether or not they should propose the foundation of the Atomic Energy Commission to the government in the second half of 1952. Those who supported this proposal intended to publicly authorize once-forbidden nuclear research under their own initiative. Those who opposed this initiative worried about the risk of possible involvement with the U.S. military strategy through nuclear research. The U.S. atomic bombing of Hiroshima and Nagasaki followed by an intense nuclear arms race between the USA and the USSR undoubtedly cast shadows over the Japanese nuclear scientists.

Because of such division, scientists were not able to take a leading role in directing the nuclear future of Japan. It was a group of politicians who took the lead by introducing the first budget for nuclear promotion in 1954. Thereafter, scientists generally played an advisory or secondary role until today.

Politicians

A group of politicians moved impressively fast setting a clear direction in favor of the use of nuclear, particularly for power generation. In March 1954, barely three months after the Eisenhower's speech on "Atoms for Peace" at the United Nations in December 1953, the first draft budget to promote nuclear research and utilization was proposed by politicians led by Member of Parliament Yasuhiro Nakasone. The budget became effective in April 1954, naturally engaging the government to that direction.

Matsutaro Shoriki, President of Yomiuri Newspaper and Nihon TV Group, also acted very powerfully towards introduction of nuclear power. He was appointed as the first Chairman of the Atomic Energy Commission in 1956 soon after being elected as Member of Parliament. He was afraid that the public protest against U.S. nuclear test in the Bikini Atoll in 1954 might lead to the spread of communism in Japan, and worked closely with the Central Intelligence Agency to mitigate anti-nuclear and anti-U.S. sentiment in Japan using every possible means including the media of his own. In a massive campaign launched by his group hailed nuclear energy as "the third fire" in contrast to combustion (the first fire) and electricity (the second fire). He strongly advocated the importation of foreign technology and equipment to short-cut the process of installing nuclear power generation in Japan.

In 1956, Hideki Yukawa, Nobel laureate in physics, opposed to this short-cut by stressing the importance of the basic research on nuclear reactor in Japan without jumping to established technology abroad. Yukawa's argument was stonewalled by Shoriki, which is symbolic of power imbalance between politicians and scientists mentioned above. His influence was used to form a cohesive alliance between politicians and the nuclear-related industry as well.

The Government

The government seemed to have been totally unprepared when the first nuclear budget was approved. Its amount was almost nominal, as little as 235 million yen (0.65 million U.S. dollars), however, only 60 million yen were spent. The officials did not know what to do with this new budget, which fell on their shoulders out of blue. The budget increased by over 30 times in four years driving the government further towards nuclear promotion. In 1956, the Atomic Energy Basic Law and related legislations paved the way for several government institutions to promote peaceful use of nuclear: the Agency for Science and Technology, the Japan Atomic Energy Research Institute, and the Nuclear Fuel Corporation.

Nuclear-related government institutions have expanded and evolved ever since. Local governments became increasingly involved with nuclear issues as the number of nuclear plants increased.

Equipment and Engineering Industry

The business community quickly responded to the initiative by politicians as a business opportunity. In March 1956, the Japan Atomic Industry Forum kicked off encompassing a broad range of companies with interest in nuclear business, notably power generation. There also emerged groups of companies such as Mitsubishi, Hitachi, Sumitomo and Mitsui. These groups were normally centered around heavy electric equipment manufactures except for Sumitomo group. The alignment was made with big foreign companies such as Westinghouse and General Electric.

However, due to the absence of nuclear research for six years, the Japanese companies took many years to establish their own technological build-up. Therefore, in the early phase of the introduction of nuclear power generation, turnkey contracts were applied. Today, Japanese nuclear equipment and engineering companies are leading the international business, but there was a long way for them to reach this stage.

Utility Industry

Japan suffered from power shortages in the early 1950s as power supply could not catch up with the speed of economic growth. The rolling blackouts were imposed frequently. Due to the limitation of sites suitable for hydro power generation, the expectation for nuclear option increased as a brand new source of energy. Turnkey contract, largely promoted by General Electric, looked attractive to power utility companies as it offered an access to matured technology and equipment at relatively reasonable costs. In a TV interview broadcasted in September 2011, a former vice president of the Tokyo Electric Power Company said that no one cared much about the location of back-up generators for cooling system of Fukushima Daiichi No.1 reactor, which is symbolic of a downside of such contract. This reactor was built by General Electric on a turnkey basis, leaving everything to the hands of the supplier until completion.

Ever since, utility companies accumulated more knowledge and experience in operating nuclear plants together with the growth of indigenous nuclear equipment manufacturers and engineering companies.

The General Public

The public opinion was generally against, or at best, extremely careful about nuclear for several years after the war. On top of the disaster in Hiroshima and Nagsaki, the exposure of the Daigo Fukuryu Maru, a Japanese tuna fishing boat, to the nuclear fallout from the U.S. hydrogen bomb test in the Bikini Atoll (March 1954) fueled the anti-nuclear sentiment in Japan. Under such circumstances, the promotion of nuclear power generation, first, needed the mitigation of the public sentiment against nuclear. Various efforts were made by the government and utility companies. Over time, Japanese public sentiment shifted in favor of nuclear power generation. The perception of "nuclear safety" has gradually been planted in the mind of people almost as a legend.

This is why the public reactions were so acute to the Fukushima nuclear disaster, which has shuttered the belief in almost legendary nuclear safety advocated by the government and utility companies over decades.

The USA

The U.S. influence was critical on nuclear power generation in Japan. Firstly, it was to the direct benefit of the U.S. to export their nuclear technology and equipment to Japan. On top of this, exporting nuclear fuels, namely enriched uranium, offered an important business opportunity to the uranium mining industry there. Second, in the Cold War era, it was imperative for the U.S. to promote the economy of countries in the Western block to demonstrate the excellence of the West vis-a-vis the communist block. To strengthen Japanese economy through introducing nuclear power was in line with U.S.

strategy.

This is why the U.S. government made a continuous and serious effort to alleviate the allergy to nuclear in Japan. There are many evidences that the CIA played an important role in Japan in support of the U.S. strategy as above. Some government officials, who were involved with the promotion of nuclear power in the late 1960s and the early 1970s, hinted that Japan was not in a position to be able to say "No" to the USA on nuclear policy in those days.

POLICY IMPLICATIONS

Nuclear energy in the Japan's energy mix had multiple policy implications.

The evolution of energy mix

The tables below show the evolution of Japan's energy mix, where the share of nuclear grew impressively. After commissioning the first commercial nuclear reactor in 1966, the number increased to 55 by the mid-2000s. Considering the following constraints, this growth could be called as "miracle":

1. Japan is the only country having suffered from atomic bombing, where the allergy to nuclear was extremely acute.
2. Its land is only 378 thousand km^2, the 64th largest in the world, with many islands.
3. It is very mountainous and only some 30% of the land is suitable for agricultural, industrial and residential use.
4. With a population over 127 million, the population density is 336 person/km^2, the 34th highest in the world.
5. It is well known for frequent earthquakes (occasional tsunamis) with various active fault lines running across the country.

Table 2-1. Japan's Energy Mix (Total Primary Energy Supply).

(%)	1952	1962	1973	1980	1990	2000	2009	2030**
Coal	49	35	18	17	17	18	21	17
Oil	11	47	78	66	57	51	46	31
Natural Gas	—	1	2	6	10	13	18	16
Nuclear	—	—	1	5	10	12	11	24
Hydro	32	13	2	5	4	3	3	13***
Renewables	8*	4*	—	1	2	2	1	

Note: * wood/charcoal ** The Basic Energy Plan (June 2010) ***Hydro+Renewables (Source: METI)

Table 2-2. Japan's Energy Mix (Power Generation).

(%)	1952	1962	1973	1980	1990	2000	2009	2030**
Coal		33	8	44	10	18	25	2
Oil	20*	22	73	5	27	9	6	11
Natural Gas		0	2	17	24	27	30	13
Nuclear	0	0	2	17	27	34	29	53
Hydro	80	44	14	17	12	10	8	21***
Renewables	—	—	—	—	—	1	1	

Note: *Total share of thermal power **The Basic Energy Plan (June 2010) ***Hydro+Renewables (source: METI)

Expectations for Nuclear Power

Why had the miracle had really taken place? The process of the introduction of nuclear power was not necessarily made in an orderly or rational manner as seen in Chapter 1. Some administrative decisions were made, in a somewhat "patch-work manner", to catch up with high expectations for nuclear power. The real driver was perceived merits of nuclear power for a resource-poor but fast-growing economy. Indeed, its GDP growth was extremely fast like China today shown as follows, when Japan was shifting its course towards nuclear.

Annual average GDP growth in Japan

1952–1961	9.7%
1962–1972	10.2%

The following reasons were publicly advocated to justify or support nuclear option. Interestingly enough, the emphasis varied reflecting the needs of time.

1. Nuclear saves foreign currencies than importing oil (the 1950s ~ the mid 1960s).
2. Nuclear is cost effective energy (the 1950s ~ the mid 1990s).
3. Nuclear contributes to the energy security (the mid 1960s ~ today).
4. Nuclear is "quasi indigenous" energy (the mid 1970s ~ today).
5. Nuclear is climate-friendly energy (the 1980s ~ today), policymakers. Japan's energy self-sufficiency will go down from 19% to only 4% without.

Among those, the "quasi indigenous" nature of nuclear power strongly appealed to counting nuclear power (note: prior to the Fukushima nuclear disaster). From climate policy perspective, nuclear energy gained renewed importance due to its CO_2-free nature. This is evident from the fact that nuclear power was given extremely high shares for 2030 in the basic energy plan issued by the government in June 2010 as shown in the above two tables. This plan is currently under outright review for obvious reasons.

POST-FUKUSHIMA DEVELOPMENTS

The impact of the Fukushima nuclear disaster has been far bigger than that of the oil shocks of the 1970s. The latter was due to the actions by the OPEC hitting oil importing countries worldwide and Japan was not alone. The former was considered to be a man-made disaster triggered by the earthquake and tsunami in March 2011, which happened in a country with extremely advanced technology and excellent safety records. The legendary confidence in nuclear safety in Japan has been utterly shattered.

Various Developments

Many things took place in parallel after the Fukushima nuclear disaster, which is self-explanatory of the wide-ranging impact of the disaster.

- Planned blackouts and rigid power conservation in the Kanto region with over 30 million people, and to a lesser extent nationwide.
- Assistance to the people evacuated from the radioactive contaminated areas.
- Radioactive decontamination of affected areas.
- Compensation for the damages from the nuclear disaster.
- Nationalization of TEPCO to avoid bankruptcy (completed on 27 June 2012).
- Procurement of LNG and heavy fuel oil for power generation to make up for the loss of nuclear.
- Feed-in tariff to promote renewable energy introduction (implemented in July 2012).
- Strengthening of nuclear safety and regulatory regime (from September 2012).
- Repositioning of nuclear in the energy mix (to be decided in August 2012).
- Restarting of nuclear power plants (the first restarting on 1 July 2012).
- Parallel investigations on the causes of the nuclear accident.

Parallel Investigations

One of the important initiatives as above is the investigation on the causes of the disaster. It ranged from a private initiative (δ) to a parliamentary one (χ) as below.

δ The Independent Investigation Commission on the Fukushima Daiichi Nuclear Accident (Founded by The Rebuild Japan Initiative Foundation: Report on 28 February 2012).

α The Investigation Committee on the Accident at the Fukushima Nuclear Power Station of Tokyo Electric Power Company (established by the Cabinet: Interim Report on 26 December 2011, Final Report expected on 23 July 2012).

β The Fukushima Nuclear Accidents Investigation Committee (established by TEPCO: Interim Report on 2 December, 2011: Final Report on 20 June 2012).

χ The Fukushima Nuclear Accident Independent Investigation Committee (established by the Parliament: Report on 5 July 2012).

Each of these contributed to unveil the complexity of the situation as well as organizational and managerial problems in multiple ways. The most impressive message through all these, so far, is the following from the report by the committee established by the parliament: "It was a profoundly manmade disaster—that could and should have been foreseen and prevented." In my understanding, this will be the most accurate analysis of the causes of the disaster, which should be reflected in all aspects of energy business and policymaking.

The New Regulatory Regime

The law to establish the Nuclear Regulatory Commission was enacted on 20 June 2012 enabling to start a new nuclear regulatory regime from September. The Commission, supported by the Nuclear Regulatory Agency, is expected to unify the regulatory process which used to be stretched across several institutions including the Nuclear and Industrial Safety Agency (NISA) under METI. The authorities given to the Commission is far more powerful compared to the current regime. The lessons form the disaster and the best practices of foreign countries seem to be duly taken into account in this initiative.

Restarting of nuclear plants

On 5 May 2012, the Tomari nuclear plant (in Hokkaodo) stopped for regular maintenance and inspection required by law in every 13 months. Since then, there were no nuclear plants operating in Japan. With the arrival of summer season at hand, there was a strong drive by the government and the business community towards restarting particularly in the service area of the Kansai Electric Power Company (KEPCO), which used to rely on nuclear about a half of its power supply.

On 8 June, Prime Minister Noda made an exceptional press statement on the need to restart No.3 and 4 reactors at KEPCO's Oi nuclear power plant by saying "The measures and system to secure safety are already in place to prevent nuclear accidents from happening even if hit by Fukushima–scale earthquakes and tsunami.—It is my judgment to restart No. 3 and 4 reactors of the Oi nuclear power plant to protect the life of the people.—". This statement let the prefectural governors agree on restarting, which led to the restart of No.3 unit on 1 July. Although enthusiastically welcomed by the business community, this decision fueled the anti-nuclear movement in Japan, which is spreading across Japan. There are a few more reactors expected to restart in the course of this year.

This in itself is very controversial as this was done before the establishment of new regulatory regime. Nor this decision did not wait for the release of the reports by the investigation committees established by the Parliament and the Cabinet.

Energy Democracy

The Japanese energy policy had been decided previously by a limited number of people, namely government officials, energy companies, academics and well-informed politicians. It was so far so good until the credibility of nuclear safety collapsed as above. For the first time, "energy democracy" has been brought forth in this country. There is no guarantee that democracy will lead us to a better place. However, a good thing is that more people pay attention to energy issues.

In this new environment, one prefecture may choose to completely close existing nuclear power plants. A neighboring prefecture may close only a half of its nuclear power plants, while the other may allow building of new plants on top of existing ones. Also, when the smart grid may network most of energy consumers, they may chose to use energy sources of their preference either from economic or philosophical reasons, from fossil energy, renewables to nuclear. This is what energy democracy is all about.

Scenario Discussion

The collapse of the credibility on nuclear safety and the rise of energy democracy made discussion on the Japanese energy future unprecedentedly dynamic and diversified. As the pre-Fukushima energy outlook has become unrealistic, the government started to renew the Basic Energy Plan from scratch.

The Fundamental Issues Subcommittee of the Advisory Committee for Natural Resources and Energy kicked off the discussion in October 2011 among members with diversified views on nuclear. As expected and very naturally, the Subcommittee was not able produce one energy scenario, therefore, in May, they came up with several scenarios on the energy mix in 2030 with nuclear share in power generation ranging from zero to 25 %, which were submitted to the Energy and Environment Council set by the Cabinet.

On 29 June, the Council proposed the following three scenarios (Table 4-1) for public hearings and comments. After nation-wide discussion on these scenarios, the government plans to decide an innovative energy and environment strategy by end-August followed by the Basic Energy Plan, the nuclear energy policy guideline and the green policy guideline by the end of this year.

Table 4-1.Three Scenarios on the energy mix in 2030.

	Nuclear (%)*	**Renewables (%)***	**Fossil Fuels (%)***	**Power Gen (Twh)**	**TFC (mkloe)**	**GHG (vs1990)**
Scenario 1	0	35	65	1000	300	≥ 23%
Scenario 2	15	30	55	1000	310	≥ 23%
Scenario 3	20-25	25-30	50	1000	310	≥ 25%
2010(Actual)	26	10	63	1100	390	≥ 0.3%

Note: * share in power generation.

Source: The Energy and Environment Council

A modest seat for nuclear in future

Due to the nature of democracy, it will be difficult to predict how the people may treat nuclear. The radioactive contamination is spreading across the country albeit minimally in remote regions, but no one is sure of its long-term effect yet. Renewable energy may increase its share faster than expected, thanks to the feed-in tariff. Technological innovation is accelerating in energy efficiency and saving, which may decrease the total energy consumption of this country much more than expected. There could be another innovation in life style as well after all these experiences.

All of these may suggest that the people's choice of the energy mix today could be different from that a few years later, which may force the government to rewrite the energy strategy adopted today. However, one for sure is that nuclear will secure a modest seat in the energy mix if the lessons form the disaster are properly taken in energy business and policymaking.

OBSERVATION

Nuclear power has established a solid position in the Japanese energy mix. In view of various constraints on the introduction of nuclear power as discussed above, this achievement is worthwhile being called as a miracle. Foreign observers might argue this as the result of the following:

a) Farsighted and thoughtful policymaking,
b) Powerful administration to implement policies,
c) High level of technology and manufacturing/engineering skills,
d) Allocation of affluent financial and human resources both in public and private sectors as national priority,
e) Harmonious collaboration between politicians, the government and the industry,
f) Public support to nuclear option in general,
g) Hard working of the Japanese people whoever involved.

Among these, (c), (d) and (g) will undoubtedly be the case. As well known, Japan is one of the top runners in nuclear power technology and the forerunner in the quest for full nuclear fuel cycle. However, others including (e) and (f) should be treated with some reservations. The government officials were unprepared when they were thrown into the world of nuclear in the 1950s, and this rather disorganized start had a lingering effect on nuclear administration afterwards. What is more regrettable is the somewhat passive role played by scientists. If they took a strong initiative in the 1950s, they would have secured more time for basic research by themselves and shed enough light on the issue of safety as they were in a better position to understand entailed risks of nuclear option than anyone else. Utility companies and the government remained complacent with the self-made belief in nuclear safety and constantly downplayed the risks. As an example, the government and TEPCO were already warned of the risk of big tsunamis hitting the Fukushima Daiichi site during the debate in the Parliament in March 2006, but both failed to take any meaningful measures.

Collaboration among key players was not always in place. Even within the government, the Agency for Science and Technology (nuclear development for improving the level of science and technology) and the Ministry of Economy, Trade and Industry (nuclear policies for energy use and nuclear safety) were often struggling for power over nuclear administration. Such power struggles within the administration, although this may not be limited to Japan, must have alleviated the efficiency and effectiveness of their performance and diverted their attention away from more important tasks on a national basis. This is the very reason to unify the regulatory regime under the Nuclear Regulatory Commission to be functional from September.

The Japanese energy structure/system, which was considered to be one of the best models, is under tight scrutiny. In other words, this is an unprecedented opportunity for Japan to develop “a new energy model”, which should be full of wisdom and new thinking to benefit all in the world. No country but Japan can do this by taking stocks from its successes as well as failures. However, in order to achieve this, all the stakeholders should contribute to this process by gathering their wisdom as it’s ongoing.

Lastly, on 10 July, NHK (the Japan Broadcasting Corporation) released the result of a questionnaire survey (6-7 July) on the scenarios as shown in Table 4-1. Very interestingly, 34% went for Scenario 1 (zero nuclear), 43% for Scenario 2 (15% nuclear) and 12% for Scenario 3 (20-25% nuclear).

REFERENCES

1. T. Arima (2008) Nuclear Power Plant, Shoriki and CIA, Shinchosha.
2. E. Chanlett-Avery (2008) "Japan's Nuclear Future: Policy Debate, Prospects, and U.S. Interests", Congressional Research Service-Order Code RL34487.
3. K. Matsui, "Energy Policy Review for 50 Years after WWII", Denryokushinposha.
4. K. Matsui (1995) "Energy Policy Review for 30 Years after the Oil Shocks", Energy Forum (2003).
5. NHK "The way towards the Fukushima Nuclear Disaster", NHK TV program (18 and 25 September 2011).
6. Richard Samuels (1987) "The Business of the Japanese State: Energy Markets in Comparative and Historic Perspective (Cornell Studies in Political Economy)", Cornell University Press.
7. The Daily Yomiuri's series, "Planning National Strategies—Resources and Energy" (April-May 2005).
8. H. Yoshioka (1999) "Social History of Nuclear Power: The Japanese Case", Asahi Shinbun.
9. "Nuclear Energy: The dream that failed", The Economist (10 March 2012).
10. Annual Report on Energy, "Agency for Natural Resources & Energy", METI, Japan (2010).
11. IEA/OECD "Energy Policies of IEA Countries: Japan 2003 Review", IEA (2003).
12. IEA/OECD "Energy Policies of IEA Countries: Japan 2008 Review", IEA (2008).
13. K. Mastui (2011) "Overcoming the Fukushima Nuclear Disaster, The Age of Global Energy Shift", Energy Forum.
14. METI "The Strategic Energy Plan for Japan", Agency for Natural Resources & Energy, METI, Japan (2010).
15. NHK "The way towards the Fukushima Nuclear Disaster", NHK TV program (18 and 25 September 2011).
16. The Independent Investigation Commission on the Fukushima Daiichi Nuclear Accident (Report on 28 February 2012).
17. The Investigation Committee on the Accident at the Fukushima Nuclear Power Station of Tokyo Electric Power Company (Interim Report on 26 December 2011).
18. The Fukushima Nuclear Accidents Investigation Committee (by TEPCO: Interim Report on 2 December, 2011: Final Report on 20 June 2012).
19. The Fukushima Nuclear Accident Independent Investigation Committee (Report on 5 July 2012).

OVERVIEW OF NUCLEAR POWER DEVELOPMENT IN CHINA

TONG JIEJUAN AND SUN YULIANG
INET, Tsinghua University
Beijing, P.R. China

ENERGY AND POWER ECONOMY

Continuous rapid economy growth in China has been taking place in the past years, along with the remarkable increase in primary energy and power consumption (Figure 1). According to the latest statistics,[1] the annual primary energy consumption in China reaches 2613 Mtoe in 2011 with an average annual increase of 9.02% over the period of 2000-2011. Electricity generation and consumption experienced even more vigorous growths in the past decade.

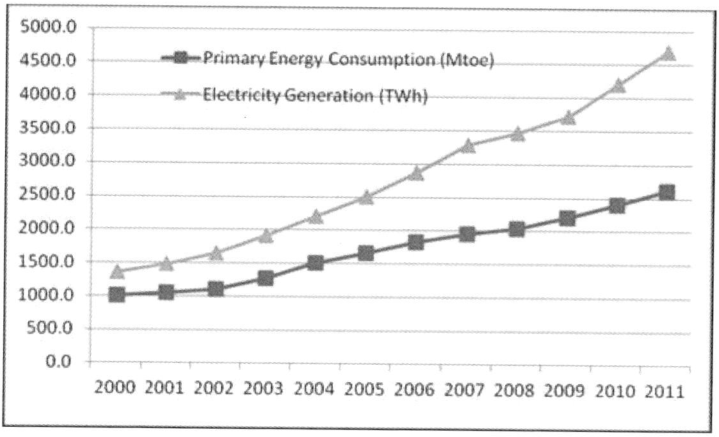

Fig. 1: Primary Energy Consumption and Electricity Generation (2000-2011).

In order to catch up with the social-economic development, China has taken a series of reformatory measures in the energy industry. Energy supply capability has been improved significantly. For example, the installed electricity generating capacity in China has been developed with an average annual increase of 9.2% and reached 1056 GW in 2011[2] (Figure 2).

According to the general social-economic development goals which Chinese government has set for the next decades,[3] the per capita GDP by 2020 will quadruple relative to that in 2000, and by the middle of this century, the per capita GDP value will increase from the current value of about $3,000 USD to $25,000 USD, about the level of some moderately developed countries. Therefore, it is expected that primary energy and electricity consumption in China will continue to grow significantly in the next decades.

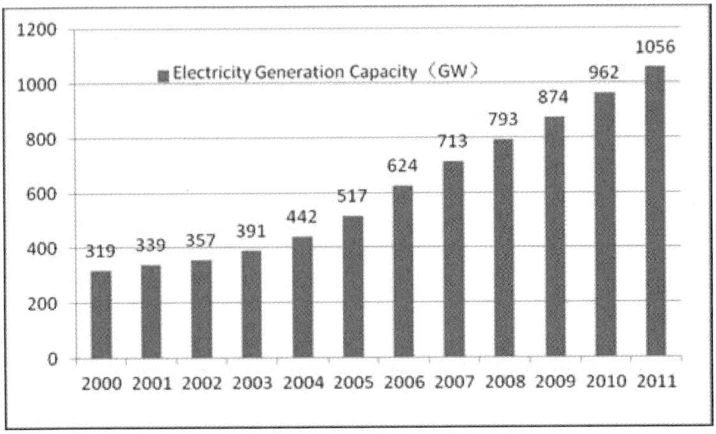

Fig. 2: Electricity Generation Capacities (2000-2011).

Besides the challenges brought by the economic development demand, the challenges from the energy restructure are notable either. One of the most striking characteristics of China's energy economy is its heavy reliance on coal (Figure 3),[1] which takes up about 70% in primary energy consumption. Hydropower contributes to 17.67%, and the rest is contributed by oil, gas, nuclear and some renewables. This coal-dominant energy structure leads to huge pressures in terms of environmental protection and carbon dioxide emission reduction.

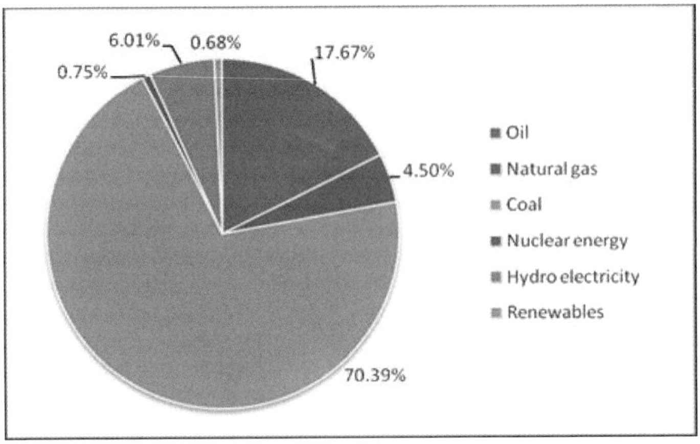

Fig. 3: Primary Energy by fuel in 2011.

These boundary conditions set by the social-economic growth and clean energy policies make the energy economy development in China rather challenging. The following two aspects would be of special significance in the mid- and long-term energy development strategies in China:

- Promotion of energy saving and overall control of total energy consumption.
- Acceleration of energy economy restructuring towards a low-carbon energy economy. The share of non-fossil energy forms is expected to reach 15% by 2020 and over 30% by 2050.

What roles should nuclear power play in the future? Before Fukushima, very few people questioned about the important role of nuclear energy. Fukushima crisis then has led to some discussions and debating. It is believed that China will resume its nuclear power development program based on enhancing nuclear safety.

PAST DEVELOPMENT AND CURRENT STATUS OF NUCLEAR POWER

China's civil nuclear power development started in 1970's. The first nuclear power plants were built in 1980's and put into operation early 1990's. But the pace of nuclear power development had not been particularly fast until several years ago when China's government shifted its nuclear power policy from "moderate" to "active" development. In 2006, China's central government approved the plan for nuclear power development until 2020, which sets up the framework for introducing and constructing Generation III advanced water cooled reactors and for constructing moderate numbers of improved Generation II nuclear power plants. That plan foresees 40GW of nuclear power to be constructed and some 18GW under construction by 2020. Under that framework, nuclear power construction has been very much active. Currently there are 15 nuclear power reactors under operation, having a total generating capacity of 12.6GW. 26 units are under construction with a total capacity of about 29.4GW. Most of these units are improved Generation II reactors, except for 4 AP1000 units and 2 EPR units. Besides, there are still around 20 nuclear reactor units under planning with the total estimated capacity of 20GW.

The nuclear power plants under operation and construction in China have already incorporated a lot of technical and safety improvement and experience feedback since the Three Mile Island and Chernobyl nuclear accidents. And Chinese nuclear power plants have gotten excellent operation records over the past years.[4] According to WANO operational performance statistics, Chinese nuclear units are generally at high international levels. The average load capacity factor over the five years is 85.7%. Some units are at advanced levels and even top ranking.

There are also advanced nuclear power technology development programs. These programs include the development of advanced light water reactor technology, sodium cooled fast reactor technology, high temperature gas cooled reactor technology and some other even more advanced technologies such as accelerator driven systems or molten salt reactors. On 21st of July 2011, the China Experimental Fast Reactor, having a thermal power of 65MW and electric output of 20MW, was constructed and successfully put into power operation. The first industrial scale demonstration nuclear power plant of high temperature gas cooled reactor with the pebble bed design (HTR-PM) is being projected in Shandong province. The project has been part of the national mid- and long-term science and technology program till 2020 which requires that by 2020 the HTR-PM demonstration plant be finished and a good industrial basis be established for future applications.

As the indispensable part of the nuclear energy development, the Chinese nuclear safety regulatory authority was established in the 1980's. Good regulatory experience and practices in the world have been studied and reflected into the regulatory system in China. In particular, the nuclear safety standards have been derived mostly from IAEA standards series which have absorbed worldwide experience in nuclear safety.

ACTIONS AFTER FUKUSHIMA AND FUTURE OUTLOOK

The Fukushima accident has resulted in profound impact on nuclear power programs in many countries in the world including China. Right after the outbreak of the Fukushima nuclear accident, the central government of China made, on the 16th of March 2011, several decisions in order to check and assure nuclear safety of the Chinese nuclear power programs. The decisions require immediate nuclear safety inspections and examinations of all nuclear power plants under operation and construction against most strict regulations, enhancement of safety management at operating nuclear units, enhancement of national nuclear safety programs, and adjustment and optimization of nuclear power development programs.

Nuclear safety inspections and examinations after Fukushima

The nuclear safety inspections and examinations lasted for 9 months in 2011 extensively, and were carried out to all the units under operation and construction. The inspection aspects covered include validation of external events used for siting evaluation, anti-flooding procedures and the actual capability evaluation, seismic design and seismic margin evaluation, prevention and mitigation measures in case of concurrence of multiple extreme natural events, evaluation and contingency guideline for station blackout, severe accident prevention and mitigation measures as well as the corresponding reliability evaluation, effectiveness of environmental monitoring and emergency response frameworks and so on.

The above mentioned inspections and examinations were implemented by several ways such as self-examination, expert walk-down, document inquiry and review, etc. The final findings and conclusions were approved for release formally by the State Council in June 2012. In order to increase the public acceptance and involvement, full text of the inspection report is published for public comments. The following conclusions are stated in the report.

Nuclear industry in China has certain advantages of "later start", benefiting from the international good experiences in design, construction and operation, and nuclear safety has been always regarded as the essential principle of the nuclear industry. Since it was founded in 1984, National Nuclear Safety Authority (NNSA) of China has set up the complete set of nuclear safety regulations and standards, which are closely connected with the international level. NNSA has fulfilled the independent inspection and review successfully. Chinese nuclear units have got excellent operation records, e.g., no operational events are serious than INES level 2. Generally speaking, all the operating nuclear units meet the existing nuclear safety regulations of China and the latest IAEA standards, having sufficient capability to deal with the design basis accidents and having certain capability for severe accident prevention and mitigation. Hence, the risks are under control and the safety of these nuclear units continuing operation can be assured.

With respect to the units under construction, the series of self-designed units are of higher safety over the similar units in the world, because the technologies adopted are the integration of continuous upgrade and optimization over the 30 years, based on the introduced international mature technology. Besides, AP1000 and EPR are the representatives of the new advanced nuclear units in the world. They have had sufficient considerations on severe accident prevention and mitigation from the design stage, the design safety levels have been improved further. In general, all the nuclear power plants under construction meet the existing nuclear safety regulations of China and the latest IAEA standards. Effective management has been implemented to the processes of siting, design, manufacture, construction, installation and testing. Quality assurance procedures are working as normal; the engineering construction meets the design requirements. Hence, overall construction quality is under control.

All the civil nuclear facilities have undergone sufficient investigations about the earthquake and flooding occurrence. It is very rare that the extreme nature event like the Fukushima nuclear accident can occur in China.

The report also identifies potentially vulnerable aspects and gives recommendations for modification actions in order to promote even higher safety of Chinese nuclear power plants.

- Recommended modification action No. 1: To meet with the functional requirements of reactor core cooling, spent fuel pool cooling as well as necessary post-accident monitoring capabilities given the situation of station blackout like Fukushima, measures such as providing portable power supply, portable pumps and the matched interface devices shall be taken.
- Recommended modification action No. 2: Recent researches and evaluations about the Fukushima shall be traced closely. Nuclear power plants shall refine the severe accident management guidelines, evaluate and improve the availabilities and reliabilities of the devices and systems used for severe accident mitigation. The possibility of hydrogen explosion shall be evaluated and mitigation devices shall be modified if necessary.
- Recommended modification action No. 3: Anti-flooding capabilities of doors, windows, ventilations and penetrations shall be examined one by one in case of beyond design basis flooding. Blocking and sealing measures shall be taken if necessary.
- Recommended modification action No. 4: Qinshan nuclear power plant is required to perform the specific modification action against flooding by dam height extension, anti-wave wall setup, water proof and drain measures implementation and so on. Before the modification action finishes, if there would be a high tide together with typhoon condition, the reactor shall be set back to the cold shutdown mode.
- Recommended modification action No. 5: All the nuclear power plants shall

start the external event probabilistic safety assessment (PSA), including seismic PSA and seismic margin analysis; Daya Bay nuclear power plant has the specific requirement to carry out the seismic-tsunami risk analysis and take the necessary modification actions.
Recommended modification action No. 6: All the nuclear power plants shall enhance the maintenance and management of seismic monitoring and recording instruments and devices to ensure the effectiveness of monitoring system; Operator actions after the earthquake shall be improved to increase the plant seismic emergency response capability.

- Recommended modification action No. 7: All the nuclear power plants shall improve the environment monitoring capability and emergency control center functions under severe accidents; Response guidelines shall be defined to deal with multiple units in the site entering the emergency state simultaneously; Emergency control capability and coordination and distribution solutions of human and material resources shall be evaluated.
- Recommended modification action No. 8: Cooperation companies such as CNNC and China Guangdong Nuclear Power Cooperation (CGNPC) shall improve the emergency response capabilities in the cooperation level. These forces shall also be part of the national nuclear emergency response management. Emergency resources and capabilities shall be shared over the region or throughout the country by effective coordination and organization.
- Recommended modification action No. 9: Information release procedures shall be refined according to the characteristics of nuclear accident. Nuclear facility information transparency shall be enhanced to increase the public acceptance.
- Recommended modification action No. 10: Recent development and trend of international nuclear safety regulations and standards shall be studied and traced, so that we can keep up with the advanced level in the world and promote the continuous safety improvement of civil nuclear facilities.
- Recommended modification action No. 11: Management to the restricted planning zones around the nuclear power plants shall be enhanced to control the population increase in the zones strictly. It is forbidden to set up and expand the big enterprises and resident living areas until the proposal would be approved by national development and evolution agency, national energy administration and NNSA.

Nuclear safety plan

Another important action after Fukushima is the formulation of nuclear safety plan[5] in China, which has been just approved by the State Council together with the nuclear safety inspection report. The nuclear safety plan sets up the overall working goals for nuclear and radiation safety by 2020, and these working goals are specified into detailed objectives, and corresponding support projects and programs are also planned for each of them.

Objective in the field of nuclear facility safety improvement

Operating units shall maintain the excellent operation performance. INES level 2 events shall be avoided; events and accidents beyond INES level 3 and more shall be prohibited definitely. New nuclear units shall have sufficient measures for severe accident prevention and mitigation. Severe core damage frequency per reactor year shall be lower than 10^{-5}; frequency of large release of radioactive material per reactor year shall be lower than 10^{-6}.

Objective in the field of nuclear accident defense

Safety modification actions shall be taken to increase the capabilities against external events and severe accident prevention and mitigation.

Objective in the field of emergency response

The capability construction of emergency control, emergency response, emergency monitoring and emergency technical support at different levels of government and organizations shall be enhanced accordingly. Engineering contingency capabilities in case of nuclear accident shall be established in a consistent and coordinative manner. Emergency resources and equipments shall be increased.

Objective in the field of nuclear safety regulation

The national R&D base for nuclear and radiation safety regulatory techniques shall be completed primarily, as well as the technical support platform, to own the relatively independent and complete capabilities of safety analysis and assessment, calculation verification and experimental verification. National network of radiation environment monitoring shall be completed, reaching the capability construction criteria by 100% in the province level and national level.

The new nuclear safety plan foresees that in 2020, safety of the nuclear facilities under operation and construction will be enhanced continuously. All the new units after 13th five year plan shall strive from the design for the practical elimination of the possibility of large release of radioactive materials.

Nuclear safety plan is actually in cooperation with the nuclear safety inspection report. It defines phased arrangements (immediate, by 2013 and by 2015) of the modification action recommendations proposed by the inspection report for the operating units, as follows. There are similar arrangements requirements also for the units under construction.

- Immediate modification actions:
  1. Implementation of anti-flooding measures of doors, windows, ventilations, penetrations based on the careful re-examination one by one.
  2. Installation of portable power supply, portable pumps and new adequate interfaces, taking into account the balanced requirements of reactor core cooling, spent fuel pool cooling, prevention of seal LOCA and necessary post-accident monitoring.

- Effectiveness assurance measures to the seismic monitoring and recording system, and the improvement of the seismic response capabilities in the nuclear power plants.

- By the end of 2013:
    1. Put into effect the flooding prevention measures of each nuclear power plant based on the possible scenario evaluations; Qinshan Nuclear power plant shall finish its specific modification project against flooding.
    2. Complete the re-evaluation to the seismic and tsunami effect on the coastal nuclear power plants, as well as the necessary modification.
    3. Develop severe accident management guidelines for each nuclear power plant and put it into effect.
    4. Evaluate the availabilities of systems and equipments used for severe accident and the possibility of hydrogen explosion, and take modification actions accordingly.
    5. Carry out safety margin analysis and evaluations under external events
    6. Carry out the research about the response solution in case of multiple units entering emergency states simultaneously.

- By the end of 2015:
    7. Carry out the external event PSA of the nuclear power plants.

In order to assure the implementation of the plan, the plan also sets up a set of management assurance measures in the following areas.

1. Accelerate the development process of the law of atomic and the law of nuclear safety, as well as the standards and guidelines, to refine the regulation foundations.
2. Optimize the management and regulatory structure and enhance the regulatory efficiency.
3. Improve policies and institutions, especially for licensing permit and nuclear safety responsibility clarification among the nuclear cooperation company, the licensee company and the professional service company.
4. Promote the nuclear safety culture framework to assure the effective quality assurance implementation with all the parties involved in nuclear matters.
5. Enhance the human resource training and education ,including establishing the shared network by government, universities, commercial training agencies and utilities.
6. Enhance the international collaboration activities in the fields of cooperation research, information exchange, experience feedback, peer review, emergency response and assist and so on.
   Expand the pubic involvement to increase the public confidence, e.g., building the information communication platforms in a open and transparent manner, increasing the nuclear knowledge spreading in basic educational level, and the public relationship maintain solutions in case of nuclear crisis
7. Ensure the financial support and investigation based on the sufficient utilization of the government leading roles and multiple investigation mechanisms.

There are still other work items underway. Obviously, a new round of discussions of safety enhancement and debates on development pace and scale of nuclear power is taking place, which is actually very much beneficial for the healthy and sustainable development of nuclear power in the future. The latest nuclear development plan has not been approved and released by the government. It seems that nuclear electricity generation capability of 60~70GW by 2020 is likely. It is generally believed that China will continue its policy of intensive utilization of nuclear energy, and nuclear power will become one of the main contributors in the Chinese energy economy, in particular in the power generation sector.

SUMMARY

China's fast economic growth in the past decades triggered correspondingly fast growth in both primary energy supply and power generation. Various studies on the mid-term and long-term economic development in China have all predicted the continued economic and energy demand growth in the coming decades. Energy supply assurance to support the economic growth has already become a tremendous challenge.

Although China is very actively implementing renewable energy programs, their contributions seem to be limited in the overall energy supply picture. Environment protection and the reduction of green-house gas emission present increasingly more rigid boundary conditions on the energy supply economy. Nuclear energy is a practical energy source which can help to greatly ease the challenge, and will be expectedly an important part in the energy economy of China.

Nuclear safety has always been the highest priority in the development of nuclear power. Nuclear safety performance in China has been very good. The excellent operational records which Chinese nuclear power plants have got during the past decades constitute a strong basis for the continuous nuclear power development. After Fukushima, a series of actions aiming at enhancing nuclear safety are being taken to draw lessons from Fukushima. A relatively large number of nuclear power plants are being operated, constructed or under planning. It is believed that nuclear power will play increasingly important roles in the Chinese energy economy.

REFERENCES

1. BP Statistical Review of World Energy, June 2012.
2. http://tj.cec.org.cn/tongji/
3. http://news.xinhuanet.com/newscenter/2007-10/24/content_6938568_3.htm.
4. Integrated safety inspection and examination report for the civil nuclear facilities in China, June 2012.
5. The 12th five year plan and the 2020's perspectives on the field of nuclear safety and radioactive pollution prevention and rescue, June 2012.

NUCLEAR POWER IN EUROPE

LADY BARBARA THOMAS JUDGE
Emeritus Chairman, UK Atomic Energy Authority
London, UK

OVERVIEW

Nuclear power in Europe accounted for approximately 15% of total energy consumption in 2011. Today however, the energy policies of the European countries vary significantly. As of January, 14 out of 27 countries had nuclear reactors. The countries that currently have reactors are: Belgium, Bulgaria, Czech Republic, Finland, France, Germany, Hungary, The Netherlands, Romania, Slovakia, Slovenia, Spain, Sweden and the United Kingdom. New reactors under construction in Finland and France, which were meant to lead a nuclear renaissance, have been delayed and are running over-budget. Specifically the situation is as follows:

Belgium

On 12 October 2009, the Belgian government decided to extend the period of operation of three reactors at the country's nuclear power plants (Doel 1, Doel 2 and Tihange 1) by 10 years (ratification by Belgium's federal parliament still needed).

In July 2012, the Belgian government agreed on a revised phase-out plan. The plan provides that the unit 1 at the Tihange nuclear power plant will operate until 2025. However Doel 1 and 2 will shut down in 2015 as scheduled in accordance with the 2003 phase-out law.

Bulgaria

New 2000-MW (2 VVER reactors) at Belene are expected to be operational by 2015 and 2016 respectively to help compensate for the closure of Kozloduy units 1-4. The project has, however, suffered delays due to difficulties in financing the construction.

Just after Fukushima, the Russian and Bulgarian governments agreed on a three-month moratorium on the construction of the Belene nuclear power plant to calculate an exact price and to fully assess seismic risks. The "stress tests" report on Kozloduy and Belene completed in January 2012 found that they both could withstand the strongest earthquake in the region. Nevertheless in March 2012, the Bulgarian government decided to abandon the construction of the Belene nuclear power plant on financial grounds.

Czech Republic

In July 2009, the Czech government adopted a strategy document that sets forth proposals for an extension of Temelin. Dukovany also opened a tender process in August 2009 for the construction of two additional reactor units at the Temelin nuclear power plant. The successful supplier is expected to be selected by 2013.

After Fukushima, the Czech government said it would carry on its new build plans. The Prime Minister and the different political parties except the Greens expressed a strong political will to complete Temelin. In July 2012, three qualified potential vendors (Areva, Atomstroyexport and Westinghouse) submitted their bids for the construction of two new nuclear units at Temelin.

Finland

A new nuclear power plant unit Olkiluoto 3 is under construction and is meant to be completed by 2013. On 21 April 2010, the Finnish government gave its "preliminary permission" to the Finnish utility TVO and to Fennovoima to build two more nuclear reactors. The decision was ratified by the Finnish parliament on 1 July 2011

The licensing process for the two new nuclear reactors to be built by TVO and Fennovoima will be carried out without any delay according to the government.

France

An EPR unit is currently under construction at a site near Flamanville in Normandy and is expected to be completed by 2014. Just after Fukushima, President Sarkozy reaffirmed the safety of the nuclear reactors; however the Socialist party won the latest elections in June 2012. During the presidential campaign, the new President, François Hollande, said he would close the Fessenheim nuclear power plant before the end of his term (2017), but no firm decision has been made yet although there has been a great deal of speculation.

Germany

The government agreed on 5 September 2010 to extend the operational duration its nuclear power plants by 12 years on average beyond 2021. The decision had been confirmed by the adoption on 28 September of a new energy strategy and had been ratified by the Parliament in November 2010.

Chancellor Angela Merkel, however, in reaction to strong political pressure from the Green Party, announced on 15 March the immediate closure, at least until June, of seven nuclear power reactors that had started operating before 1980. On 6 August 2011, eight reactors were declared permanently shut down. On 30 May, the government adopted the decision to phase out all nuclear by 2022 (ratified by Parliament on 30 June) and on 6 June it made amendments to the country's Atomic Energy Act to effect this decision.

Hungary

The government announced on 16 February 2009 a proposal to double the capacity of the country's sole nuclear power plant at Paks. It was ratified by the Parliament on 31 March 2009. The new reactors should be completed by 2025.

Italy

On 23 July 2009, the Italian Parliament adopted a bill bringing to an end a ban on the use of nuclear energy which had been in place since 1987. On 15 February 2010, the Italian government adopted a decree defining the criteria and procedures for the construction of nuclear power plants in the country by 2020. In January 2011, there was a ruling of the constitutional court allowing a referendum to decide on whether there should be construction of nuclear power plants.

Thereafter the government approved on 23 March a one-year moratorium on the construction of the country's first nuclear power plant by 2020. On 19 April 2012, an amendment was tabled that indefinitely puts on hold new build plans. Then on 12 and 13 June, Italian citizens voted against nuclear power in a referendum.

The Netherlands

In January 2006, the Dutch government decided to prolong the life of the country's sole nuclear power plant, Borssele, for another twenty years. On 17 February 2011, the Economics Minister, Maxime Verhagen, sent a letter to the parliament allowing the construction of a new nuclear unit to begin in 2015.

After Fukushima, the Dutch government said it would carry on with its new build plan. The stress tests' report of the Netherlands concluded that Borssele meets all the safety requirements. However in January 2012, Delta, the Dutch energy utility, decided mainly on financial grounds to postpone its new build plan for 2 to 3 years.

Poland

13 January 2009, the government adopted a resolution on nuclear energy that indicates that by 2020 electricity will be generated by one or two nuclear power plants. The current problem in Poland is that financing is difficult, but the Poles seem determined to go forward.

Romania

The second nuclear reactor of Cernavoda nuclear power plant was inaugurated on 5 October 2007. Plans to complete Cernavoda 3 and 4 are also under way and the government intends to go forward with its new build plans. The investors for Cernavoda 3 and 4 should be selected very soon.

Slovakia

On 3 November 2008, Prime Minister Fico and Fulvio Conti announced that the construction of units 3 and 4 at Mochovce due to be completed in 2012 and 2013 respectively had begun. In December 2008, the Slovak government chose Czech utility CEZ as its strategic partner in proposals to construct its fifth reactor.

Slovenia

On 20 November 2009, the Economy Minister, Matej Lahovnik, announced that the planned second reactor at the Krsko nuclear power plant (NEK) would be completed between 2020 and 2025. On 14 January 2010, the Slovenian utility, GEN-energija, sent an application to the Economy Ministry for a permit to build a second unit at the Krsko nuclear power plant.

Spain

When the PSOE (Socialist Party) won the general elections, it had made a political statement to phase out nuclear power, but no calendar or specific strategy had been fixed. On 2 July 2009, there was a decision to grant an operating licence to the Garoña nuclear power plant in Northern Spain for a further four years. On 15 February 2011, there was ratification by the Spanish Congress of a Sustainable Economy Law, in which the reference to a maximum operational duration of 40 years for nuclear power plants had been deleted

The conservative government that came into office in November 2011 is more favourable to nuclear. In July 2012, it overturned the decision of the previous socialist government that had decided to grant an operating licence to the Garoña nuclear power plant in Northern Spain only until 2013 opening up the possibility for it to operate until 2019.

Sweden

In February 2010, the Swedish government put forward a draft law that would allow the construction of a maximum 10 new nuclear units in the country to replace existing units as they are shut down. It was ratified by the parliament on 18 June. This brings to an end the nuclear phase-out policy that was first introduced in 1980.

The Prime Minister said that the government would not reconsider its decision to replace nuclear reactors at the end of their lifetime.

Switzerland

In February 2007, the Swiss government adopted a new energy policy strategy, which emphasised the necessity to replace existing units with new ones to avoid a future energy gap. On 15 November 2010, there was a positive assessment of the Swiss Federal Nuclear Safety Inspectorate on the suitability of 3 sites for the construction of two new nuclear power plants.

After Fukushima the government had suspended the approval process for the construction of the two new nuclear power plants in order to review safety standards. The Swiss Federal Nuclear Safety Inspectorate has also been required to carry out safety evaluations at Switzerland's five existing nuclear power plants, which they passed successfully. On 25 May 2011, the government decided not to replace ageing nuclear reactors and to phase out nuclear by 2034.

While parliament has agreed to this strategy it has added a backdoor which would allow reconsideration of the decision if "new generation reactors" with significant improvements in safety, efficiency and waste production would become available. Even though government and parliament have agreed to the nuclear phase out it will still need public support in a referendum held presumably in autumn 2014.

United Kingdom

On 10 January 2008, the British government published a Nuclear White Paper, which announced that a new generation of nuclear power plants would be built in the UK. The new coalition government adopted a revised draft National Policy Statement on energy, which recognised the essential role that nuclear has to play in ensuring the UK's energy supply along with other low-carbon energy sources by 2025. It also confirmed eight sites as suitable locations for nuclear new build.

The coalition government (Conservatives and Liberal Democrats) is very committed to pushing forward with a new fleet of nuclear power station in the UK. There is also strong cross-party support, with Labour committed to a new nuclear build programme as part of long-term energy policy for the UK. The new secretary of state for energy, Liberal Democrat Ed Davey, has already indicated that he will follow the government plan to take forward new nuclear build.

NUCLEAR POWER IN THE MIDDLE EAST FOLLOWING FUKUSHIMA

ADNAN SHIHAB-ELDIN
Kuwait Foundation for the Advancement of Sciences
Safat, Kuwait

ABSTRACT

This paper comments on the significance of Fukashima both technically and with regard to public opinion. On the former, it is argued that the culture of the industry and resultant history are overwhelmingly more important indicators of the safety of nuclear power. Although in the wake of Fukashima, safety improvements are being made, in the resultant politics, it is probable that costs are being increased to address combinations of events of negligible probability and attention diverted from more proactive assessments. However, the effect on public opinion has been decisively negative and lasting.

The MENA (Middle East and North African countries) context of rationale, cost and political environments are discussed. The impending crisis of domestic use of hydrocarbon resources in the Gulf States is presented and the alternative path of sustainable energy is demonstrated using the example of Kuwait. It is suggested that the concentration on supply solutions, whilst expedient, is not optimum and that the much more challenging, and ultimately more significant issue of demand control through tariff increase and sustainable energy policy must be addressed. The requirement for a nuclear contribution to the resultant reduced energy mix in these countries and to address more fundamental energy shortfalls and security issues elsewhere in the region remains strong. Nuclear power programs are summarized and the most significant in UAE, Saudi Arabia, Turkey, Jordan and Egypt, in order of certainty, discussed.

1 LESSONS OF FUKASHIMA

It has now been admitted[2] by the Japanese Government that the Fukashima accident was caused by failure to design for known tsunami conditions. It is also probable that seismic design, known to be inadequate, was a contributory factor in accident response.

The tsunami design basis chosen was 5.7 M while 20M events had occurred in the previous century. TEPCO Internal reports to this effect and practical experience of the vulnerability of emergency generators to flooding were available before the accident. The event on 11th March was 14 M. and Fukashima was designed in the late 1960's before rigorous seismic design was possible.

In 2006, NISA (Nuclear and Industry Safety Agency) required a general seismic re-evaluation. No significant assessments were performed. Reports submitted were limited. It was known that appropriate seismic reinforcements of safety related pipe work were not in place at Fukashima. Following an earthquake in 2007, the more recent Kashiwazaki Kariwa plant was shut down for 21 months for reassessment and modification.

Thus both TEPCO and, more seriously, NISA were fully aware of the possibility of core damage consequent on flooding, inadequacy of seismic design and of the need for safety improvements but that no action was taken. The latter organization did not monitor nuclear safety. A succession of inadequate disclosure and falsification of information by the Japanese nuclear industry is well publicized.

Kiyoshi Kurokawa, chairman of the commendably frank and courageous report by the National Diet of Japan Nuclear Accident Independent Investigation Commission has stated:[2]

> *"The accident was a profoundly man made disaster that could and should have been foreseen and prevented."*

> *"Its fundamental causes are to be found in the ingrained conventions of Japanese culture."*

He explains that this culture led to a mindset, which in extreme placed corporate interest above safety and which managed to avoid learning the lessons of Three Mile Island and Chernobyl.

1.1 Safety and Regulatory Significance

To a first order, Fukashima is a uniquely Japanese failure in both the safety culture and regulation essential for the safe use of Nuclear Power. In the U.S. and Europe, the Three Mile accident in 1979 proved the integrity of containment of light water reactors, despite failures of instrumentation and operation leading to a partial meltdown. It also transformed safety culture and regulation, resulting in orders of magnitude improvement in reliability and availability. Figure 1 shows fleet averages of availability and significant incidents for the 104 U.S. reactors. The latter values are per reactor year and defined as events with a core damage probability of 10^{-5} or higher. There is therefore a strong trend of improvement in both operational performance and safety despite the aging of the fleet. There are no discernable radiological health effects from operations and incidents.

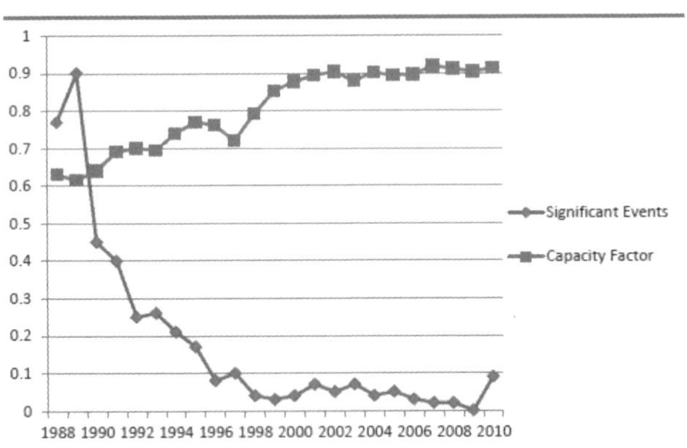

Fig. 1: Availability and Significant Events.[4]

Had these standards of safety culture and regulation been applied in Japan, safe shut down at Fukashima would have been achieved. However this has not resulted in complacency in the world nuclear industry. It has recognized the risks of the inevitable balance between safety and cost, and the possibility of design shortfalls, and has undertaken reassessment of all reactors by stress testing. Its position is well expressed by Mike Weightman, UK Chief Inspector of Nuclear Installations.[3]

> *"We see no reason to curtail operation of nuclear power plants. Once work is complete proposed improvements will be considered in line with our normal regulatory approach"... "We need to continue to adhere to the principle of continuous improvement No matter how high the standards are, the quest for improvement should never stop."*

This culture of safety-in-depth against realistic operational, fault and emergency design conditions has been rigorously practiced for three decades making increasing use of probabilistic assessments in a strong independent regulatory environment. It is unlikely that major functionally justified changes or safety improvements will result from Fukashima. However, in the prevailing climate of opinion, there is a strong possibility of major cost escalation due to counter-measures against deterministically imposed combinations of events of negligible probability. There is also a danger that excessive emphasis on earthquake and flooding will obstruct the quest for other causes of accident by the more proactive normal safety culture.

While it is unlikely to materialize in the foreseeable future, an enforceable international safety regime, (similar to international safety regime in commercial air transport) could contribute tangibly to assure public opinion and policy makers. The concept was raised by France and others and discussed briefly within IAEA following the accident, but was opposed strongly by most other nuclear states.

1.2 Public Relations and Political Significance

Public opinion is conditioned by perceptions of nuclear weapons, explosions and fear of radiation and waste disposal, exacerbated by the sometimes alarmist an uninformed media coverage. It is influenced by anti nuclear movements, initially focused on weapons but now extended to nuclear power which they maintain is an unsafe and because of sustainable energy unnecessary technology propounded by an untrustworthy industry and establishment. These views are strengthened by major accidents but are intolerant of the facts of those accidents. Despite inadequate instrumentation and gross operator failure, Three Mile Island proved the safety of containment of an early PWR without public harm. Both Chernobyl and Fukashima were solely caused by willful disregard of the safety in depth culture routinely practiced in the U.S., Europe and South Korea. They are also intolerant of the fact that existing nuclear power plants in these countries are the safest, cheapest and least polluting source of electricity currently available.

Also, although the largest technically feasible and economically justifiable renewable energy component in the energy mix is required, very large-scale deployment is beset with difficulty, and its cost trajectories and the extent of deployment remain uncertain. For the next two decades, a renewable share of around 20-30% appears to be a limiting practical ceiling on economic (cost of storage) and technical (grid stability) for most countries.

These influences have already achieved have their objectives. Public opinion is shown in Figure 2. Before Fukashima, in all countries, only a large minority considered that nuclear power was a safe and important source of electricity and that new plants should be built.

Fukashima is catastrophically and permanently damaging to the vital role of nuclear power in the world economy and in reducing carbon emissions. Recognition of this is evident in the following statement by the Japanese Government, immediately prior to IAEA Ministerial conference on nuclear safety in June 2011,[1] evidently seeking to mitigate the damage:

> *"Japan takes it very seriously and with remorse that this accident has raised concerns around the world about the safety of nuclear power generation and above all, Japan feels severe regret for causing anxiety among people all over the world about the safety of nuclear power and release of radioactive materials."*[1]

This statement and those from[2] quoted above are too late. Japan is deeply respected by the public worldwide because of its excellence in high public profile technology such as cars and consumer electronics and because of its perceived integrity. The facts of the gross design and regulatory failures will remain incredible to the public; leaving a lasting perception that safe nuclear power is beyond the capability of even such a respected nation.

The response of Germany, similarly respected, is seen to endorse this perception and is equally damaging. The hasty acquiescence to anti-nuclear public opinion for political expediency, abandoning the considered views of the executive which produced it is not recognized.

Actions by the industry, intended to recover public opinion may have the opposite effect. Emphasis on Generation III+ designs with passive or multi-redundant systems, and the declared expectation that the reassessment process and stress tests following Fukashima will yield major safety improvements, leaves the public perception that the previous safety process and existing reactor stock are unsafe.

Mitigation of socio-political impacts of accidents is essential but overt emphasis implies an establishment accepting their possibility and the requirement for as remote siting as possible. In fact, size of exclusion zones is dependent on the severity of the accident postulated. The only basis on which Nuclear Power can be responsibly proposed is that it is engineered and regulated in a manner which renders the probability of such accidents negligible. The history of U.S. and European reactors and their siting relative to major centers of population supports this basis.

Post Fukashima opinion is also shown on Figure 2. With the exception of U.S. and UK, public support in nations with nuclear power is sharply reduced. The position of the segment supporting continued use of existing plant but opposing new, despite its lower risk, appears inconsistent.

Few opinion surveys in the Middle East are available. That for the UAE, conducted for the Emirates Nuclear Energy Commission by an independent polling organization is remarkable.

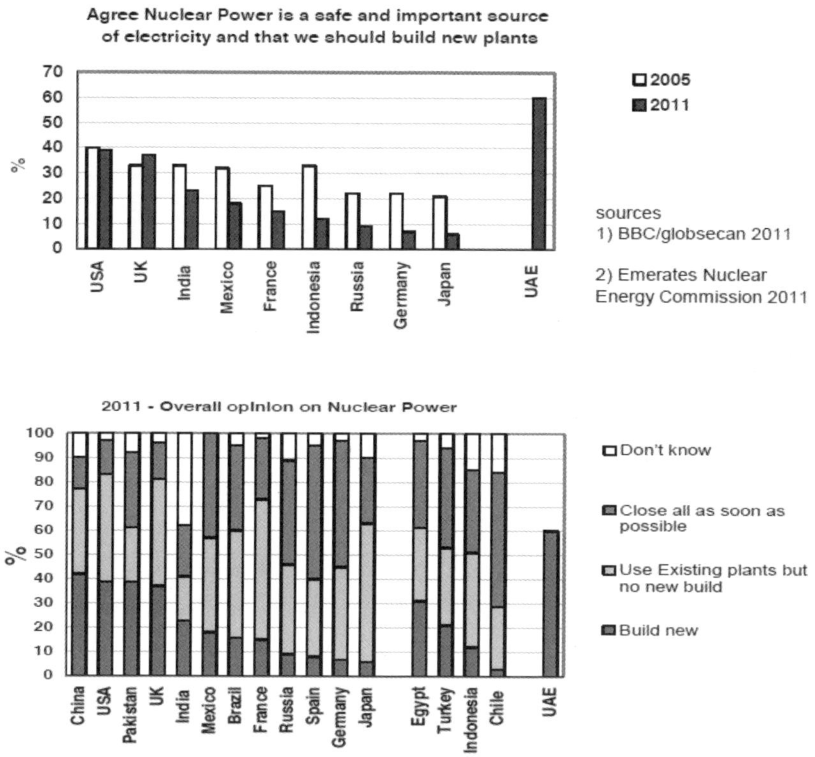

Fig. 2: Public Opinion Before and After Fukashima.

GCC states are monarchies with either no or limited participatory democracies. Kuwait and Bahrain are most advanced with legislatures in which a majority of members are elected. Other countries in the wider MENA region, have, both monarchies with elected legislation (Jordan and Morocco) but with limited power, and pseudo "democratic" republics, with presidents of chiefs for life. The storms of the Arab spring are sweeping the region promising transformative but bumpy and long pathways to true democracies. It is notable that in the Gulf states nuclear programs are certain where political power is most strongly vested in ruling monarchy elites (the UAE and Saudi Arabia) and have been abandoned in response to public opinion by governments in countries with strong legislatures (Kuwait and Bahrain). In Jordan, public opinion may yet have the same effect. Its significance in Egypt is as yet uncertain.

2 RATIONAL FOR NUCLEAR POWER IN MENA

2.1 The Gulf States

The primary rationale is to counter the domestic use of national hydrocarbon resources and consequent reduction of national income. With the exception of Bahrain, the GCC are rentier states with national income dependent on hydrocarbon exports—ranging from about 70% in the UAE to 93% in Kuwait. Historically, these revenues and small populations provided budget surpluses which allowed governments to provide citizens with material benefits, including heavily subsidized electricity and water, transport fuel, public sector employment, welfare, free education and housing without taxation. This has become the basis of a de facto social contract between the government and its people, and the basis of popular support for the former. Tariffs are shown in Figure 3.

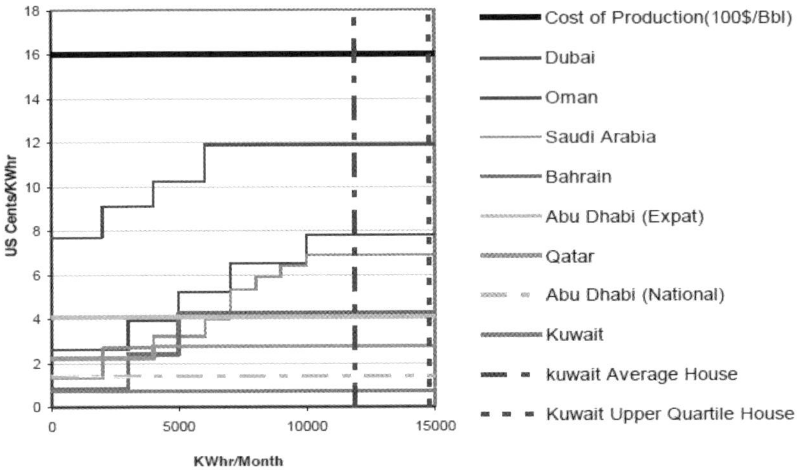

Fig. 3: GCC Tariffs relative to cost of production @100 4/Bbl.

These levels of subsidy lock in a consumer base of buildings, equipment and vehicles chosen with little consideration of efficiency and eliminate conservation in their operation. Together, with the absence of taxation, they have also eliminated the financial instruments normally used to incentivise efficiency and conservation.

These effects, more than severity of the environment, are responsible for the situation shown in Figure 4. Although they are not industrial economies, the three wealthiest Gulf States, including Kuwait have the highest per capita fossil generated electricity demand, and therefore emissions in the world. Resultant consumption of domestic hydrocarbon resource is shown in Figure 5.

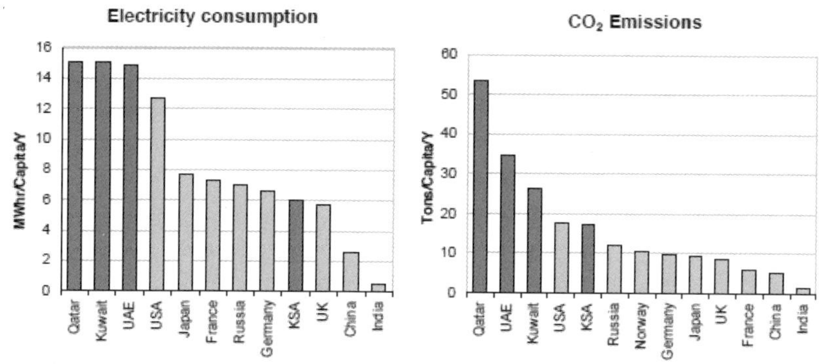

Fig. 4: Per Capita Electricity and Carbon Emissions.[5,6]

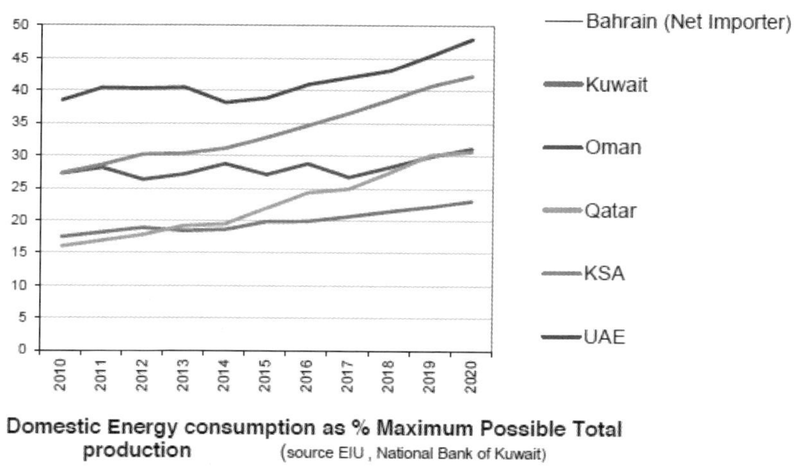

Fig. 5: Domestic Hydrocarbon Consumption as a % of Total Production.[7]

The decade of rising oil prices and increasing budget surpluses from 1998 to 2008 diverted attention from these issues and encouraged the postulation of ambitious national development plans, but now they are a major cause for concern.

Saudi Arabia is the most serious case. Despite exceptional sustained diversification efforts, the economy remains almost 90% dependent upon oil exports. Raising oil production is increasingly difficult and more costly. Per capita demand is rising rapidly because of its present relatively low levels and urbanization. A recent study by an international independent research body[8] gives the following forecasts of the development of the energy economy under present policies.

The implications are declining revenue and inability to meet the increasing expectations of a rising young population leading to economic unsustainability and, in the new Arab trend to expectations of democratization, political unrest. Because of the social contract and the resultant political difficulties of demand management, attention is concentrated on supply. Major programs of Nuclear and Renewable Energy have been announced, as yet without official detail—nuclear first but with increasing emphasis on renewables post-Fukashima and as their costs reduce.

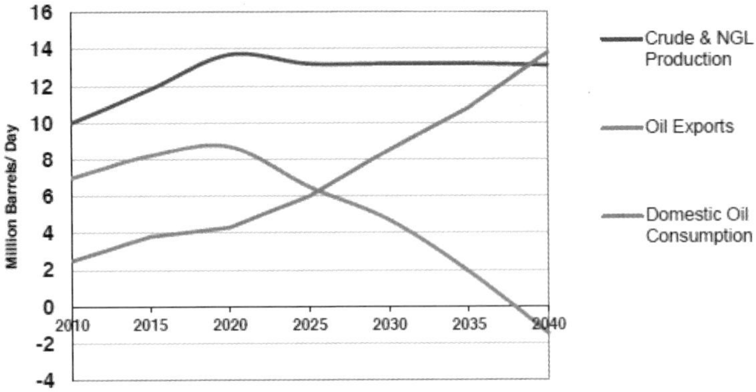

Fig. 6: Oil Balance in Saudi Arabia on Business as Usual Trajectory.[8]

Although this is an extreme case, it illustrates the general problem, which to various extents and timescales, faces all the Gulf States, and the policies being adopted.

The economic competitiveness of nuclear power is discussed below in 3.1. Financing has always been problematic, because of high capital cost, risk of delays and cost overruns, slow returns on investment and decommissioning commitments. Now these difficulties are exacerbated by the consequences of Fukashima, both in cost escalation and investor confidence. The MENA region's policy of following the world trend to merchant power industry is now almost impossible to apply to a nuclear power program because of these superimposed risks. However, the advantages of nuclear power in reducing domestic hydrocarbon consumption, and providing low cost electricity for operating lives of more than 40 years as now proven in the USA and France, remain attractive as a contribution to economic sustainability of the Gulf States. Combinations of a majority of direct state funding, together with loan guarantees and mitigation of risks to encourage private sector investment, is one of the optimum uses of present fiscal reserves.

2.2 Energy Security

This is a significant consideration in the Gulf States, particularly for Kuwait, which to avoid the use of high value exports at high official demand scenarios will rely on LNG imports for over 30% of its energy requirements. However, it is the primary rationale for Bahrain and Jordan, which are entirely dependent on imports, Turkey and Egypt.

2.3 The Knowledge Based Economy

All of the MENA Countries recognize the imperative to transition toward knowledge based economy and envisage nuclear technology as a contribution to doing so. Some, notably Iran, regard it as a matter of national prestige. Many of the countries, except the Gulf States, operate research reactors, Egypt for 40 Years. However, all with power programs intended to lead to Power will outsource training and have entered into suitable international agreements.

3 ALTERNATIVE PATHS

3.1 The Effect of Fukashima on Generation Economics

Overnight costs must increase but the extent is yet uncertain. Figure 7 compares nuclear with fossil and renewable technologies and shows the effect of up to 30% increase on the anticipated 4.5 \$/W pre Fukashima. The effect of various levels of state participation are indicated by discount rates of 5%, (state ownership), 8% (major state participation, loan guarantees and absorption of major risks) and 12% (IPP). Because of the present world economic situation, and the climate of opinion on nuclear power, the IPP model is considered unlikely. Major state participation, either by the host or the technology providing country, is considered the only probable financing model, justified as discussed above and now the model for the UAE Barakah plant.

Unlike the U.S., competitiveness of CCGT is set by the Middle East /Asia LNG market in which, for the foreseeable future, prices will be near fuel oil parity. The Fossil results are for an IPP with 10% discount rate and fuel prices benchmarked to crude. Assuming a 30% cost escalation, fully state supported nuclear power is competitive with CCGT in MENA above a crude benchmark of 50\$/Bbl and Joint Ventures above 70\$/Bbl. For some countries, including Kuwait and Saudi Arabia, CCGT may not be possible because of gas and LNG shortage. Reheat steam plant shown is then the most efficient plant available for heavy fuel oil firing but is uncompetitive. The RE results use a consensus of IEA and NREL costs, an IPP model with a 10% discount rate and resource data for Kuwait. RE competitiveness is time dependent. By 2020 Wind is so with all alternatives other than State supported nuclear and CCGT with benchmark crude@ < 100 \$/Bbl. The same is predicted to apply to utility PV by 2051.

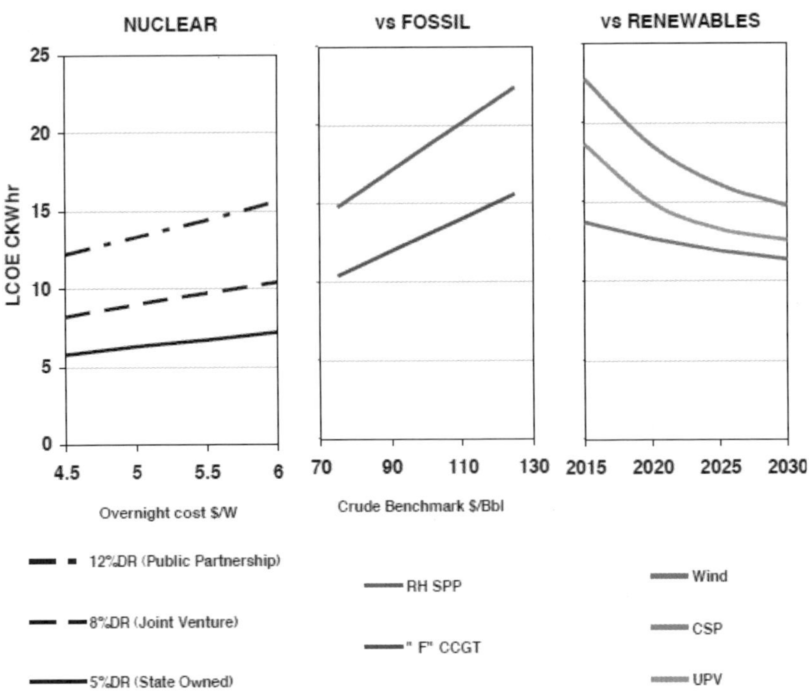

Fig. 7: Comparison of Generating Costs.

3.2 Conservation

Because of previous neglect of conservation resulting from state subsidies of electricity, potential for conservation, particularly in the Gulf States is almost certainly the largest in the world. Residential is the largest demand sector. Figure 8 shows estimates of technically feasible savings on grid energy consumption of a typical Kuwaiti villa with rooftop PV. Recent unpublished studies show that using current technology, on existing buildings retrofitting windows and sealing, use of efficient energy appliances and partial use of efficient lighting has potential to save 36% and rooftop PV a further 14%. Envelope improvements, and further use of efficient lighting in new buildings included in a recent building code, could save a further 8%. District Cooling is transformative, projecting a further 17% saving. Thus, including 14% due to PV, savings of 50% on air cooled and 75% on district cooled buildings are now possible. With some building envelope development, increased appliance and A/C plant efficiency and design to accommodate larger PV arrays, within less than a decade further savings of more than 10% are possible. Saving potential in large buildings with lower ratios of roof and exterior wall to occupied floor area, which already utilize water cooled AC, are less.

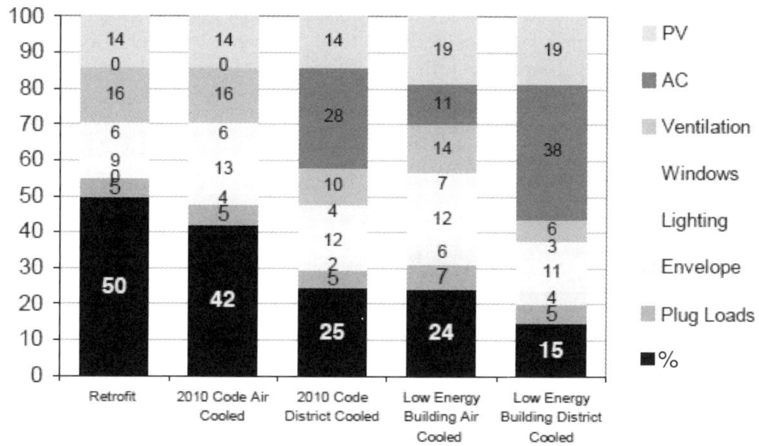

Fig. 8: Potential % Saving on Energy Demand of Kuwaiti Villas.

Difficulties in realization of such potential are the absence of consumer incentive and the levels of national coordination and resource required.

The consumer considers only first cost. Unless tariffs are greatly increased, retrofits and the additional cost of energy efficiency measures mandated in codes for new buildings must be state subsidized. However, irrespective of efficiency of design, tariff increase is the only consumer motivation to operate his building efficiently and is ultimately inevitable. In the present political climate, this is a major challenge to governments.

Conservation requires the declaration of saving targets in law and national development policy. It must be included in the mandates of all relevant government entities and the planning and technology requirements accepted. Cash incentives (rebates) for consumer saving of excessive use of electricity have the potential to be a win-win solution, meriting serious consideration.

Human resource requirements are large to design, implement and audit sustainable energy throughout national economies. This has larger potential for the national knowledge base than nuclear power.

These challenges are the most important for national sustainability; they offer the largest potential for reduction in domestic energy consumption, for industrial diversification, quality employment and development of national capability. They are fundamentally more appropriate that the concentration on supply measures presently apparent in the Gulf States. They are the only solution to the problem discussed in 2.1 and, must be confronted in the near future.

3.3 The Example of Kuwait

Figure 9 is a typical example of the uncertainty of energy planning in the Gulf

States. It shows three demand scenarios: A optimistic official estimates including national development plans funded by fiscal reserves implying more than 60% increase in per capita consumption, B based on estimates of national income, without significant conservation initiatives and C, as B assuming half of the conservation potential shown in Figure 8 is achieved. Two fuel availability scenarios covering uncertainty in realization of domestic gas resources are shown. The supply and demand scenarios range from requirements of between 50 and 15% imports in case A, depending on domestic gas realization, to balance at certain fuel availability in case C.

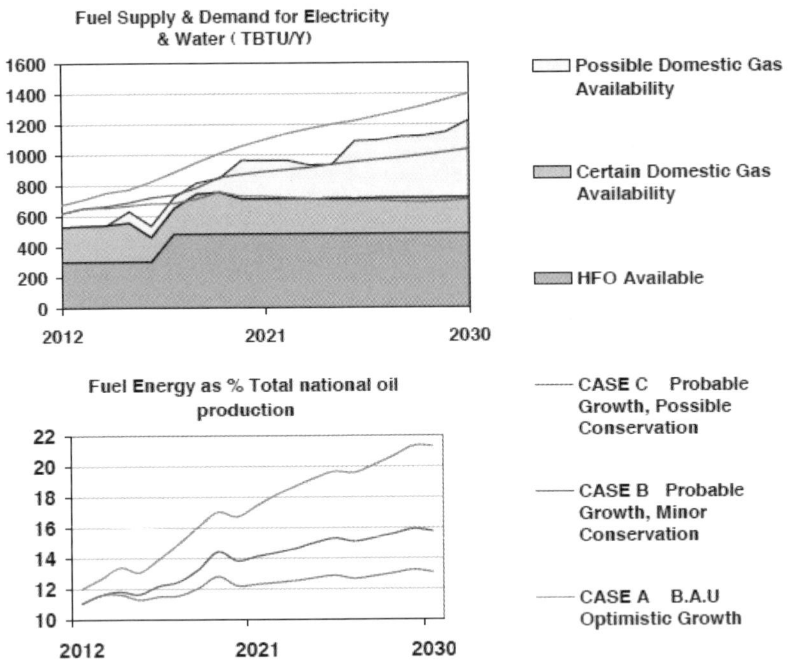

Fig. 9: Kuwait Example—Energy Supply and Demand.[9]

Case A was the primary justification for the now terminated Kuwait Nuclear Program. A secondary justification extending to cases B & C depended on the generation economics as discussed in 3.1.

4 MIDDLE EAST NUCLEAR PROGRAMS

Programs are shown in Figure 10 and summarized in Table 1.[10,11] The most significant are discussed below.

4.1 United Arab Emirates

The program was initiated in 2008 to counter forecast 50% shortfalls in availability of gas which is used almost exclusively for power generation in the UAE. Plans are for 5 GW of nuclear power by 2020 with up to 20 GW by 2032, depending on growth, a consortium led by Korean electric Power Company and involving Samsung, Hyundi & Doosan has been selected to build 4 AP 1400 reactors. A site at Barakah, 300 Km West of Abu Dhabi was chosen. Licenses for component construction were issued in 2010, for site preparation, followed by ground breaking in 2011 and for construction of the first two units in 2012. The licensing process included review post Fukashima.

The program is on schedule to commission the 4 reactors between 2017 and 2020 the first less than 10 years after commencement. There is expectation that the consortium will have a major role in operation for 60 years. The program sets a world standard and model for fast tracked, efficient implementation. International expertise is used in key posts in the national regulator and nuclear corporation whilst development of national resource is pursued in parallel. Policy is based on full cooperation with the non proliferation treaty, an IAEA safeguards agreement and the Additional Protocol. First fuel loads are provided by the consortium. Long term arrangements for uranium supply and fuel manufacturing are being pursued. In country enrichment is banned by the law which established nuclear infrastructure. In country HLW repositories are being considered. The UAE has been complimented on its good patience, safety culture, independence of regulation and human resource development by the IAEA. The financing model has evolved to increasing state participation and support of equity partners.

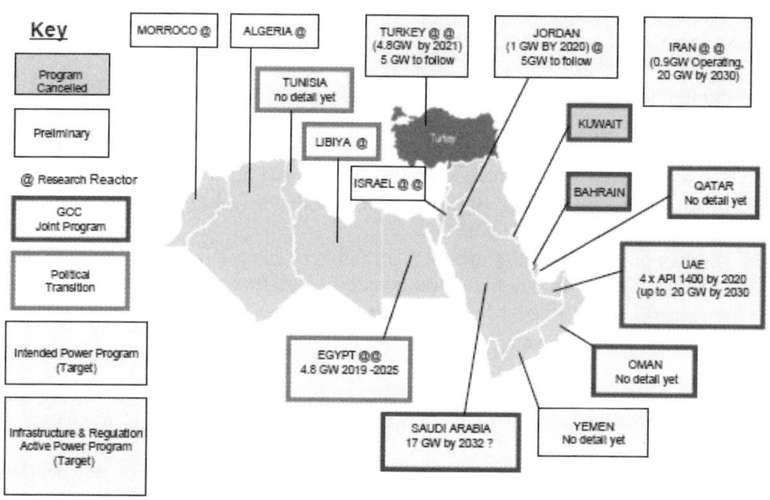

Fig. 10: Middle East Nuclear Power Programs.

4.2 Saudi Arabia

Because of the imperative of the problem outlined in 2.1 above, the current availability of huge financial resources and an absolute monarchy this will certainly be the largest MENA program. However, it was initiated only in 2009 with a royal decree establishing the King Abdullah Centre for Atomic and Renewable Energy (KACARE) in 2010 as the responsible entity. Multiple international agreements have been signed, notably with Argentina for small modular reactors with emphasis on desalination, and with China including fuel fabrication. High level strategy and definitive plans are still under consideration. Statements are that generating capacity will double to over 100 GW by 2032, that 2 reactors will be built by 2022 and 16 by 2032. Intent of 40 GW of solar and 4GW from waste and geothermal have been stated. The remit of KACARE is an energy mix, evidently of the order of 15:45:40, nuclear, fossil and RE.

4.3 Turkey

Rationale is energy independence and to meet demand growth. At present, 33% of the energy slate is Russian gas and 17% Iranian. Turkey is industrializing rapidly and, per capita, demand has trebled in 2 decades. Nuclear power has been considered to be a key component of economic growth since 1970 but the program has been characterized by delayed and cancelled tenders partly due to concerns about finance. In 2008, only Atomstroyexport bid for 4 x 1200 MW reactors for the Akkuyu site on the Mediterranean with required COD 2017-2009. This was cancelled after lengthy delays by the Government but cooperation agreements with Russia were signed. In 2010, Rosatom, as part of an aggressive policy to establish Russia in the international nuclear industry offered, and Turkey signed, an intergovernmental agreement to Build, Own and Operate the Akkuyu plant using the AES 2006 reactor and to provide100% initial equity in a Turkish company established for the purpose. The licensing process is on course for COD in 2020.

Attempts to develop a further 4 x 1400 MW plant at Sinop on the Black sea have seen bids by KEPCO in 2010 for construction, using the AP1400 and part ownership, cancelled over electricity sales guarantees and a subsequent bid by TEPCO using ABWR cancelled as a result of Fukashima. A further round to find a BOO partner involving Mitsubishi, Areva, Candu Energy and a possible re-entry by KEPCO is now in progress. Two further sites have also been identified.

Turkey is pursuing development of its relatively small Uranium resources and the Rosatom Akkuyu agreement includes the possibility of fuel fabrication in country but Turkey has signed IAEA safeguards agreements and the additional protocol.

4.4 Jordan

Jordan is dependent on imported energy, 10% of electricity is imported from Egypt and Syria. The fuel slate for generation of the balance is 25% gas from Egypt with the remainder; heavy fuel oil and light distillated from the GCC. Energy requirement is predicted to double by 2030. Jordan has also considered Nuclear power and developed infrastructure for more than 40 years. Throughout, it has sought to link development of its uranium resources with power. A joint venture ventures with Areva have been formed for this purpose but a similar arrangement with China was terminated in 2011. Non proliferation safeguards and an additional protocol are in place. Current nuclear power strategy is for first plant by 2020 and 30% nuclear generation by 2030, with intent to become a net exporter. It has International agreements with France, Russia, and Canada involving power and desalination and uranium resource development. In 2009, a contract for a 5 MW university research reactor was signed with a consortium of Korean universities and Daewoo. Construction will start in 2012 and operation is expected in 2016, to serve as the basis of training for the national program.

In 2008, a CANDU reactor to utilize domestic natural uranium was considered. From 2009, Jordan has been evaluating bids for 2 x 1 GW reactors, finally short listing an Areva MHI consortium and Atomstroyexport. It has also evaluated bids for investor operators for the plant from potential Spanish, Russian and Korean Partners. These are based on 70:30 Debt Equity ratios and 45 year power purchase agreements. There is also the possibility of an asset exchange involving part of the Uranium resources.

Separately, Rosatom has offered B.O.O. of 4 x 1200 MW reactors for longer term development in a deal similar to that in Turkey.

The program is affected by a shortage of suitable sites. Jordan only has about 30 Km of coastline on the Gulf of Aqaba. This location has been rejected because of seismicity and inland sites proposed. Initially, a site 40 Km N of Amman, with cooling from a municipal wastewater treatment plant was chosen. This has now been changed to one 100 KM from Amman. Cooling arrangements have not been described. Jordan has a growing anti-nuclear movement and in 2012 the lower house of the Legislature suspended the power program and uranium resource development pending re-evaluation of costs, quantity and quality of uranium resources and safety studies by the Jordan Atomic Energy Commission.

Despite the clear rationale and decades of commitment, this development superimposed on uncertainties of finance, siting, linkage to uranium production and processing make the program increasingly uncertain.

4.5 Egypt

From a low base of per capita consumption, high demand growth rates are inevitable. Fuel slate is predominantly domestic gas which is expected to be depleted in 20 years. The incentive for nuclear power is strong and Egypt has pursued it for more than 40 years. Research reactors have been operated and infrastructure established over this period. A succession of plant projects including desalination for sites west of Alexandria and on the Gulf of Suez, have been attempted since 1964 but they were interrupted because of Chernobyl. The program was restarted in 2004 by cooperation agreements with Russia and China. Since then it has been subject to continuous change. Proposals for a 1000 MW cogeneration plant at Al Dabaa, 250 Km W of Alexandria were developed in 2006. In 2008 preconstruction contracts were placed for a 1.2 GW plant with COD in 2017 on a site to be determined. This was extended in 2010 to 4 x 1200 MW by 2025.

Since then Egypt has been dominated by political change. The rationale for a nuclear program is clear, strengthened by expectations of economic development following that change. However, its future depends on newly empowered public opinion and the availability of finance, possibly via the partnering policies now established by Russia and China in Turkey and Jordan.

4.6 <u>GCC</u>

Saudi Arabia was one of the main proponents of a GCC program. Given its, and the UAE's independent programs and the supply limited situation of all member states, the future of a GCC program, other than cooperation on nuclear emergencies and probably, regional grid development to provide system security under large reactor trip conditions is uncertain.

The GCC Secretariat continues to pursue the preparation of feasibility studies for a regional program. The Saudi nuclear program, could be the nucleus in which other Member states participate in investments and manpower requirements and share production proportionally, leaving regulation to Saudi Arabia

A regional HLW managed repository, accepting waste from GCC and outside, probably located in Saudi Arabia, eventually followed by a reprocessing facility is an attractive but very long term possibility.

5 CONCLUSIONS

In general, MENA has so far responded rationally, apparently recognizing the economic and environmental benefits of nuclear power and its long term record of safety and reliability in South Korea U.S. and France rather than the specifically Japanese events at Fukashima and the political expediency in Germany.

The significance in MENA of these latter events depends on the empowerment of public opinion to affect national policy. It is notable that in two of the most democratized countries, Kuwait and Bahrain, nuclear intent has been dropped by the executive before encounters with the legislature. Programs in UAE and Saudi Arabia, with effectively absolute rule by heredity elites, and that in Turkey are certain, Jordan is encountering increasing opposition in addition to siting difficulty. Egypt depends on as yet unknown political response.

The UAE has established a very impressive benchmark for implementation by a developing country, making maximum use of international expertise in executive roles, whilst simultaneously developing national resource, rejecting lengthy programs of prior development of national resource, research reactors and involvement in fuel production.

Build Own and Operate Policy by Russia in Turkey and Jordan demonstrates an aggressive intent to develop as a dominant provider of nuclear technology and may be very significant in MENA.

The Gulf States have chosen to address the looming energy supply/demand imbalances which threaten national sustainability and ultimately, political unrest, predominantly by attention to alternative supply sources. Although this is expedient because of the political dangers of breaking or severely modifying the social contract for massively subsidized energy and the challenges of national coordination and financial instruments to implement sustainable energy policy, it is shortsighted. The latter policy has greater long-term potential for reducing demand, and creating

Industrial diversification and quality employment than any solely supply side measure and must be addressed. Because of locked in demand and implementation timescales of sustainable energy, the rationale for early construction of a nuclear component of the energy mix remains for many of the MENA countries, but the longer-term extent of deployment now being proposed appears unlikely.

Table 1.

Country	Pre Fukashima	Post-Fukashima
UAE	Rationale: Partially compensate fuel (predominantly gas) shortfall 50% in 2020, increasing thereafter Strategy: 5 GW by 2020/ Up to 20 by 2032. Standardize technology Fast track. Outsourced expertise. Parallel indigenous development. First plant state financed with small contractor equity Subsequent by JV's (Government/investor/ contractor) 2008-2010 Regulator, Nuclear Energy Corporation, Safety Review board 2009 KEPCO Selected from 3 bids, 4 x AP1400. COD 2017-2020 2010 Site: Barakah 300 KmW of Abu Dhabi	2011 public outreach, polls 2012 Construction license delayed by additional safety reviews 2013 Possibility of early follow on plant depending on demand
Kuwait	Rationale: Minimize use of crude and GO to meet 50% fuel state shortfall by 2030	"Kuwait is no longer pursuing NPP". Fuel issue not addressed.
KSA	Rationale: unsustainable domestic oil and gas consumption(25% production in 2009, 60% predicted in 2032 for B.A.U) Strategy: 2 reactors by 2022, 16 by 2032. Interest in small reactors for desalination. No plant/ site details yet 2009 Royal decree stating NPP essential to meet energy demand and prevent depletion of hydrocarbon resources and establishing King Abdullah City for Nuclear and renewable energy KACARE	Evidence of shift of emphasis to Solar
Bahrain	Rationale: Minimal hydrocarbon resources, net importer, urgent requirement for alternative generation, energy security	2012 Minister "Plans abandoned"
Qatar	Rationale: Maximize gas availability for petrochemicals and GTL. Constraints on gas production. 8% electricity demand increase 2008 Feasibility studies with EDF 2010 Cooperation with Rosatom. Interest in Russian Floating Nuclear Power and desalination plants	2012 studies: viability at crude > 55$/bbl. No changes in policy announced but go ahead not imminent
Oman	2009 Cooperation agreement with Rosatom for technology transfer and preliminary studies Various statements of intent but no firm proposals	

Table 1. *(Continued)*

Country	Pre Fukashima	Post-Fukashima
GSS Joint	Rationale: Shares costs, technology, human resource requirements. Allows Optimum siting for emergency exclusion zones and cooling conditions Optimum use of GCC grid to ensure electrical network stability under reactor trip	2012 Consultant studies on emergency response. No detailed proposals
	2007 IAEA study shows Economic viability. 2010 Consultant studies on siting, knowledge sharing, safety and regulation	No changes in policy announced
Turkey	Rationale: Energy independence. (Present Energy Slate 33% Russian gas, 17% Iranian gas, 25% Coal, 25% Hydro. Demand growth 8% Several projects proposed , tendered and cancelled since 1970 2010 Intergovernmental agreement for Rosatom to B.O.O. 4 x VVER 1200. For Mediterranean site at Akkuyu. Includes Fuel Supply and spent fuel return. 100 % Initial Rosatom equity, retaining 51%. Construction start 2013 C.O.D 2018 -2021 2010 Toshiba/ TEPCO Bid for 5 GW at Black Sea site at Sinop	2012 TEPCO Sinop bid withdrawn. 2012 Renewed discussions with France, South Korea, Japan and China on Sinop No changes in policy
Iran	Rationale: Conserve hydrocarbons for export and petrochemicals Strategy: 23 GW by 2030 1975 KWU contract for 2 x 1.2 GW PWR at Bushehr 1979 50 & 85% completion, KWU withdrew 1995 Atomstroyexport to replace 1 reactor with 915 MW VVER. 2005 Fuel Supply and spent fuel return agreement with Russia 2011 Completion & commencement of operation	2011 Grid connection Operation by Russians under IAEA supervision No changes in policy. Plans for unit 2 construction
Jordan	Rationale: Energy security. Fuel (Gas 28%, HFO 36%, GO 36%), 11% of Electricity all imported. Concerns re Egyptian gas. Uranium reserves. Interest as fuel vendor with enrichment partner Strategy: 1GW by 2020, 5 by 2030, Part financing by operator partners. 2007 Atomic Energy commission (JAEC), Regulator (JNRA) 2009 Contract for 5MW research reactor for technology centre 2009 Areva/MHI, SNC Laval. Atomstroyexport bids for 1 GW. 2010 Concern re. seismicity of Aquaba site. Inland location 40Km N of Amman chosen. Cooling from wastewater plant 2011 Investor operator Bids received 2011 Rosatom offer BOO for 4 reactors	2012 Selection of contractors for 1GW. Start 2014, COD 2020 2012 Site moved to100 Km N of Amman 2012 Political opposition. Suspension by parliament pending studies by JAEC 2012 Possible Delays associated with Syrian Civil War

Table 1. *(Continued)*

Country	Pre Fukashima	Post-Fukashima
Egypt	Rationale: Energy security. Fuel slate (Gas 72%, HFO 16%, 14 % hydro). Gas expected to be depleted in 20 years. Demand growth 7%. 1956 Atomic energy Commission. 1976 Nuclear Power Plant Authority. 1976 -1986 Various projects and tenders. Program cancelled because of Chernobyl. 2004 ± 2008 Cooperation agreements with Russia & China 2008 Preconstruction contract for 1.2 GW plant COD 2017 2010 .plan extended to 4 plants COD 2019- 2025 2011 delayed because of political instability	No policy changes announced Muslim brotherhood statements supportive of nuclear power program
Israel	No Civil Nuclear Power Program	
Tunisia	1981Agency for Nuclear Safety, Centre for Radiation Protection, 2008 Nuclear law revised with IAEA 1990 Atomic Energy Commission 2006 Cooperation agreement with France, extended in 2008	Evaluating 0.6 to 1 GW plant No policy changes announced
Algeria	2008 various international cooperation agreements 2009 Declared intent to build reactor in 2020 and then 1 per 5 years	No policy changes announced
Morocco	Study with Atomstroyexport for building first plant in 2017 Study with China on 10 MW , 2MG/D desalination plant 2010 2 x 1 GW reactors under CDM to be tendered 2014	No policy changes announced
Yemen	Developing proposals with IAEA for 300 MW reactors in 2025 -2030 and 1.5 GW about 2035	No policy changes announced

The serious and advanced plans for nuclear power in all MENA region countries have not adequately addressed the fuel cycle challenge: planning and implementing redundant approaches to assured practical nuclear fuel cycle services. The UAE, and Kuwait before it abandoned its nuclear plan, pursued multiple approaches including participation in and support to regional and international arrangements for assured fuel procurement and services. Such arrangements need to be pursued and encouraged by all those who believe in the role of safe and secure nuclear power in the future energy mix of the MENA region.

REFERENCES

1. Report of Japanese Government to the IAEA ministerial conference on nuclear safety, June 2011.
2. Report of National Diet of Japan Fukashima Nuclear Accident Independent Investigation Commission, June 2012.

3. "Japanese earthquake and tsunami: Implications for the UK Nuclear Industry," Interim Report, Office of Nuclear Regulation, May 2011.
4. "Forging a New Safety Construct. "The ASME task force on response to Japanese Nuclear Power Plant Events, June 2012.
5. EIA World Fact Book 2006.
6. U.S. DOE Carbon Dioxide Information Analysis Centre 2008.
7. "The GCC in 2020: Resources for the Future," *Economist Intelligence Unit*, 2010.
8. Laln, G. and Stephens, P. (2011) *Burning oil to keep cool, the hidden energy crisis in Saudi Arabia* Chatham House, Dec 2011.
9. MEW Kuwait Strategy Planning Analysis.
10. IAEA 2010 International Prospects for Nuclear Power.
11. World Nuclear Association, Emerging Nuclear Countries, Accessed August 2012.

OUTLOOK FOR NUCLEAR POWER IN THE UNITED STATES

ROBERT J. BUDNITZ
Lawrence Berkeley National Laboratory, Earth Sciences Division
Berkeley, USA

[This paper summarizes a verbal presentation by the author in Erice on 21 August 2012. Like that presentation, this paper is not a comprehensive discussion of the topics covered. Rather, it is an overview summary which contains the author's personal opinion on each of the topics covered, and is intended to introduce the subject to an audience of technically trained experts in other areas who may or may not have been introduced to these topics before.]

ABSTRACT

There are 104 nuclear power units operating in the U.S. today on 67 sites. The fleet is generally operating well and is far safer than was the case 20-30 years ago. About 30 new nuclear units are in various stages of being proposed for construction in the U.S., although probably not all of them will be built. A number of issues confront nuclear power today in the U.S., and this paper provides the author's personal opinion about each of them. Among the topics are natural gas prices; Yucca Mountain's demise and the future of nuclear waste management; Fukushima lessons-learned; license extensions out to 60 years or beyond; safety and reliability enhancements; the advent of more risk-based utility decisions and NRC safety decisions based on PRA; the NRC's design-certification process and process for new-plant combined operating licenses; and the possible advent of new SMR (small modular reactor) designs.

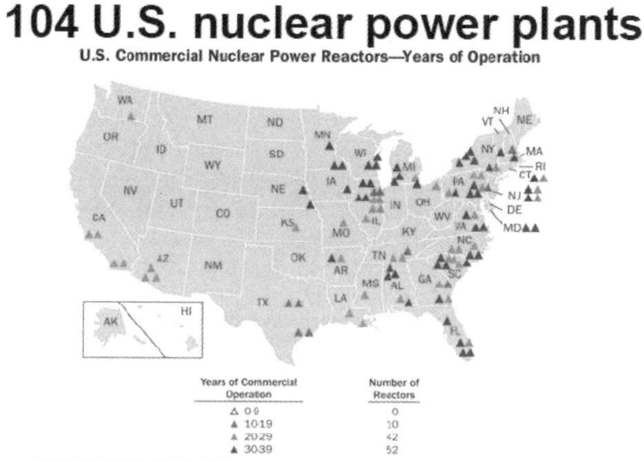

Fig. 1: Nuclear reactors operating in the U.S.

Figure 1 shows the locations of the 104 nuclear plants operating in the U.S. today. All produce more than 500 MW (electric), and some are as large as about 1300 MW. The U.S. has about one quarter of the nuclear power reactors operating in the world today. The youngest of these was designed in the 1970s, so some of the technology of the operating power reactors is very old. For example, the control rooms generally use analog electronics, although many of them have recently begun to switch over in a major way to digital technology.

Figure 2 shows the locations of the 30 or so proposed new U.S. nuclear plant units as of today, from the U.S. Nuclear Regulatory Commission (NRC) website. They are proposed for 19 sites, 5 of which are new sites. (The others are planned to be built adjacent to existing reactors on existing sites.) Many of these have signed contracts to begin design or to procure major equipment, and four of them have received NRC licenses and are in the midst of actual construction. Whether all of these will eventually be built is not known today. My own "guess" is that perhaps half of them will incur very long delays, or may even never be built, due to a variety of economic and other factors to be discussed below.

There are several major issues confronting nuclear power in the U.S. today and the bulk of this discussion will be to provide the author's personal views on each of them.

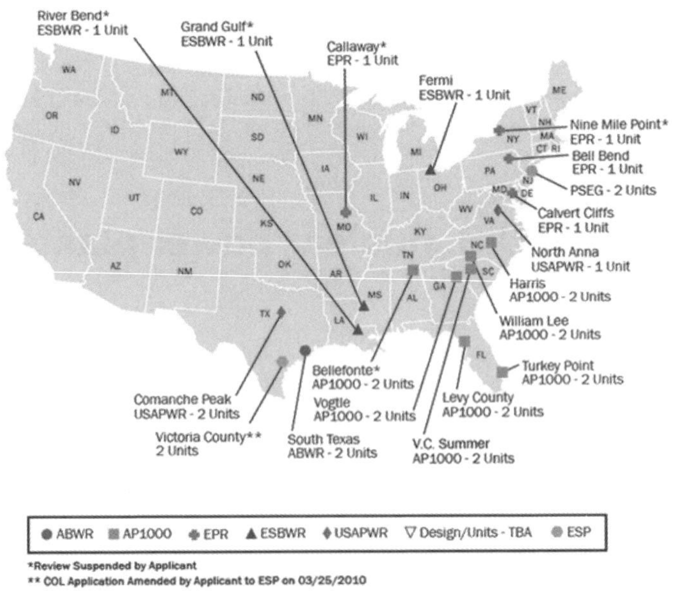

Fig. 2: Proposed new U.S. nuclear reactors.

The topics to be covered are as follows:

- Natural gas prices—will they hamper plans for building new nuclear plants?
- Current plants can be viewed as "cash cows"
- Yucca Mountain's demise
- Fukushima lessons-learned (flooding protection, station blackout protection)
- License extensions out to 60 years and possibly beyond
- Capacity factors and outage management continue to improve
- Safety and reliability enhancements
- The NRC's design certifications and new-plant combined operating licenses
- The advent of more risk-based utility decisions and NRC safety decisions based on PRA
- SMRs (small modular reactors?)

<u>Natural gas prices—will they hamper plans for building new nuclear plants?</u>

The short answer is "yes." Recently, natural gas prices in the U.S. have plummeted, and predictions are that they will remain low for a long while, perhaps a decade or more. If sustained, this is very likely to make a huge difference in the current plans by many electric utilities to maintain their current positions vis-à-vis the new nuclear plants now on the drawing board (see Figure 2). Nobody knows yet how much the impact will be, and to predict it would be folly now, but this is the single largest uncertainty factor today as to whether most or only a few of the proposed new nuclear plants will be built. My own feeling is that a significant fraction of the proposed new plants will end up incurring major delays.

<u>Current plants can be viewed as "cash cows."</u>

This is certainly the case today. Most of the current plants are old enough that their capital cost has been paid off by now, so the only day-to-day cost "on the books" is their operating costs: labor, fuel, replacement parts, etc. These costs are quite low in aggregate, meaning that the cost to the utility and hence to the ratepayers of operating them is now very low and highly competitive with other options. This fact, which helps the local economy wherever the plants now run, is a major factor in the current U.S. economic picture vis-à-vis these operating plants. It means that even what might seem like sizeable investments to keep them running, such as multi-hundred-million-dollar steam generator replacements, are seen as economically viable. It also means that there is a huge financial incentive to keep these plants running for as long as they are technically able to do so safely.

<u>Yucca Mountain's demise</u>

The proposed Yucca Mountain deep geological repository for disposing of the country's nuclear waste is currently "dead in the water," having been stopped by the current Federal administration. Efforts by some to revive it may or may not succeed, and this is a highly political matter beyond my ability to predict. In 2010, the President appointed a "Blue Ribbon Commission" to recommend what to do next, and that commission's report was issued a half year ago. It contains a number of very useful policy recommendations, most of which are important to do whether or not the Yucca Mountain repository is revived. If it is, it will incur a delay of perhaps a decade compared to the "old" (Yucca Mountain) schedule; if it is not, finding another site and developing it, technically and politically, probably means a delay of 20 years or more compared to the "old" schedule, in my opinion. In the meantime, all of our reactor waste is still on the reactor sites where it always has been, although much of it is now in dry cask storage (or soon will be), which is both adequately safe and adequately secure in my opinion for a period of several decades.

Fukushima lessons-learned (flooding protection, station blackout protection)

After the Fukushima accident in Japan, both the industry (through its industry organizations—NEI, INPO, and EPRI) and the NRC embarked on a series of lessons-learned studies. These have already resulted in some backfits and other changes, and more is to come. The major short-term insight on the U.S. side is the conclusion that the root cause of the accident at Fukushima was the result of a very poor siting decision, and we have tentatively concluded that none of our sites is vulnerable in the same way. Nevertheless, several other potential vulnerabilities have been identified, and each is being worked on by both the industry and the NRC. The NRC has ordered each plant to do a re-evaluation of their site's vulnerability to flooding and to large earthquakes, and to study the capability of the design itself to withstand potentially larger floods and larger earthquakes than previously contemplated. All of the plants are also reviewing their ability to withstand an extended "station blackout" (loss of all electricity) and to identify ways to improve their capabilities. The stations are also reviewing their operating procedures and plans to cope with a major beyond-design-basis event, and their abilities to obtain offsite assistance in the event it is needed. A few studies and changes related to other issues are also underway now. In my opinion, there is excellent cooperation between the NRC and our industry on all of these issues, which is very nice to report.

All of the above is going on now. As mentioned, there are several longer-term initiatives too, whose study and implementation will take a few years. In the meantime, everyone (in the U.S. and around the world) is watching as the Japanese learn their own "lessons." Which lessons will apply where, and how, is still for the future, and many of them are likely to be highly targeted to one or another specific reactor design feature or site.

License extensions out to 60 years and possibly beyond

Each U.S. reactor license was issued originally for 40 years. Based on extensive technical studies (measurements, analysis, inspections, etc.), 71 of our 104 operating reactor units have already been granted an extension to 60 years by the NRC, and essentially all of the others have applied to the NRC for these extensions, or soon will. The technical issue now confronting everyone is whether there is a valid technical basis for an even longer extension. The studies are just beginning. There is still time to do these studies, because the first of our operating units has recently passed its 40th-year mark, and all the others are "younger." Of course, much of the rest of the world is working on this issue too and is watching the U.S.

Capacity factors and outage management continue to improve

The long-term trend of vastly improved capacity factors and improved management of refueling outages has made a remarkable difference to our U.S. plants. (See Figures 3 and 4.) Their economic viability is far better, and the enhanced reliability of operation has also been a major benefit in improving their safety performance a lot too. This is a source of pride in the industry, and in the NRC also, and it is justified. There is no one single factor (or even 2 or 3 factors) that accounts for this long-term improvement. *It is instead the result of a myriad of small incremental improvements in every facet of operations.* It is surely a sign of how much can be done by long-term concerted effort on everyone's part.

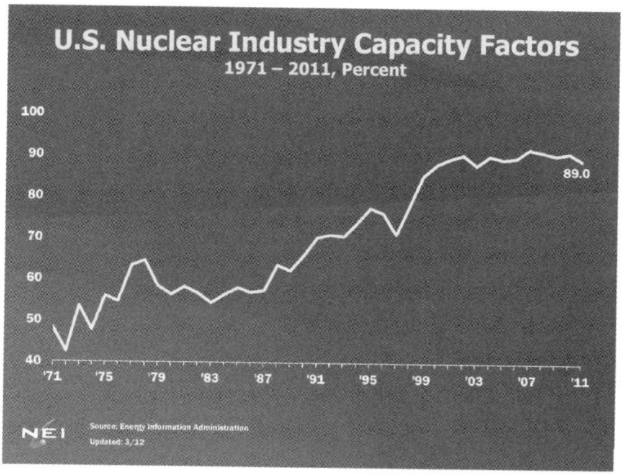

Fig. 3: U.S. Nuclear Industry Capacity Factors 1971–2011.

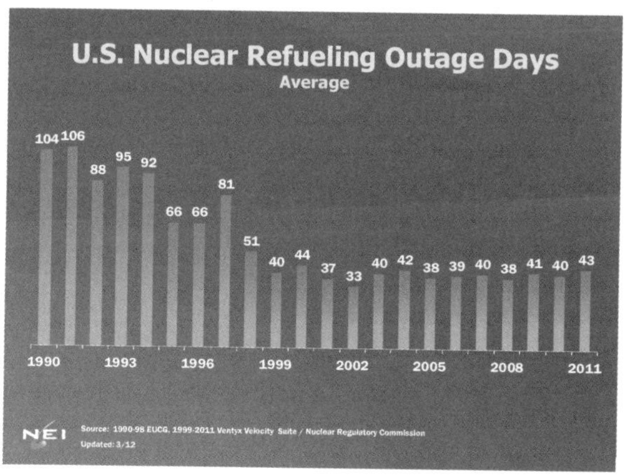

Fig. 4: U.S. Nuclear Industry Refueling Outage Days, 1990–2011.

<u>Safety and reliability enhancements</u>

There is no doubt, on a statistical average basis, that the U.S. nuclear plant fleet is much safer and much more reliable than it was two or three decades ago. (This does *not* mean that accidents can't or won't happen—a single outlier or poor performer could have an accident any time, as we saw in Japan last year.) But the statistics and their interpretation speak for themselves.

The gains have been impressive, as shown by numerous measures of reactor performance over the last two decades. The number of so-called "significant events" per reactor year for U.S. reactors since 1988 is shown in Figure 5. Figure 5 also shows the NRC's definition of the term "significant event." Back in 1979, at the time of the Three Mile Island accident, there were about 4 such events per nuclear unit per year. By 1988 this had been reduced to less than one (0.9). By 2006, that number had dropped to 0.01, that is, only one event that year throughout the U.S. for our 104 operating units. For the 5-year period 2006-2010, the average was about 3 per year country-wide. These data show a safety improvement in this figure-of-merit by about two orders of magnitude.

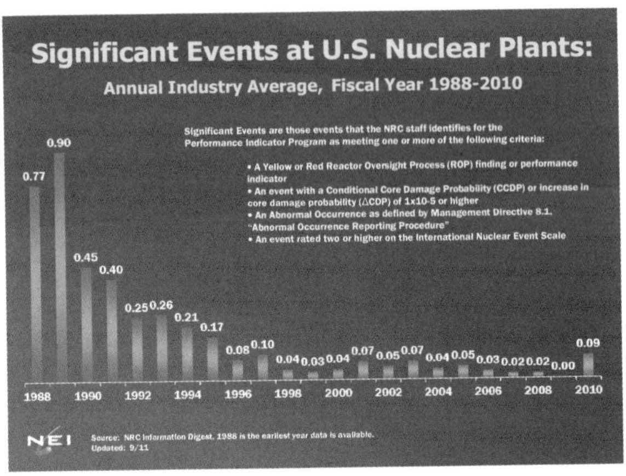

Fig. 5: Significant Events at U.S. Nuclear Plants.

Another indicator of safe operation is the number of scrams while the reactor is critical, shown in Figure 6. If any of a large number of abnormalities occurs, a signal is generated automatically to shut down the plant, or "scram" it, by inserting the control rods. The operators can also scram the reactor if a signal is not automatically generated. In 1985, there were about 4 scrams per nuclear unit per year. By 2006 that number had dropped by an order of magnitude to 0.32, or only about one scram per year for every three units. Note that a scram does not necessarily mean that the reactor is in trouble. It is often due to a minor issue that needs to be fixed.

The forced outage rate, or the percent of time that a plant was shut down as a result of repairs required in the wake of an event, is shown in Figure 7. The average outage rate for the U.S. fleet has dropped from about 10% in 1998 to less than 1% in 2010. The smaller outage rate is largely because the number of forced outages is fewer, and also because the repairs are less complicated, so the plant can come back online more quickly.

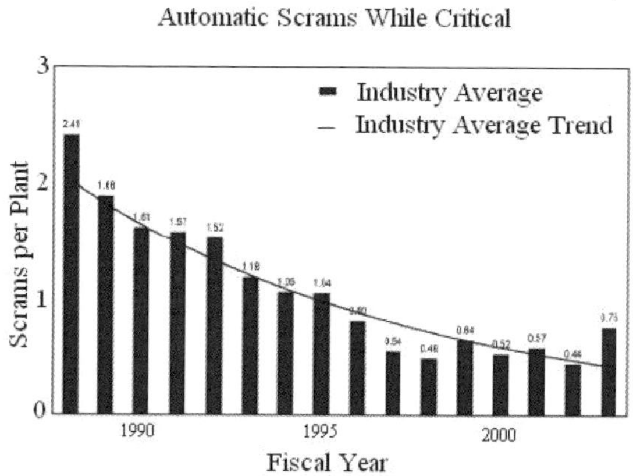

1980	7.3 scrams/year
1985	4.2 scrams/year
2006	0.32 scrams/year

Fig. 6: Automatic Scrams While Critical, 1980–2006.

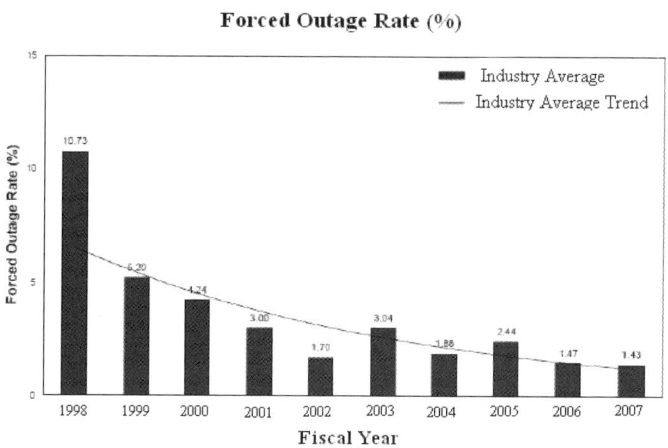

Fig. 7: Forced Outage Rate, 1998–2007.

The overall safety improvements in the U.S. reactor fleet can be attributed to a number of factors. In my opinion, almost all reflect the development and importance of a culture of safety:

- Learning from experience: an industry-wide reporting system and no-fault reporting.
- Analysis: major efforts to analyze each event for its causes and implications.
- Maintenance: concentrating on the important things and designing for easier maintenance.
- Operator errors: simulator training and better procedures.
- Industry-wide peer-to-peer inspection visits, task forces.
- Design changes: eliminating design flaws and working toward a "forgiving" design.
- NRC: Risk-informed enforcement actions (less attention to minor events).

The NRC's design certifications and new-plant combined operating licenses

The NRC staff is diligently working on the review of several new reactor designs, to provide a so-called "design certification." Several designs have been through the NRC process, and the NRC is now reviewing several other design submittals, or on certifying modifications to designs that have already been certified. All of these are for what we call "large LWR" reactors, which represent evolutionary improvements over the existing fleet. Some of the new design features, however, are almost "revolutionary," such as the features that provide much more passive means of reacting to accidental upsets and bringing the plant to a safe shutdown state without active systems or active operator intervention.

Essentially all of the new reactors being ordered or discussed in the U.S. (discussed above—see Figure 2) plan to use these new "design certification" plant designs, and will go through a new NRC one-stop licensing process to obtain a so-called "combined operating license." This new NRC process, which is still being tested for the first time now, is intended to provide the same full NRC safety review and the same full opportunity for public involvement, but in a fashion that expedites the procedures and decreases the risk of unnecessary delays. In my opinion, this change in NRC's regulator procedures is one of the major new factors in making possible an expansion of nuclear power in the U.S. of other factors also fall into place.

The advent of more risk-based utility decisions and NRC safety decisions based on PRA

Over the last decade or so, both the NRC and the U.S. operating plants have increasingly used PRA (probabilistic risk assessment) methods to analyze plant performance and to improve both safety and reliability. This has been a major initiative, and some experts credit this as the single largest technical contributor to the increased plant reliability and safety that has occurred. (I am among those who hold to this view.)

The NRC itself has been using "risk-informed" and "performance-based" approaches to its regulations for a long while. Among the most successful areas where this has been used is the NRC's reactor oversight process, used by the resident inspectors and the regulatory staff to track and understand how the plants are performing. It supports a so-called "graded approach" to NRC's regulatory oversight and their enforcement actions. In this approach, those areas that contribute most to safety are given most weight in terms of both time spent and importance attached to them. Areas judged to represent lesser risk are correspondingly given somewhat less attention. PRA methods and results are used to support the technical judgments. This has been remarkably successful by all accounts. This same approach is now being used in many other areas of NRC regulation, and the benefits are apparent to all.

Recently, a two-year effort by a specially appointed NRC "task force" produced an important report, NUREG-2150, "*A Proposed Risk Management Regulatory Framework*." This report is intended to provide a decade-long blueprint for how the NRC might move even more extensively into using a risk-type philosophy to modify how the agency regulates, not only in the reactor area but across its many other responsibilities too. How this will play out in the future is hard to predict in detail, but in my opinion the NRC is sure to be relying more and more on this philosophical approach throughout its work.

The industry on its side is also using the same ideas more and more. Each operating nuclear plant now has an up-to-date PRA which is used to help the plant decide how to carry out a whole host of different activities. On-line maintenance is deployed using PRA insights; outages are managed using PRA-based analysis; operator training and maintenance schedules are affected by risk-informed ideas; engineering analyses of the more traditional kind, such as thermal-hydraulic, metallurgical, fuel-performance, and reactivity analyses, are now routinely informed by PRA analyses.

The industry and the NRC have worked cooperatively to develop a methodology standard for performing PRA, published as an ANS-ASME standard. This standard is now used extensively by the industry and the NRC to help assure that PRA analyses relied on for decision-making have acceptable quality.

This is all to the good, and the trend will only continue.

<u>SMRs (small modular reactors</u>)

There is a lot of industrial activity today, both in the U.S. and around the world, to develop so-called SMR (small modular reactor) designs. There are many different design concepts, some of which are smaller versions of the latest large LWRs, some of which are LWRs but of very different designs, and some of which use different concepts such as gas-cooled designs or liquid-metal-cooled fast-spectrum designs. It is beyond my scope here to discuss these in detail. However, it is important to note that the NRC is actively engaged now in trying to adapt its technical regulatory positions, many of which were developed explicitly to regulate large LWRs, to enable licensing of these several new SMR design ideas.

Some of these designs are for individual nuclear units in the range of 200 to 300 MW-electric, and a few are larger, but some of them are far smaller, even as small as 20 to 30 MW-electric in size. Of course, the idea is typically to deploy several on a single site, which is why the term "modular" is used.

Why is there so much industrial interest in developing smaller reactors? There are several reasons. First, it's much easier to keep a smaller core cool, largely thanks to its higher surface-to-volume ratio and certain passive safety features. Some small cores can almost cool themselves without any intervention. Because the cooling problem is less severe, the plant would not require as much auxiliary equipment, so the design could be simpler, there would be fewer things that could go wrong, and the plant would be easier to understand and operate. The general opinion industry-wide, which I share, is that many of these features will make these designs far safer, and also far easier to analyze to provide high assurance that they are safer.

There are other reasons to favor small plants. They would not require the utilities to raise as much initial capital for their construction. Additional reactors could be added more easily for increased capacity, and the waste inventory per plant would be reduced. Also, when one unit goes off-line for any reason, planned or unplanned, the impact on the electrical grid would be far less.

The main reason that smaller reactors may not be widely deployed soon would be the economics. The economies of scale that are gained in the capital cost with a bigger reactor are a major factor. LWR plants of half the size of the new ones we are starting to build are estimated to cost 75% as much to build if built the same way (with mostly field fabrication.) Although many manufacturers and constructors would prefer to design and build standardized and modular nuclear plants about one fifth to one tenth the size of the current generation, and many utilities would prefer to deploy them, the cost per megawatt of doing so using the traditional approach could be two or more times higher. This has led to some innovative design and manufacturing ideas that go well beyond the "traditional approach," with a major emphasis on *factory fabrication* and *transporting major modular components from factory to site*, to bring the costs down. Anther major cost driver is staffing, and some of these innovative SMR designs claim to be able to run with much less in the way of operating crews and maintenance staffs, and even of security staffs. Whether these cost savings will be sufficient to allow them to penetrate into the marketplace is still not known.

I predict that this area will be one of great activity, both technical and regulatory, over the next few years. Where it will come out in the end is anybody's guess.

ACKNOWLEDGMENT

The author is an employee at the University of California's Lawrence Berkeley National Laboratory, but none of his funded projects has supported the development of this paper, which was written on his own time. However, broad general support from LBNL is gratefully acknowledged.

An earlier version of this paper by the same author, covering some of the same scope, was published recently by the American Institute of Physics as a chapter whose full citation is as follows:

"Topics in Nuclear Power," by R.J. Budnitz, chapter in Physics of Sustainable Energy II: Using Energy Efficiently and Producing it Renewably, D. Hafemeister, D. Kammen, B. Goss Levi, and P. Schwartz (editors), published by the American Institute of Physics (conference proceedings, 5-6 March 2011, published 2012).

The sections of this paper on safety and reliability enhancements and on small modular reactors have borrowed significantly from the AIP paper. Acknowledgment is hereby made to the AIP for the use of this material.

ATTRIBUTES OF SUSTAINABLE PLANETARY-SCALE NUCLEAR FISSION ENERGY SYSTEMS

ROBERT PETROSKI
Nuclear Innovation Engineer, TerraPower LLC
Bellevue, USA

LOWELL WOOD
Hoover Institution, Stanford University, University of California and
Lawrence Livermore National Laboratory, Stanford, USA

INTRODUCTION

Global population is forecasted to grow to approximately 10 billion within the next several decades. Providing this population with a developed-world standard of living will require a large increase in total global energy production. For example, if all 10 billion people use a time-averaged 10 kW per person (the present day U.S. *per capita* primary energy supply), this necessitates a global energy supply rate of 100 TW, a 6-fold increase over present. Supplying this greatly increased amount of energy is an important planetary challenge, with major implications for human health, quality of life, the environment, and global climate.

Present-day energy systems, which are centered almost entirely—approximately 80%—on combustion of fossil fuels, are currently unsuitable for meeting this challenge for several reasons. First is one of basic resource availability: at 6X their present-day usage rate, conventional petroleum resources will last between one and two decades, while unconventional petroleum and other fossil fuel resources (gas and coal) will last less than one century. Second is the effect that fossil fuel emissions have on human health: for example, the WHO attributes *one million additional deaths per year to particulate emissions from coal power production* (with roughly half of these occurring in China). Third is the impact that fossil fuel energy production has on the environment and global climate: greenhouse gas emissions from ever-more intensive utilization of fossil fuels may change the climate significantly and thus represent a significant issue-set with respect to global prosperity, stability, and security.

The attributes needed for a sustainable global energy system follow from these challenges facing present-day energy systems. First, such a system should have sufficient resources available to last for *multi-century* to *millennium* timescales. Such a long-term outlook is important for guaranteeing that future human civilization has sufficient time in which to transition to "ultimate" energy sources having no resource limitations, such as full-scale/baseload-capable renewables, fusion energy, space-based solar, or something currently quite unforeseen. Such basic energy transitions have historically taken on the order of 50-100 years (e.g., from wood to coal, then from coal to oil and gas), and transitioning to ultimate energy sources may take longer, e.g., if such energy sources are capital-intensive to harvest or require major technological development. Second, such a system should have minimal impacts on human health, by limiting emission of pollutants and by avoiding major accidents (e.g., dam breaks, deleterious pulses from processing plants, and large-scale radioactive releases from nuclear plants). Third, a sustainable energy system must either limit or mitigate its greenhouse gas emissions in order to avoid global climate changes significantly above natural variability. In order to increase energy production by a factor of six while maintaining greenhouse gas emissions at no more than their present day rates, over 80% of the new energy system must be carbon-neutral; more prudent emission rates will require carbon neutrality from at least 90% of any such new energy system. This fact sets a fourth requirement: a sustainable energy system must be implementable *at full-scale* and *in the near-future*. In order to operate at full scale, a sustainable energy system must be capable of satisfying a large majority of all energy end uses, and it must be *genuinely* affordable in order to be adopted widely, particularly in the developing world. Finally, because of the long time needed for major transitions in energy systems, buildup of new systems (and phasing out of present ones) must *get seriously underway in the quite near term* in order to meet sharply-increasing energy demands likely within the next several decades.

This list of requirements is stringent, and it is virtually certain that a mix of energy technologies will be needed to satisfy it. The single most cost-effective approach is likely to involve significant degrees of energy conservation and increased efficiency, which have the potential to relax the required build-out magnitudes and timescales, and in some cases may be implemented at *negative* cost under reasonable discount-rate assumptions.

Of necessity, fossil fuels will continue to play an important role in the relatively near-term, and can be overall performance-enhanced with more extensive adoption of pollution abatement technologies, although longer-term use of such fuels seemingly will eventually require new technologies to sequester or mitigate greenhouse gas emissions in order to avoid major climate impacts. (To give a sense of scale, burning all of the Earth's accessible fossil fuels would release on the order of 5000 gigatonnes of carbon, which would increase atmospheric CO_2 concentrations by roughly a factor of five, assuming half of all released CO_2 is absorbed into the oceans during the release interval.)

Renewable energy sources (solar, wind, geothermal, hydro, and biomass) will likely play an increasing role in areas where they can be economically harvested. It will be challenging to obtain a majority of planetary energy from renewable sources because of limited site options for hydro and geothermal, and low energy density for solar, wind, and biomass, which increases land/sea area requirements and reduces economic density as well. In addition, solar and wind energy are both intermittent, and would need to be coupled with large-scale energy storage and/or very long-range transmission in order to provide a large fraction of total energy demand.

Finally, there is nuclear fission energy, which has the potential to satisfy the above sustainability criteria and furthermore has the advantage of already having been demonstrated at global scales (~1 TW). This paper discusses the technological and institutional advances that are needed for nuclear energy to become *planetary-scale sustainable*, and sketches a limiting scenario in which a system of nuclear reactors supply all of the 100 TW needed to provide primary power to an energy-replete human civilization comprised of 10 billion people.

IMPORTANT NEAR-TERM ATTRIBUTES OF NUCLEAR ENERGY SUSTAINABILITY

Before discussing the resource availability and range of applicability of nuclear energy (which are covered in the section on long-term sustainability), it is important to first address perhaps the most significant challenge to the sustainability of nuclear energy: nuclear reactor safety. As illustrated most strikingly by Chernobyl, severe nuclear accidents can have significant and long-lasting impacts on public health and the environment. Even less catastrophic accidents, such as at Fukushima Daiichi, can have remarkably large economic and social consequences, and may greatly erode public confidence in and tolerance of nuclear energy. In order for nuclear energy to be considered sustainable, reactor accidents that result in significant releases of radioactivity must be exceedingly rare, occurring globally as something of the order of once-per-lifetime events. With potentially thousands of nuclear reactors operating around the world, an accident rate of approximately 1×10^{-6} per reactor-year is required to make nuclear accidents rare in this context.

This high degree of reactor safety is best achieved by applying the principles of "*manifest safety*." The idea of "manifest safety" is the combination of two simple concepts: engineering to a very high degree of safety, and making the safety characteristics of a system transparent and fully known to everyone. Making safety *manifest* is necessary for two reasons: first, because it allows professional peer-and-public verification of the safety status of a nuclear facility, and second, because it is a prerequisite to public knowledge and acceptance of such safety. Therefore, manifest safety has two components: a technological one and an institutional one.

Technological aspects of manifest safety

Improving nuclear safety is a topic as old as nuclear energy itself, so only the topic's most salient points will be very briefly summarized here. The key to achieving a high level of safety is to have defensive measures at each stage in any possible accident sequence, and furthermore to layer such defenses at each stage so that it takes a combination of many simultaneous, independent failures to defeat a given stage—an approach referred to as "defense in depth." To the greatest extents possible, *such defenses should operate entirely independently of operator inputs and instead should appeal entirely to physical mechanisms and natural phenomena* (e.g., motions of masses under gravity, passive thermal conduction, circulation of differentially-buoyant fluids, etc.) *for the totality of their essential functions.*

The key stages of accident interdiction are guarding against accident initiators, ensuring reactor core shutdown and afterheat removal, and preventing significant radiological releases to the biosphere. The first stage prevents accidents from occurring in the first place, and includes stringent construction, operation, and maintenance practices supported by a pervasive safety culture, as well as sensors and inspections that fully indicate the state of the plant at any moment. Accidents occurring due to external initiators (most notably earthquakes, but including extreme meteorological conditions and inundation) can be effectively precluded by apt use of remote seismic and weather monitoring, appropriate structural engineering for the site of interest, and seismic isolation/dampening for critical components and systems.

The second stage, ensuring core shutdown and afterheat removal, prevents an accident sequence from progressing to the point of core damage. The major technological advancement in improving defenses at this stage is the development of *passively safe* reactors, i.e., reactors that are able to automatically shut down and remove afterheat with no operator inputs and no electrical power. This is accomplished by providing redundant, automatic shutdown systems, as well as adequately large thermal masses (e.g., graphite in the core, large tanks of water, or the atmosphere itself) to serve as heat sinks, and means to flow afterheat to the heat sinks based on conduction and natural convection. Passive safety is quickly becoming a standard feature in new nuclear power plants, such as those centered on the Westinghouse AP-1000 reactor, and is a defining characteristic in most other advanced reactor designs. Its benefits are borne out by probability risk analyses, which show that reactors with passive safety systems have core damage probabilities that are two orders of magnitude lower—i.e., 100X smaller—than those without (Westinghouse, 2007), allowing levels as low as 1×10^{-6} per reactor-year to be achieved.

The third stage is aimed at preventing radiological releases to the biosphere in the event that the core is damaged. This can be achieved in a definitive manner by underground siting of the reactor core, allowing the core to be totally isolated from the biosphere at a moment's notice. It is also possible to design reactor containments with predefined "escape routes" for e.g., volatile material that would tend to escape during a severe accident involving breach-of-containment, and to employ filters and cold traps along these routes to retain and thereafter sequester radioactive materials. Escape of solid or molten fission products can be prevented by employing one or more "core catchers" — featured in some modern Russian reactor designs and the European Pressurized Reactor beneath the core that are able to trap and immobilize any downfalling molten core material.

Individually, many of these safety measures could have prevented entirely or at least mitigated substantially the three major accidents in the history of civilian nuclear power: the severe incident at Three Mile Island, the Chernobyl catastrophe, and the Fukushima Daiichi disaster. Implemented together, they have the potential to ward off nuclear power plant accidents in all but the most devastating natural or military circumstances and to successfully mitigate severe consequences of most if not all of them that nonetheless might occur.

Institutional aspects of manifest safety

The second key aspect of *manifest safety* is fully communicating the safety characteristics of any and all nuclear power reactors to the global nuclear power community and to the general public, which will rationally insist on proper accounting for associated benefits and costs. Making safety manifest is enabled to a degree never before possible by modern data recording, storage, processing and transmission capabilities—the costs of all of which continue to decrease exponentially with time at Moore's Law rates—and by nearly ubiquitous and instantaneous public access to all such data through the swiftly-advancing capabilities of the Internet.

As a basic aspect of communicating *manifest safety*, the measures just described must be implemented in a fully transparent manner, from design to construction and on through operation and eventual decommissioning, in order to support unceasing professional-peer and general-public reviewing of the safety postures of each and every nuclear power plant, around the world.

Secondly, integrated results from this ongoing safety reviewing by peer-professionals must be effectively conveyed to the public. One important element of such communication likely would be a universally accessible website operated by professionals of nearly universally-recognized expertise and objectivity that provides up-to-date risk information regarding each power reactor anywhere. This information would be based on probabilistic risk assessment analyses that provide order-of-magnitude forecasts of core damage and radioactive release probabilities, taking into account current conditions of each plant such as its design and construction, its operational history and status, equipment condition, and anticipated probability of major external events such as earthquakes, inundations, or attacks. An example of a similar service already exists in the form of the U.S. Nuclear Regulatory Commission website, which gives regularly updated safety status information for every operating reactor in the United States. This information can be made even more impactful if it can be made more transparent to the public, and more easily verifiable by peer-professional reviewers (e.g., by also making available the data and methods used in the analyses).

Information of such quality, timeliness and completeness would make it adequately clear to the public, as well as to regulators and political decision-makers which plants are most susceptible to which types and levels of risks and for what reasons. For example, such a service could have revealed the relatively severe susceptibility of the Fukushima Daiichi nuclear power station to tsunamis, which had been known prior to the 2011 Tohoku Earthquake (Sakai, 2006; Epstein, 2011) but which facts were not disseminated to degrees sufficient to politically compel preventive actions.

A third component of adequately communicating all possibly-pertinent safety information would be comprehensive, quantitative studies of total health impact of different energy sources, where such health consequences are measured in the widely-used unit of disability-adjusted life-years (DALYs). Such a component could provide objectively-defensible results – based on transparent choices-made regarding modeling parameters – with respect to both DALY gains (e.g., arising from improved levels of health and life-expectancy directly and indirectly resulting from greater levels of low-pollution electrification and more economical societal energy-intensity) and DALY losses (e.g., due to anticipated pollution/radiation releases during fuel mining/extraction and deployments, reactor operation, and waste management) to <u>both</u> the energy-sourcing professional community <u>and</u> to the energy-consuming general public and its political leaderships.

These ongoing studies in turn would enable informed public debate regarding which energy technologies in which particular modalities are the most sustainable, economically preferable and overall desirable from health and welfare perspectives. The nuclear power component of such studies and debates likely would benefit greatly from ongoing research quantifying the fundamental health impacts associated with low total doses and low dose-rates of ionizing radiation, such as (Olipitz, 2012). These MIT studies reported that <u>*no*</u> genetic damage could be detected even after continuous irradiation at 400 times natural background dosing-rates, up to 10 rad total-dose levels, in *any* of several independent "standard assays" for DNA damage—which dosing-rates and total doses both were greatly in excess of exposures of the general population seen in the aftermath of the Fukushima Daiichi disaster, as well as most all of those general population exposures seen with less certainty after the Chernobyl catastrophe (there being no general population exposure seen in the course of the Three Mile Island incident).

Implementing manifest safety

The ultimate goal of manifest safety is to continually improve the safety posture of nuclear facilities worldwide, and so responsibility for implementing it should begin with nuclear regulators, as part of their ongoing commitment to the public to ensure safety. As seen most recently in the aftermath of the Fukushima disaster, the public rightfully demands transparency from both its nuclear utilities and regulators as a basic prerequisite of their acceptance of nuclear energy. Nuclear regulators can address this demand by being making it part of their central mission to communicate their findings to the public.

The independence and competence of nuclear regulators worldwide can be greatly bolstered through international cooperation and oversight, the beginnings of which can be seen in entities such as the Western European Nuclear Regulators' Association. Achieving manifest safety will also require international cooperation and openness among nuclear reactor vendors and operators so everyone can keep watch over each other, share institutional safety knowledge, financially co-insure each other's operations, and assist each other when needed, analogously to the U.S. Institute of Nuclear Power Operations or World Association of Nuclear Operators. Coinsurance premiums and refunds based on degree of risk and risk uncertainty can be used to financially incentivize utilities to improve both safety performance and the timely and accurate availability of safety information. This insurance funding can also be applied to resourcing an accident remediation fund, which would act as an additional assurance that any negative impacts to the public caused by nuclear energy operations anywhere would be financially compensated, and that all externalities would be paid-up fully, promptly and reliably.

Organizing and encouraging international cooperation and peer monitoring on safety matters would seemingly be well-suited as a basic program goal for the International Atomic Energy Agency, which already has an international status and overall mission that are compatible with such an undertaking. In doing so, the IAEA would have the opportunity to play a leadership role in helping to bring about manifest safety, by developing best practices for manifest safety as well as the capability to offer evaluation and recommendations for a site's manifest safety characteristics. While the above discussion has focused on the safety of nuclear power plants, it can be similarly applied to guarding nuclear fuel cycle facilities (enrichment and reprocessing plants) against the risk of surreptitious diversion of fissile material.

Economic aspects of near-term sustainability

Outside of basic requirements for nuclear safety, near-term adoption of nuclear fission systems will depend in large part on the economic competitiveness of nuclear power plants, including their degree of economic risk. Currently, there are examples of passively-safe light water reactors being built at what appear to be competitive costs (Bloomberg News, 2010), but a complete and reliable perspective on cost will only become available after a number of these modern nuclear construction projects are seen through to completion. Similarly, costs of new nuclear reactor designs (such as various types of modular reactors) will be difficult to evaluate reliably until such systems can be built at scale. Thorough implementation of manifest safety may ultimately aid significantly in achieving lower costs, since the emphasis on rigorously quantifying and validating safety performance will also reveal those safety measures which are most effective in realizing the requisite degrees of safety at the lowest costs. Likewise, wide deployment of highly standardized designs will allow manifest safety to be more readily and pervasively achieved while also minimizing aggregate costs.

IMPORTANT LONG-TERM ATTRIBUTES OF NUCLEAR ENERGY SUSTAINABILITY

By instituting the principles of manifest safety and developing technologies that continue to improve the economics of nuclear energy, the important near-term requirements for a sustainable nuclear energy system can be effectively addressed. However, expanding this system to supply even a majority of global energy and extending it to do so for millennia will require new technologies to be deployed. This section describes a planetary-scale nuclear energy system supplying ~100 TW that uses three such technologies: uranium harvesting from seawater, manifestly safe breeder reactors, and deep borehole disposal of nuclear waste. In addition, it considers a variation that uses traveling wave reactors (TWRs) instead of standard breeder reactors in order to eliminate the need for any fuel reprocessing.

System using standard breeder reactors

Employing standard breeder reactors and reprocessing plants, it is possible to fission nearly 100% of all uranium (i.e., with some small fraction being not recovered during fuel-reprocessing). Complete fissioning of one kilogram of uranium is sufficient to provide 8.1×10^{13} J of thermal energy, so producing 100 TW for one year corresponds to fissioning of **40,000 metric tonnes** of uranium. This uranium use-rate is actually smaller than the *current* global uranium demand of ~70,000 MT per year (World Nuclear Association, 2012) which however provides a far-smaller amount of energy—eloquent testimony to the vastly improved uranium energy efficiency achievable with breeder reactors.

Harvesting this rate of uranium from seawater would involve efficiently extracting the uranyl ions from approximately 1.4×10^{13} m^3 of seawater per year, corresponding to an average flow rate of 0.43 Sverdrups (i.e., 430,000 m^3 per second) through a high-efficiency extraction system. Such flow rates of seawater are readily available from natural currents; for example, the Florida Current carries about 30 Sv through the Straits of Florida, so only a small fraction— ~1% —of even this modest current would need to have its dissolved uranium harvested to satisfy all of the energy needs of an energy-replete, fully-populated human civilization. Uranium can be harvested from seawater using specialized adsorbent fabrics or yarns (e.g., comprised of graft-polymerized polyethylene augmented with specialized side-groups) that can selectively bind to uranium. Using the early adsorbent technology that's already demonstrated to be able to harvest 2 g of uranium per kg of adsorbent (Tamada, 2006), and assuming a total collection cycle-time of two months (i.e., somewhat greater than that demonstrated to be sufficient), the amount of adsorbent material needed to be deployed at any given time would be roughly 3,400,000 MT. Masses of this magnitude are not at all unprecedented for deployment at sea; for example, oil supertankers individually capable of carrying over 500,000 MT have been built and operated.

With this early technology, adsorbent materials have been demonstrated to be reusable at least six times, so the average adsorbent production rate would need to be at most 3,400,000 MT/y. (By way of comparison, current global production of oil is approximately 4,000,000,000 MT/y, over 1000 times larger.) The minimum of 12 g of uranium produced from each kg of adsorbent (at least 6 use-cycles x 2 gm-U extracted/cycle) can produce 1 x 10^{12} J of thermal energy, equivalent to approximately 40 tons of coal. Meanwhile, more recent ORNL work has demonstrated adsorbent materials capable of adsorbing seven times more uranium per unit mass, which would reduce the amount of absorbent needed proportionally—and researchers in the field anticipate further progress in absorption performance *and* in cycle-life. (Oak Ridge National Laboratory, 2012)

Producing new fuel for the reprocessing variant of this reference 100 TW nuclear energy system would require quite substantial reprocessing facilities, which would recycle used seed and blanket fuel into fresh seed fuel, if it is to employ standard breeder-reactor technology. The total reprocessing capacity needed would be larger than the amount of uranium needed, because fuel exiting the breeder reactors would only be incompletely burned. If the average discharge burnup level (taking into account both seed and blanket fuel) is taken to be 9%, then the total reprocessing capacity would be (40,000 MT/y)/ 9% = 450,000 MT/y. This corresponds to roughly 600 commercial-size reprocessing plants with capacities of 800 MT/y, each servicing about 50 reactors of GWe scale. This number is inversely proportional to average burnup, so breeder reactors that achieve higher burnups would require less reprocessing capacity.

On the back end of the nuclear fuel cycle, waste will be produced mostly in the form of fission products and used cladding hulls from the reprocessing facilities. Some small amount of actinides will be lost during reprocessing as well and would be disposed of as waste. Finally, radioactive facility wastes created during reprocessing operations may also need disposal. Estimates for waste volumes from a pyrometallurgical reprocessing facility are 0.3 m^3 of ceramic waste (containing most of the fission products) and 0.05 m^3 of metal waste (containing some fission products, zirconium from the fuel-alloy, and stainless steel cladding hulls) per MT of fuel processed (Ackerman, 1997). The volume of low level wastes generated is expected to be small compared to these high level wastes, because pyrometallurgical processes are able to reuse their process fluids. Given these waste volumes, the total volume of waste generated each year would be roughly 160,000 m^3/y.

These wastes can be disposed of in deep boreholes, which involves drilling a comparatively deep (~5 km sub-surface) borehole into the Earth's crust employing already-widely-used petroleum extraction technology and equipment, stacking canisters of waste at the bottom of the borehole, then plugging the uppermost ~3 km of the hole with rock and cement (as was routinely done for several decades of underground nuclear weapons testing). Deep boreholes are advantageous because they isolate wastes very far from the biosphere, do not rely on engineered barriers, and are small in size compared to a mined repository.

Assuming a deep borehole capacity of 200 m^3 and a factor-of-two increase in volume to account for packaging, this corresponds to roughly 1600 new boreholes—or borehole branches, if directional drilling is exploited to minimize total drilling effort and costs—per year to accommodate the wastes generated by a 30,000-reactor global nuclear energy system. Assuming a rather generous borehole-to-borehole spacing of 100 m and no usage of the trunk and branch option, these 1600 boreholes required for annual global disposal of all energy production wastes would occupy an area of about 16 km^2, which is about 0.5% of the total area of the Nevada Test Site—so that an NTS-sized area would suffice for two centuries of deep borehole disposal of all nuclear energy system wastes from energizing the entire planet. As another way to think of this issue, each borehole would be capable of storing roughly 20 years' worth of waste from each reactor, so only 2 or 3 unbranched boreholes (or a single lightly-branched one) would suffice over the operational lifetime of a GWe-scale reactor to dispose of all of its radiological wastes, adding only a very modest additional area to the 'footprint' of a plant.

This energy system exhibits several synergies between its various technologies. First, the comparatively small amount of uranium required by breeder reactors means that the procurement of uranium at present market prices would involve only a very small fraction—less than 1%—of the present-day cost of electricity. This means that more expensive sources of uranium, such as that adsorptively extracted from seawater, can be used in an eminently economical fashion. Second, waste volumes are minimized because of the high fuel burnups achieved in breeder reactors, and because most of the fuel constituents are recycled to form new fuel. This synergizes well with borehole disposal, since the amount of waste that can be disposed of in a borehole is primarily limited by borehole volume, unlike in mined repositories which are limited by heat load from radioactive decay processes, due to their much smaller surface-to-volume ratios. Therefore boreholes are a natural choice for low volume, high specific heat-rate breeder reactor waste, in contrast with comparatively high volume, low specific heat-rate LWR waste. Third, the relative permanence of borehole disposal is more applicable to a breeder reactor system than to an LWR one. This is because the disposal of potentially useful plutonium fuel is no longer a significant concern when using breeder reactors, due to the facts that they are capable of producing additional fissile fuels (i.e., uranium-233 and plutonium-239) as needed *and* that they intrinsically employ inputted uranium and thorium with such exceptionally-high mass efficiency.

<u>System using traveling wave reactors</u>

Simply put, a traveling wave reactor is a special type of breeder reactor with an exceptionally efficient neutron economy *that does not require a reprocessing plant to supply it with fresh fuel,* and *is able to operate using solely natural or depleted uranium as fuel reloads* (Teller, et al. 1996; Ellis, et al. 2010). After a TWR breeds new fissile plutonium-239 from uranium-238, it can continue to directly burn this freshly-bred fuel in a single pass, without needing the plutonium to be reprocessed and formed into new fuel. To start up this breed-burn process in a TWR, it is sufficient to provide a one-time investment of "fissile starter fuel," such as low-enriched uranium (~15% uranium-235) from an enrichment plant, after which no further fuel enrichment is ever required—even for 'successor reactors' whose fuel-loads can be directly taken from a 'predecessor reactor' reaching the end of its service lifetime.

Because its fuel never undergoes reprocessing, a TWR does not completely fission 100% of natural uranium. Instead it fissions about 20% of it, after which the used fuel can be sent to a deep borehole for disposal. As a result of this, a nuclear energy system using TWRs would require a factor of 5 more uranium input as a standard breeder reactor system of the same power and energy rating which is still a factor of 30 less uranium than an equivalent fleet of LWRs. The amount of seawater processing and uranium adsorbent materials required would therefore scale proportionally. Meanwhile, the number of boreholes (or branches) required would actually decrease by a factor of two, to ~800 per year for an all-planet 100 TW primary energy-supply system, because entire intact fuel assemblies can be disposed of, with no volume added due to reprocessing.

The same synergies between seawater extraction of uranium, breeder reactors, and borehole disposal are present in the TWR system as were in the one based on 'standard' type breeders. The increased burnup attainable with TWRs (~20% of fuel-loaded) relative to LWRs (~6% of fuel-loaded, or ~0.6% of mined/extracted uranium, when mass inefficiencies in isotopic enrichment is considered) makes power from a TWR energy system far less sensitive to the price of uranium, and therefore able to utilize uranium extracted from seawater—even when the employing contemporary, early technology—in an economically-practical manner.

Meanwhile, the TWR energy system benefits fully from borehole disposal, because of its greatly-reduced waste volumes and its indifference to the irretrievable disposal of bred-but-unburnt plutonium in the fuel-waste stream. Finally, the TWR is unique among breeder reactors in *its ability to sustain full energy production operations indefinitely with no additional fissile material inputted from an enrichment or reprocessing facility.*

Sustainability of nuclear energy systems

Figure 1 shows the duration and power scales over which the two energy systems just overviewed can be sustained on an "energy scenario map", along with corresponding results for light water reactors (LWRs) and fossil fuels. In Figure 1, the starting point for all scenarios is one that corresponds to a global civilization using 100 TW of primary power. From this starting point, the lower bound in Figure 1 represents a scenario in which energy use stabilizes at the aforementioned 100 TW level, while the upper bound in illustrates a scenario in which global energy use grows steadily at a rate of 3% per year, continuing the long-term global trend from the 20th century. The area between these two bounds includes all scenarios that have intermediate or variable growth rates between none and the singularly-great one of the past century.

As indicated by the dashed horizontal line in Figure 1, a level of energy use exceeding 10 PW is likely bounding for a fundamental physical reason: it begins to approach-from-below the total incoming energy-rate the Earth receives from the Sun (~170 PW). For terrestrially-sourced energy such as fossil fuels, nuclear fission/fusion, as well as geothermal energy, adding energy significantly beyond this scale to the energy budget of the Earth's surface would require a corresponding major increase in the Earth's surface average temperature (or its concatenated spatially-, time- and spectrally-averaged emissivity or albedo) in order to radiate this 'extra' energy-rate into space, or novel means to sink remarkably immense quantities of energy into the oceans or back into the solid Earth. For non-geothermal renewable energy of classic types, the power received from the Sun is a hard limit on how much can be harvested, since these forms of energy (wind, solar, hydro) are ultimately derived from the Sun (neglecting the small-scale exception of tidal energy). Expanding mankind's energy supply far beyond this level would require extraterrestrial energy production (and use), such as collecting solar energy in space. In such a situation, humanity will no longer be solely dependent on terrestrially-sourced energy, and standard terrestrial notions of resource sustainability would no longer apply.

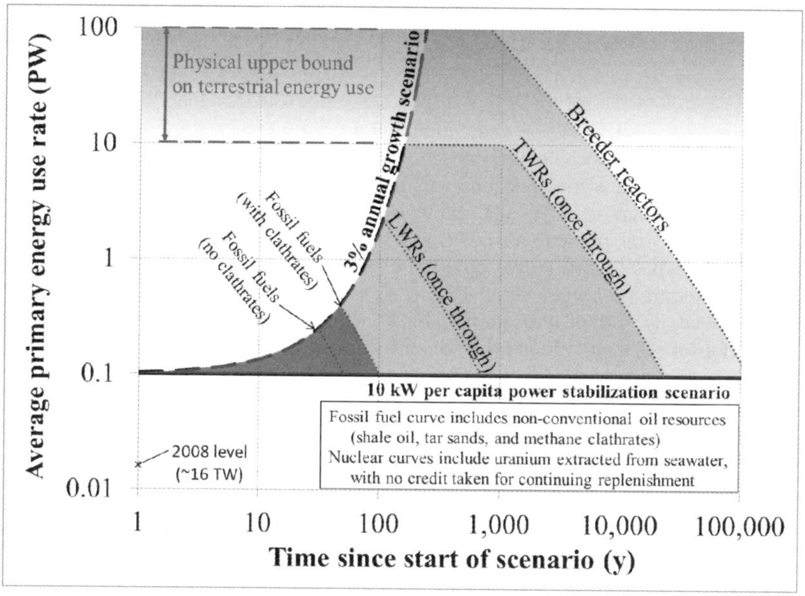

Fig. 1.

The shaded areas on this energy scenario map correspond to which energy-use scenarios are accessible by different energy systems, which depends on the total energy available to each. For example, in the breeder reactor system, there is sufficient uranium in seawater (an astonishing 4,600,000,000 MT) to power a 100 TW civilization for over 100,000 years, or allow civilization to grow by 100X above the reference 100 TW one—to 10 PW—over a period of roughly 5,000 years. The space covered by TWRs covers less of the map, but still easily supplies enough energy to last for many millennia and remarkably large multiples in total energy use. (There is a horizontal limit at 10 PW in the TWR scenario corresponding to roughly how many TWRs (~3,000,000) can be started up using the uranium present in seawater, if one uses only enriched uranium as the initially-invested "starter fuel" in each one; this limitation is inapplicable to other TWR scenarios, for example if used TWR fuel is used a second time to start up new "daughter" TWRs.) In contrast, fossil fuels contain much less total energy and thus cover a much smaller portion of the map. Even including all speculative sources of fossil fuels (oil sands, shale gas and shale oil, and under-ocean methane clathrates), fossil fuels are unable to sustain a 100 TW energy system for more than a single century. Indeed, Figure 1 very clearly shows how nuclear systems are qualitatively distinct from fossil fuels from the consideration of planetary-scale sustainability of power supply.

For both types of nuclear energy systems described above, the extraordinarily-high energy density and innately-clean nature of this type of energy mean that both perform exceptionally well with respect to environmental impact metrics such as land use, habitat preservation, and greenhouse gas and pollutant (e.g., SO_2, NO_x, fly-ash, carbonaceous and metallic particulates...) emissions. On the front end, uranium mining or extraction from seawater can be made to have very little environmental impact, simply because of the relatively modest amounts of uranium required and the isotropically-negligible consequences of extracting it from seawater. The extremely high energy value of uranium fuel implies that the costs of habitat protection and restoration involved in its use have minimal impact on the economics of nuclear energy, and therefore even a conventional uranium *mining* operation is far more capable of paying for high-quality land remediation than other types of mining might be.

As discussed previously, only a very small (~1% of a single ocean current) fraction of oceanic currents would need to be processed to harvest sufficient uranium for a 100 TW planetary energy economy, so it will be possible to use spatially-diffuse collectors that do not disturb local habitats and can readily avoid protected areas. Uranium adsorbent materials and their mooring/ship attachments naturally would need to be appropriately engineered so as to not impact ambient marine life to untoward extents. Meanwhile, the gradual rate of uranium accumulation means that uranium collection modules can move slowly—if indeed they are moved at all—and with correspondingly minimal fuel consumption.

As for the nuclear facilities themselves, 30,000 large nuclear power plants sourcing 40 TWe from 100 TW of primary power would occupy an area on the order of 30,000 – 100,000 km^2, a very modest aggregate area for providing the entire globe's energy needs of all types. Taking the continental United States as an example, its share of ~1000 reactors would occupy an area of 1,000 to 3,000 km^2, equal to 0.013-0.04% of its total land area. In comparison, the total area occupied by *present day* U.S. coal mining operations *alone* occupies approximately 8,000 km^2 (Serchuck, 2000), not including areas disturbed by past mining. As another comparison, using the equivalent land area for high-efficiency solar collectors would produce about 20 times less peak power and 100 times less energy in the time-average than the land area-equivalent nuclear power system, due to the diffuse and intermittent nature of solar energy—and wouldn't be baseload-capable. Land area required for borehole disposal of all plant wastes only increases the required area by a factor of 0.015-0.05% of the base plant area per year, implying the feasibility of colocation of plants and waste disposal boreholes—which may be a significant operational convenience.

As noted previously, the waste produced by nuclear reactors is small in volume due to the energy density of nuclear fuel, and may therefore be completely localized. Nuclear energy produces no dispersed atmospheric pollutants such as the sulfur and nitrogen oxides, particulate, soot, and heavy metal smokes that are released from burning coal or petroleum. Since pyrometallurgical reprocessing of relevance to conventional breeder reactors reuses its process fluids, it does not add significant volume to the nuclear waste stream, as noted previously. The absolutely-small volume of the resulting nuclear waste-streams means that it can all be disposed of geologically, e.g., via emplacement in deep boreholes.

Similarly, the lifecycle greenhouse gas emissions of nuclear energy are very small. A metastudy by Sovacool (2008) of the *contemporary* nuclear energy system cites values of greenhouse gas emissions between 1.4 and 288 g CO_2 equivalent/kW-hr of electricity produced; in comparison, fossil fuel combustion results in 500–1000 g CO_2 equivalent/kW-hr. The huge range in LWR values—and its peculiarly-large upper-end—is due to differences in both the study methodologies employed and the scenarios studied. The upper-end values are largely due to use of legacy gas diffusion enrichment plants powered by fossil-fuel generated electricity. In contrast, modern centrifuge enrichment plants are able to use 50 times less electricity per unit of enrichment—and CO_2 emissions from enrichment plants can be eliminated entirely by utilizing nuclear- or renewable-sourced electricity. Both of the nuclear energy systems considered above—the standard breeder and the TWR advanced breeder—greatly reduce mining and enrichment energy requirements below those of the current LWR-centered nuclear energy system, and therefore can achieve greenhouse gas emissions-levels several hundred-fold below those of fossil fuel power-equivalents, levels easily sufficient to essentially halt atmospheric build-up of CO_2.

As discussed in the introduction, the range of applicability of a given technology is an important aspect of its sustainability. Currently, nuclear energy is used almost entirely for production of electricity, accounting for 20% of U.S. electricity production and about 13% of worldwide electricity-sourcing (IAEA, 2012). As a source of baseload electricity, nuclear energy can be used to provide a large fraction of total electricity demand, e.g., as in France, which obtains nearly 80% of its electricity from nuclear reactors. It would be possible to increase the share of nuclear electricity considerably beyond a baseload level (while using a minimum of energy storage or variable energy sources) by introducing major uses of electricity that are time-insensitive and energy cost-dominated, such as charging secondary batteries associated with electric ground vehicles at night, aluminum production, heat-pumping such as making ice/chilled-brine or heated water at night for the next day's cooling and heating uses, and water desalination/purification and distribution. As computing data centers spread around the globe, it may also be possible to distribute computational loads such that the majority of all computing and associated telecommunications—presently consuming about 2% of all planetary electricity and predicted to rise to ~10% by mid-century—takes place at locations where electricity is locally cheap.

Desalination and water distribution can offset variations in electricity demand to an excellent degree in regions without abundant and conveniently-positioned fresh water. As an example, U.S. *per capita* electricity consumption presently averages around 40 kW-hr per day, and around one-third of this—13 kW-hr—is needed to fill in the nighttime valley in energy demand. Meanwhile, U.S. *per capita* fresh water consumption is around 4.5 m^3/day, which would require ~10 kW-hr to produce from seawater using reverse osmosis, and an additional ~3 kW-hr to pump over distances required for long-range irrigation. Therefore, it appears to be a fitting synergy for a nuclear reactor to supply electricity for residential and commercial use in the daytime, and then have its extra capacity at night used toward the production and distribution of fresh water.

Nuclear energy can also be used for many traditionally non-electric applications. It can be used for efficient residential and commercial heating via heat pumps (effectively air conditioners operated in reverse), or from providing a heat source for powering district heating and cooling systems. Plug-in hybrids and electric cars and trains allow electricity to be used for transportation, and nuclear electricity and/or high-temperature heat can be used to produce hydrogen from water, a versatile energy carrier applicable to large scale transportation including aircraft. For very-large scale transportation, i.e., very large ships, with 20-300 MWt power plants, it is not unreasonable to apply nuclear reactors directly to power their propulsion, as major military powers have done for several decades now and as the Soviets and Russians have done for a good-sized fleet of nominally-civilian icebreakers for use in their Arctic waters.

Overall, nuclear reactors are already able to meet a large fraction of global energy demand by sourcing a major fraction of electricity production, which will continue to grow in importance as human civilization naturally becomes increasingly electrified. Nuclear energy clearly has the potential to satisfy ever larger portions of humanity's total energy demands, especially as new technologies and practices are developed to continue to drive down costs of employing cheap, high-grade nuclear heat. Indeed, it's not easily feasible to identify energy-use sectors that aren't intrinsically amenable to being supplied with nuclear-derived energy.

CONCLUSION

By concatenating uranium extraction from seawater, manifestly safe breeder reactor technology, and borehole disposal of nuclear waste, it is possible to create a nuclear energy system that is *planetary-scale sustainable*, i.e., one that's capable of providing such an immense amount of energy at such high power levels that it can be counted upon to fuel most—and potentially all—of human civilization throughout essentially all future scenarios of serious interest. In this regard, nuclear energy is *qualitatively* different from other consumable energy resources and must be objectively deemed to be fully sustainable in all respects. In the near term, the ability to thereby develop and demonstrate *manifest safety* with the current generation of new nuclear reactors likely will be essential for laying the groundwork for a sustainable future based on nuclear energy.

Civilization's current reliance on fossil fuels for sourcing of primary energy is not just innately challenging due to long-term resource availability, but also may be problematic from a relatively short-term ecosystems perspective, because our current net rate and rate-of-growth of greenhouse gas emissions into the atmosphere, if sustained through even the present century, seemingly shadows the thermal structure and major meteorological features of the biosphere of future generations over the next several centuries, until most of the associated CO_2 pulse becomes sunk into the oceans. Furthermore, these impacts are likely distributed across all populations around the globe, so that means to properly remediate them by "making whole" any and all persons that are arguably negatively affected to one-or-another extents would be immensely difficult to put into practice.

A nuclear energy infrastructure, properly and *sustainably* implemented, is one of relatively few options which are reasonably and reliably accessible that can potentially fully address these significant, comparatively near-term challenges. Therefore, it seems important that we as creators and residents of a global civilization begin to lay the technical and institutional groundwork that will enable a sustainable, full-scale nuclear energy system to become an operational reality in the adequately near term, while concurrently doing likewise with respect to other practical forms of sustainable energy supply at-scale.

REFERENCES

1. Ackerman, J.P., Johnson, T.R., Chow, L.S.H., Carls, E.L., Hannum, W.H., Laidler, J.J. (1997) "Treatment of Wastes in the IFR Fuel Cycle"; *Progress in Nuclear Energy*, Vol. 31, No. 1/2, pp.141-154.
2. Bloomberg News, "Nuclear Boom In China Sees Reactor Builders Risk Their Know-how For Cash"; available http://www.bloomberg.com/news/2010-12-02/china-nuclear-boom-sees-reactor-builders-risk-know-how-for-cash.html; Dec 2, 2010.
3. Ellis, T., Petroski, R., Hejzlar, P., Zimmerman, G., McAlees, D., Whitmer, C., Touran, N., Hejzlar, J., Weaver, K., Walter, J., McWhirter, J., Ahlfeld, C., Burke, T., Odedra, A., Hyde, R., Gilleland, J., Ishikawa, Y., Wood, L., Myhrvold, N., Gates, W. (2010) "Traveling-Wave Reactors: A Truly Sustainable and Full-Scale Resource for Global Energy Needs", *International Congress on Advances in Nuclear Power Plants* 2010, ICAPP 2010, v 1, p 546-558, San Diego, CA.
4. Epstein, W. (2011) "A Probabilistic Risk Assessment Practitioner looks at the Great East Japan Earthquake and Tsunami," Tokyo Institute of Technology, Ninokata Laboratory white paper, April 19, 2011.
5. International Energy Agency, Electricity/Heat in World in 2009, available http://www.iea.org/stats/electricitydata.asp?COUNTRY_CODE=29, 2012.
6. Olipitz, W., Wiktor-Brown, D., Shuga, J., Pang, B., McFaline, J., Lonkar, P., Thomas, A., Mutamba, J.T., Greenberger, J.S., Samson, L.D., Dedon, P.C., Yanch, J.C., Engelward, B.P. (2012) "Integrated Molecular Analysis Indicates Undetectable DNA Damage in Mice after Continuous Irradiation at ~400-fold Natural Background Radiation", *Environmental Health Perspectives*, http://dx.doi.org/10.1289/ehp.1104294, April 26, 2012.
7. Oak Ridge National Laboratory (2012) "ORNL technology moves scientists closer to extracting uranium from seawater"; ORNL news release; Aug. 21, 2012; available online: http://www.ornl.gov/info/press_releases/get_press_release.cfm?ReleaseNumber=mr20120821-00
8. Sakai, T., Takeda, T., Soraoka, H., Yanagisawa, K. (2006) "Development of a Probabilistic Tsunami Hazard Analysis in Japan", *Proceedings of ICONE 14*, International Conference on Nuclear Engineering, July 17-20, Miami, FL.
9. Serchuck, A. (2000) "The Environmental Imperative for Renewable Energy: an Update," *Renewable Energy Policy Project*, April 2000.
10. Sovacool, B.K. (2008) "Valuing the greenhouse gas emissions from nuclear power: A critical survey", *Energy Policy* 36 (2008) 2940-2953.

11. Tamada, M., Seko, N., Kasai, N., Shimizu, T. (2006) "Cost Estimation of Uranium Recovery from Seawater with System of Braid Type Adsorbent", *Transactions of the Atomic Energy Society of Japan,* v 5, n 4.
12. Teller, E., Hyde, R., Ishikawa, M., Nuckolls, J., Wood, L. (1996) Completely Automated Nuclear Reactors for Long-Term Operation II.", Proceeding of ICENES'96, International Conference on Emerging Nuclear Energy Systems, June 24-28, Obninsk, RF.
13. Westinghouse (2007) *AP1000 Design Control Document*, Chapter 19.
14. World Nuclear Association (2012) "World Nuclear Power Reactors & Uranium Requirements", available http://www.world-nuclear.org/info/reactors.html, May 2012.

ECONOMICS OF NUCLEAR POWER IN LIBERALIZED POWER MARKETS[1]

CARMINE DIFIGLIO[2]
Office of Policy and International Affairs, U.S. Department of Energy
Washington, USA

BRENT WANNER[3]
Directorate of Global Energy Economics
International Energy Agency

ABSTRACT

Projections of electricity markets such as the International Energy Agency's (IEA) *World Energy Outlook* (*WEO*) or the Energy Information Administration's (EIA) *Annual Energy Outlook* (*AEO*) rely on estimated levelized costs for nuclear power and other power generation technologies. These levelized cost estimates show nuclear to be a very competitive power generating technology in most regions of the world. This paper analyzes unpublished data from the *2011 WEO* to assess the competitiveness of nuclear power in liberalized and non-competitive power markets. We consider alternative estimates of nuclear capital costs (Risk Scenario) that might well be considered before investing in nuclear power in competitive markets without government guarantees. The Risk Scenario also assumes lower natural gas prices that could result from higher shale gas recovery in the United States and new shale gas development in Europe. The Risk Scenario causes the relative cost difference of nuclear power vs. natural gas to change by 5 cents/kWhr. Instead of nuclear power holding a competitive advantage over natural gas, in the Risk Scenario, natural gas gains a competitive advantage over nuclear power, even if a \$50/ton CO_2e charge is assessed.

The analysis suggests that nuclear power plants will not be built in liberalized power markets without government guarantees. If nuclear power plants built with government support or those built by state-owned power industries prove that nuclear power plants can be reliably built at a competitive cost, power plant economics will be a significant positive factor for the nuclear industry, especially taking into consideration national commitments to reduce greenhouse gas (GHG) emissions. If costs continue to increase over the estimates being used by the IEA and EIA, power-plant economics become a significant negative factor for the nuclear industry that may not be offset by pricing CO_2 emissions.

These findings do not apply to reactor projects in many Asian countries. Nuclear power is estimated to enjoy a more favorable cost advantage than in the United States, Europe or Japan. In addition, private financing is less important in many Asian countries and delays due to licensing problems are less likely. Given the high electricity demand

[1] The data and views expressed in this paper are those of the authors and are not endorsed by the U.S. Department of Energy, the United States government or the International Energy Agency.
[2] Deputy Assistant Secretary for Policy Analysis , U.S. Department of Energy: carmine.difiglio@hq.doe.gov.
[3] Directorate of Global Energy Economics, International Energy Agency: brent.wanner@iea.org.

growth expected in Asia, more competitive costs, and government support, it is not surprising that over 60% of currently-planned nuclear reactors are in Asia.

INTRODUCTION

Estimating the future mix of electric power generation technologies usually begins with a comparison of the costs of competing forms of electricity generation. Most major energy models employ a "bottom-up" methodology that takes into consideration estimated capital, fuel costs and operating costs of alternative power-generation technologies. These are usually expressed as a "levelized-cost." Levelized cost represents the wholesale cost to produce each kilowatt hour (kWhr) of electricity over the economic life of the power-plant investment. While energy models consider a variety of factors, including the dispatchability of power in relationship to daily and seasonal power demand, there is a tendency to estimate high future market shares for power technologies with the lowest levelized costs. Given the estimated levelized costs of nuclear power compared to natural gas, coal and renewables, one would expect the projected growth of nuclear power to be greater than most projections might imply. This is partly explained by different public policies among countries. Modelers must restrict model projections in countries that have national policies that are adverse to nuclear power development. However, there may be other reasons why current cost estimates may imply a stronger future for nuclear power than may reasonably be expected. This paper considers the factors that might explain an apparent mismatch between estimated costs and realistic expectations for nuclear power expansion, even in markets that price GHG emissions.

LIBERALIZED POWER MARKETS

For the purposes of this paper, power markets are divided into two general types: traditional and liberalized. Traditional power markets are uncompetitive arrangements that adhere to the concept of a natural monopoly. A natural monopoly is a condition in which a single firm can provide a product or service more efficiently than multiple firms. Since a power company must provide power-lines to their customers, it would be inefficient to have multiple power lines from different power providers. For this and other reasons, integrated power companies arose that encompassed all aspects of providing power to consumers, from generating electricity through distribution to consumers. These companies have traditionally been regarded as natural monopolies that should enjoy exclusive access to customers. These natural monopolies have usually been organized as a state-owned company or a private company with an exclusive right to serve that must seek approval from government regulators when making large investments or setting prices.

Utilities and regulators use estimates of levelized cost to acquire a "level-playing-field" comparison of alternative power plant investments that have substantially different combinations of capital cost, fuel cost, operating cost, and plant life. Levelized cost estimates the per-kilowatt hour (kWr) cost of a technology derived from the present value of building and operating a generating plant, assuming a particular duty cycle, amortized over its economic life. It was a tool that allowed State regulators to determine whether to approve investments in new power plants by private utilities. Levelized cost estimates allowed the regulatory authorities to evaluate whether new power plant investments would be prudent and the likely effect they would have on consumer costs over long periods of time. If approved, the utility could proceed with the power plant investment and the utility's banks and shareholders would be confident that the power plant would generate revenues sufficient to pay the loans and provide expected dividends.

The U.S. utility industry was initially unregulated. Thomas Edison's first power plant began operation in 1882 (New York City). By the beginning of the 20th century there were about 3,000 unregulated U.S. electric firms in addition to almost 1,000 municipal systems. State regulation of power plants began in 1907 in New York and Wisconsin. The private providers actively sought State regulation as the competition among power providers was counterproductive and the private companies needed some protection against the municipal providers. Under state regulation, there were fewer power companies as the investor-owned utilities became much larger. A collapse of utility stocks during the Great Depression led to the Public Utility Holding Company Act (PUHCA) that gave significant powers to the Securities and Exchange Commission (SEC) to control the structure of investor-owned utilities. In 1935, the Federal Power Commission was created to oversee Interstate transmission of electricity, including rates.

Between 1935 and 1970 U.S. electricity consumption grew at an annual rate of 7.9% requiring predictable power plant additions. Under the U.S. regulatory regime, the investor-owned utilities would acquire permission from their State regulator to build the plants. They were assured that the cost to build, fuel and operate these additional plants would be considered when setting tariffs and they were assured that these investments would provide a normal rate of return to the utilities' investors. The cost of carrying excess generating capacity was also included in the rate base. Consequently, the investors' returns, within limits, did not depend on whether actual electricity demand fulfilled expectations when the power plant investments were planned.

Levelized cost calculations are also considered by state-owned power companies. In each type of market, investor-owned regulated utilities or state-owned utilities, the investment in a new power plant is substantially less risky than private sector investments without guarantees. Until the 1970s in the United States, there was little likelihood that a power-sector investment would adversely affect a privately-held utility. Regardless of less-than-expected electricity sales, capital cost overruns or increased fuel costs, retail electricity rates would be adjusted to make the investment whole.

These arrangements were taken for granted until the mid-1970s energy crisis. High oil costs required retail rate increases that led to a flattening of electricity demand ending decades of predictable electric power growth. This particularly affected the nuclear power industry as plans for 113 U.S. nuclear power plants were scrapped between 1972 and 1984. At the same time, many players in the U.S. power industry and non-governmental organizations (NGOs) began to propose new ways to organize the power industry recognizing that it consisted of three distinct services: generation; transmission; and local distribution. A dozen U.S. States implemented regulatory reforms that allowed competition for industrial and private consumers. The reforms also affected markets that remained regulated. While regulated utilities could still build power plants and have them admitted to the rate base, they often have to show that their plants would provide power at lower costs than contracting for power from independent power producers. Several regulated utilities also purchase power from unregulated private companies using competitive contracting procedures.

These reforms can be generally characterized as a "liberalized power market." Any party considering a power-plant investment in a liberalized power market realizes that it carries the risks of any business investment, substantially riskier than a corresponding investment by a non-competitive regulated utility. In addition, regulated utilities in States that have not implemented significant regulatory reform often evaluate power plant investments as if they were engaged in a competitive market since these utilities are less confident that State regulators will add the full cost of new power-plant investments in the rate base if there is excess generating capacity or if the cost of the added generation is higher than expected.

IEA'S *2011 WEO* ESTIMATES OF LEVELIZED COSTS

Table 1 shows the estimated long-run marginal cost of four power-generation technologies in the United States, the European Union, China and Japan. These estimates underlie analysis used in the *2011 WEO's* World Energy Model. There is a wide range of estimated levelized costs for renewable energy since a variety of power-generating technologies are included in this group, from inexpensive wind power to more-costly solar power. The costs shown for natural gas only reflect combined-cycle base load plants. Peaking plants were not included since these would not be considered to be alternatives to nuclear power plants. The estimated costs for coal plants include a variety of coal power technologies with different operating efficiencies. The estimated nuclear costs reflect a Generation III+ pressurized-water or boiling water plant. Plants of this type include the Westinghouse AP1000, the General Electric Economic Simplified Boiling Water Reactor and similar designs from other manufacturers. IEA's *WEO* capital cost estimates were informed by *Projected Costs of Generating Electricity, 2010 Edition* published by the IEA in collaboration with the Nuclear Energy Agency (IEA/NEA 2010).

The levelized costs for nuclear power are estimated to be very similar in the United States, the European Union and Japan. However, the levelized cost for nuclear power is substantially lower in China (about one-half). The IEA/NEA estimates that the overnight cost of building a Generation III+ reactor in China is about one-third lower than in the United States ($2,302/kWe vs. $3,382/kWe).[4] After taking into account the differences in estimated construction times and financial risks, the IEA's estimated levelized cost is about half as much in China compared to the United States (4 cents/kWhr vs. 8 cents/kWhr). Construction times are estimated to be lower in China due to an entirely different structure of licensing and oversight than exist in the United States. Likewise, financial risk to the State-owned power industry is substantially lower in China than in the United States unless the power-plant investment has Federal loan guarantees and backing from a State Public Utility Commission.

Renewable energy technologies have a wide variation of estimated levelized costs. The lowest costs are for wind power with favorable wind energy environments. When the best wind resources are taken, the cost of wind power increases as less-productive wind sites are employed or if wind has to move offshore. Other renewable

---

[4] p. 59, IEA/NEA 2010.

technologies such as solar or geothermal tend to be at the higher end of the estimated costs for renewable energy and, consequently, can be the most expensive sources of electricity. In addition, these estimated levelized renewable electricity costs do not include natural gas peaking plants or energy storage systems that are typically required as back-up for intermittent renewable energy production.

Table 1: Long Run Levelized Cost (U.S. cents/kWhr)[5] in 2035.

	Nuclear	Renewables	Natural Gas	Coal
United States	7.5 – 7.8	7.7 – 12.8	8.5 – 8.8	8.1 – 8.7
EU	7.3 – 7.6	7.8 – 15.9	10.4 – 10.8	9.6 – 11.4
China	3.8 – 4.1	4.8 – 9.5	9.6 – 10.8	5.7 – 6.9
Japan	7.9 – 8.2	7.6 – 15.3	12.1 –12.9	10.3 – 11.6

Nuclear power is estimated to have the lowest estimated levelized costs in the United States, European Union, China or Japan. If competitive levelized costs were the dominant factor determining future power investments, the share of nuclear power should increase in these countries and world-wide.[6] Competitive nuclear costs are not relevant in countries that have polices to restrict or prohibit new nuclear reactors. Opposition to nuclear power becomes stronger and more widespread with nuclear accidents. Fortunately, nuclear accidents are infrequent but, when they occur, they can reverse a country's policies on nuclear power. The Fukushima accidents have brought about a near-total loss of confidence in the Japanese nuclear power industry. As a result, a country that had been committed to increasing its fleet of nuclear power plants is now considering a phase out of its existing nuclear fleet. Nonetheless, nuclear power remains an acceptable power option in most countries. The cost estimates in Table 1 suggest that nuclear power should be an attractive power investment in these countries.

Nuclear power economics improve if there are GHG taxes or a GHG cap-and-trade system. Table 2 shows IEA levelized cost estimates with a \$50/ton CO_2 charge added to the cost of natural gas and coal power plants.

Table 2: Long Run Levelized Cost (U.S. cents/kWhr)[7] with a \$50/tonne CO_2 Charge in 2035.

	Nuclear	Renewables	Natural Gas[8]	Coal w CCS[9]	Coal w/o CCS
United States	7.5 – 7.8	7.7 – 12.8	10.2 – 10.5	8.5 – 11.8	11.8 – 13.1
EU	7.3 – 7.6	7.8 – 15.9	12.1 – 12.5	10.2 – 13.3	13.3 – 16.1
China	3.8 – 4.1	4.8 – 9.5	11.4 – 12.6	6.3 – 9.4	9.4 – 11.6
Japan	7.9 – 8.2	7.6 – 15.3	13.7 –14.7	NA[10]	14.0 – 16.5

---

[5] Unpublished data from the World Energy Model used for the IEA *2011 WEO* (2008 U.S. cents).

[6] Based on IEA data, the relative levelized cost estimates presented for the United States, European Union, China and Japan are not significantly different than those estimated for India, Brazil and Russia.

[7] 2008 U.S. cents.

[8] The carbon charge for natural gas (1.7 cents/kWh) is for a combined cycle gas turbine without CCS and is based on a plant efficiency of 59% and fuel carbon content of 2.3 tCO_2 per toe.

[9] The carbon charge for coal varies by technology, for an ultra supercritical coal plant without CCS it is 3.7 cents/kWh and is based on a plant efficiency of 47% and fuel carbon content of 4.0 tCO2 per toe. CCS plants are estimated to have the lowest costs with a CO_2 charge if favorable assumptions are made about CCS deployment.

[10] The geology in Japan does not favor carbon storage.

We have estimated the carbon charge for natural gas plants assuming efficient base-load natural gas plant instead of a less-efficient peaking plant. A base load natural gas plant is the alternative to a nuclear reactor investment. Without the carbon charge, U.S. natural gas plants are estimated to produce electricity at 0.7 to 1.3 cents/kWhr higher cost than a nuclear plant. With the $50 carbon fee, this increases to a much more significant advantage of 2.8 to 3.5 cents/kWhr (natural gas becomes 31% to 40% more expensive than nuclear power). The advantage of nuclear power over natural gas with a carbon charge is estimated to be higher in the European Union, China and Japan: a $50/ton CO_2 charge increases the cost penalty of natural gas compared to nuclear to as high as 71% in the European Union, 232% in China and 86% in Japan.

Table 3: Cost Advantage of Nuclear Power Relative to Natural Gas with and without a $50/tonne CO_2 Charge in 2035.

	Cost Advantage of Nuclear over Natural Gas w/o Carbon Charge		Cost Advantage of Nuclear over Natural Gas with Carbon Charge	
	U.S. cents/ kWhr	% Cost Increase Over Nuclear	U.S. cents/ kWhr	% Cost Increase Over Nuclear
United States	0.7 – 1.3	9 – 17	2.4 – 3.0	31 – 40
EU	2.8 – 3.5	37 – 48	4.5 – 5.2	59 – 71
China	5.5 – 7.0	134 – 184	7.3 – 8.8	178 – 232
Japan	3.9 – 5.0	48 – 63	5.5 – 6.8	67 – 86

As shown in Table 2, we have estimated the effect of a carbon charge on two different types of coal plants: with and without carbon capture and storage (CCS). A $50/ton CO_2 charge is sufficient to make CCS plants more competitive than coal plants without CCS. If CCS could be deployed at costs estimated for 2035 by the IEA, a $50/ton CO_2 charge only adds about 0.4 cents/kWhr to the estimated levelized cost of a coal plant with CCS. Since this does not constitute a significant cost increase, we will not discuss how a carbon charge affects the competitiveness of coal plants with CCS. For the remainder of this paper, we will only compare the competitiveness of nuclear plants to coal plants without CCS.[11] As shown in Table 4, a $50/ton CO_2 charge causes nuclear power to have cost advantages over coal at least as high as 51% (minimum U.S.) and as high as 205% (maximum China). Given this carbon charge ($50), natural gas and coal cease to be competitive power technologies compared to nuclear in these four countries, and, barring special circumstances, world-wide.

[11] As with cost estimates for nuclear power, cost estimates for CCS are highly uncertain and CCS faces deployment obstacles that may be has great or greater than those for nuclear power.

Table 4: Cost Advantage of Nuclear Power Relative to Coal (no CCS) with and without a $50/tonne CO_2 Charge in 2035.

	Cost Advantage of Nuclear over Coal w/o Carbon Charge (no CCS)		Cost Advantage of Nuclear over Coal with Carbon Charge (no CCS)	
	U.S. cents/ kWhr	% Cost Increase Over Nuclear	U.S. cents/ kWhr	% Cost Increase Over Nuclear
United States	0.3 – 1.2	4 - 16	4 – 5.6	51 – 75
EU	2.0 – 4.1	26 – 56	5.7 – 8.8	75 – 121
China	1.6 – 3.1	39 – 82	5.3 – 7.8	129 – 205
Japan	2.1 – 3.7	26 – 47	5.8 – 8.6	71 – 109

PROJECTED POWER GENERATION IN THE *2011 WEO*

The *2011 WEO* provided three projection scenarios: 1) Current Policy Scenario—assumes no change to current national polices; 2) New Policy Scenario—assumes the adoption of policies that have been announced to address energy security and environmental goals; and 3) 450 Scenario—assumes policies sufficient to maintain world-wide atmospheric GHG concentrations at 450 parts per million CO_2e.

The New Policy Scenario assumes that, by 2035, the EU, Korea, Australia and New Zealand have adopted a $45/ton CO_2 charge and China has adopted a $30/ton CO_2 charge. The 450 Scenario assumes, by 2035, CO_2 charges of $120/tonne or $90/tonne have been adopted in the OECD, China, Russia, Brazil and South Africa. The 450 Scenario is a very ambitious end-state driven scenario that would require immediate uptake of strong climate policies in the twenty or so countries that are responsible for the bulk of world-wide CO_2 emissions. The types of changes to the world energy economy implied by the 450 Scenario appear less likely with each passing year (Difiglio, 2010). Consequently, for the purposes of this paper, we only consider the *2011 WEO* Current Policy and New Policy Scenarios.

Electric power generation is expected to grow from 20 thousand TWhr to 36 to 39 thousand TWhr (New to Current Policy Scenarios respectively). As might be expected, the *2011 WEO* projects substantial additions to coal, natural gas, renewables and nuclear power generation capacities through 2035. However, despite the favorable costs estimated for nuclear power, nuclear power is estimated to have a smaller share of world-wide electricity supplies than in 2009 in either the Current Policy or New Policy Scenarios. The share of nuclear power is only expected to increase in the 450 Scenario with its very high carbon charge.

Table 5: IEA *2011 WEO*—World-Wide Projected Share of Electric Generation Technologies in 2035 (percent).

	Fossil Fuels	Nuclear	Hydro-Electric	Other Renewables
2009	67.1	13.5	16.2	3.2
Current Policy Scenario	66.5	10.3	13.1	10.1
New Policy Scenario	56.5	12.8	15.2	15.4

The projected share of nuclear power is also estimated to be smaller in OECD countries in the Current Policy Scenario and the New Policy Scenario as shown in Table 6.

Table 6: IEA 2011 WEO—OECD Projected Share of Electric Generation Technologies in 2035 (percent).

	Fossil Fuels	Nuclear	Hydro-Electric	Other Renewables
2009	60.7	21.6	12.7	5.1
Current Policy Scenario	55.3	17.7	11.1	15.8
New Policy Scenario	46.3	20.9	12.0	20.8

The *2011 WEO* lists several factors causing a cut back in nuclear plant builds. These include: government reviews of nuclear policy following the Fukushima accidents; early retirement of German nuclear plants by 2022; no lifetime extension of nuclear plants in Switzerland; delays in capacity additions in China; and difficulties in securing financing for nuclear power plants. The *2011 WEO* also cites an increase in expected construction costs, compared to the *2010 WEO*, of 5% to 10%.

ASIA

As shown in Tables 3 and 4, the advantage of nuclear power over natural gas or coal is substantially higher in China than in the United States, the EU or Japan. This cost advantage is also shared by other Asian economies. In addition, the structure of the power markets in Asia, the importance of government investment instead of private investment, government policies favoring nuclear power, less likelihood of delays due to regulatory problems and a relative absence of political opposition all make it likely that nuclear power will expand in Asia. Consequently, of the currently planned nuclear reactors, 130 are in Asia, 54 are in Western and Eastern Europe, 8 are in the Middle East, 11 are in North America and 4 are in South America (World Nuclear Association, 2012).

EXPANDING LIBERALIZED ELECTRICITY MARKETS WOULD HELP MEET WORLD-WIDE GROWING ELECTRICITY DEMAND

Electricity generation is estimated to increase by 81% to 96%[12] between 2009 and 2035 (IEA, 2011). The estimated growth outside the OECD is much higher (138% to 164%). Large investments will be needed to finance this expansion of electric generating capacity. The IEA estimates that $9.8 trillion will be required to finance new power plants (New Policies Scenario). Of this, $5.5 trillion will be required outside of the OECD. It is not likely that these capital requirements can be met without a predominate role for the private sector. In many of the countries outside of the OECD, several challenges need to be overcome in order for the private sector to have an effective role.

First of all, there has to be a likelihood of profits. Inefficient state-owned power systems make an attractive business model more difficult. When power is sold for less

[12] For the New Policy Scenario and Current Policy Scenario respectively. The additional efficiency investments in the New Policy Scenario reduce the need for additional electric generating capacity compared to the Current Policy Scenario.

than it costs, an investor would then depend on government payments in addition to electricity sales. When you add uncertainty about government policies and regulations, it is unlikely that sufficient capital can be raised without power-sector reforms. Power-sector reform reduces government fiscal responsibilities, improves the reliability of electricity supplies to industry and households, and can improve electricity access to the poor (Difiglio, 2010b).

A successful pattern of power-sector reform involves separating the electric utility from the relevant ministry and obliging it to operate according to commercial principles. The second step involves developing economic regulation of the power market that is applied transparently by an autonomous regulator. Third, independent power producers can help the reform process by demonstrating the benefits of private investment and management. The forth step is unbundling or restructuring the electric power supply chain to enable the introduction of competition, Fifth, the unbundled electricity segments are privatized under dispersed ownership. Finally, the development of competition in the generation and supply segments follows from the development of power exchanges (World Bank, 1993).

Non-OECD countries may adopt this pattern of reform to achieve a more efficient and reliable power sector and to reduce federal fiscal liabilities. Such reforms may be required in order to raise the $5.5 trillion needed to build the power plants that will be needed outside of the OECD. If they are, the scope of liberalized power markets will grow and, with their growth, reduce the opportunities to secure State financing or guarantees for nuclear power plants.

UNCERTAIN NUCLEAR POWER PLANT COSTS

Cost estimates for Generation III+ power plants have been based on vendor estimates rather than actual experience (Kessides, 2010). These estimates might underestimate future nuclear power plant costs compared to estimates for other power technologies that are based on numerous actual projects and contracts. Generation III plant designs emerged in the 1990s. The cost estimates for these designs were typically in the neighborhood of $1,000 per kWe. As these designs have evolved, current U.S. or European cost estimates have grown to be several times higher. Plant cost estimates in the United States currently range from $2,500/kWe to $10,000/kWe. If the outlier cost estimates are deleted, current U.S. cost estimates range from $4,200/kWe to $6,600/kWe (Thomas, 2010). Unfortunately, there is no assurance that cost escalation will not continue. Past experience in the U.S. nuclear power industry is not encouraging. For example, between 1974 and 1975, fourteen power plants were estimated by utilities to cost, on average, $1,263/kWe. Actual costs averaged $4,817 (281 percent overrun). While the cost overrun for these 1974-75 plant start ups were higher than earlier plants, the average cost overrun for all U.S. nuclear power plants was 207 percent (Schlissel, D., Biewald, B., 2008).

TECHNOLOGY LEARNING

Predictions of future energy technology costs usually employ the concept of technology learning. The general principal is: for each doubling of capacity of a new energy technology, cost goes down by a fixed percentage (IEA, 2000). Since the nuclear industry is well established, additions to capacity would fall well short of any opportunity for further technology learning unless the Generation III and III+ designs were judged to be sufficiently different than Generation II reactors to be new technologies. Most estimates of future nuclear technology costs do not view them as such and do not anticipate cost reductions in Generation III reactors. For example, the cost estimates for a European Generation III+ in the *2011 WEO* remain constant at \$3,800/kWhr from 2015 through 2035.

An analysis of the cost trajectory of the French nuclear reactor fleet indicates that construction costs, in constant dollars, have increased over time (Grubler, A., 2010). While this is consistent with U.S. experience (described above), it may be surprising since the French reactor fleet is generally regarded as the most successful and well-managed nuclear power deployment, in part, by avoiding the U.S. experience of building many different designs each requiring individual regulatory assessments.

Going forward there may be reasons for nuclear reactor costs to systematically increase. About 60% of the cost of a nuclear reactor stem from on-site engineering as opposed to the major equipment items (reactor vessel, steam generators and turbines). Reactor cooling is also an important cost component that has increased. In particular, environmental regulations to protect fish and other species have made it more likely that cooling towers would be required when reactors are located on rivers.

With high Asian economic growth that is not expected to lessen, the demand-supply balance for construction materials is likely to produce higher prices. For example, metallurgical coal prices are now at record levels despite a weakening world economy. As with petroleum, the supply of mineral resources is not likely to keep up with high world demand growth at constant prices.[13] Likewise engineering and labor costs are likely to increase with high economic growth. Compared to the opportunities to reduce the costs of manufactured products, large nuclear power plants appear to face the likelihood of higher costs, not lower costs.

IMPORTANCE OF CAPITAL COST ESCALATION AND OTHER RISKS IN A LIBERALIZED POWER MARKET

Risks of cost over-runs are less likely to deter nuclear power plants investments in uncompetitive power markets or state-owned enterprises. A costly power plant in these markets will not necessarily have adverse consequences on investors as the higher cost can be averaged into the existing fleet of power plants. A private investor in an uncompetitive State-regulated power market would still receive the normal rate of return. A state-owned company likewise passes on the incremental costs to ratepayers or the government incurs higher electric subsidy costs.

A liberalized power market presents an entirely different picture to an investor. Delays and cost over-runs not only threaten to diminish their rate of return but put the investment itself at risk. This risk is compounded by five factors: 1) nuclear power has high capital costs per kilowatt of electrical generating capacity (kWe) compared to

[13] While technological breakthroughs, such as hydraulic fracturing, can significantly reduce the cost of mineral resource extraction, these are likely to be the exception rather than the rule.

natural gas or coal power plants; 2) nuclear power has large economies of scale and require large investments for a unit of investment (1 gigawatt or more); 3) having high capital costs, nuclear power is more vulnerable to increases in material costs and construction delay; 4) construction delays are more likely due to uncertainties in the regulatory process, especially for new plant designs; and 5) because of potential public and NGO opposition to nuclear power, construction delays can also be caused by delays in issuing local permits, lawsuits and political pressure. Because of these risk factors, financing is not feasible without government guarantees. Equity financing is typically not an alternative to debt financing as it is difficult to ask shareholders to forgo dividends and accept risk that banks were unwilling to take.

Recognizing these barriers to nuclear power plant investments, the U.S. government established government subsidies and risk-sharing for nuclear power plants in the Energy Policy Act of 2005. These included production tax subsidies for the earliest new nuclear plants, guarantees against regulatory delays and a loan-guarantee program to attract private financing. While no new nuclear plants have been built take advantage of the production tax subsidies, the U.S. Department of Energy (DOE) has issued two loan guarantees for two 1.1 gigawatt Westinghouse AP1000 nuclear reactors at the Vogtle generating plant. Like all of the several loan-guarantee applications DOE received, these new plants would be built at an existing nuclear plant campus. The project sponsors include the Georgia Power Company. Notably, the Georgia Power Company does not operate in a liberalized power market. The Georgia Power Company is fully regulated by the Georgia Public Service Commission. Competition is limited to customers with manufacturing or commercial loads of at least 900 kW. The Vogtle plants received $8.33 billion of DOE loan guarantees, or approximately $3,800 per kWe capacity. The plants are estimated to cost $11.9 billion or $5,000 per capacity (Thomas, 2010). A project such as Vogtle should enjoy low interest rates. The project is guaranteed in two ways: first, the DOE loan guarantees ensure that the banks will receive 76% of the total estimated plant cost; and second, the Georgia Public Service Commission will likely guarantee any remaining exposure to the Georgia Power Company.

Because of the high cost to build a single nuclear power plant, along with the likelihood of delays and cost overruns, Moody's Investors Service has characterized a decision to build a nuclear power plant as a "bet-the-farm" risk. "From a credit perspective, companies that pursue new nuclear generation will take on higher business and operating risk profile, pressuring credit ratings over the intermediate- and long-term...Of the 48 issuers that we have evaluated during the last nuclear building cycle (1965-1995), two received ratings upgrades, six went unchanged and 40 had downgrades...the average downgrade fell four notches" (Moody's Global). It is notable that these credit downgrades affected utilities that were mostly operating in a state-regulated environment where regulators have effectively transferred the utilities' risks to ratepayers. Moody's concludes that utilities that pursue new nuclear power plants will experience credit downgrades and should: 1) rely on strategic partnerships; 2) increase reliance on equity financing; 3) moderate dividend policies to increase cash flow; and 4) adopt a "back-to-basics" focus on core electric utility operations. These recommendations are aimed at utilities that "can expect regulators to support the financial health of the utilities they regulate and will authorize recovery of investments and costs over a long time frame" (Moody's Global, 2009). In a liberalized power market where power companies and investors would not enjoy such protections, the risk of credit downgrade would be much higher.

LEVELIZED COSTS AND MODELING FUTURE NUCLEAR POWER

Considering the difference in risk between traditional and liberalized power markets, it may be necessary to reconsider how levelized cost estimates are used when modeling the uptake of power-generation technologies. Current levelized cost estimate that do not account for possible cost overruns may be perfectly appropriate when modeling the uptake of nuclear power in uncompetitive power markets. Governments or government regulators can consider their long-run need for additional power and be willing to accept problems that may be encountered with the first few nuclear plants. Governments may also be committed to nuclear power development as a matter of climate or energy security policies. If climate policies are a high priority, it is reasonable that governments would assume additional financial risk to increase the share of nuclear power as modeling studies show how difficult it would be to avoid high concentrations of greenhouse gases in the atmosphere without nuclear power (IEA, 2010). Likewise, for countries that do not enjoy indigenous fossil fuel resources, avoiding reliance on energy imports can be an important policy consideration.

A private investor considering a nuclear power plant in a liberalized power market will carefully consider the risk of construction delays and cost over-runs. Current estimates of levelized cost may not be given much weight by such an investor. If contingencies for overruns and delays are built into the capital cost estimate, the estimated levelized cost for nuclear power could be significantly higher. If these estimates showed the power-plant to be uncompetitive, it is not likely that the investment would be made.

In order to develop a levelized cost estimate that would be appropriate in a liberalized power market we have developed a Risk Scenario and, as before, compare the Risk Scenario power sector levelized costs with and without a \$50/ton CO_2 charge. This type of Risk Scenario would certainly be considered by utilities and investors before building nuclear power plants in competitive power markets.

RISK SCENARIO

Considering the history of cost overruns for the vast majority of nuclear power plants, it is reasonable to assume that there is a high likelihood that any particular nuclear power plant project might be more costly than planned. For example, Lew Hay, Chairman and CEO of Florida Power and Light said "although suppliers keep quoting overnight costs of \$2500 to \$3500 per kilowatt, I believe the all-in costs are likely to be much higher—possibly twice as much once you factor in the owners' costs such as land, cooling towers, etc., interest during construction and cost escalation due to inflation and cost overruns. And of course, we have to have a contingency as well."[14] Florida Power and Light has, in 2007, estimated that a nuclear plant would have overnight costs of \$3,108/kWe to \$4,540/kWe and total plant cost of \$5,492/kWe to \$8,081/kWe. Entergy Corporation

[14] Biannual General Meeting of the World Association of Nuclear Operators, Chicago, September, 2007.

Chief Executive J. Wayne Leonard has said regarding a decision to build a new nuclear power plant "Utilities do not want to take that risk...it's risk we don't control." Leonard added that to make the economics of nuclear work for Entergy, he would need to see "double-digit natural gas prices and carbon blow-out prices" starting at $25 per ton and escalating toward $50. He added, since nuclear vendors don't want to assume the risk of a cost overrun they have put construction costs too high for most companies.[15] These statements and available data for a number of nuclear plant bids show that vendors and utilities are assuming significantly higher capital costs for new Generation III+ reactors than the $3,800 to $4,000 per kWe being used as long-run capital costs for U.S. and EU nuclear plants by the IEA. That said, it does not mean that $4,000/kWe is not a reasonable cost estimate for what new power plants in these markets may eventually cost. Due to government guarantees and climate policies, some plants are likely to be built. These plants could establish a nuclear industry capable of building near-identical plants at predictable prices within an efficient regulatory process. Nonetheless, the predictable price could be considerably higher than the current estimates used in the IEA modeling studies.

A Risk Scenario should account for several factors. There is very limited experience constructing or operating any of the Generation III+ reactor designs and no experience for some of the proposed designs. The limited experience building new reactor designs is not encouraging as projects in Finland, Japan and elsewhere have experienced delays and cost overruns. It should be expected that reactor costs will increase further to meet stricter safety requirements especially with regard to on-site spent-fuel management in light of the Fukushima accidents. Based on various vendor offers (Thomas, 2010), past cost overruns, factors likely to increase future construction costs and the likelihood of construction delays (resulting from regulatory changes, etc.) nuclear capital costs could increase capital costs by 50% to 100% higher than the $3,800/kWe to $4,000/kWe estimated for the EU and U.S. by the IEA. Using the 50% figure, the capital cost for a Generation III+ nuclear plant in the United States and European Union is assumed to be $2,000/kWe higher in the Risk Scenario. This increases nuclear capital costs to $6,000/kWe in the United States and $5, 800/kWe in the EU.

NATURAL GAS PRICE RISK

The competitiveness of nuclear power would also be reduced if natural gas prices were significantly lower than assumed in the *2011 WEO*. Since 2005, the United States has greatly expanded its natural gas production by fracturing natural-gas bearing shales. There are uncertainties whether this technology will be as widely adopted in other countries with promising shale resources. There are "above-ground" and "below-ground" uncertainties whether European countries with shale resources (including Poland, France, Sweden, Austria, Germany and the Baltic States) will repeat the U.S. experience. There are also different estimates of the sustainability of U.S. production. Higher shale-gas production, along with lower natural gas prices, would make nuclear less competitive. Should U.S. shale productivity be at the high end of the plausible range, U.S. natural gas prices would be significantly lower than estimated in the *2011 WEO*. Likewise, European

[15] Reuters Global Energy Summit, Houston, May, 2010.

natural gas prices would be significantly lower if European shale plays turned out to be productive and a variety of "above-ground" constraints were overcome. Consequently, for the Risk Scenario, we have lowered the estimated 2035 natural gas price in the United States from $8.56/mmBTU to $5.99/mmBTU.[16] For the European Union we have lowered the 2035 natural gas prices from $12.08/mmBTU to $8.70/mmBTU.[17]

RISK SCENARIO RESULTS (HIGHER NUCLEAR COST & LOWER NATURAL GAS PRICES)

Table 7 provides the revised levelized cost estimates for the Risk Scenario showing a higher estimated cost for nuclear and a lower cost for natural gas. Table 8 shows the estimated levelized costs with a $50/tonne CO_2 charge.

Table 7: Long Run Levelized Cost (U.S. cents/kWhr) (Higher Nuclear Cost & Lower Natural Gas Prices Scenario) (2035).

	Nuclear	Renewables	Natural Gas	Coal
United States	10.8 – 11.1	7.7 – 12.8	6.9 – 7.0	8.1 – 8.7
EU	10.6 – 10.9	7.8 – 15.9	8.4 – 8.5	9.6 – 11.4

Table 8: Long Run Levelized Cost (U.S. cents/kWhr) with a $50/tonne CO_2 Charge (Higher Nuclear Cost & Lower Natural Gas Prices Scenario) (2035).

	Nuclear	Renewables	Natural Gas	Coal w CCS	Coal w/o CCS
United States	10.8 – 11.1	7.7 – 12.8	8.6 – 8.7	8.5 – 11.8	11.8 – 13.1
EU	10.6 – 10.9	7.8 – 15.9	10.1 – 10.2	10.2 – 13.3	13.3 – 16.1

Table 9 provides the estimated cost advantage of natural gas over nuclear for the United States and the European Union. In the United States, instead of enjoying a 9%—17% cost advantage over natural gas (or 31%–40% with a $50/tonne CO_2 charge), nuclear is now estimated to be more costly than natural gas by an average of 4 cents/kWhr (or 2.3 cents/kWhr with the carbon charge). Consequently, natural gas enjoys a cost advantage of 54% to 61%. Even with a $50/tonne CO_2 charge, natural gas retains a substantial cost advantage (24%-29%).

Table 9: Risk Scenario—Cost Advantage of Natural Gas Relative to Nuclear with and without a $50/tonne CO_2 Charge (2035).

	Cost Advantage of Natural Gas over Nuclear w/o Carbon Charge		Cost Advantage of Natural Gas over Nuclear with Carbon Charge	
	U.S. cents/ kWhr	**% Cost Increase Over Natural Gas**	**U.S. cents/ kWhr**	**% Cost Increase Over Natural Gas**
United States	**3.8 – 4.2**	**54 – 61**	**2.1 – 2.5**	**24 – 29**
EU	**2.1 – 2.5**	**25 – 30**	**0.4 – 0.8**	**4 – 8**

[16] $5.99/mmBTU is the EIA estimated 2035 price for natural gas from its *2012 Annual Energy Outlook*: High Estimated Ultimate Recovery case. See Figure 105.

[17] $8.70/mmBTU is the estimated European natural gas price in the Unconstrained Shale-Gas Scenario from a DOE study: World-Wide Unconventional Gas Potential: Energy Security and GHG Emissions. See Difiglio 2012a.

In the European Union, instead of enjoying a 37%–48% cost advantage over natural gas (or 59%–71% with a \$50/tonne CO_2 charge), nuclear is now estimated to be more costly than natural gas by an average of 2.3 cents/kWhr (or 0.6 cents/kWhr with the carbon charge). Consequently, natural gas enjoys a cost advantage of 25% to 30%. With a \$50/tonne CO_2 charge, natural gas retains a slight cost advantage (4%–8%).

Considering that natural gas plants have significantly lower capital costs per kWe capacity, investors' rationale to build a nuclear plant pretty much disappears if a natural gas plant is estimated to undercut the nuclear plant by 4 cents/kWhr. With these cost estimates, and taking into account the various risks facing any specific nuclear plant project in a liberalized power market (licensing problems, public opposition, changing government policies, spent fuel disposal, etc.), the "bet-the-farm" risk becomes a "lose-the-farm" certainty. These costs constrain nuclear builds to governments that value energy security, have a high priority to reduce their greenhouse gas emissions or have other priorities that nuclear power would support.

The Risk Scenario also reverses the estimated competitiveness of coal and nuclear when there is no carbon charge. As shown in Table 4, nuclear had a 4%–16% cost advantage over coal in the United States and a 26%–56% advantage in the European Union. As shown in Table 10, with the Risk Scenario, coal now has a 24%–37% cost advantage over nuclear. In the European Union, the cost advantage of coal is minus 10% to plus 15% (a lower- efficiency coal plant would be less competitive than a nuclear plant, hence the minus 10% estimate). With the Risk Scenario, nuclear power plants are not only uncompetitive with natural gas plants but do poorly against coal plants.

Table 10: Risk Scenario—Cost Advantage of Coal Relative to Nuclear with and without a \$50/tonne CO_2 Charge (2035).

	Cost Advantage of Coal over Nuclear w/o Carbon Charge (no CCS)		Cost Advantage of Nuclear over Coal with Carbon Charge (no CCS)	
	U.S. cents/ kWhr	% Cost Increase Over Nuclear	U.S. cents/ kWhr	% Cost Increase Over Nuclear
United States	2.1 – 3.0	24 - 37	0.7 – 2.3	6 – 21
EU	minus 0.8 to 1.3	minus 10 to 15	2.4 – 5.5	22 – 52

CLOSING COMMENTS

We do not intend to represent the Risk Scenario as a projection of what nuclear power plants will cost. It is intended to show the high economic uncertainty facing nuclear power in competitive power markets. Nonetheless, the Risk Scenario is as likely to be an accurate projection of future nuclear power costs as the costs assumed in the *2011 World Energy Outlook*. More significantly, from the standpoint of current investments in nuclear power, investors are unlikely to rely on cost estimates as low as \$4,000/kWe when evaluating nuclear power projects.

Very strong climate policies could support higher nuclear plant costs. These climate policies may be in the form of carbon charges, as considered in this paper, or regulations such as those that have recently been proposed by the U.S. Environmental Protection Agency that prohibit new coal power plants without CCS.

While the economic obstacles appear to be formidable in the United States, U.S. policies continue to support nuclear power development with production subsidies and guarantees. If U.S. plants built with government support establish stable or declining nuclear cost expectations, the future of nuclear power in the United States could be strong. However, until positive experience is achieved building nuclear plants in the U.S., investors are unlikely to build U.S. nuclear power plants without Federal guarantees.

The low nuclear costs estimated for China suggest that China, and other countries outside of the OECD, may succeed in building large nuclear plant fleets. State-supported construction in China and elsewhere could help demonstrate that nuclear plants can be built at competitive and predictable prices without delays or other problems.

REFERENCES

1. Difiglio, C. (2010) "Challenge of Reducing Greenhouse Gas Emissions; The IEA 450 Scenario," *International Seminar on Nuclear War and Planetary Emergencies, 40th Session,* World Scientific Publishing Co., World Scientific Publishing Co.
2. Difiglio, C. (2012a) World-Wide Unconventional Gas Potential: Energy Security and GHG Emissions, *International Seminar on Nuclear War and Planetary Emergencies, 44th Session,* World Scientific Publishing Co., World Scientific Publishing Co.
3. Difiglio, C. (2012b) "Financing Power Sector Investments," *Energy for Development: Resources, Technologies, Environment*, F.L. Toth (Editor), Dordrecht: Springer.
4. EIA 2012 (2012) *Annual Energy Outlook*, Figure 105, Energy Information Administration Washington, DC.
5. IEA (2000) *Experience Curves for Energy Technology Policy*, International Energy Agency Paris.
6. IEA (2010) *Energy Technology Perspectives 2010, Scenarios & Strategies to 2050*, International Energy Agency Paris.
7. Grubler, A. (2010) "The costs of the French nuclear scale-p: A case of negative learning by doing," *Energy Policy.*
8. IEA (2011) *2011 World Energy Outlook*, International Energy Agency Paris, 2011.
9. IEA-NEA (2010) *Projected Costs of Generating Electricity, 2010 Edition*, International Energy Agency and the Nuclear Energy Agency, Paris.
10. Kessides, I. (2010) "Nuclear power: Understanding the economic risks and uncertainties," *Energy Policy*, April 2010.
11. Joskow, P.L., Parsons, J.E. (2009) "The economic future of nuclear power", Daedalus.
12. Moody's Global (2009) *New Nuclear Generation: Ratings Pressure increasing*, Moody's Investors Service, New York.
13. Schlissel, D., Biewald, B. (2008) *Nuclear Power Plant Construction Costs*, Synapse Energy Economics, Inc., Cambridge.
14. Thomas, S. (2010) *The Economics of Nuclear Power: An Update*, Heinrich Boll Stiftung Foundation, Berlin.

15. World Bank (1993) *The World Bank's role in the electric power sector: Policies for effective institutional, regulatory and financial report* (Policy Paper), World Bank, Washington, DC.
16. World Nuclear Association (2012) WNA Reactor Data Base, 25 June, 2012.

SESSION 5

ENERGY & SUSTAINABILITY IN CITIES

ENERGY, CITIES, AND GLOBAL SUSTAINABILITY

ROBERT N. SCHOCK
World Energy Council, Senior Advisor
London, UK

"In the fair city of this vision, there were airy galleries from which the loves and graces looked upon him, gardens in which the fruits of life hung ripening, waters of hope that sparkled in his sight. A moment, and it was gone."(Charles Dickens, A Tale of Two Cities, 1859)

ABSTRACT

Arguably, the most acute energy emergency today is associated with the rapid growth of cities, most notably in Asia and Africa. The populations of all cities are growing rapidly, and two-thirds of the global population will live in large cities by 2050. Economic, political, and social entities are important elements of cities, which are complex systems in themselves, but energy issues tend to be afterthoughts. Energy in itself is a large and complex system and is the "glue" that holds an entire city system together, and energy must be integrated equally into the mix, in planning as well as execution. Energy's full integration provides an opportunity to substantially contribute to discussions and lead to city solutions going forward, and by extrapolation to the sustainability of the entire planet. Issues such as energy-efficient technologies, policies, regulations and their enforcement, energy services, and communication all need to be part of the solution. These problems are amenable to the new science of modelling and controllability of complex networks, with a real promise of new insights and the identification of unanticipated patterns and key control points using statistical physics and non-linear dynamics.

In working on energy and cities, it is important to interact with the entities already involved in other parts of the system. Productive interactions logically begin with the mayors of cities. Public–private partnerships already exist and the private sector must be engaged. The financial community is critical and capital investment firms must also be involved. Last but not least, national governments have a role to play by offering long-term technology development and financial incentives.

INTRODUCTION

Economic health, social stability, security, public health, and respect for the environment are integral parts of a very complex global system. To these we must add energy, often taken for granted, and on which the entire system depends: energy that is available, reliable, affordable and accessible. Success or failure at supplying and using energy effectively and efficiently has huge implications for nations, geographic regions, and people. Although the individual components of this complex global system are reasonably well studied and understood, the collective behavior of the ensemble and the ability to control it, does not respond to simple explanation and has not been studied sufficiently as a complete system. This paper proposes that we begin to change this situation.

WHY CITIES?

The most acute and rapidly gathering global energy emergency is around the growth of cities. Cities are the largest and fastest growing entities on the planet. Over the next 20 years, the world will see the addition of the equivalent of 10 new cities of 7 million people *every year* (World Energy Council, 2010), most in Asia and Africa (see also Grubler, 2012). Population densities in large cities are already highest in Africa and Asia (Figure 1).

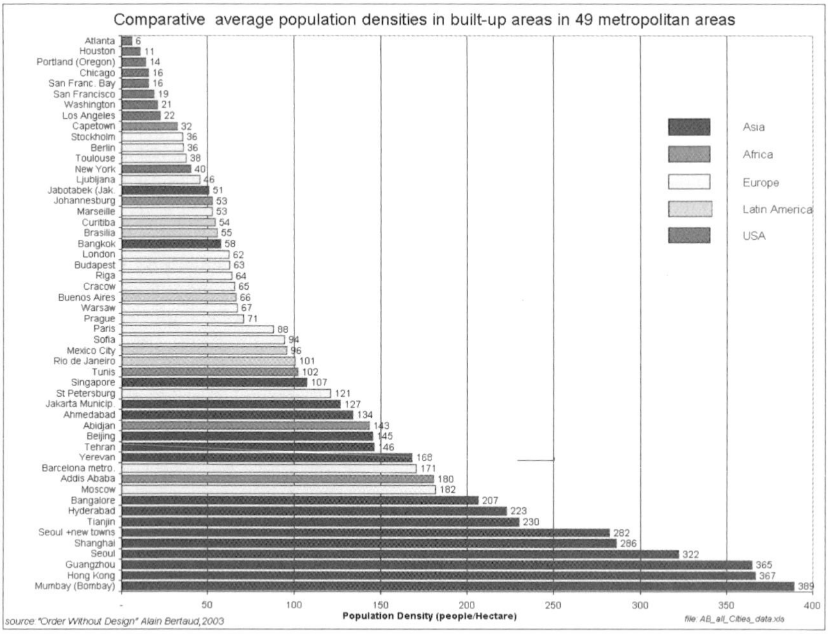

Fig. 1. Population densities in large cities (World Energy Council, 2010).

The United Nations (Table 1) estimates that by 2050 urban populations will comprise 67% of the total global population, up from 52% in 2011 (United Nations, 2011). This amounts to an average of over 60 million additional people in cities, every year.

Table 1. Total and urban population forecasts, UN, 2011.

	2011	2050	Annual Change Rate (1970-2011)
World Total	6.97 billion	9.31 billion	+ 1.55%
Urban	3.63 billion	6.25 billion	+ 2.41%
% Urban	52%	67%	

Alarmingly, the rate of city growth is expected to be highest over the next 20 years in very large cities. The largest projected growth is in the largest cities (Figure 2) where 630 million people are expected to be living in cities having a population greater than 10 million by 2025, compared with slightly more than 350 million people in cities this large today.

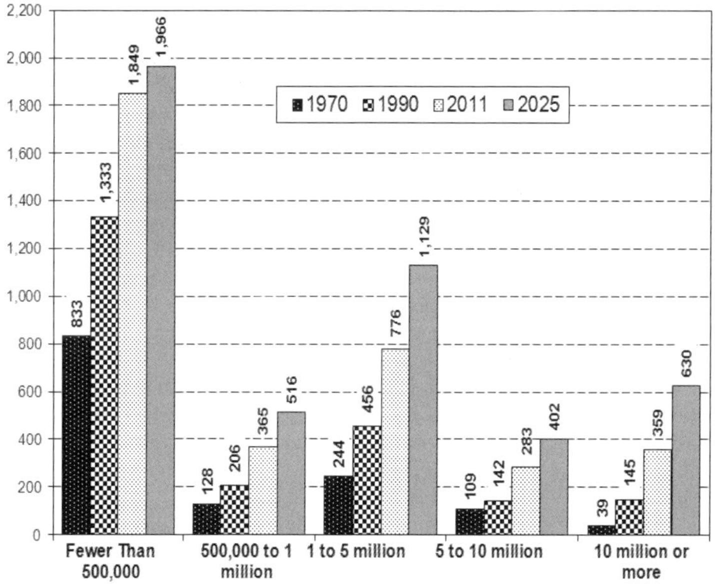

Fig. 2: Population by city size, in millions (United Nations, 2011).

The reason for this alarming projected growth of cities is straightforward—people and their children can find work and take advantage of improved health care and education not available in the rural countryside, even though they may have to live in a city slum instead of a rural community. Because of this extraordinary growth in cities, innovation in their energy supply and use systems can have the most immediate global impact. Most importantly, from the viewpoint of opportunity, these new cities are being built now (Bettencourt, 2012) and how they are built and organized will put systems in place that will exist for years to come. Finally, solutions to the problem of energy and sustainability in cities will have broad ramifications to the sustainability of the planet as a whole.

Another reason cities are so important is that cities contribute disproportionately to the global economy, in total and on a per capita basis. More than 50 percent of global GDP now comes from the 600 largest cities having 22 percent of the population (McKenzie, 2011). Because of more opportunities for the exchange of ideas and access to more capital, populations in cities clearly do more with less. Furthermore, most cities have well-established political systems that depend on national governments to varying degrees, but in many cases very little. By looking intensively at cities, we can work the problem of energy from the bottom-up, rather than the usual top-down, and give fresh input to economic, social, and political dynamics. Arguably, this is the key to understanding the complex system affecting cities' efficiency, most notably energy efficiency, but also food and water. How growing cities approach the provision of services and the livelihood of their citizens, both in urban infrastructure and energy-using technology, presents a major opportunity for implementing energy system transformations (see for example, an article in *Scientific American*, September 2011).

Managing this complex economic–social system, and not just in cities, involves science and technology, economics, and politics. To deal with the problems of cities, these different disciplines provide a template for working together in areas such as water, food, security, limits of development, climate, but beyond into the critical areas of politics, economics, and policies, all of which are necessary for action.

The logical place to begin is to examine some of the best examples of success within cities in critical areas such as buildings, transport, water, food, waste, public health, policies, and especially the interactions between them (Kretschner and Kollemberg, 2011; Lemann, 2011). Examples of city successes in the developed world need to be examined side-by-side with cities in the developing world. It has been suggested that it would also be useful to contrast mono-cultural cities (e.g., Tokyo, Mexico City, and Delhi) with those that are more multi-cultural (e.g., Toronto, New York, and Johannesburg). Comparing policies in mono-cultural cities with those in multi-cultural cities would point to lessons that can be practical for everyone.

CONTROLLABILITY OF COMPLEX SYSTEMS

The complex systems that are cities and the interactions between their individual elements need to be identified and studied if we are to understand and better control the outcome and achieve societal goals. The networks that compose cities are amenable to forecasting and control if treated as a system. Energy systems must be brought into the equation early, rather than later when plans are being implemented. The problem may be tractable with the developing science of modelling and controlling networks involved in complex systems. Much work has been done in biological, social, and communications systems (See for example, Newman, 2010; Liu et al. 2011; Motter and Albert, 2012).

All city systems have a strong social component. Control theory uses statistical physics and non-linear dynamics to identify unanticipated patterns and avoid policies based on preconceived ideas, thus maximizing chances for success. A framework for controlling complex, self-organized systems has been lacking (Liu et al. 2011). Models are now used to identify a set of "driver nodes" with time-dependent controls to guide the dynamics of the entire system. The application of control theory can help cities guide the behaviour of the complex energy–economic–security–social–environmental system to a desired result.

To get some idea of what cities are dealing with in terms of the complex set of interactions involved in achieving a sustainable system for their inhabitants that is secure and affordable, examine just the electricity grid, shown in stylized form in Figure 3.

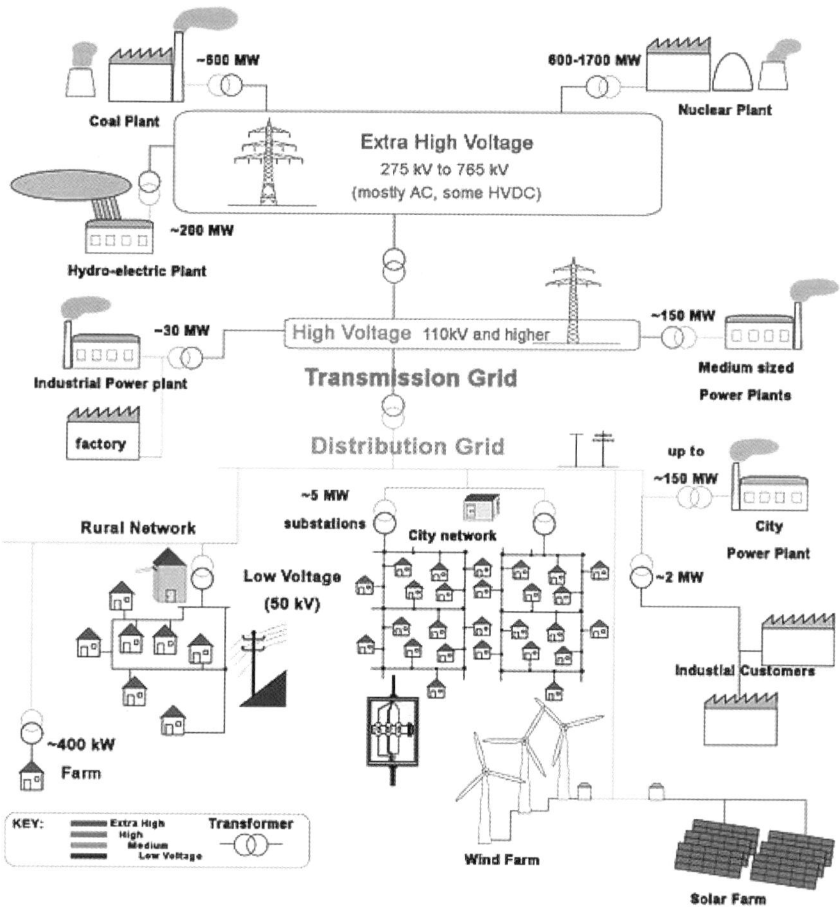

Fig. 3: Representation of an electric grid in modern developed countries (la.wikipedia.org/wiki/Fasciculus:Electricity_grid_schema-_lang-en.jpg).

It should be emphasized that this schematic representation includes neither the diverse end uses of electricity in rural homes, nor in industrial uses, both of which are integral to the grid. Adding all the appliances and manufacturing components in typical urban homes and industry easily doubles the numbers of interconnections or nodes in Figure 3. Liu, et al. (2011) has identified 5,855 links in the power grid that serves Texas, which is separate from the larger United States grid. The key is to identify those nodes (elements) that can guide the behaviour of the entire system. This part of the system that supplies electricity to cities is at the heart of what are termed "smart grids," which attempt to vastly improve the balance of electricity supply and demand in real time using a modern electronic command and control elements built into the elements of the system (Schock and Sims, 2012; Razdan, 2012).

Taking the electric grid in Figure 3 and adding other energy use systems (e.g., appliances, industry, and transportation), water and food distribution makes the city system begin to look something like the figure below (Figure 4). Some connections are stronger and have more influence than others. Using physics and mathematics to examine these systems will generate new insights and identify unanticipated patterns, particularly of control, and eliminate biases from preconceived ideas. These new tools hold the promise of allowing us to acknowledge all the elements of a complex system, such as a city, without approximations to make the system more tractable (Zichichi, 2011).

Fig. 4: A network of scientists (Porter et al. 2009).

POLICIES

Any examination of the areas important to the success of large cities, especially the central role of energy, must ultimately include policies. Policies typically involve a number of key areas. Using complex network theory can help identify linkages where policy can be most effectively implemented. City operations where energy efficiency increases can be made are obvious candidates; but when examining the entire system, this may not be true. Building standards, power generation, and transportation quantity and flow are areas where the optimized implementation of policies may have a large effect. There are regulations, whose effectiveness depends to a large degree on the capabilities for enforcement. The provision of services where cities influence the behaviour of individual people or businesses is equally important. Water, waste disposal, utilities, parks, recreation, and even food supplies are critical to the success of the whole city as an enterprise. Finally, there is the area of communication with citizens, corporations, other cities, and national governments, which offer an exploration of policies that can have profound effects.

COLLABORATORS

To be effective in this endeavour, the World Federation of Scientists (WFS) must begin to involve entities already working on parts of the problem. In government, mayors of various cities and the C40 group (www.c40cities.org/) of the largest cities are already working the issues around governance while focusing on climate (The C40 represent almost 300 million people). There is the Energy Sector Management Assistance Program managed from the World Bank (www.esmap.org/esmap/). For transportation, see the work of the Deutsche Gesellschaft für Internationale Zusammenarbeit (www.gtz.de/en/themen/28263.htm). Public–private partnerships already exist and the private sector must be involved. See for example Siemens (2012) (www.usa.siemens.com/sustainable-cities/?stc=usccc025507). The financial community is critical and capital investment firms must eventually be involved. Last but not least, national governments have a central role to play. These are examples and there are many others.

NEXT STEPS

There should be a series of expert workshops, co-sponsored by WFS. The first should involve people knowledgeable about city energy networks and experts in complex network theory. The goal is to create network maps with nodes and linkages, and then make a preliminary attempt to identify control nodes. The electrical grid is a reasonable place to start.

Subsequent workshops can then investigate the controllability of other energy networks (transport, fuels) and the relationships between the energy system and other systems such as water and food. At the appropriate time, officials from key cities should be brought into the process who have been working the larger issues (e.g., Toronto, Shanghai, Cape Town, Buenos Aires, New York, Delhi, Tokyo), from global organizations (e.g., World Bank, UNDEP, UNEP, ESMAP, C40, OECD, ICLEI), from national and regional political appointees (e.g., California, New York, Ile de France, Russia), and also from the financial community (e.g., Goldman Sachs, Barclays, Deutsche Bank),. Papers derived from these workshops and meetings will be published by the World Federation of Scientists in print and on group websites. Finally, interaction with city policymakers is essential to disseminate the results to achieve more sustainable cities.

SUMMARY

At present rates of population growth and migration, an unprecedented number of large cities will appear in the next 20–30 years with consequences for social stability, energy, food and water systems, public health, and social stability and security. This presents us with a unique opportunity to ensure that these cities are built using the latest cost-effective technologies to ensure the city infrastructure will be energy and environmentally sustainable and encourage economic and social security for decades. This will require that city energy supplies and their use are integrated with evolving political and economic services, and that scientific work on the controllability of complex networks is fully utilized.

ACKNOWLEDGMENTS

I am grateful for discussions with William Barletta, Luis A.M. Bettencourt, Carmine Difiglio, William Fulkerson, Thomas Johansson, and Anil Razdan.

REFERENCES

1. Bertaud, A. (2003) *Order Without Design*, http://alain-bertaud.com/
2. Bettencourt, L.A.M. (2013) *Complexity, Cities and Energy*, in International Seminars on Planetary Emergencies; 19-24 August, Erice, Italy.
3. Grubler, A., Convening Lead Author (2012) *Urban Energy Systems*, Chapter 18, in *Global Energy Assessment*, Cambridge University Press.
4. Kretschmer, F. and Kollenberg, M.E. (2011) *Can Urban Agriculture Feed a Hungry World?* Spiegel Online, 22 July 2011, www.spiegel.de/international/zeitgeist/0,1518,7775754,00.html.
5. Lemann, N. (2011) "Has the Celebration of Cities Gone Too Far?" *The New Yorker*, June 27, 2011.
6. Liu, Y.-Y., Slotine, J.-J., and Barabási, A.-L. (2011) *Controllability of Complex Networks*, Nature 473, 167-173, http://leonia.zoology.washington.edu/publications/LiuEtAl11.pdf.
7. McKinsey Global Institute (2011) *Urban World: Mapping the Economic Power of Cities*, www.mckinsey.com/mgi.
8. Motter, A.E., and Réka, A. (2012) "Networks in Motion," *Physics Today*, April, pp. 43-48.
9. Newman, M.E. J. (2010) *Networks: An Introduction*, Oxford University Press.
10. Porter, M.A., Onnela, J.-P., and Mucha, P.J. (2009) *Communities in Networks*, Notices of the American Mathematical Society, 56, p. 1082.
11. Razdan, A. (2012) "Smart Grids for Sustainable Mega Cities", in *International Seminars on Planetary Emergencies*; 19-24 August, Erice, Italy.
12. Schock, R.N. and Sims, R., Convening Lead Authors (2012) *Energy Supply Systems*, Chapter 15, in *Global Energy Assessment*, Cambridge University Press.
13. Scientific American Editors (2011) *Cities: Smarter, Greener, Better*, September.

14. United Nations (2011) Wold Urbanization Prospects (rev. 2011), 318pp, http://esa.un.org/unup/pdf/FINAL-FINAL_REPORT%20WUP2011 _Annextables_01Aug2012_Final.pdf.
15. World Energy Council (2010) *Energy and Urban Development*, www.worldenergy.org/publications.
16. Zichichi, A. (2007) *Complexity at the Fundamental* Level, 42nd Course of the International School of Subnuclear Physics, Erice, Italy, 29 August, v. 42, pp. 251, World Scientific Publishing.
17. Zichichi, A. (2011) *Science, Culture and the Planetary Emergencies*, 44th Seminar on the Planetary Emergencies, Erice, Italy, 20 August, World Scientific Publishing.

SMART GRIDS FOR SUSTAINABLE MEGA CITIES AND A DELHI INNOVATION

ANIL RAZDAN
Former Secretary Power, Government of India,
PTC India Ltd., New Delhi, India

A RAPIDLY URBANIZING WORLD

- For the first time in history, the proportion of world population living in urban areas reached 50% in 2008.
- Will reach 60% in 2030 if present trend continues.
- Developed countries had attained 50% urbanization in the 1950's. Developing countries will attain that level in a decade.
- The least developed have most of their population in rural areas and will face the fastest urbanisation in the near future.
- Between 2007 and 2050 the world population growth will be 2.5 billion and the cities will absorb 3.1 billion.
- Urbanisation in developing countries not linked to economic growth will happen even otherwise.

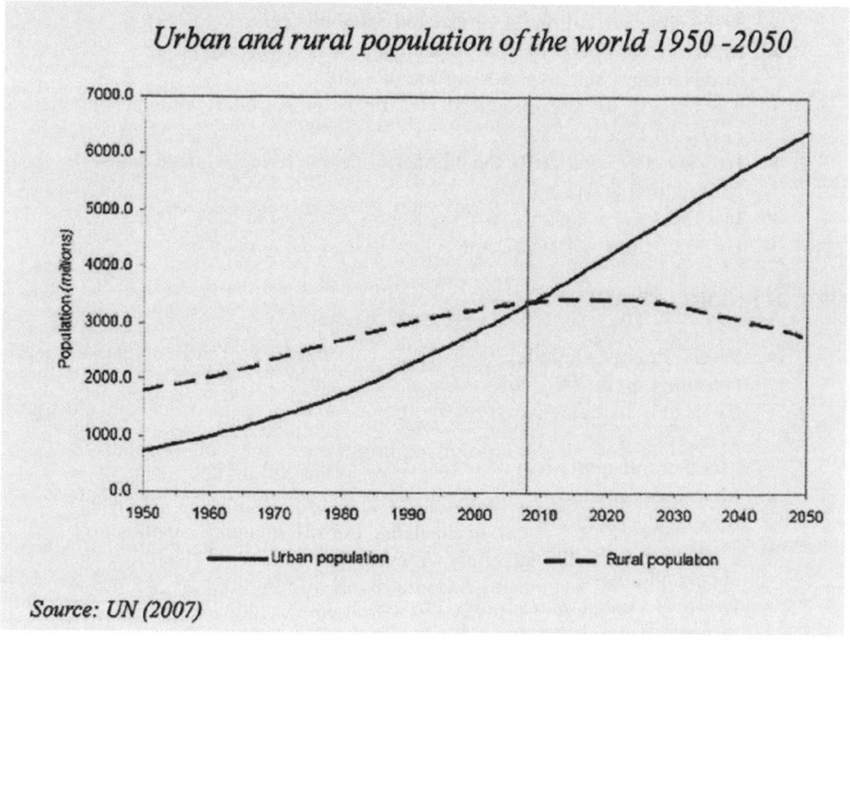

Source: UN (2007)

THE CHALLENGES OF URBANISATION

- Populations in developing countries move to cities for better wages, amenities, health care and education.
- People prefer to be poor in city slums rather than in a remote rural area for better economic opportunities.
- Besides housing, this presents vast challenges of meeting energy demands of electricity, transportation and cooking fuels.
- Need to assure ENERGY Availability, Accessibility and Acceptability (3As).
- In the context of developing countries, 4Es are relevant—Energy, Environment, Efficiency and Equity.
- Challenge becomes greater on account of multilevel governance and financial jurisdiction: local, state and federal.
- Poverty and nascent democracy restrain harsh enforcement measures.

MEGA CITIES

- In 1975, there were 179 cities with more than 1 million inhabitants. In 2007, there were 430.
- MEGA CITIES, defined by UN as an urban conglomeration of more than 10 million, constitute only 4% of urban population.
- Mega cities are cultural, political and economic centres.
- Most visual part of the urbanising world depicting the advantages, disadvantages and disparities of life in a city.
- The biggest 30 cities produced 16% of the global GDP while accounting for only 4% of the population.
- Between 1975 and 2010, the number of people living in mega cities rose from 53.2 million to 318 million.
- In 1950, there were only two mega cities: Tokyo and New York.
- By 2025, there will be 27, and Asia will have 15 of them.

URBAN ENERGY SURGE

- World Energy Outlook 2008: Cities consumed two thirds of global energy consumption in 2006, will consume three fourths by 2030.
- Cities in transition move from traditional bio mass fuels to LPG and electricity as incomes rise.
- Mode of transport often determined by density and spread.
- Metro services reduce oil consumption and pollution, but demand more and reliable electricity.
- Vertical rise of cities raises electricity demand for climate control (AC), water supply and lifts.
- However, mega cities often provide a better opportunity for innovation through better finances, political clout, media attention and visibility for federal support.

CHARACTERISTICS AND BURDENS OF MEGA CITIES

- Given their economic importance, each mega city must remain highly competitive in terms of infrastructure in a globalized economy.
- Serious energy constraints in developing countries made more acute in high demand mega cities.
- Climate change and local pollution concerns make capacity addition from conventional fuels more difficult.
- Business, transportation, high rise buildings, sensitive IT-based information and communication system, traffic signals and large hospitals place a huge demand for electricity.
- The presence of ultra rich with energy intensive life styles adds to the demand.
- The integration of neighboring villages in extended limits of mega cities leads to pockets of lower quality electricity service.
- Energy poor settlements and erstwhile rural areas without an urban culture create serious problems of power theft and governance.
- Metro transportation has an increased burden of secure and steady power.
- Working population leaving and returning home at the same time in a service oriented economy leads to peak demand.
- Space constraint for developing renewable energy infrastructure to meet renewable purchase obligations.
- Energy Demand Management becomes a critical issue to avoid disruptive black-outs and shortages.
- It is crucial and necessary to involve consumers in demand response management through differential tariff.
- Developed countries retro fitting smart grids.
- Danger of mega cities emerging in the developing world through sheer population size without modern logistics to ensure safe and secure electric supply.

Solution

We have to find a technology solution which minimises power demand, gives quality supply, is remotely controlled and involves participation of consumers.

As life in a mega city is controlled by electricity, the installation of Smart Grid is no longer a luxury or a brand idea. It has become a necessity.

DELHI: A MEGA CITY WITH CHALLENGES

- From a small town of 0.4 million in the early twentieth century, Delhi is home to nearly 17 million people today.
- The National Capital Region NCR has a population of over 48 million.
- Highest per capita income in India.
- By 2012, the population of Delhi 23 million NCR will be 64 million.
- 3.63 million households and an electricity consumer base of 3.33 million.
- Peak Power Demand met in 2004-05: 3490 MW, and in 2012: 5600 MW.

- Aggregate Technical and Commercial Power loss in 2002: 52%, and in 2011: 18%.
- Vast improvement in distribution control system post-Public Private Partnership in 2002.
- Further reduction in losses hampered by some urban and rural pockets having high power and material theft and malpractices.
- Quality of power in villages poor on account of long distribution feeders.
- Regulator has imposed Renewable Power Purchase obligations. No local renewable source except solar.
- Space constraint for any solar farm in congested urban area.
- Number of private vehicles has risen from 2.75 million in 1997-98 to 5.30 million in 2007-08.
- Large scale introduction of Metro Rail Service to provide Mass Rapid Transport to decongest roads and reduce vehicle pollution has increased dependence on electricity.

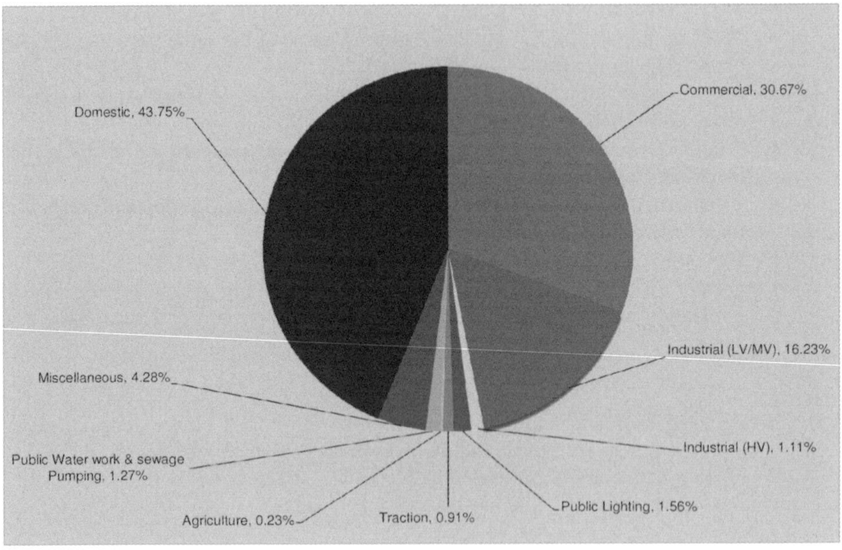

Delhi: Pattern of Electricity Consumption in 2007-08 Source: Central Electricity Authority

Yearwise Status of AT&C Losses for Private Distribution Companies in Delhi (In Percentage).

Distribution Company	Opening Levels of AT & C Loss*		2002-03	2003-04	2004-05	2005-06	2006-07	2007-08	2008-09	2009-10	2010-11
BYPL	57.2	Target	56.45	54.70	50.70	44.65	39.95	34.77	30.52	26	56.45
		Achievement	66.89	54.29	50.12	43.88	39.03	29.82	20.02	24.32	21.95
BRPL	48.1	Target	47.55	46.00	42.70	36.70	39.10	26.69	23.46	20	
		Achievement	47.40	45.06	40.64	35.53	29.92	27.17	20.59	19.64	18.82
NDPL	48.01	Target	47.60	45.35	40.85	35.35	31.10	23.03	20.35	18.67	
		Achievement	47.79	44.86	33.79	26.52	23.54	18.31	15.41	15.16	

*AT & C Loss: Aggregate Technical and Commercial Loss.

MAKING DELHI SUSTAINABLE

- Conserve scarce water through waste water recovery and rain water harvesting.
- Building designs to be energy efficient.
- Solar reflecting roofs and water heating.
- Encourage Roof Top Solar power.
- Widespread use of CFL/LED lighting.
- Subsidised power tariff for low power consumption.
- Strong public campaign for consumer awareness to install energy efficient labeled appliances.
- Scrapping coal-based thermal power stations replaced with gas fired stations.
- Erect a world class Metro Rail network to substitute petrol/diesel driven vehicles.

In 2012, Delhi Metro has 160 KM corridor carrying 2.2 million passengers daily with an average trip length of 15 Km.

By 2017 nearly 4 million passengers.

By 2021nearly 250 km corridor to carry 10.8 million passengers daily with an average trip length of 15 Km.

Issue: How to optimise power demand and availability

POWER SECTOR MODERNISATION VITAL FOR DELHI

- Delhi has become highly dependent on a secure and reliable power supply system without adequate generation of its own.
- Delhi is part of Northern Region (NR) which has met a peak demand of 40,000 MW this summer. Peaking shortage in NR has been 12-14% though Delhi has virtually no peaking shortage.
- Delhi is part of an interconnected NR, WR, ER, NER grid of about 80,000 MW.

- Owing to over drawls and grid indiscipline by some constituents there were two successive grid collapses on 30th July and 31st July 2012 after almost 10 years leading to huge disruption for some hours.
- Large base load Coal Thermal Power means large off peak surplus. Need to curtail it through Smart Grid analytics and solutions.
- Distribution and Transmission networks need urgent modernisation and intelligent grids, as well as islanding of Delhi.
- Delhi being a national capital needs to immediately develop a smart grid to ensure safe supply, discipline consumers, optimise power requirement and improve quality of supply.

KEY CHALLENGES IDENTIFIED BY A DELHI DISTRIBUTION COMPANY

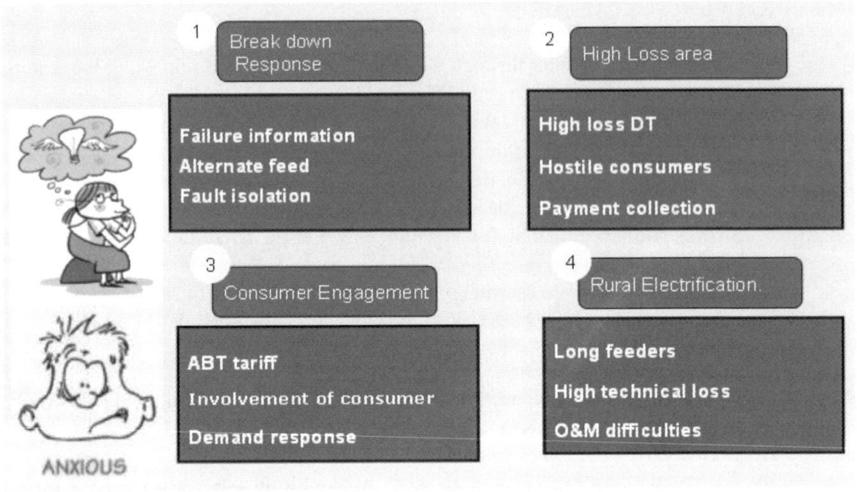

WHAT IS A SMART GRID?

Smart Grid is the integration of intelligent IT technologies, basic electricity principles, and processes into transmission and distribution networks, with the use of smart meters for the mutual benefit of grid controllers, distribution companies and consumers, thereby optimizing the use of minimal installed capacity and ensuring system stability with self healing characteristics.

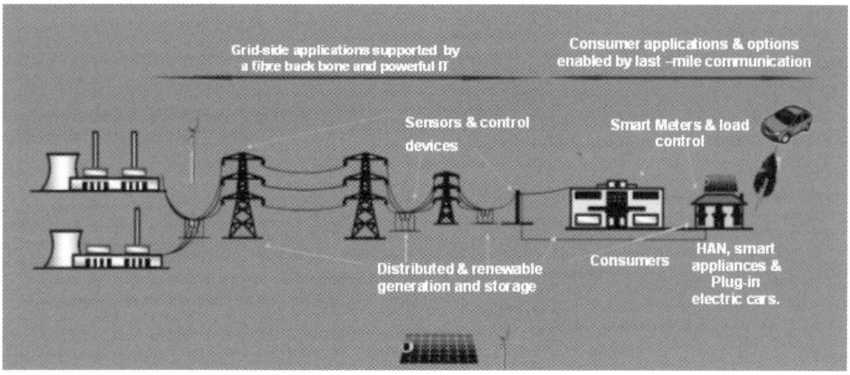

SMART GRID CHARACTERISTICS

20th Century Grid	21st Century Smart Grid
Electromechanical	Digital
One-way communications (if any)	Two-way communications
Built for centralised generation	Accommodates distributed generation
Radial topology	Network topology
Few sensors (limited to Zone SCADA)	Monitors and sensors at all levels of T&D network
"Blind"	Self-monitoring & timely
Manual restoration	Semi-automated restoration and, eventually self-healing
Prone to failures and blackouts	Adaptive protection and islanding
Check equipment manually	Monitor equipment remotely
Emergency decisions by committee and phone	Decision support systems, predictive reliability
Limited control over power flows	Pervasive control systems
Limited price information	Full price information
Few customer choices	Many customer choices
Electromechanical	Digital
One-way communications (if any)	Two-way communications

Source: Energy Networks Association

SMART GRID—KEY COMPONENTS

Customer Applications	**Grid Applications**
• In home displays (access consumption data from appliance and not meter) • Direct load control via smart appliances • Home automation (load control of high priority smart appliances)	• Volt-VAR Control – Fault detection, isolation and restoration – Sub-station and feeder monitoring and diagnostics – Wide area measurement
• Smart Meters – Remote meter reading – Real time consumption and power factor data – Two way communications to: ▪ Set supply capacity limits ▪ De-engerisation and re-engerisation ▪ Provide direct load control of customer appliances (via home area networks) ▪ Remote service checking ▪ Other utility meters (e.g., water) via HAN	• Integration of renewables and distributed generation • IT Systems and back office – Enhanced system control, network planning, operations and maintenance

RURAL DISTRIBUTION CHALLENGES IN DELHI

To enable improvement in reliability and quality of rural electricity in Delhi it was decided by the distribution company to design a Smart Grid Rural Electrification Network to meet one of the four identified challenges.

<u>Problems in Identified Villages</u>

- Technical loss 15%, AT&C loss 50%, low voltage, long feeders.
- Material theft.
- Billing and Collection difficulties. Hostile and aggressive consumers
- Difficult physical disconnection.
- Long breakdown: need to make fault intimation, isolation and alternate feed instantaneous.
- Fault forecasting.

<u>Solution</u>

Install Renewable Energy Distribution with Smart Grid as a win-win solution. It will also be cheaper to install solar panels in rural area to meet Renewable Power Obligation

SMART GRID MATURITY MODEL

Is a Management tool developed for electric utilities under the aegis of Carnegie Mellon University, USA, to:

- Map the current state of smart grid deployment and capability within an electric utility.
- Help in establishing future strategies and work plans for Grid modernization.
- Help organizations to bridge gaps between Smart Grid strategy and execution.

- Share vision of the smart grid with internal and external stakeholders.
- Prioritize options and measure success of Smart Grid Implementation.

The Delhi distribution company model has been validated by Carnegie Mellon USA which believes that the company is at a right position to start Smart Grid implementation.

PROPOSED SYSTEM ARCHITECTURE

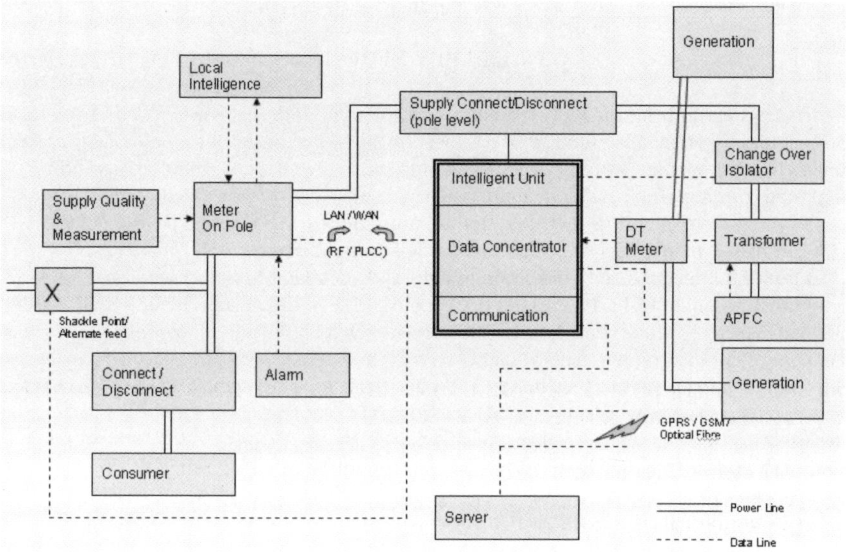

SMART NETWORK—PHASE ONE COMPONENTS

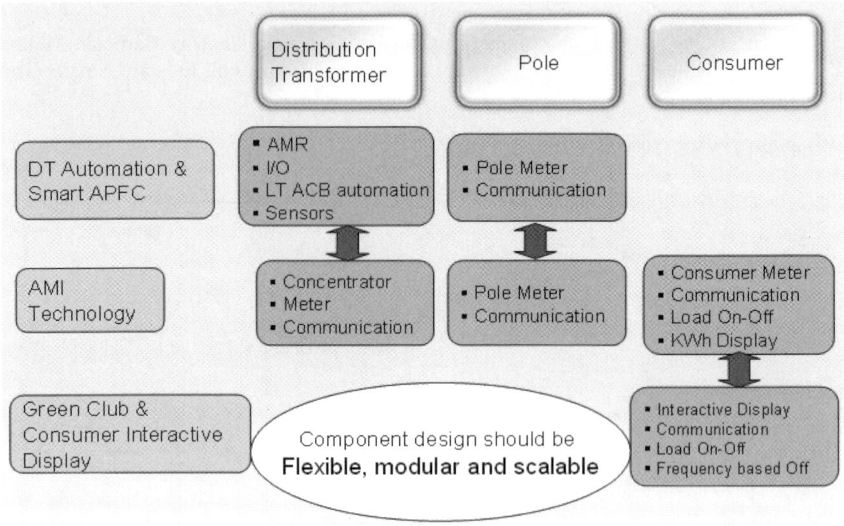

PROPOSED SITE FOR RENEWABLE DISTRIBUTION WITH SG. NAME OF SITE – VILLAGE PANDWALA KALAN, NAJAFGARH

- DT electrical parameters
  - Nos of DTs…03
  - Rating…Each 400KVA
  - Feeder length (HT)…12~14KM
  - Consumer tagging…Yes

Generation capacity
50 KW Solar/Solar Wind

Cost
Rs. 5 Million
Approx U.S.$ 0.1 Million

- Consumer brief
  - Nos of consumers…1000
  - DT loss (%)…>50%
  - Any shackle joint…YES

System Cost
Rs. 45 Million
<US$ 1 Million

Large power theft to be checked using Energy Gap Method between DT and consumer meters.

- Infrastructure
  - Communication…GSM/ CDMA
  - Road…YES

Alarm to be sounded for theft with automatic disconnection.

Village community to isolate power thieves.

AMI SOLUTION—FEATURES

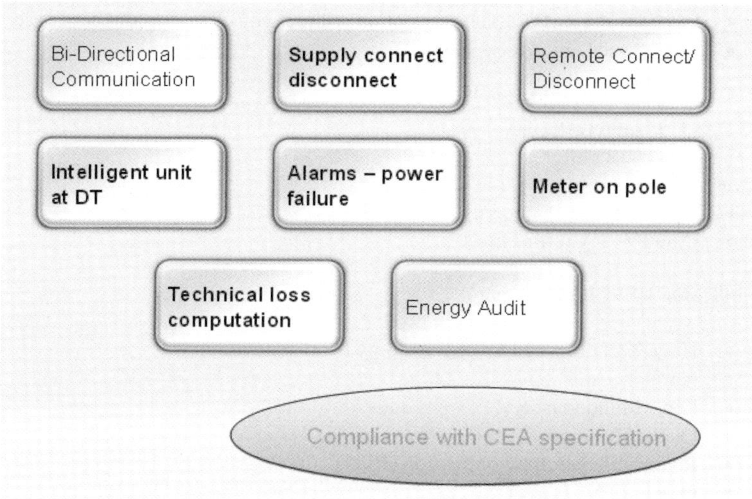

AMI-REMOTE CONNECT/DISCONNECT

- The command can be executed on-line.

- Can disconnect/reconnect customer based on defaults/ request.
- Server keeps history of Connection/ Disconnection in Database.
- Provides Pre or Post payment option without any change in the hardware.
- Logic based auto-disconnect
  - Load limiter
  - Peak load limiter
  - Load type limiter

Negatives

- Cyber Security Concerns.
- Data Privacy (crime etc.).

BENEFITS OF REPLICABLE MODEL

- Huge T&D loss can be contained by local generation.
- No fuel cost.
- Least Maintenance.
- Quality Power.
- Renewable purchase obligation (RPO).
- Quick ROI.
- Reliable and quality power.
- Mitigating Climate change concern and carbon credits.

COMPLEXITY, CITIES AND ENERGY

LUÍS M.A. BETTENCOURT
Santa Fe Institute
Santa Fe, USA

The world is spiky: population, wealth, innovation, energy consumption are concentrated in cities! Florida and Stolarick, 2005.

Night lights over Europe, 1997.

THE PROBLEM: ENERGY AND CITIES

- Most people live in cities. In a few decades the world will be fully urban.
- Most of the great cities in the world are being built NOW.
- Planning and energy choices made NOW will be locked in for >50 years.

Cities can support radically new technological & organizational solutions.

"The **methods** of theoretical physics should be applicable to all those branches of thought in which the essential features are expressible with numbers."

Paul A.M. Dirac
Nobel Lecture, 1933

MAP OF OUR DATA COVERAGE AND ANALYSIS

August 2012

Bettencourt et al. 2007, 2008, 2009, 2010

GENERAL SCALING OF METROPOLITAN GDP ACROSS NATIONS

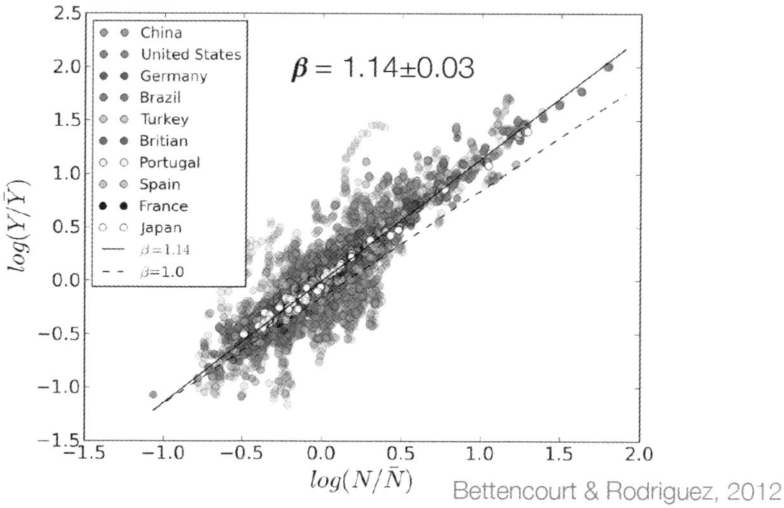

SMOKING GUN: INFRASTRUCTURE VOLUME AND SOCIAL OUTPUTS ARE RECIPROCAL

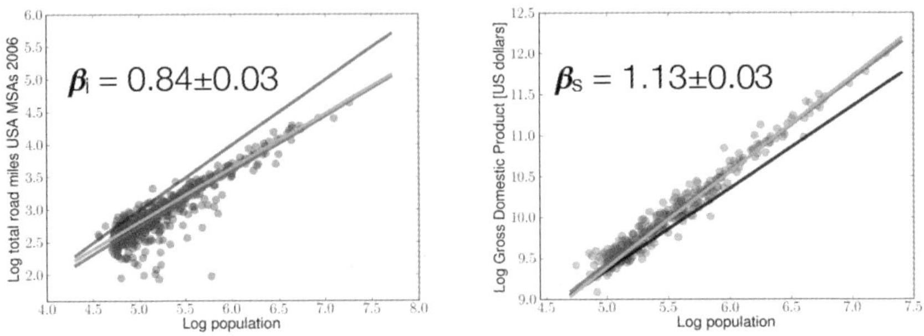

Volume of Infrastructure

$$\sim N^{\beta_i}$$

$$\beta_i = 1 - \delta$$

$$\delta \simeq 0.15$$

Social Outputs

$$\sim N^{\beta_s}$$

$$\beta_s = 1 + \delta$$

THE ORIGINS OF SCALING IN CITIES

Cities obey four general principles:

1. **Mixing Populations**

$$A(N) = aN^{\alpha}, \quad \alpha = \frac{D}{D+H} \sim \frac{2}{3} \quad \text{US MSAs}$$

2. **Decentralized infrastructure networks**

$$A_n(N) \sim \rho^{1/D} N \sim N^{5/6}$$

3. **Bounded human effort**

$$G = (ga_0 l) \text{ is } \sim \text{constant of city size.}$$

4. **Social outputs reflect local human interactions**

$$Y(N) \sim G \frac{N^2}{A_n(N)} \sim N^{7/6}$$

Bettencourt 2012, submitted

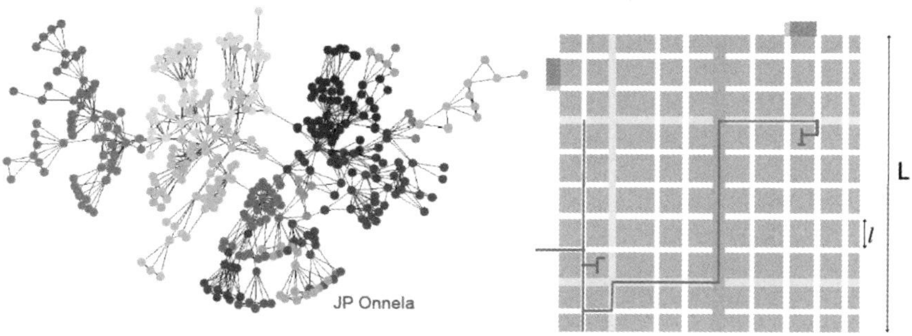

Spatial and temporal social proximity facilitates functional integration at increasing scales.

People can be the infrastructure but infrastructure cannot be the people.

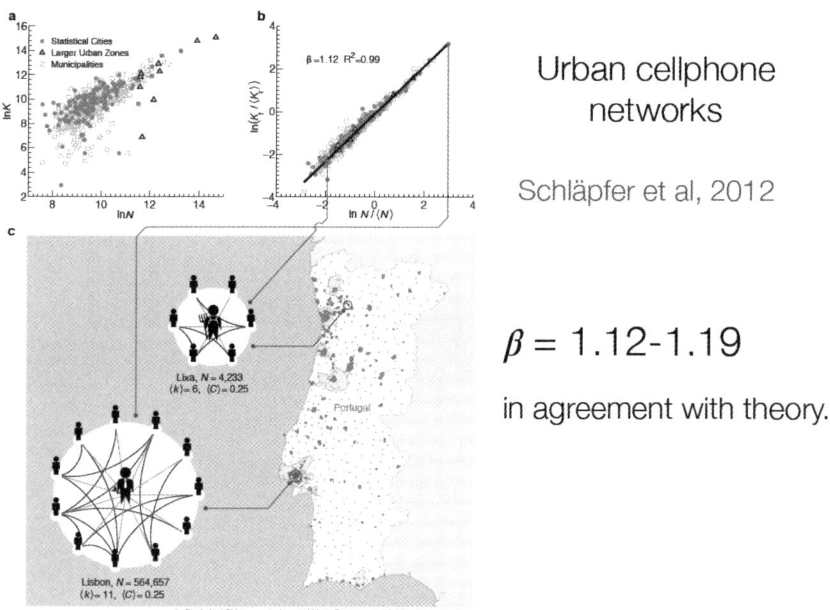

ENERGY CONSUMPTION IN U.S. CITIES

Household consumption vs. Transportation

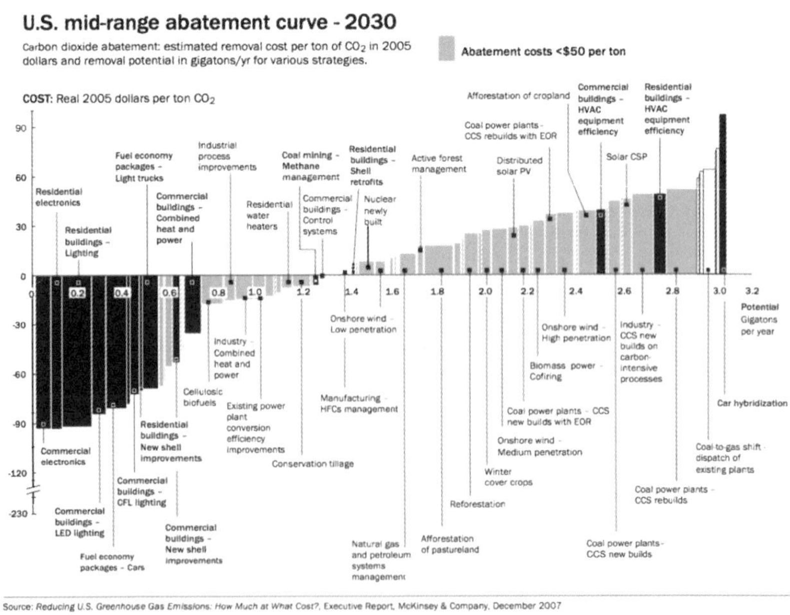

RESIDENTIAL CONSUMPTION (Similar for Japan, Germany, China)

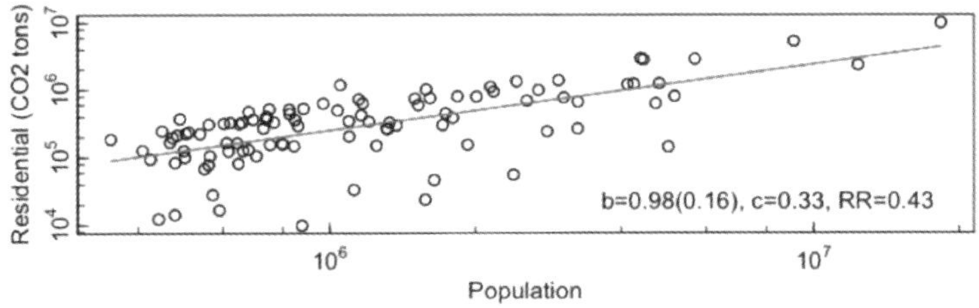

Household Climate Control

$$\Delta E = ekS\Delta T$$

efficiency / external surface

thermal conductivity

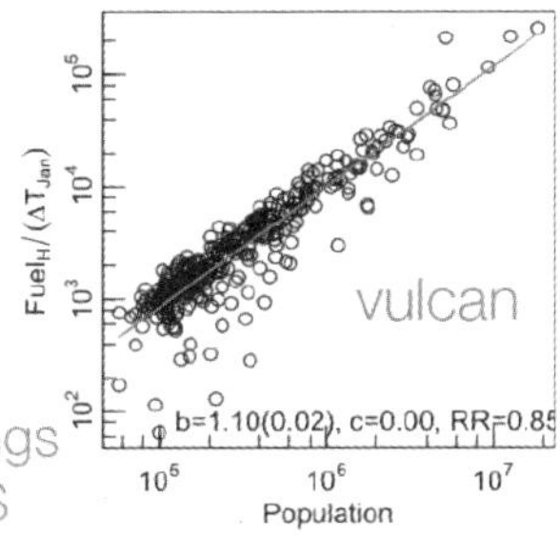

-New York City must deal with its buildings

-San Diego is the greenest city in the US

TRANSPORTATION! (Similar for Japan, India, UK)

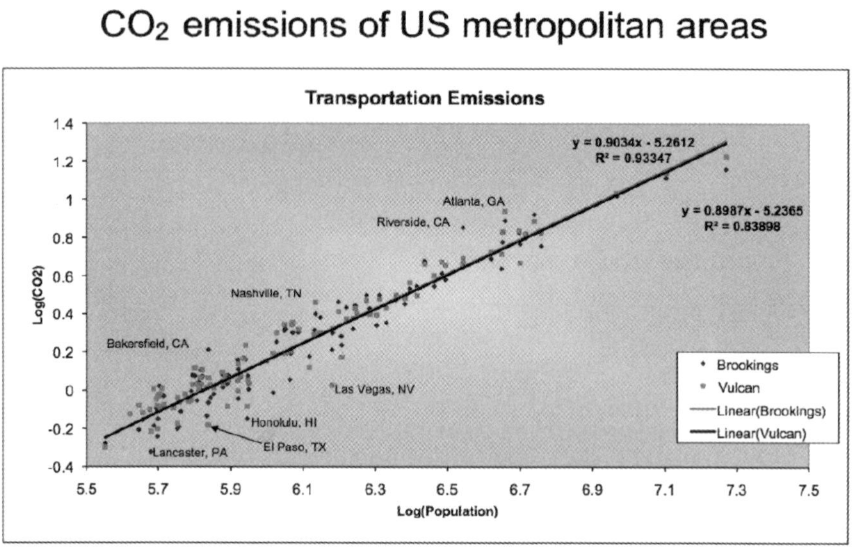

WEALTH, TIME, SCALE AND TRANSPORTATION CHOICES

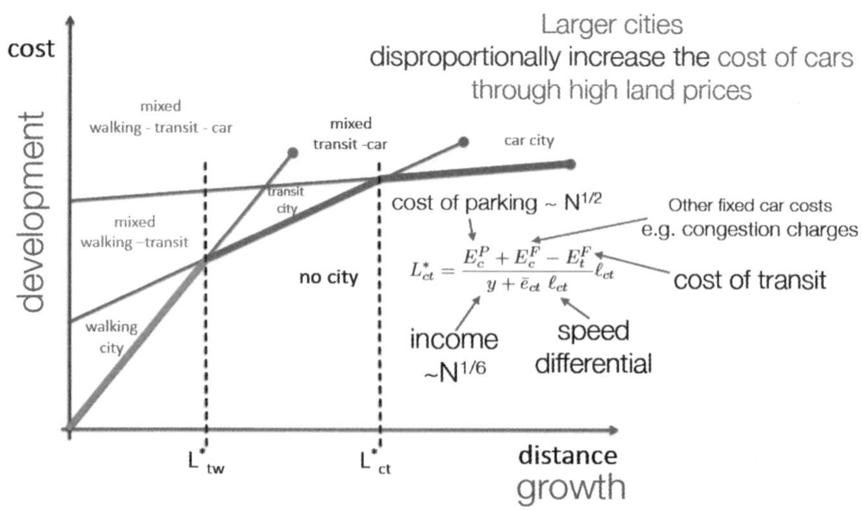

<u>Cities are centers of:</u>

- Consumption
- Innovation
- Efficiency

Solutions to societal problems rise and fall by the needs and uses in cities.

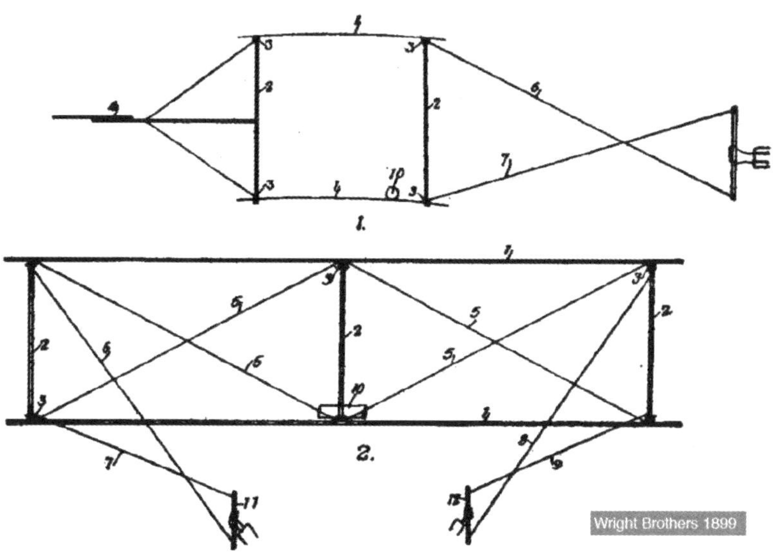

Wright Brothers 1899

"Whatever nature has in store for mankind, unpleasant as it may be, Ignorance is never better than knowledge."

Enrico Fermi
1901-1954.

Thanks to:

Clio Andris (MIT, SFI)
Andres Gomez-Lievano (ASU)
Marcus Hamilton (SFI)
José Lobo (ASU)
Nathaniel Rodriguez (UI, Bloomington)
Horacio Samaniego (LANL, Valdivia)
Deborah Strumsky (UNC Charlotte)
Geoffrey West (SFI)
HyeJin Youn (SFI)
Rockefeller Foundation
John Templeton Foundation
James S. McDonnell Foundation
National Science Foundation
Bryan J. and June B. Zwan Foundation

http://www.santafe.edu/research/cities-scaling-and-sustainability/

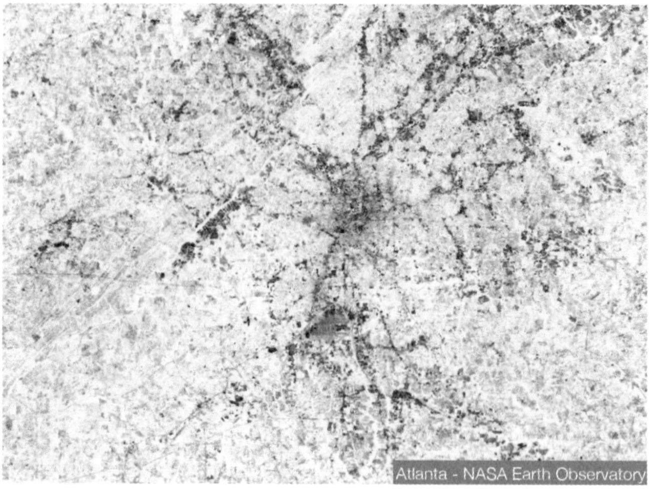

TRANSPORTATION ENERGY IN INDIA

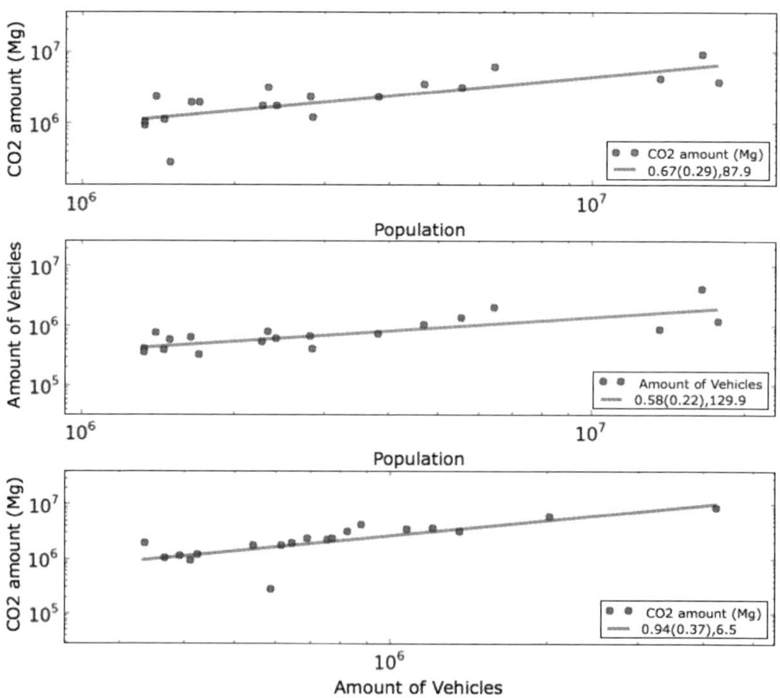

PREDICTIONS AND CONSEQUENCES

Urban Scaling Relations	Model (D=2,H=1)	Model (D, H)	Effect
Land area $A = aN^{\alpha}$	$\alpha = \frac{2}{3}$	$\alpha = \frac{D}{D+H}$	spatial densification
Network volume $A_n = A_{n0}N^{\nu}$	$\nu = \frac{5}{6}$	$\nu = 1 - \delta = \frac{D^2+DH-H}{D(D+H)}$	growth of infrastructure
Network length $L = L_0 N^{\lambda}$	$\lambda = \frac{2}{3}$	$\lambda = \alpha$	area filling networks
Average network width $\bar{S} = \bar{S}_0 N^{\bar{\sigma}}$	$\bar{\sigma} = \frac{5}{6}$	$\bar{\sigma} = 1 - \delta$	widening of roads
Interactions per capita $y = Y_0 N^{\delta}$	$\delta = \frac{1}{6}$	$\delta = \frac{H}{D(D+H)}$	increased interactions
Socioeconomic rates $Y = Y_0 N^{\beta}$	$\beta = \frac{7}{6}$	$\beta = 1 + \delta = \frac{D^2+DH+H}{D(D+H)}$	acceleration of social rates
Power dissipation $W = W_0 N^{\omega}$	$\omega = \frac{7}{6}$	$\omega = 1 + \delta$	increased congestion
Land Value $P_L = P_0 N^{\delta_L}$	$\delta_L = \frac{3}{2}$	$\delta_L = 1 + \alpha + \delta$	increased land rents

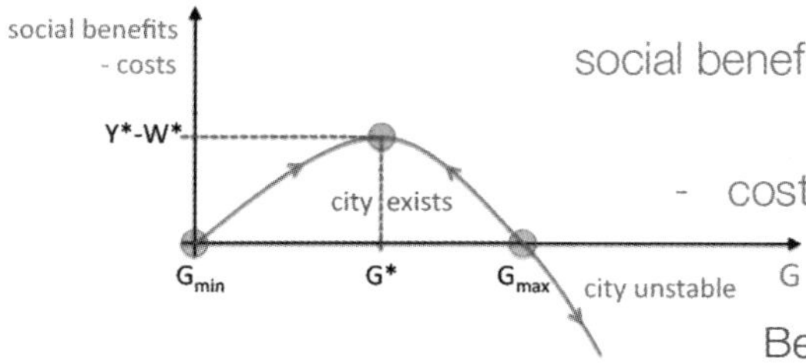

social benefits: $Y(N) \sim G \dfrac{N^2}{A_n(N)} \sim N^{7/6}$

- costs: $W(N) \sim W_0 \dfrac{I^2}{A_n(N)} \sim N^{7/6}$

Bettencourt 2012, submitted

SESSION 6

GLOBAL FOOD PRODUCTION & FOREST DYNAMICS

ENERGY IN GLOBAL AGRICULTURE AS THE HUMAN POPULATION PEAKS

LOWELL WOOD
Hoover Institution, Stanford University, University of California and Lawrence Livermore National Laboratory, Stanford, USA

PHILIP ECKHOFF
Intellectual Ventures
Bellevue, USA

INTRODUCTION

Provision of adequate nutrition to a population that is forecasted to grow to >9 billion individuals by the year 2050 is a major challenge facing humanity. When considering requisite major improvement in the nutrition level of the existing population, net global agricultural production will have to at least double over the next four decades. This growth necessarily will be driven primarily by increases in agricultural productivity rather than by major increases in land under cultivation. If dietary preferences continue to develop globally towards those more characteristic of developed world settings, global food productivity may have to rise even further.

A major component of previous improvements in agricultural productivity and a key requirement for further increase is the availability and cost of energy. Energy is a key input for chemical fertilizer generation and utilization, planting, harvesting, transportation, storage, water supply, and many other aspects of modern agriculture. Unfortunately, the areas of the globe with the largest nutritional shortfalls today—and the largest projected shortfalls in 2050—are also the regions with the least effective access to energy today.

For much of human history, human-supplied power, fueled by food intake, was the primary source-term in agricultural energy accounts. The available food supply determined not only the supported population, but the anthropometrics of the supported population and the fraction of the population required for agricultural work. These factors would settle into an equilibrium determined by population, population anthropometrics, labor market, and food supply. Animal labor was the other key input, but this too required either diversion of food supply or allocation of land for grazing. Baseline human metabolism depends on body size, and thus available food calories determine how much excess energy for work will be available for a given body size.[1] A lower food supply supports smaller people—who are less absolutely productive agricultural workers—so that they will have sufficient excess energy to perform the work of bringing in the harvest. Smaller, less-nourished body sizes also spent relatively more time-and-energy in coping with acute and chronic illnesses, further limiting productivity. Net, low excess *per capita* calories clamped the amount of agricultural work that could be performed for nearly all of history prior to the 20th century.

As agricultural productivity increased, human anthropometry was able to improve as well, resulting in more productive body sizes, healthier populations, and a lower fraction of the population required to perform essential agricultural labor. As development continued, several key factors improved agricultural productivity, made the food supply more robust, and changed the sources of energy central to agricultural output. Early mechanization made available human and farm animal energy more efficient in driving crop production. As mechanization progressed, human and farm animal energy, which were in turn dependent on agricultural land and the food supply, were replaced to ever-greater extents by fossil fuels. Existing soil levels of fixed nitrogen, which had been dependent on natural processes such as lightning, crop rotations with legumes, or mined and transported nitrates, were supplemented drastically with outputs of the Haber-Bosch ammonia-synthesis process.[2] This technology has made giant strides in efficiency, but such fertilizer production and distribution requires energy inputs. These inputs have enabled tremendous increases in food productivity per cultivated hectare over the past 9 decades, so much so that over half of humanity is presently 'carried' nutritionally by Haber-Bosch process outputs.

Modern agriculture is extensively and intensively intertwined with energy production and availability. Energy is required for fertilizer production in Haber-Bosch plants throughout the world, as well as for the transportation and distribution of fertilizer of all types. Energy, primarily as fossil fuels, is required for transportation of crops from sites of production through processing points and on to sites of consumption. Separation of these sites is often inevitable due to the differences in agricultural productivity between the most productive regions and the regions with the highest populations, with efforts to move production closer to the population centers often resulting in higher net energy cost per output-unit than the nominal transportation savings. Regardless, energy remains a key requirement for transportation of crops. Once transported, crops and agricultural products must be stored until use, e.g., via both standard refrigeration of perishables and the operation of modern granaries to buffer food supply both prior to and following processing steps. Fossil fueled-energized mechanization is now synonymous with high-productivity agriculture in most locales. Energy also is often required in substantial quantities for the sourcing and transportation/distribution of agriculturally-required water.

In the present context of doubling agricultural production, severe energy challenges arise. First and foremost, energy availability is lowest and prices are often highest in areas with the most severe malnutrition. These regions are often also the ones with the highest anticipated population increases during the next four decades. Moreover, these areas generally have low present-day agricultural productivity and tend to face the worst issues with food security and stability. The concatenation of these challenges is formidable.

Agriculture is a key driver of human demand for water, and as water becomes scarcer and more in demand with growing populations and stepped-up agriculture, supplying of water will be a key driver of agricultural energy requirements. Sourcing may have to resort to such energetically-expensive means as desalination or similar enhancement of local low-quality water supplies, and more water will need to be transported to regions at risk of drought, or without access to water supply levels required for near-optimal productivity.

Increasing agricultural production demands energy of several other major types. More fertilizer will be produced as it becomes more utilized in developing world agricultural areas currently working at subsistence levels, as intensive fertilizer usage can result in order-of-magnitude yield-gains. Many regions of sub-Saharan Africa have difficult (e.g., weather-impaired), costly and long-latency transportation, so that fertilizer typically is not readily available and is much more expensive—often, 2X or more—at farmgates than at source-points in best-case. These transportation challenges doubly impact farmers in such areas, as they do not receive major fractions of the ultimate sales-prices of their crops due to high costs involved in moving products from farmgates to urban marketplaces or ports, moreover in transiently market-saturated circumstances arising from minimal crop-storage capabilities. Construction of all-weather roads requires petroleum, and improved transportation of fertilizer and crops requires major energy sources as well.

We explore a variety of routes towards meeting these energy demands in the contexts of doubling global agricultural production over the next four decades, especially in areas most in need of such large and swift gains. Precision planting and water and fertilizer usage combined with deployment of sharply performance-enhanced, locally-optimized seedstocks will attain and maintain productivity gains with significantly lower inputs and thus less energy, capital and labor than would otherwise be required. Waste and water management in cities and transportation of food to cities will be key drivers of overall use-efficiency as urbanization advances globally, and strategies for mitigating energy demand for each sector will all benefit the food supply outlook. Finally, new cultivar strains have the increasingly-tangible potential to require less water and fertilizer while concurrently providing large yield-gains, thereby engendering notably nonlinear energy savings. [We are aware of a number of thoughtful studies[3-5] documenting the feasibility-in-principle of adequately nourishing ~50% more people by large-scale redistributions of *presently-employed* quantities of fertilizers and water supplies concatenated with major shifts in dietary practices, e.g., abandonment of preferences for meat intakes in favor of high-yield cereal grains and uniform distribution of foodstuffs. However, our present focus is on identification of ways and means for practically evolving human food generation and use in less centrally guided and implicitly coercive manners, according greater respect to national sovereignties, unequal resource distributions, extant and evolving dietary tastes of individuals and sub-populations, etc.—i.e., taking note of the salient features and working within the constraints of the quite non-ideal world in which we find ourselves.]

Facilitating the increases in agricultural productivity indicated as required over the next four decades will have dramatic impacts on human health and well-being. Nutritional status is a key factor in many types of disease and chronic illness.[6] Most notably, adequate macro and micro-nutrition for humans in their first half-dozen years of life is essential for cognitive development, and has been demonstrated to dramatically improve the health, productivity and overall outlook for future individuals and nations alike; seeing to such nutrition will provide the single greatest return-on-investment in agricultural productivity in the developing world, especially when the relatively very modest demands—of the order of 2-3%—placed on total human food supply are considered. Finally, mitigating stresses on food supply, water, and energy can dramatically bound major triggers for large-scale conflict, as the global population swiftly crests.

THE CHALLENGE OF ADEQUATE NUTRITION

All of human civilization rests on an agricultural base, the source of all food and most fiber. Farming is the traditional occupation of 70-90 percent of humanity, and 50 percent of all people still live in rural areas, engaged in predominantly agricultural occupations. The fraction of the human population involved in agriculture has declined remarkably over the past two centuries as productivity gains have allowed a smaller agricultural labor force to feed humanity. As the global population grows beyond 9 billion individuals by 2050 --with a UN-estimated eventual peak at approximately 10 billion—still more food will need to be produced to feed the planet.

Unfortunately, these required gains are not starting at food adequacy today, as extensive malnutrition still persists in many human populations.[7] This current malnutrition leads to stunting and wasting of growing young people—approximately a

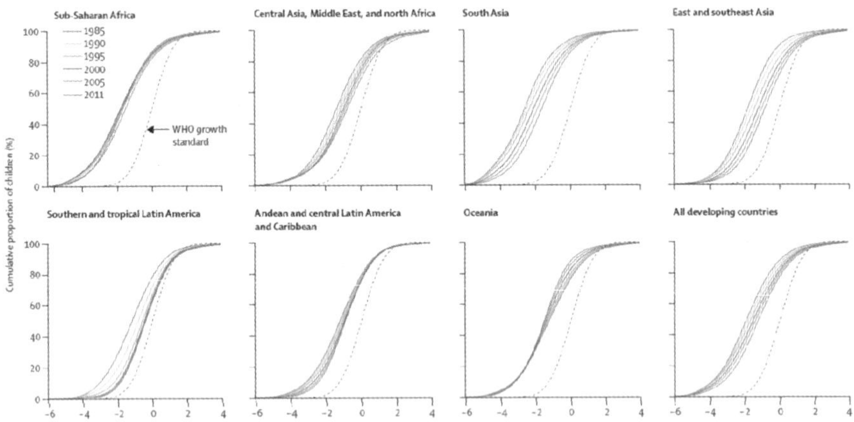

Fig. 1: Trends in height for age Z-scores (HAZ) by region. Each panel plots estimated distributions of HAZ at 5-year intervals. The more of the distribution below -2, the higher the prevalence of malnutrition in a region's children. The more of the distribution below -3, the higher the prevalence of severe malnutrition. South Asia started with the worst malnutrition in 1985 but has made dramatic progress in each 5-year interval. East and southeast Asia have made tremendous progress as well. Southern and tropical Latin America are actually approaching the WHO child growth standards. Sub-Saharan Africa, on the other hand, had a decade of increasing malnutrition from 1985 to 1995 and is just now getting back to the HAZ distribution of 1985. Figure from[7].

half-billion of whom have already been irrevocably ruined and another billion of whom are at risk between now and 2050, if current conditions persist. There has been tremendous progress in certain regions of the world since 1985, especially East Asia and South Asia, but sub-Saharan Africa is just recently catching up to its nutritional standards of 1985 after over a decade of decline through the mid-1990's.[7] Figure 1 shows the trends in height-for-age Z-score (HAZ) by region over the past 3 decades relative to 2006 WHO child growth standards.

South Asia still dominates in total numbers of children with HAZ and WAZ (weight-to-age Z-score) below 3 standard deviations of the mean of the WHO standard distributions, but this region has made great progress in both mean Z-scores and in total numbers of children most at risk since the mid-'80s. East Asia has raised its mean Z

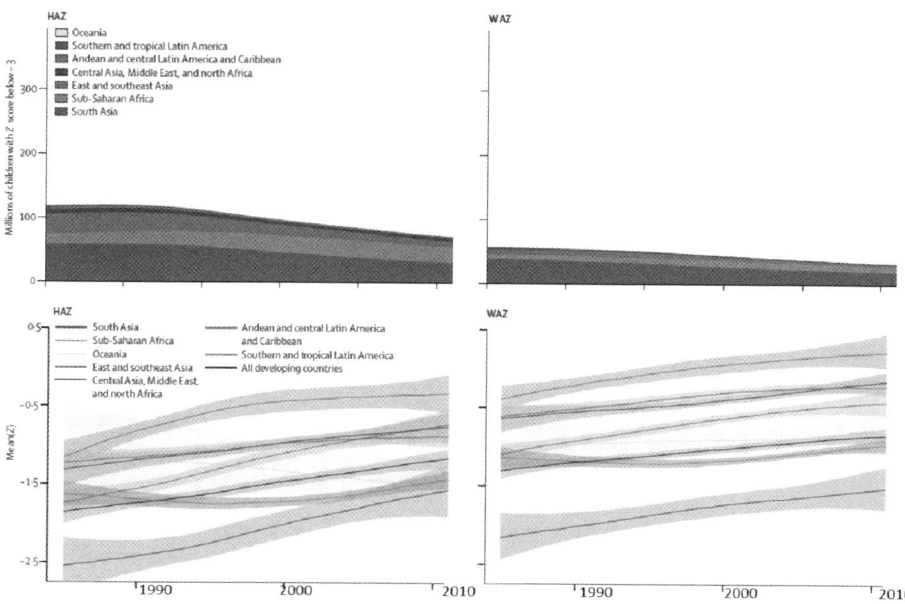

Fig. 2: Trends in mean Z-scores and numbers of severely malnourished children by region. South Asia still has more children with HAZ and WAZ below -3 than any other region, but the prevalence has been decreasing even faster than populations have been growing. So even the absolute number of severely malnourished children is decreasing. In contrast, the prevalence in sub-Saharan Africa is similar to 1985 but populations have increased, resulting in an increase in the absolute number of severely malnourished children. East Asia used to represent a sizable fraction of the absolute number of severely malnourished children in 1985, but the situation has changed dramatically for the better. Figure from[7].

scores from ones statistically similar to sub-Saharan Africa to scores closer to Latin America. The mean Z-scores in sub-Saharan Africa are similar to those in 1985, and the increase in population there means that the total number of children with Z-scores below -3 has actually risen while the globe's corresponding numbers have fallen. These trends and patterns are displayed in Figure 2.

The first Millennium Development Goal—MDG-1—focuses on a two-fold reduction in malnutrition (specifically defined as reducing the prevalence of children with WAZ < -2 by a factor of 2 or attaining a low target prevalence), but the estimated probability of attaining MDG-1 is quite low across much of sub-Saharan Africa and South Asia (Figure 3).[7]

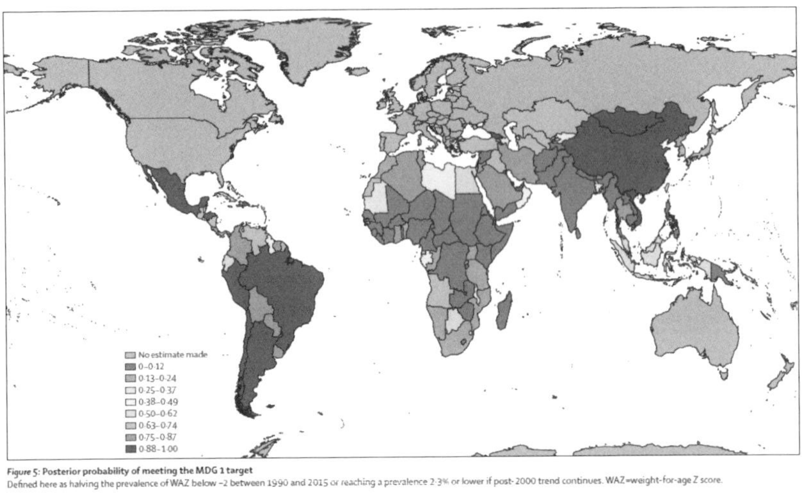

Fig. 3: Estimates of the likelihood of attaining MDG-1 by 2015. Figure from[7].

In addition to improving nutritional standards for existing populations and increasing food production for future populations, agriculture is necessarily responding to a large-scale evolution in dietary preferences. Dietary preferences across the developing world are changing towards the less crop-efficient, more meat-intensive diets characteristic of the developed world. The diets of more economically developed countries exhibit higher total per capita caloric inputs to diets and more upgraded dietary protein sources such as eggs, milk, and meat. Figure 4 exhibits trends over the past half-century in total dietary caloric demand and protein demand for a variety of economic categories.[8]

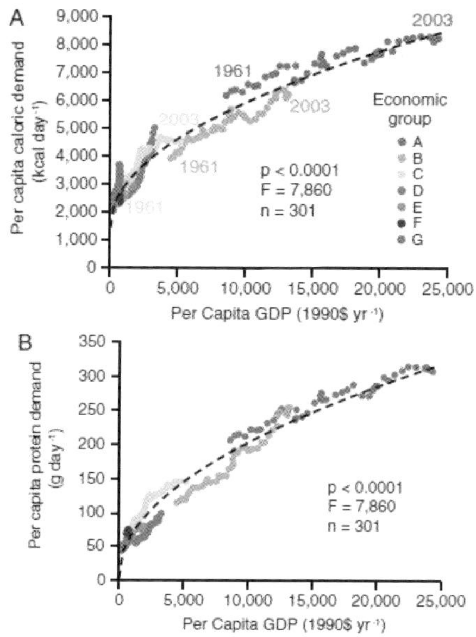

Fig. 4: Trends in caloric and protein demand with changes in GDP. Figure from[8].

As much of the readily arable land of our planet has already been enlisted in crop production, these necessary gains in crop production will have to come primarily from gains in agricultural productivity. Increasing total land area under cultivation will become increasingly difficult, and urbanization in many regions may be expected to even reduce total arable land. Concatenating all these factors, agricultural production will have to at least double in the next four decades, and the average productivity per unit of cultivated area will have to more than double.

THE CHALLENGE OF AGRICULTURE PRODUCTIVITY IS A CHALLENGE OF ENERGY SUPPLY AND UTILIZATION

Cheaper and more plentiful energy has been a major driver of previously documented major improvements in agricultural productivity. Energy is a key component driving generation and transport of fertilizers, crop planting and tending, water supply, harvesting, transportation, and processing and storage. Of special note is the water supply, as agricultural uses dominate the human-controlled global water budget, currently accounting for ~69% of human water use.

Unfortunately, the regions with nutritional shortfalls and thus behind their commitments on attainment of MDG-1 often have less effective energy access. Table 1 shows the annual electric power consumption *per capita* (in kW-hr/year) for a variety of countries, with data from IEA.[9]

Table 1: Electric power consumption per capita (kW-hr/year). Data from[9].

Country	Consumption	Country	Consumption	Country	Consumption
Haiti	36	Bangladesh	252	China	2631
Ethiopia	46	Cameroon	271	Argentina	2759
Tanzania	86	Pakistan	449	Malaysia	3614
Benin	91	Mozambique	453	Libya	4170
Nepal	91	India	571	UK	5692
DRC	104	Zambia	635	Russian Fed.	6133
Togo	111	Vietnam	918	Germany	6779
Kenya	147	Algeria	971	France	7468
Senegal	196	Zimbabwe	1026	Belgium	7903
Angola	202	Botswana	1503	Australia	11113
Cote d'Ivoire	203	Namibia	1576	USA	12914
Yemen	219	Thailand	2045	Norway	23550

Transportation and mechanization are other drivers of energy consumption in agriculture, and here too effective access to energy is more limited in regions most in need of productivity increases. Table 2 shows the distribution of vehicles, diesel consumption, and the cost of gasoline normalized by local GDP (PPP) for a range of countries.

Table 2: Data for motor vehicles per 1000, diesel fuel consumed, and the cost of gasoline normalized by per capita GDP indicate that normalized costs are higher and availability is lower for transportation in countries requiring the largest increases in food supply. Data from[10-12].

Country	Motor vehicles per 1000[10]	Diesel fuel consumed (road sector, kg oil equiv. per capita)[10]	Cost of gallon of gas in June 2010[11]	Gallon of gasoline normalized by *per capita* GDP (PPP) per day[12]
Eritrea	11 (2007)	6 (2009)	9.59 USD	4.76
Senegal	22 (2008)	41 (2009)		
Kenya	23 (2009)	21 (2009)	4.31 USD	0.9
India	18 (2009)	26 (2009)	4.25 USD	0.42
Pakistan	13 (2009)	38 (2009)	3.02 USD	0.39
Nigeria	31 (2007)	4 (2009)	1.62 USD	0.23
Indonesia	79 (2009)	44 (2009)		
Brazil	209 (2008)	149 (2009)	5.69 USD	0.18
U.K.	523 (2009)	351 (2009)	6.60 USD	0.067
France	598 (2009)	470 (2009)	6.04 USD	0.063
U.S.	802 (2009)	384 (2009)	2.85 USD	0.02

TRENDS IN ENERGY USAGE OVER THE HISTORY OF AGRICULTURE

For most of human history, human-supplied power, fueled by food intake, was the primary source of agricultural energy. People prepared fields, planted, tended, harvested, and threshed their crops with energy supplied by ingestion of fractions of their previous agricultural output. Available food supply determined both the number of the supported population *and* its anthropometrics, which determined the labor effectively available to produce the next season's food supply. Energy exertion in various activities can be measured as a multiple of the **B**asal **M**etabolic **R**ate. Including the necessary functions of eating, digestion, hygiene, etc,, a bare-survival diet requires approximately 1.27 x BMR.[6] Work, e.g., the production and processing of food, requires additional food energy beyond this modest BMR-multiple (see Table 3 for typical required BMR multiples for various agricultural activities central to food and fiber production).

Table 3: Energy requirements of representative agricultural activities. Data from[13, 14], collected by[6].

Activity	Multiple of BMR (males)	Multiple of BMR (Females)
Sleeping	1.0	1.0
Standing or Eating	1.4	1.5
Strolling	2.5	2.4
Walking, normal	3.2	3.4
Walking with 10 kg	3.5	4.0
Walking, uphill	5.7	4.6
Sitting and sewing	1.5	1.4
Milking cows	2.9	
Hoeing		5.3-7.5
Collecting and spreading manure	6.4	
Uprooting sweet potatoes	3.5	3.1
Weeding	2.5-5	2.9
Binding sheaves	5.4-7.5	3.3-5.4
Plowing	4.6-6.8	
Threshing		4.2
Felling trees	7.5	

Improved agricultural productivity has a wide variety of benefits that accrue over intervals from years to generations. More available food provides more available energy to produce next year's food supply. Resulting better nutrition for children reduces stunting and wasting and endows them with healthier, more labor-productive bodies as adults. Wages and labor participation increase as function of body-height, and wages increase as a function of BMI up to 27 kg/m^2.[6] Conversely, low BMI limits ability to do physical labor and is associated with increased health risks, both acute and chronic.[15] In addition to the impacts on stunting and wasting, child nutrition has a profound role in determining successful schooling:[16] starving children learn slowly and retain poorly. The *increases* in energy available for all types of work per consumer (in kCal/day) over the past 3 centuries in 3 relatively developed countries are shown in Table 4.

Table 4: Increase in energy available for work per consuming unit (kCal). From Table 1.3 in[1]. Note the significant 'bottoming out' in above-minimum food energy in the USA during the 19th century, which is more extensively documented elsewhere.[6]

Year	France	England & Wales	United States
1700		720	2313
1705	439		
1750		812	
1785	600		
1800		858	
1840			1810
1850		1014	
1870	1671		
1880			2709
1944			2282
1975	2136		
1980		1793	
1994			2620

Severe undernutrition during childhood can lead to stunting and excess disease as adults—and consequently lower labor productivity, especially in manual labor such as agricultural work. A less productive adult labor force will have difficulty feeding the next generation, especially when conducting less mechanized agriculture. Improving nutrition thus has a beneficial ratchet effect from generation to generation, as body sizes and general health increase and more work per capita is sustainable.[1] The feedback of improved available nutrition in Table 4 on population anthropometrics is seen in Figure 5, and Table 5.

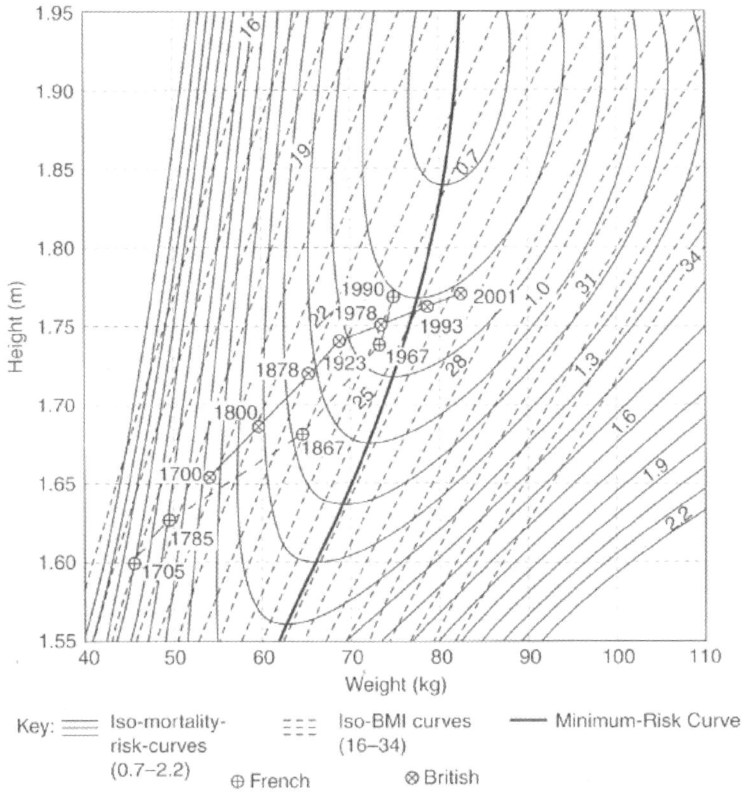

Fig. 5: Development of French and English body heights and weights across a mortality risk Waaler surface with iso-BMI and iso-mortality curves. Both England and France started the 18th century with body heights that were significantly shorter and weights representing lower body-mass indexes for a given height. The iso-mortality curves show constant rates of mortality at different height and BMI combinations. The minimum-risk curve plots the optimal weight for a given height in terms of minimizing mortality risk. England and France have both demonstrated an approach towards this minimum risk line while increasing heights and achieving even lower overall mortality rates. Figure 2.4 from[6].

Table 5: Changes in estimated average male heights at maturity (cm). Data from Table 2.5 in[6].

Quarter-Century	France	Great Britain	Hungary	Norway
18-III		165.5	167.4	165.6
18-IV	163.0	168.6	166.6	165.5
19-I	163.7	167.9	163.1	166.6
19-II	164.3	171.2	163.5	167.4
19-III	164.7	167.2	162.3	168.7
19-IV	165.4	168.0	163.8	169.6
20-I	166.3	168.2	165.4	171.0
20-II	168.0	170.0	168.4	173.8
20-III	171.2	175	170.7	177.6
20-IV	174.7	176.6		179.5

For much of human history, rather than a continually improving ratchet, there were long intervals of stagnation at similar levels of caloric availability, agricultural production, stunting and wasting, and adult anthropometries. The introduction of body-external sources of agricultural energy constituted a basic shock to this system and helped accelerate progress. Global exploration introduced new cultivars e.g., potatoes which improved the ratio of required energy inputs to attainable caloric outputs. Mechanization allowed fossil fueled-engines to augment and substitute for human and animal muscle-power. In the most mechanized agricultural settings, dramatic progress means that grandchildren of a ~40 acres' USA single-farmer today are >1000 acres single-farmers. This change in area per farmer—entirely unprecedented in all of human history—enabled the farm population-fraction in the more developed countries to evolve from >0.5 to <0.03 in less than a single century—and this while per capita total calorie intakes (including those via agricultural animals) were accelerating drastically (see Figure 4)

Another important source of agricultural efficiency was the imports of nitrates mined in South America. These significantly boosted agricultural efficiency per hectare and per unit of human work.[2] The availability of these nitrates was limiting, but eventually Haber-Bosch ammonia-synthesis process enabled production of nitrogen fertilizers through fossil fuel energy and mass inputs. Nitrogen fertilizers provide a tremendous increase in agricultural output per hectare— 4-10X—when employing the full set of modern "best farming practices." The transition to nitrogen fertilizers turned ever-more-readily-available fossil energy into sustained, large-scale improvements in farmer-efficiency, especially during the last two-thirds of the 20th century.

These chemical fertilizer-based gains were then further leveraged by the accelerating availability of improved seeds and much more intensive irrigation: the Green Revolution. Today there has been both very extensive proliferation and capacity-growth of Haber-Bosch plants worldwide. They presently consume ~2% of all fossil fuel energy and ~6% of all natural gas, and they support food generation sufficient to 'carry' 40-60% of the planet's present human population. Figure 6 illustrates the improvements over time in agricultural productivity coinciding with the increase in use of nitrogen fertilizers. Recently, there have been improvements in the efficiency of utilization of nitrogen fertilizers as seen in Japanese rice yields (top left). China's extraordinary commitment to the production and consumption of nitrogen fertilizers over the past few decades has drastically increased that country's agricultural productivity and has driven the tremendous decrease in average East Asia malnutrition over the past quarter-century that's exhibited in Figures 1-2.

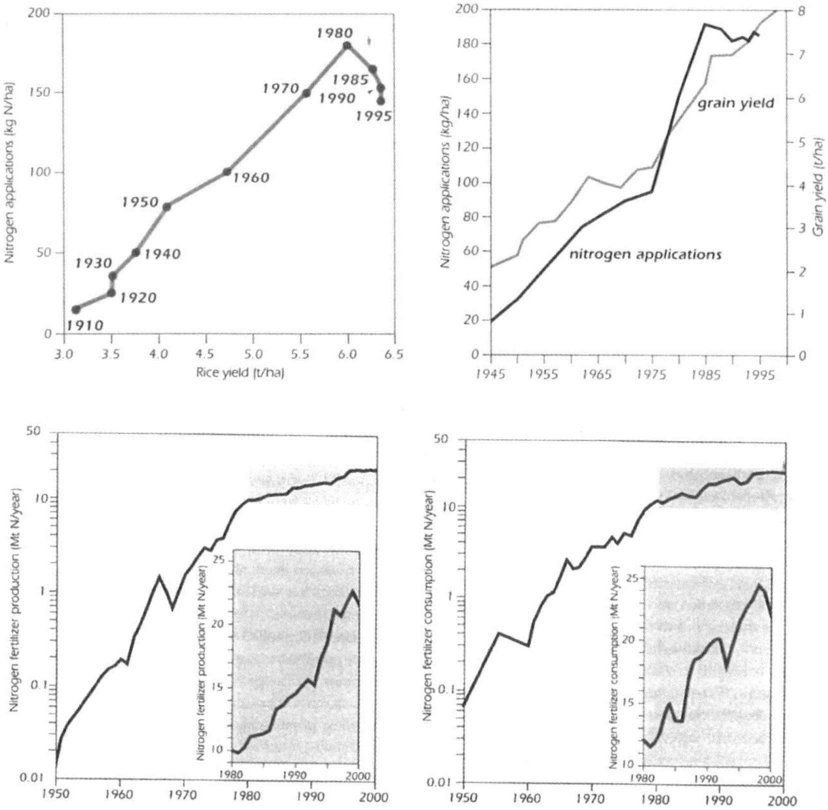

Fig. 6: (left top) Nitrogen fertilizers and rice yields in Japan demonstrate recent improvements in yield efficiency, as rice yields has gone up as average nitrogen applications have peaked[2] *(top). Use of nitrogen fertilizers dramatically increased winter wheat yields in England over the past 65 years*[2] *(bottom). Tremendous* increases *in Chinese food production and consumption of nitrogen fertilizers coincided with dramatic* decreases *in numbers of under-nourished children, as seen in Figures 1-2.*[2]

In summary, energy consumption is a critical component of developed world agriculture. Energy is required for sourcing and distribution of water, production and transportation of fertilizer, multi-modal mechanization, transportation of foodstuffs, and storage and processing and use-efficiencies of agricultural products. Energy use provides enhanced structure *and* improved nodal efficiencies of modern food webs. These gains are further leveraged by gains in 'seed quality,' improving cultivar efficiencies along multiple axes.

THE ENERGY CHALLENGE FOR DEVELOPING WORLD AGRICULTURE

Energy access and utilization has enabled irrigation, fertilization, mechanization, and transportation and storage of crops to revolutionize agricultural productivity in the developed world, but comparable gains have been scarce in the developing world. Much of the developing world still experiences acutely limited availability of electricity, as outlined earlier in Table 1. In addition, agricultural mechanization and commodities transport are often severely constrained by availability and costs of liquid fuels, as illustrated in Table 2.

Fertilizers, which have driven much of the last century's improvements in agricultural production, exhibit highly-variable availability and cost. N-based fertilizer costs may stabilize as natural gas becomes cheaper, but K- and (particularly) P-based fertilizer costs will likely grow substantially, as K- and P-based fertilizers have modest reserves and few suppliers. These issues raise rational concerns that currently ample-to-generous returns-on-investment (ROI) to farmers associated with moderate levels of fertilizer-usage will diminish.

In addition to the shortages of consumables such as electricity, liquid fuels, and fertilizers, developing world agriculture is often seriously undercapitalized. There are often pervasive shortfalls in all-weather road availabilities. Remediation will require energy, equipment and materials to construct and surface asphalt roads—the cheapest roads having acceptable qualities—"up to farmgates." These transportation challenges are compounded by frequently-severe shortages of crop-storage facilities adequate for local prevailing conditions, another consequence of undercapitalization. These combine to create sometimes-major losses and inefficiencies in getting crops to market/export.

In some countries, the food supply was increased during recent decades by expansion in the area under cultivation. As outlined earlier, future such expansions are likely to be quite limited, and such arable land-area limits compel improved productivity as populations grow. Limited arable land can be further decreased by urbanization. Available arable land is close to use-saturation in most densely populated regions, including those with widespread under-nutrition. This has impelled improving crop yields per hectare through fertilizers—especially NH_3-based ones—as well as use of new cultivars and more extensive irrigation. The increasing use of fertilizers as arable land *per capita* decreases is illustrated in Figure 7. Those countries which have not yet followed this path will soon do so out of necessity, unless they opt for the always-perilous path of importing ever-larger fractions of their food supplies from world markets not distinguished for their price-stability.

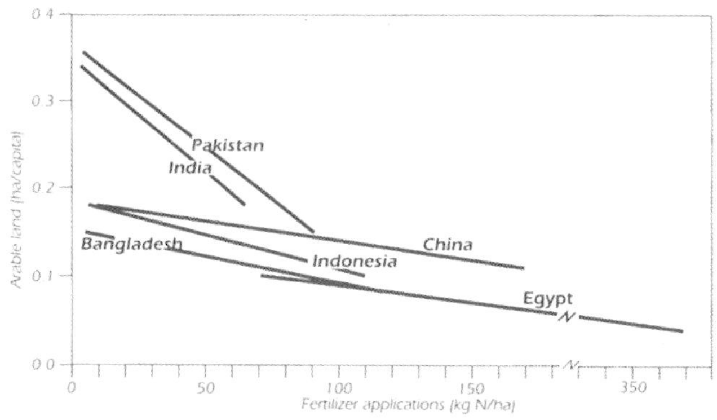

Fig. 7: Increasing nitrogen application trends as arable land per capita decreases.[2]

With substantial gains being available as agricultural practices advance towards best-in-class,[2,3,17] as seen in Figure 8, the world definitely has the ability in principle to meet the mid-21st century agricultural production targets that we've sketched. Much of these gains will require new energy expenditures, as well as more efficient utilization of water, fertilizers, and energy. The rest of these gains will require improved capitalization of developing world agriculture together with the corresponding energy expenditures. In the next section, we outline pathways towards meeting these energy needs and estimate the total amounts of energy required.

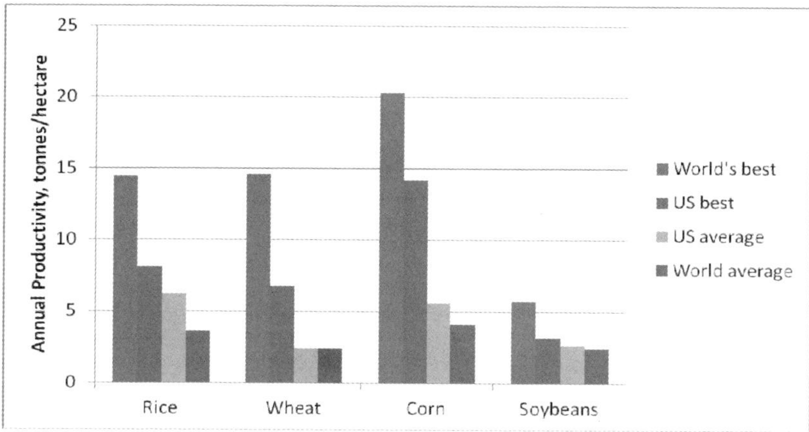

Fig 8: Available productivity gains when moving from world average to world's best.[2]

PATHWAYS TOWARDS SATISFACTION OF AG-ENERGY NEEDS

As fertilizer demands increase to support food production needs, more efficient use or production will be required. More precise use of fertilizers will save on production and transportation costs, while making usage more economical—and thus accessible—in developing world contexts. Remote sensing can inform timing and quantities-used. Improved time-release packaging and release can improve the fraction of fertilizer that ends up in valued agricultural plant matter rather than being leached out to then constitute polluted run-off. Precision placement-on-demand is comparatively technology-intensive, but may rather drastically improve the efficiency of fertilizer use.

An obvious pathway to explore is improving the efficiency of N-based fertilizer production. However, the best Haber-Bosch plants are approaching stoichiometric requirements in N-fertilizer production efficiencies, as seen in Figure 9 (left).[2] It is important to move the global average up towards best-demonstrated, but as the right panel of Figure 9 illustrates, large gains are available by moving produced fertilizer more effectively and efficiently into plants' tissues. Improvements in state-of-the-art production efficiencies may saturate, and improved usage to defeat leaching into run-off and N-'evaporation' will provide gains up to a point. But in the end, it will be necessary to use significantly more energy and generate much more N-based fertilizer in order to attain the requisite doubling of food production.

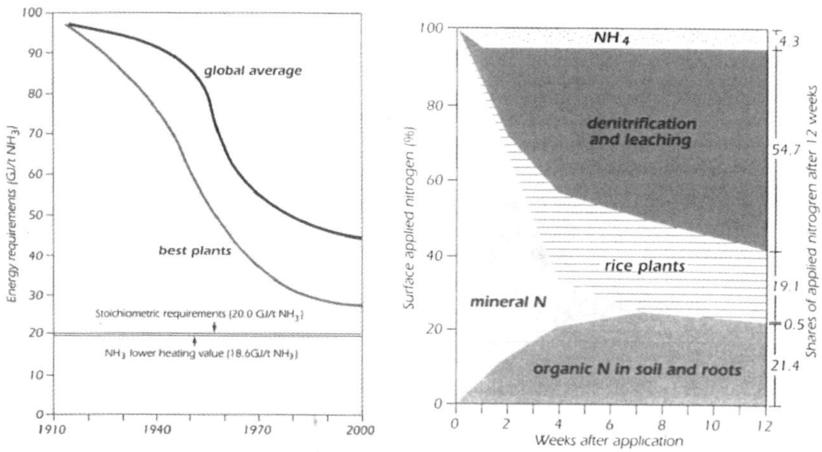

Fig 9: Best plants approaching limits, but global average lags (left). The energy requirements of the early Haber-Bosch plants were very high, but there has been consistent progress in reducing input energy for a given output of NH_3. Given the exothermic reaction, the energy requirements are driven by the stoichiometric requirements for hydrogen supplied by natural gas (CH_4). The best plants are approaching this limit, and the global average has made progress but remains more than a factor of two away from optimal. Much of applied nitrogen fertilizer presently does not make it into plants (right). When fertilizer is first applied, most of the nitrogen is in the form of mineral N. As time continues, this nitrogen either is incorporated into rice plants or organic matter in the soil or is leached away. More precise spatiotemporal application can reduce the amount lost to leaching. Figures from[2].

Water sourcing and irrigation in water-stressed and water-unstressed regions offers tremendous benefits for agriculture. Sufficient irrigation enables maximum carbon-fixation and growth for a plant, and reliable irrigation limits crop failures due to too-scant or -infrequent rains. As such, agriculture dominates the water budget—the 'water footprint'—of today's global civilization, as illustrated in Figures 10 and 11. Irrigation is a major driver of agricultural productivity, but extending irrigation has energy requirements for sourcing and transport. The easier case is when water is available but merely under-utilized due to lack of capital equipment. In some parts of the world, including much of Africa, there are available water resources from aquifers, as indicated in Figure 12. In these cases, water is available for sourcing to growing fields, and enhanced capitalization will allow pumping of water to these fields. Drip irrigation and more advanced means of minimizing evaporation losses will enable further capitalization to significantly improve water utilization efficiencies for several varieties of crops.

Table 6. Global water footprint of production (1996–2005)

	Agricultural production					
	Crop production	Pasture	Water supply in animal raising	Industrial production	Domestic water supply	Total
Global water footprint of production, Gm^3/y						
Green	5,771*	913†	—	—	—	6,684
Blue	899*	—	46†	38	42	1,025
Gray	733*	—	—	362	282	1,378
Total	7,404	913	46	400	324	9,087
Water footprint for export, Gm^3/y	———	1,597	———	165	0	1,762
Water footprint for export compared to total, %	———	19	———	41	0	19

*Mekonnen and Hoekstra (21, 22).
†Mekonnen and Hoekstra (23).

Fig. 10: Blue (surface and ground water), green (rainwater), and gray (variably-polluted and brackish freshwater) water footprints for agriculture, industry, and in total. The water footprint of production represents the water consumed from different sources in different sectors of the economy, and the water footprint of production is dominated by agricultural production. Rainwater is critical for crop production and animal pastures. Exports of products that require water inputs represent substantial international flows of water resources. Table 1 from[18].

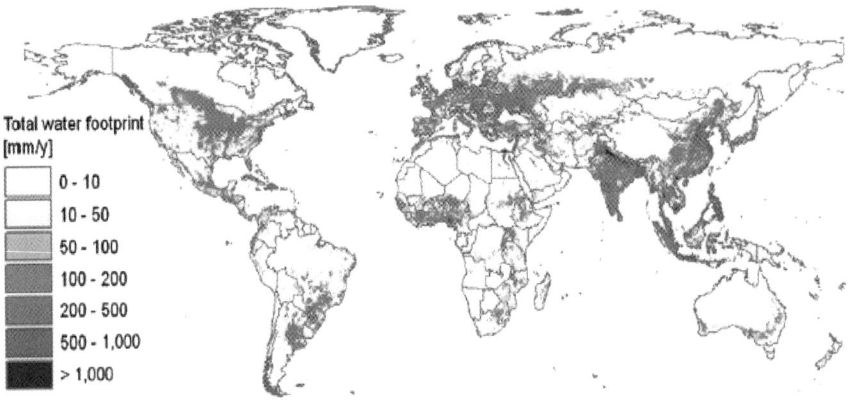

Fig 11: Water footprint of humanity, 1996-2005. These calculations from[18] *calculate the mm of water used per year in production and supporting the residential population. Higher footprints can be driven by agricultural production as seen in the central United States or by large populations, or by both. In areas in which the water footprint exceeds replenishment rates, groundwater stores will be gradually depleted. Figure from*[18].

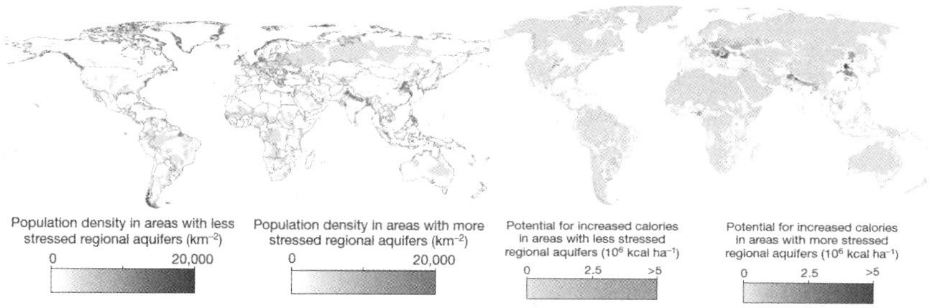

Fig. 12: Some parts of the world exhibit unstressed aquifers which could provide easier water sourcing. Many of the world's aquifers are stressed, including many in areas which could provide more calories if water supplies were available. Measures of stress are determined in[19] *by calculating a 'groundwater footprint' of how much area is required to replenish aquifers given their utilization rates. Aquifers with groundwater footprints larger than their actual area will become depleted over time and are classified as stressed. Many regions of the world including the central United States, north-west India, and north-east China could produce even more calories of food than they do at present. These increases in production would require additional water from aquifers are already stressed and do not represent a long-term sustainable source of increase. Large scale water transport would be required to sustain food production gains in these areas. Other parts of the world have available food production increases without overstressing the local aquifers. Figures from*[19].

The more difficult and energy-intensive problem is water sourcing in stressed regions. As seen in Figure 15, the aquifers in the highly-populated regions of India and China are already stressed. China and India have both made tremendous strides in reducing malnutrition, primarily through reasonably intensive use of (especially, nitrogen-based) fertilizers, irrigation, and improved cultivars. Reliable water sourcing will be required to sustain and further extend these gains. Other regions of the world that may require improved water sourcing include the central United States and south-central Canada, which is already a critical grain-producing region for the planet.

Once easily available freshwater is fully-utilized, water sourcing will involve upgrading of brackish and polluted water and desalination. Energy requirements for water sourcing from saltwater begin with the basic thermodynamics of desalination, which requires 1.06 kW-hr/m^3 or 3.8 kJ/L. Including pump efficiencies, the typical minimum energy-expense will be ~1.5-1.6 kW-hr/m^3,[20] and the overall water-plant requirements will be even higher due to pretreatment and post-treatment energy expenses. In a triumph of technology, current energy efficiencies of reverse osmosis of seawater have made tremendous improvements towards thermodynamic optima over the past few decades, as seen in Figure 13. This thermodynamic minimum energy naturally depends on the water recovery fraction and on the salinity of the feedwater. These also represent major improvements over early desalination technologies relying on thermal means. Other desalination options include wave-energy driven, multi-stage flash distillation and non-photoelectric solar power, but these typically have non-competitive capital and/or operating energy costs when compared to the best reverse osmosis-based plants.

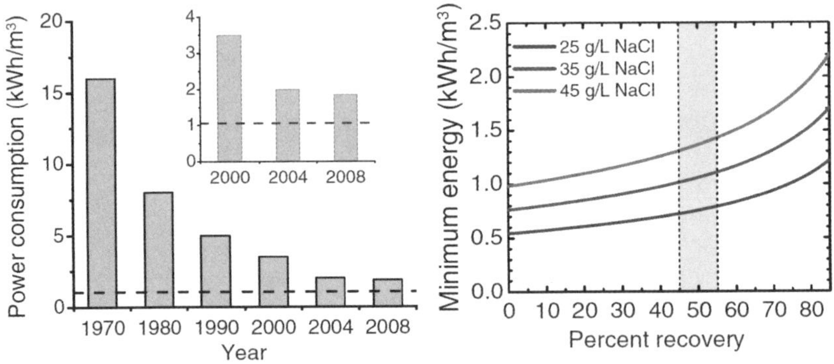

Fig 13: (left) Progress in efficiencies of reverse osmosis stage towards thermodynamic optimum. Thermodynamic minimum energy requirements are plotted as a function of salinity and percent recovery (right). Higher salinity feedwater (e.g., from the Persian Gulf) requires more desalination energy. Higher water recovery fractions naturally require more energy per unit volume. Figure from[20].

Once water is sourced, it must be transported for use, and there are other efficiency gains potentially available for obtaining, moving, and using water. Precision 'irrigation' enhances water delivery to roots and lowers the amount of water required for a given agricultural output. Drip-fed irrigation is the "best practice" today; sub-surface sourcing may soon be preferred. Spatially-precise delivery of water and temporal delivery precision will allow water to be delivered just where it is needed, when it is needed. The appropriate timing and location is driven by current soil moisture, recent and near-future weather (e.g., precipitation, temperature, humidity and wind-speed), current condition of the crop plants, and stage of the growing season. Remote sensing can inform each of these to a possibly-surprising extent, and expansion of the use of remote sensing data will play a critical role in enabling these spatiotemporal precision improvements.

Locally-optimized selection of cultivars can also improve water usage efficiency. Cultivars best-adapted to local climate, soil, pests and tastes-in-foodstuffs can provide improved effective caloric production for a given set of energy, water, fertilizer, and labor inputs. Photosynthetic performance-improved cultivars can be rationally expected to express better CO_2 fixing and transport efficiencies. Since plants typically transpire ~100X their tissue mass-gain in water while conducting their CO_2-'harvesting' activities, efficiency gains in CO_2 fixing and transport immediately map into superior water use efficiency and lower water demand. In a striking recent example, over a 20% gain in soybean harvest has been seen with a single-gene-modification resulting in enhanced intra-leaf CO_2 transport.

It is possible to provide an upper-bound estimate on how much water sourcing will be required. Plants grow best when water supplies on root-systems typically are evaporatively diminishing: long, cloud-free, sun-high-overhead days. Unsurprisingly, irrigation to keep roots water-replete greatly increases plant food and fiber productivity. A quantity of 0.5±0.1 meters of water-supplied to soil-surface over later growing season typically "makes all the difference" in a crop plant's success in seed-filling, enabling maximum CO_2-harvesting and carbon-fixation. Substantial economization from this figure is of course available via more efficient water-supplying, such as drip-irrigation and sub-surface sourcing. These approaches greatly reduce evaporation losses but require capital investments. As there are ~1.6 gigahectares of arable land now under cultivation, approximately 10^{13} m^3 of water are needed annually for full irrigation coverage. When averaged over a growing season's approximately 10^7 seconds, that globally-aggregated demand requires a flow of ~10^6 m^3 /sec: 1 Sverdrup (Sv).

This figure is roughly the aggregate global riverine-flow, or 4X the Amazon's—delivered over the growing season. Three-times lower flows suffice, if reservoir- or soil-storage of water is feasible: e.g., soil-storage allows fields to accumulate water year-round and then use it in 'burst mode' when roots of growth-surging plants access the stored water. If 1 Sv is all obtained via reverse osmosis from seawater, roughly 5 MJ/m^3 of hydraulic work is required, or 5 TW for a third-year interval. This energy requirement is several times the current USA electrical generating capacity, or that of ~10^3 of the largest power stations. In the aggregate then, meeting this demand would require a time-averaged build-out of 2 large stations per month for 40 years. If pumped from lakes and rivers(-deltas), as little as 1^+ TW may suffice to move this 10^4 km^3 of water, corresponding to a 100 m 'lift' against gravity in order to enable a ~1,000 km horizontal 'run' at typical mid-flow-speed riverine rates. Upgrading brackish or polluted water will have intermediate energy requirements and will depend on water quality and net-availability for a given location. Although these average power levels—especially the fractional-year usage ones—are high, they're modest in relative terms, e.g., when noting that ~40 TWe would be required to give all ~10 billion people inhabiting the Earth a few decades hence the electricity supply presently enjoyed by Americans.

Energy requirements can also be estimated based on how much fertilizer will be needed. Many modern cultivars—e.g., the bellwether ones of the Green Revolution—have been 'tuned' via breeding to be "hungry but productive." These cultivars 'demand' lots of N/P/K fertilizer *and* water—but typically grow a few-fold more food when they're replete with these, relative to 'natural' conditions. These improvements are exemplified in the huge productivity gains in the cereal grains which "snatched the human brand from the Malthusian burning" in the later 20th Century: the "Green Revolution." These gains can be attained with ever-more-efficient time utilization of fertilizers, and there is a significant preference for sourcing-just-as-needed. As soil fixed-N storage capacity is often only a fractional-season's worth of 'best' (e.g., maize) plant-demand, iterated providing of time-appropriate masses are indicated in order to ensure adequacy over the growing season and to limit losses to leaching. The world's agricultural sector presently uses ~110 million metric tonnes (MT) of fixed nitrogen annually, corresponding to an *average* of ~7 kg-N/hectare over the 1.6 gigahectares of arable land (albeit with only a fraction of it being fertilized).

Currently, approximately 140 MT of natural gas— ~6% of total global production—are used in Haber-Bosch ammonia synthesis. These applications have resulted in incremental production of 50±10% of total present human food-intake. Use of fertilizers makes tremendous economic sense, since many agriculturally-important plants gain close-to-linearly in food productivity up to ≥100 kg-N/ha-year. Thus plants can readily return ~10X more food-value than the corresponding N-cost (when adequate water, P and K are also available). In the context of these available gains, several-fold gain in fixed-N usage is indicated *with present cultivars*—and improved strains will likely more fully-leverage far higher (2-3X) fixed-N application. For example, some SE Asian rice farms already use ≤300 kg-N/ha-year—with much 'wasted' to runoff water and into the air, to be sure—as the ROI from incremental crop yields are still attractive at that fixed-N cost and usage-level. Therefore, 30±10% of total natural gas production is likely to be used in 2050 for N-fixation—*combined* with ~2X gains in efficiency of usage-application. Efficiencies of usage-application getting more applied nitrogen into plant tissues will also have collateral environmental benefits. As N-based fertilizer uses increases, there will also be associated major demands for stepped-up sourcing of P- and K-based fertilizers and agricultural water, if there aren't marked improvements in the efficiency-of-utilization of these often-limiting inputs.

It is not enough to provide adequate water and fertilizer (and other ingredients of "best farming practices") if there are too few roads of adequate quality to and from farmgates due to undercapitalization. Getting materials to and from farmgates and markets is overall productivity rate-limiting in much of the developing world at present. The lack of accessible transportation crimps overall agricultural productivity and continually threatens smallholder farmers' margins. Getting crops to markets without undue losses—and also getting 'bulk' ag-materials (e.g., ~20 T/km^2 of fertilizer per season) to farms is keenly needed. In view of present crop losses and the need for improved net production, paved/all-weather roads durably supporting moderate-speed/multi-ton powered vehicles are urgently required. Most road-building materials can be sourced reasonably locally, but asphalt-base must come from distant refineries. A useful approximation is that approximately 1 kg-asphalt base is required per cm of paved "1½" lane roadway. If the 'pitch' of farmlands roads is to be ~USA standard (i.e., by-sections), then a requirement for 2 km of roads for every ~4 km^2 of farmed area is implied. As approximately 1 gigahectare—10^7 km^2—of present global farmland (~60% of present total) still needs paved roadway systems embedded in it, ~5 M km of roads must be paved, demanding ~$5x10^{11}$ kg or $5x10^8$ tonnes of asphalt-base, i.e., $3.5x10^9$ barrels. This asphalt-base corresponds to of the order of a half-year's global-total petroleum production 'by-product.' This number could be reduced if significant fractions of the road network "capillary bed" remained unpaved, but this would leave the corresponding portions of the network sensitive to weather excursions and associated farm productivity and crop losses.

Once roads are in place with these capital and petroleum expenditures, transport energy expenditures will be required on an annual basis to utilize these investments. For many decades, motorized "heavy lift" transport has been a centerpiece of modern agricultural systems. These transportation costs include moving farming materials from piers and manufacturing facilities to farmgates in ~1 gigatonne per year quantities a few decades hence. Even more significantly in terms of mass flows, crops must then be transported from farmgates to storage and processing plants and marketplaces in ~10 GT/year quantities in 2050. The corresponding energy requirements are varied, depending on transport means.[21] Approximately 0.03 MJ/tonne-km is required when materials-moving is done via modern marine transport, ~0.3 MJ/tonne-km when done via diesel locomotive-drawn train, and ~3 MJ/tonne-km when done via modern heavy diesel-engined truck. Thus a 100 km *truck-haul* from farmgate to rail-loading point 'costs' roughly as much energy as a ~1,000 km *train-haul* to pierside, or as much as ~10,000 km *ocean-transport* to another pier: ~300 MJ/tonne. For example, wheat-or-rice for export probably costs ≤1 GJ/tonne to transport from farmgate to consumption-point, when using modern engine technologies. This expenditure is small compared to the food-energy 'value' of typical cereal grains of ~17 GJ/tonne. Moving ~10 GT/yr of food from world's farms to the world's dining-tables in 2050 will likely involve $\sim 10^{19}$ J/yr, or a time-average ~0.3 TW of shaft horsepower, or $\sim 7 \times 10^8$ tonnes-diesel fuel/year, or 5 billion barrels-of-oil-equivalent annually—a lot in absolute terms, albeit not so much relatively speaking (world petroleum production presently averages ~30 billion barrels/year, and the petroleum-fraction of interest for road-building is refinery residue or 'waste,' otherwise useful only for specialized applications such as marine diesel engine-fueling).

Finally, available gains can be maximized by limiting the wasting of crops. A first step includes construction and operation of granaries of adequate quantity and quality. Several gigatonnes of global-aggregate capacity are indicated, i.e., several cubic km are needed to store a substantial fraction of a year's crops. There are mass and energy requirements to build, maintain, and operate, including temperature and/or humidity control and pest and vermin-proofing. Improved transportation—farmgates to and from markets—will allow as much as possible of the developing world farmer's output to result in monetary benefit to the farmer and in nutrition in the world's population. Improved transportation will deliver foodstuffs and will leverage capitalized cheap and plentiful refrigeration for perishable commodities.

More efficient limitation of crop wastage can be achieved by telecommunications informing farmers and intermediaries of best-accessible markets for crops. Telecommunications will also inform markets of crop availabilities, and remote sensing-driven prediction of crop yields and advance sales may provide more economic stability to farmers on the margin. Undergirding by crop insurance will minimize farmers' often risk-intolerant behaviors, e.g., *"Can't risk usage of pricey fertilizers or seeds, due to resulting financial wipe-out if the rains don't come and the crop fails!"* Concatenating these steps will maximize efficiency for a given energy expenditure.

SUMMARY: ENERGY USE CAN ENABLE ENHANCED AGRICULTURE TO ACHIEVE NECESSARY PRODUCTIVITY GAINS

Nutrition of adequate quantity *and* quality is essential for proper infant and child development and educateability—*and* labor-productivity of adults. Poor early-life nutrition scars the human (epi)genome *for generations*: e.g., the 'Dutch Famine Syndrome'.[22] Starving young children are forever "at the margin" re food availability, just when their brain development is at the greatest—and subsequently irredeemable—

risk of permanent damage: there's *no* 'catch-up' opportunity "later on" for the human brain. For adults, even 500-1,000 calories/day of food-intake 'above metabolic minimum' translates into *huge* impacts on their capacity for labor output and their resistance to both infectious and degenerative diseases.

"Baked into the cake" *global population increases*, *remediating current widespread under-nutrition* and escalating demands for *higher-quality diets* concatenate to require at least a doubling of crop production by 2050. Realities and constraints of the world-as-it-is and will-be duly considered, this crop-doubling will require *significantly* more than 2X increases in total energy inputted to food production: substantial large-scale inhomogeneities in distribution and utilization of key agricultural resources—e.g., soil qualities and water-availability—are virtually certain to persist and less marginal ROI is inevitable as the energy-intensivity of agriculture grows.

Paths forward for achieving the production increases and improving energy efficiency in doing so include improvements in water sourcing and distribution and gains in spatiotemporal precision of fertilizer use. Enhanced timing and location of fertilizer and water usage can be achieved via remote sensing and telecommunications. Greatly enhanced capitalization of African agriculture in particular will be required to enable efficient irrigation and fertilizer use over that large and potentially quite agriculturally-productive continent. Capitalization will also result in improved transportation, storage and processing of foodstuffs, and less waste of foodstuffs in moving from growing-fields to stomachs.

Greatly stepped-up energy sourcing and utilization where it's most needed are *sine qua non* in each of these high-priority areas. "Business as usual" will result in major food-production shortfalls and consequent widespread under-nutrition during the next few decades. It's essential to be forever-mindful that the *Malthusian Wolf* is never all *that* far from Humanity's Door!

There is no way around the fact that humanity "has its work cut out for it" re food production advances over the next several decades. Greater than two-fold gains in total food production are clearly required by 2050, which translates into a sustained time-average growth of >1.8%/year on the present base. It is sobering to realize that these are larger absolute productivity gains—by far!—than during *any* time previously, and this is to be accomplished after the "easy" gains have already been attained.

For achieving these agricultural production targets, even 'ordinary' means may suffice—*if* exercised vigorously and consistently and soon. Greatly stepped-up R&D expenditures *and* infrastructure investments may both be required—perhaps to Green Revolution levels. These expenditures and capitalization will serve as 'safety nets' underneath greatly-improved food storage and distribution infrastructures—*especially* in imperiled areas and regions. As outlined above, these exercises will require increased energy usage in-the-aggregate, and dramatically increased energy usage in the presently-underserved regions. Beyond 'ordinary means,' a resort to 'extraordinary' means may assure worldwide success even in the face of multi-axially 'pressured' farmlands. Easily envisioned pressures arise from population growth/urban expansion, erosion and salination, irreversible chemical-nutrient extraction, bio-fuel cropping, and monoculture risk-mitigations. These 'natural' pressures are multiplied by politicoeconomic 'complexities' and governance issues already too apparent in many regions.

'Extraordinary means' will provide possibly-crucial 'insurance' against likely shortfalls in food production performance-gains involving 'ordinary' means.

The time to act is now. Delays in attaining and thereafter maintaining ~2%/year planetary-averaged food productivity gains will erode extant margins to possibly disastrous extents. Well-thought-out, purposeful, sustained, adequately-resourced programs must commence soon. Success will prevent another *billion* of our fellow humans from being *irretrievably crippled*—physically and cognitively—by sustained poor nutrition *in our own times*, and will forestall Malthusian catastrophe of quite unprecedented magnitude over the next four decades—and perhaps significantly beyond.

ACKNOWLEDGEMENTS

We gratefully acknowledge many useful discussions with our colleagues at Intellectual Ventures and at the Bill and Melinda Gates Foundation.

REFERENCES

1. Fogel, R.W. (2004) *The Escape from Hunger and Premature Death, 1700-2100.* New York, NY: Cambridge University Press.
2. Smil, V. (2001) *Enriching the Earth.* Boston: Massachusetts Institute of Technology.
3. Smil, V. (2002) *Feeding the World.* Boston: Massachusetts Institute of Technology Press.
4. Mueller, N.D., et al. (2012) "Closing yield gaps through nutrient and water management." *Nature.* **advance online publication**.
5. Foley, J.A., et al. (2011) "Solutions for a cultivated planet." *Nature.* **478**(7369): p. 337-342.
6. Floud, R., et al. (2011) *The Changing Body: Health, Nutrition, and Human Development in the Western World since 1700.* New York: Cambridge University Press.
7. Stevens, G.A., et al. (2012) "Trends in mild, moderate, and severe stunting and underweight, and progress towards MDG 1 in 141 developing countries: a systematic analysis of population representative data." The Lancet. **380**(9844): p. 824-834.
8. Tilman, D., et al. (2011) "Global food demand and the sustainable intensification of agriculture." *Proceedings of the National Academy of Sciences.* **108**(50): p. 20260-20264.
9. International_Energy_Agency, *Energy Statistics and Balances of Non-OECD Countries and Energy Statistics of OECD Countries*, in *IEA Statistics*2012, OECD/IEA.
10. International_Road_Federation, *World Road Statistics and data files*, 2012.
11. Schepp, D. (2011) "Gas Prices Around the Globe: Who Feels the Most—and Least— Pain at the Pump" in *Daily Finance.*
12. International_Monetary_Fund (2012) *World Economic Outlook Database, April 2012.*

13. Durnin, J. and R. Passmore (1967) *Energy, Work, and Leisure*. London: Heinemann.
14. FAO/WHO/UNU (1985) *Energy and Protein Requirements: Report of a Joint FAO/WHO/UNU Expert Consultation. Technical Report Series no. 724*. Geneva: WHO.
15. Kimhi, A. (2003) "Socio-economic determinants of health and physical fitness in southern Ethiopia". *Economics and Human Biology*, **1**: p. 55-75.
16. Behrman, J., H. Alderman, and J. Hoddinott (2004) *Malnutrition and Hunger*, in *Global Crises, Global Solutions*, B. Lomborg, Editor. Cambridge University Press.
17. Cassman, K.G., P. Grassini, and J. van Wart (2010) *Global Food Security in a Changing Climate: Crop Yield Trends and Yield Potential*, in *Handbook of Climate Change and Agroecosystems: Impacts, Adaptation, and Mitigation.*, D. Hillel and C. Rosenzweig, Editors. Imperial College Press: London. p. 37-51.
18. Hoekstra, A.Y. and M.M. Mekonnen (2012) "The water footprint of humanity." *Proceedings of the National Academy of Sciences*. **109**(9): p. 3232-3237.
19. Gleeson, T., et al. (2012) "Water balance of global aquifers revealed by groundwater footprint." *Nature*. **488**(7410): p. 197-200.
20. Elimelech, M. and W.A. Phillip (2011) "The Future of Seawater Desalination: Energy, Technology, and the Environment." *Science*. **333**(6043): p. 712-717.
21. Smil, V. (2010) *Prime movers of globalization: The history and impact of diesel engines and gas turbines*. Cambridge, MA: MIT Press.
22. de Rooij, S.R., et al. (2010) "Prenatal undernutrition and cognitive function in late adulthood." *Proceedings of the National Academy of Sciences*. **107**(39): p. 16881-16886.

QUALITATIVE ANALYSIS OF FOREST DYNAMICS MODELS USING ELLIPTIC FUNCTIONS

MIKHAIL J. ANTONOVSKY
Carbon Dioxide Division, Institute of Global Climate and Ecology
Moscow, Russia

"...Mathematics gives the deep description of the Nature, but any attempt to express the Nature, based upon philosophical Principe or an intuitive mechanical analogy never brings to the seriously scientific results."

Paul Dirac,
Nobel Price on Physics, 1933

The mathematical approach to description and investigation of the biological and ecological phenomena has attracted the attention of a large number of researchers in the 20th century. E. Wittecker, one of the outstanding mathematicians of the Great Britain of the 20th century, has written on the survey of the works of Vitto Volterra, one of the most famous scientist's of the 20th century, the President of the Accad Lincei: "Biologists have criticized Volterra for his extremely careful workout abstract mathematical models, that are far off complexity of the Nature. Now this principle is a main one which is based on the triumph of physical sciences". Lately several influential physical phenomena were discovered on the result of investigation of corresponding mathematical models.

In some sense the continuation of the mathematical approach to biological and ecological problems is the research of Robert May, President of the Royal Society (2000-2005), see, for example, his book "Theoretical Ecology: Principles and Applications".

Another example is the article of M.Ya. Antonovsky et al. "Forest—Pest Interaction Dynamics: The simplest Mathematical Models", Theoretical Population biology, vol. 37, no. 2 (1990), 343-367 pp. I have received almost 600 letters from all continents and from almost all universities of the USA.

In this report I would like to introduce (apply) another mathematical device to the analogue problems of forest dynamics. From one side it establishes the connections with the models of classical and quantum mechanics through the equations of the type of Korteveg de Vries. From the other side we are convinced that some of the models of forest dynamics are of Hamiltonian one that permits us to apply the methods of classical mechanics.

In an assumption that the solution of model has the shape of a running wave we have shown that explorated model is equivalent to the ordinary non-linear differential equation that has exact analytical solution. This solution is written in terms of the elliptic Weirstrass function with the parameters of initial models. It in their turns permits us to find effects of periodicity and quasi-periodicity (see Appendix 3). Let us stress that far from every model is a Hamiltonian one; it is a special property (the same rare enough that the equation has exact analytical solution).

Picturesquely saying the Hamiltonian is a way. But how the "car" moving on—it is uniformization. Mathematics tells about the possibility, the selection of necessary is turn of forester. The result of Weirstrass is that elliptic curves have such uniformization.

When we found such one to one presentation we can recognize the hidden periodicity and quasi-periodicity, which could not see on phase plane.

These mathematical models of Forest Dynamics could be considered as the tools for monitoring of the latent phenomena in Forest Dynamics. First we found the effects mathematically and then—in reality, creating appropriate notions.

Let us consider the Frame of our Report:

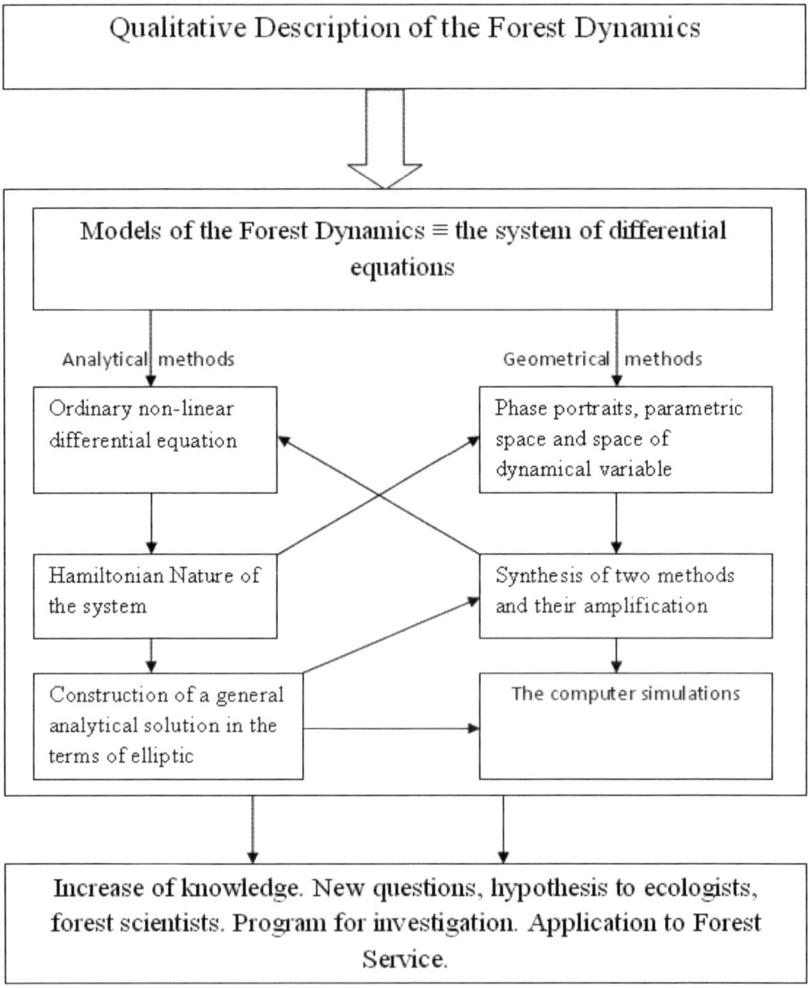

In the beginning, we shortly describe each block of this scheme and then we present the results on the base of this scheme.

First of all, we cite some of our works which we use in this report:

1. M.Ya. Antonovsky et al. (1990) "Forest-Pest Interaction Dynamics: The Simplest Mathematical Models", *Theoretical Population Biology*, Vol. 37, No. 2, 343-367 pp.
2. M.Ya. Antonovsky (1975) "Impact of the factors of the environment on the dynamics of populations (mathematical models). In: *Comprehensive Analysis of the Environment*. Proc. Soviet-American Symp. Tbilisi, USSR, 1974, 218-230 pp., Leningrad, Hydromet.
3. M.Ya. Antonovsky, E.A. Aponina, Yu.A. Kuznetsov (1991) "On the stability analysis of the standing forest boundary". WP-91-010, Laxenburg, Austria, International Institute for Applied Systems Analysis.
4. Yu.A. Kuznetsov, M.Ya. Antonovsky et al. (1994) "A Cross-diffusion model of forest boundary dynamics", *Journal of Mathematical Biology*, 32, 219-232.
5. M.Ya. Antonovsky, V.M. Buchstaber, J.Yu. Bunkova (2012) "Wave processes in Forest Ecosystem" (in publication, ДАН, 2012).
6. M.Ya. Antonovsky, M.D. Korzukhin, V.K. Mackjavichus (1989) "Periodically behavior of age-distributed population of trees", WP-89-012, Laxenburg, Austria, International Institute for Applied Systems Analysis.
7. M.Ya. Antonovsky, M.D. Korzukhin (1983) "Mathematical modeling of economic and ecological-economic processes" in: *Integrated Global Monitoring of Environmental Pollution*. Proceedings, 2nd International symposium, Tbilisi, USSR, 1981, pp. 353-358, Hydromet.
8. R.A. Fleming, M.Ya. Antonovsky, Yu.A. Kuznetsov (1987) "The Response of the Balsam Fir Forest to a Spruce budworm Invasion...", WP-87-71, International Institute for Applied Systems Analysis, Laxenburg, Austria.

Let us remember that the goal of this report is to show that the study of mathematical solutions of corresponding formalized forest models which we offer to the professional forest people could be interpreted on real phenomena in real forest ecosystems. As it is well known, in the forest ecosystems permanently precede the physical, chemical and biological processes. For example, age waves on ecosystems arise due to concurrence for the light, for the water (roots and crowns), for minimal elements of soil and so on. Moreover the forest fires, aging, transition in categories "to be ready" to be burned. For example, at Boreal Belt each lot of forest burned down with big probability once at 120 years. For example, in 2010 and 2012 were burned several millions of ha of high-quality forest. Beside, arise outburst of vermin, illness and other internal causes of disappearing of some kind of trees. The waves of heat and waves of cold hit on mechanism of productivity and so on. Moreover, the year cycle also create some type of wave motions. Thus, phase space of boreal belt is some tor in which there exist periodically and quasi-periodically motions with different frequents and amplitudes, including solutions, for example, age groups. In fact, as we have stressed, if one short period (1 year) of changing physiological processes in trees, the other, like successions, and even their abrupt, fire and so on, create large circle of tor. Mathematical findings could give to forester the possibility to think if it is possible to organize the forest ecosystem move to such states from which it can go to chaos—destruction from fire, strong wind, outbreak of parasites, and so on.

The phase picture of the state of given part of forest and doesn't let to go to another state with the help of forestry or possibility to transfer into other states of equilibrium.

Let us note that the global and local forests are extremely important for mankind. From one side forest mainly define Biodiversity on the Globe, from the other side; forest is a time bomb and even the weapon of mass destruction: the forest fire last 5 years.

As I have told above, this report is based on our works[1-8]. At the beginning we construct mathematical models (phenomenological) of special-temporal dynamics of trees population, and investigate them. The first such a model we have proposed in the work [1] and it has the following form:

Let u—the number of "young" trees (or density), υ—the number of "old" trees (or density) on some surface of a region. Here for simplicity we consider only two age classes. In the work[6] we consider situation with arbitrary number of age classes. Parameters ρ, f and h are coefficients of reproduction, ageing and mortality of "old" trees respectively, while γ (υ) is a mortality rate of "young" trees. Our hypothesis that there exists some optimal value of "old" trees density under which the recruitment of "young" trees is maximal, it is possible to choose $\gamma(\upsilon)=a(\upsilon-b)^2+c$ with positive parameters a,b,c. Our model is phenomenological

$$\begin{cases} \dot{u} = \varsigma\upsilon - \gamma(\upsilon)u - fu \\ \dot{\upsilon} = fu - h\upsilon \end{cases} \tag{1}$$

It is clear that any spatial distributed generalization of model (1) should have space-homogeneous solution corresponding to two stable state of model (1). To account for spatial effects, one has to consider in greater detail the process of regeneration which was described in model (1) in a phenomenological way by reproduction coefficients. It is well known that the regeneration usually proceeds through four main stages involving seeds, namely: seed production, seed transport, seed deposition and seed establishment. The seed transport (in many cases) is the main spatial effect that should be taken into consideration in the forest modeling. We take a hypothesis that there is no prevailing wind direction and, therefore the Gaussian (diffusion) model can used for describing seed dispersion for a general treatment of wind-induced seed dispersal. Let introduce an additional w-variable (density of air-borne) seeds into the model:

$$\begin{cases} u_t = \delta\beta w - \gamma(\upsilon)'' - f(u) \\ \upsilon_t = fu - h\upsilon \\ w_t = \alpha\upsilon - \beta w + Dw_{xx} \end{cases} \tag{2}$$

where α,β and δ are seed production, deposition and establishment rates respectively, and D is the diffusion coefficient along a space variable X. The system (2) is of "reaction-diffusion" type. It is possible to derive an approximate system of equations for "old" and "young" tree classes densities by eliminating the seed variable from (2) under the hypothesis of different time scales for dispersing seeds and trees dynamics.

$$\begin{cases} u_t = \rho \upsilon - \gamma(\upsilon)u - fu + K\upsilon_{xx} \\ \upsilon_t = fu - h\upsilon \end{cases} \tag{3}$$

Note that model (3) has a "cross-diffusion" term while (2) has a classical diffusion term. By a linear change of variables, time and space scale we will transform system (3) into the following dimensionless form

$$\begin{cases} u_t = \rho \upsilon - (\upsilon - 1)^2 u - su + \upsilon_{xx} \\ \upsilon_t = u - h\upsilon \end{cases} \tag{4}$$

The analysis of the model (4) begins with the search for spatially non uniform stationary solutions:

$$u(x,t) = U(x),\ V(x,t) = V(x)$$

$$\begin{cases} \dfrac{d^2V}{dx^2} + \rho V - (V-1)^2 U - SU = 0 \\ U - hV = 0 \end{cases}$$

The second equation is algebraic and allows one to find U(x) if V(x) is known. Thus the problem of finding out the stationary solutions of eqs (4) reduces the analysis of the equation:

$$\frac{d^2V}{dx^2} + \rho V - h(V-1)^2 V - shV = 0,$$

which can be rewritten as a system of two first-order differential equation (where ´ indicates space derivatives); introducing W=V' we obtain:

$$\begin{cases} V' = W \\ W' = shV + h(V-1)^2 V - \rho V \end{cases} \tag{5}$$

Solutions of (5) define profiles of stationary solutions of system (4).

This system differs only by space scale from the corresponding standing wave system for model (4).

Therefore, we can use those results in our present analysis. From (5) we obtain (6)

$$V'' = W' = shV - h(V-1)^2 V - qV \tag{6}$$

Equation (6), as it is known, (see Appendix 2) has analytical solution:

$$V = -2\frac{\wp(t) - a}{\wp'(t) + b\Upsilon(t) + C} \tag{7}$$

where a,b,c – are constants.

Here $\wp'^2(t) = \varphi\wp(t)^3 - g_2\wp(t) - g_3$

where g_2 and g_3 – constants (see Appendices 1,2).

It is said that our model is uniformized by function (7).

We go back to the system above

$$(6) \qquad \begin{cases} V' = W \\ W' = shV + h(V-1)^2 V - \rho V \end{cases}$$

Let us show that our system is a Hamiltonian system, i.e. there exist such function H=H(V,W), that

$$\begin{cases} V' = \dfrac{\partial H}{\partial V} \\ W' = \dfrac{\partial H}{\partial V} \end{cases}$$

We look for H(V,W) in the form [which is prompted by system (1)]. To this Hamiltonian corresponds a conservation law

$$\frac{1}{2}W^2 + \mathrm{P(V)} = \mathrm{const},$$

where $\frac{1}{2}W^2$ corresponding to Kinetic energy of the system, and P(V)—potential energy. The proving of this assertion consist in finding of P(V):

$$W' = -\frac{\partial}{\partial V}\left(\frac{1}{2}W^2 + P(V)\right) = -\frac{\partial P(V)}{\partial V}$$

Hence, according to the system (6) we obtain:

$$\frac{\partial P(V)}{\partial V} = -shV - h(V-1)^2 V + \varrho V, i.e$$

$$P(V) = -\int (shV + h(V^2 - 2V + 1)V + \varrho V)dv$$

$$= -\frac{1}{2}(sh - \varrho)V^2 - \frac{1}{4}V^4 h + \frac{2}{3}hV^3 - \frac{1}{2}hV^2 = \text{polynomial of 4 deg}$$

Obtaining P(V) we can study lines of level of the Hamiltonian
H(V(t),W(t))=const

What is sapid sense of Hamiltonian? The phase plane with the coordinates (V,W) is subdivided on the line of level of the function H(V,W). The motion at phase plane defined by dynamical system (6) is the motion along the lines of the level correspondingly.

Thus we obtain, that

$$H(V,W) = \frac{1}{2}\left(W^2 + [\varrho - (s+1)h]V^2 - h\left(\frac{V^4}{4} - \frac{2V^3}{3}\right)\right)$$

We use the well known fact that the equation H(V,W)=const is the equation of elliptic curve and the equation H(V,W)=const under the condition V=W' gives elliptic curve. For each point of the trajectory we have H(V(t), W(t)=c.

The standard Weierstrass model of elliptic curves is

$$W^2 = 4V^3 - g_2 V - g_3$$

Let us stress, that the main interest in both methods—phase portraits and elliptic description how to combine the potential of both methods to reach the best description of the phenomena is study. We try to demonstrate both cases, using the development we presented above [see equation (6)].

The parametric and phase portraits of Hamiltonian of the system (6) (see Figure 1) that describes forest standing boundary. On the line Q we will obtain it below the system has a heteroclinic orbit connecting E_0 and E_2 see calculations below).

Lines D_1 and D_2 on Fig. 1 are bifurcation lines equilibrium of the system (6) in region 1 have the following coordinates

$$E_0 = (0,0) \text{ and } E_{1,2} = (V_{1,2}^0, 0), \text{ where } V_{1,2}^0 = 1 \pm \left(\frac{\varrho - sh}{h} \right)^{1/2}.$$

Equilibriums E_0 and E_2 are saddles while equilibrium E_1 is a center. Bounded nontrivial orbits of the system (6) either are closed or connects saddles. The equation of line Q on which a separatrix connecting the saddles exists and can be found analytically from the condition that isoline H=0 goes through saddle E_2:

$$q - \left\{ (qh) : q = \left(s + \frac{1}{9} \right) h \right\} \tag{*}$$

For parameter values satisfying (*) the equation ( ) can be explicitly written

$$V(x) = \frac{3}{4} \left\{ 1 + \exp\left(-\frac{4}{3} \left(\frac{h}{2} \right)^{1/2} x \right) \right\}^{-1}$$

Therefore, we have established a possibility for the existence of a standing boundary in the forest modeled by a system (**), but only for a specific relation between parameters (ϱ, s, h).

(**)

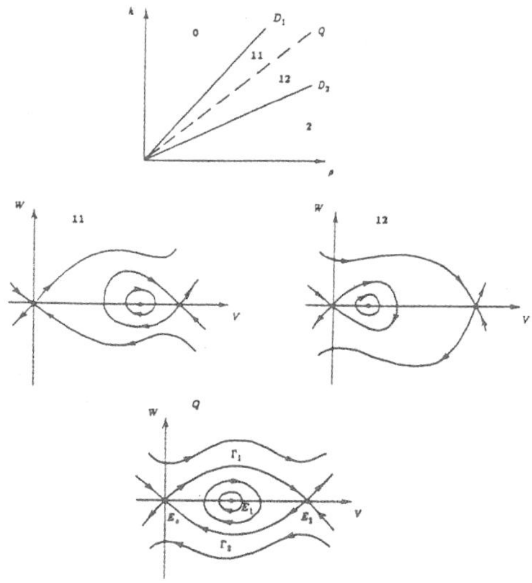

Fig. 1. (See Article 2 for the meaning.)

Let us now introduce an insect pest into our initial model and consider two extremely simple situations.

1. The pest feed only on the "young" trees.
2. The pest feed only on the "old" (adult) trees.
3. The general case was considered by on of my summer student (Samarskaya, E.A.: forest-pest interaction dynamics in temporal and spatial domains. WP-89-16, Laxenburg, Austria, International Institute for Applied Systems Analysis, 1989).

Let z is insect density, ε the mortality rate of insects and the terms with xz and yz represent the insect-forest interaction.

Thus for the case where the pest feeds on "young" trees we obtain the dynamical model

$$\begin{cases} \dot{x} = \varrho y - \gamma(y)x - fx - Axz \\ y = f\dot{x} - hy \\ z = -\varepsilon\dot{z} - Bxz \end{cases} \qquad (A_1)$$

while for the case where the pest feeds on adult trees we obtain the (equations) dynamical model

$$\dot{x} = \varrho y - \gamma(y)x - fx$$

$$\dot{y} = fx - hy - Ayz \qquad (A_2)$$

$$\dot{z} = -\varepsilon z + Byz$$

The main tools for our investigation are the bifurcation theory of dynamical systems and the numerical methods of this theory.

By a linear transformation of variables, parameters, and the time the system (A_1) will transit into

$$\dot{x} = \varrho y - (y-1)^2 x - sx - xz$$

$$\dot{y} = x - hy \qquad (AM)$$

$$\dot{z} = -\varepsilon z + Bxz$$

where the previous notations are preserved for new variables and parameters which have the same sense as in initial model. The new parameters can be represented in terms of the old one as

$$\varrho := \frac{f\varrho}{a^2 b^4};\ s := \frac{f+c}{ab^2};\ h := \frac{h}{ab^2};\ \varepsilon := \frac{\varepsilon}{ab^2};B := \frac{B}{ab^2}.$$

In the first octant (i.e., where the variables take on biological possible value), the system (AM) can have from one to four equilibrium points: E_0, E_1, E_2, E_3 (see Antonovsky et al.[1]).

We show the parametric portrait of system (AM) and the phase portraits of the model (AM).

This presentation is useful device for exact notions in a forest science. For example, the notion of probability of outbreak of insect could be formulated only on the base Figure 2.

The parametric and the phase portraits of system (AM): only 3 of 9 possible states of the system (AM) could be ready for outbreak. It means that probability of outbreak in the system described by (AM) is 1/3.

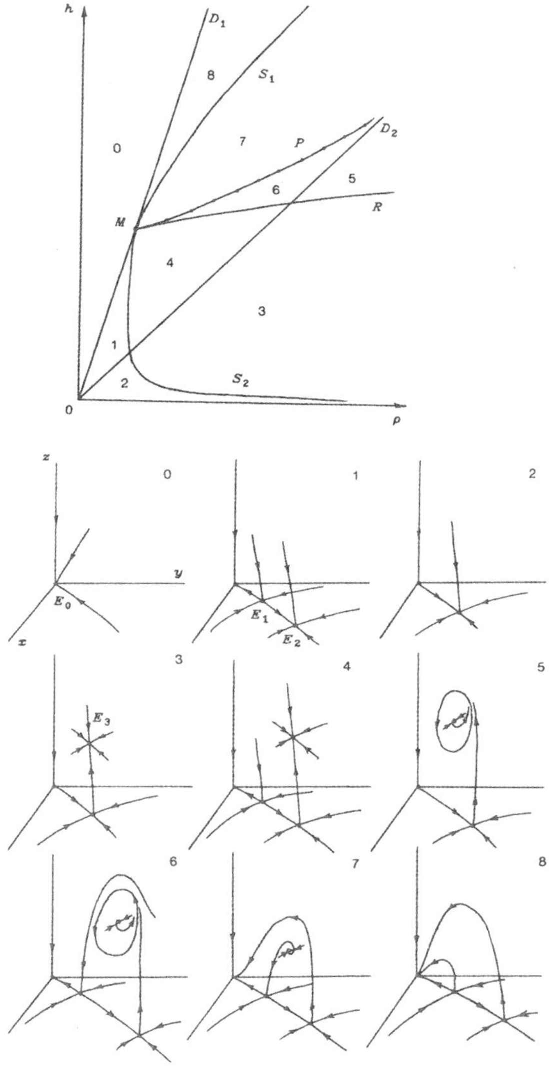

Fig. 2. (See Article 1 for the meaning.)

Appendix 1.

For the sake of convenience of readers we will refresh some notions that we are using and that can be found in any mathematical reference book. The elliptic functions are meromorphic double periodic functions. Two important notions—primitive periods and parallelogram of the periods (on complex plane). If Ω and Ω′, so called primitive periods of an elliptic function, then parallelogram with vertexes O, Ω, Ω+Ω′, Ω′ don't has the points of periods nor inside, neither at the sides, not taking into account vertexes. It arises some net of parallelograms that cover all complex plane. The Weierstrass function $\wp$(u) by definition is

$$\wp(u)=\frac{1}{u^2}+\sum\nolimits'_{m,m'}\left\{\frac{1}{(u-2mw-2m'w)^2}-\frac{1}{(2mw+2m'w)^2}\right\}$$

It turns out that $\wp$(u+2W′)=$\wp$(u) and $\wp$(u+2w)=$\wp$(u), i.e. $\wp$(u) is an elliptic function of the second order. It is the main function in the Weierstrass theory.

As it is well known in special case of degradation elliptic functions transit into trigonometric functions. Let us remember that

$$\sin u = u\prod{}'\left(1-\frac{u}{m\pi}\right)e^{\frac{r}{m\pi}}$$

$$\text{ctg u}=\frac{d}{du}\ln \sin \text{u}$$

$$\frac{1}{\sin^2 u}=-\frac{d}{du}\text{ctg u}$$

$\wp$(u)=$\wp$(u;w,W′)=$\wp$(u,g_2,g_3), g_2=g_2(w,W′); g_3=g_3(w,W′), and

$\wp''(u)^2=4\wp^3(u)-g_2\wp(u)-g_3$ – differential equation for Weierstrass functions.

If w=∞ and w′=∞, then $\wp(u)=\frac{1}{u^2}$, $g_2=g_3=0$. In this case we get the general solution of our problem and instead of $\wp$ and $\wp'$ we substitute these values. Contently it means that the conditions when periodic solutions disappears. In case when only one periodic equal w′=∞ and ∞=finite, then

$$\wp(u)=-\frac{1}{3}\left(\frac{\pi}{2w}\right)^2+\left(\frac{\pi}{2w}\right)^2\frac{1}{\sin^2\frac{\pi u}{2w}}$$

Discriminant $g_3^2 - 27g_3^2$ of Weierstrass functions will be considered for different values.

Appendix 2. (See in any math reference book.)

Let us consider well known equation of mathematical physics $-\dot{f} = f''' + 6\delta ff' + 6\varepsilon f^2 f',\ \ f = f(t,z)$ when δ=1, ε=0 it became the equation of Korteveg de Vries (KdV); when δ=0, ε=*1* we obtain modified equation of Korteveg de Vries (mKdV). The problem of solutions of this equation in the form of traveling wave

$$f(t,z) = \phi(z - 4ct)$$

Is reduced to the solution of equation

$$4c\phi' = \phi''' + 6\delta\phi\phi' + 6\varepsilon\phi^2\phi' \tag{4}$$

The general solution of this equation with the initial conditions

$$\phi(0) = 0,\ \ \phi'(0) = 0,\ \ \phi''(0) = 1$$

is the function

$$\phi(z) = \frac{3}{6\gamma(z) - 2c} \qquad \text{(W)},$$

where $\wp(z) = \wp(z; g_2, g_3)$, $g_2 = \frac{\gamma}{3}c^2 + \delta$, $g_3 = -\frac{8}{27}c^3 - \frac{1}{3}\delta c + \frac{1}{\gamma}\varepsilon$, where c is a velocity of travelling boundary, parameters g_2 and g_3 through the parameters of initial model (ϱ,h,s), i.e. c=φ_1(ϱ,h,ε), σ=φ_2(ϱ,h,s), ε=ϱ,h,s,f,…).

Thus the traveling wave of the initial system of the equation (B) will be expressed through Weierstrass function. It is important to see all four transformations of the initial parameters. It is worth to note that the shape of equation (4) is preserved under the transformation

$$\phi(z) \to \psi(z) = \frac{1}{\alpha^3}\phi(\alpha z) + \beta,$$

where $\alpha \neq 0$ and β are free parameters. Under this transformation we have

$$(\mathrm{c},\delta,\varepsilon)\rightarrow(c\alpha^2+\frac{3}{2}\delta\alpha^5\beta-\frac{3}{2}\varepsilon\alpha^8\beta^2,\ \delta\alpha^5-2\varepsilon\alpha^8\beta,\ \varepsilon\alpha^8);$$

$$(x_0,x_1,x_2)\rightarrow\left(\frac{x_0}{\alpha^3}+\beta,\frac{x_1}{\alpha^2},\frac{x_2}{\alpha}\right).$$

It takes place Lemma. General solution of equation (1) with initial condition

$\phi(0)=1$, $\phi'(0)=1$, $\phi''=x$ is given by formula:

$$\phi(t)=f_{EI^{(t)}}=-2\frac{\wp(t)-\frac{1}{4}x^2+\frac{2}{3}c}{\wp'(t)+x\wp(t)-\frac{1}{3}cx-\frac{1}{2}\wp},$$

where $\wp(t)=\wp(t,g_2,g_3)$ – Weierstrass function $g_2=\wp_x-\varepsilon+\frac{4}{3}c^2$;

$$g_3=\frac{1}{4}x^2\varepsilon-\frac{1}{3}c\wp x-\frac{\wp}{27}c^3-\frac{1}{4}\wp^2.$$

In conclusion, I would like to stress again that the goal of mathematical modeling is to find and present the mechanism of observed effects. This is exactly what we did in this report on example of age waves on Forest.

SESSION 7

WATER POLLUTION & TERRORISM

THE TERRORIST THREAT TO WATER AND TECHNOLOGY'S ROLE IN SAFEGUARDING SUPPLIES

DAN J. KROLL
Hach Homeland Security Technologies
Loveland, USA

INTRODUCTION

The potential for terrorists to access and compromise our drinking water supplies is a widely recognized infrastructure vulnerability. Such an attack has the potential to cause substantial damage and could result in mass casualties. The threat of a waterborne attack is especially troublesome due to the low level of technical knowledge required to orchestrate an attack and the ready availability and low cost of the materials needed for such an operation. Many parts of the water supply networks that we rely upon are not capable of being hardened in a physical mode with locks and fences due to their widespread geographic footprint and many access points. The preclusion of physical deterrents as a means to harden the system requires reliance upon monitoring and early detection of such an attack as the only means to ameliorate the potentially dire consequences.

WHY ATTACK WATER?

In general, terrorism's goal is to focus attention on a group's plight or demands and sway the public into pressuring their government into making the policy changes that the terrorist organization seeks. Columbia University's Brigitte Nacos states that there are three principal components which terrorism strives to exploit: "the news media, the public and governmental decision makers."[1] Brian Jenkins in his seminal 1975 report on terrorism prepared for the Rand Corporation referred to "terrorism as theater".[2] He made the statement that, "Terrorists want a lot of people watching, not a lot of people dead."[3] While this may have been true at one point, in today's world of global media it is necessary for terrorists to continually escalate the size and scope of their attacks to maintain their audience's attention. Dozens of attacks occur every week around the world but few garner any significant attention from the general public as they are not covered outside of local media due to the fact that they are not spectacular enough. In order to attract the attention and publicity that they crave and to generate the general fear needed to accomplish their goals, terrorists are forced to raise the levels of their attacks.

Hijackings, hostage takings, kidnappings, car bombings, and suicide bombings have all played a role in the terrorist's play book. Victims and targets are not chosen at random but are carefully selected to generate the maximum of shock value and fear. Schools, shopping centers, bus and train stations, and restaurants and nightclubs have been targeted both because they attract large crowds and because they are places with which members of the civilian population are familiar and in which they feel safe. One goal of terrorism is to destroy the public's sense of security in the places most familiar to them.[4] An attack on water could hit at the population where they feel the most secure. An attack on the water supply could find them at work, school, vacation and, most importantly, at home. Whereas water is necessary to life, no one can completely avoid contact with it and easily regain a sense of safety after such an attack occurred.

The public tends to exhibit an ability to rationalize away their fear of potential terrorist actions. Terrorism is something that happens to someone else. The thought is either I don't use the sort of facilities that are commonly attacked or don't frequent areas where a terrorist attack is likely to occur, or, if I do, the attack will not happen to me. This rationalization becomes more difficult for a waterborne event. We all use water for drinking, cooking, bathing, cleaning and a variety of other uses. This sort of attack can occur at any time and any place including our own homes. The psychological repercussions of such an attack could be extreme.

HOW SEVERE IS THE THREAT?

From source to tap our water supplies are vulnerable. Water is vulnerable to various types of attack from a denial of service by blowing up supply lines to an intentional contamination event. Intentional contamination of the water supply is the most serious of the possible scenarios. Studies conducted by the U.S. Air Force and Colorado State University have shown that drinking water systems are highly vulnerable to contamination.[5] These studies show a few gallons of highly toxic material was enough, if injected at a strategic location via the proper method, to contaminate an entire system supplying a population of 100,000 people in a matter of a few hours. Material and significant contamination was not relegated to only the areas surrounding the introduction point. Material flowed through each neighborhood and then reentered main trunk lines, thus, making its way to the next area until the contaminant had permeated the entire system.

Using computer simulations, when a military nerve agent material was used over 20% of the population was determined to have received a dose adequate to result in death and when a common chemical was used in place of the warfare agent the result was a casualty rate of over 10%.[5] Thousands of deaths could result from this very inexpensive and low-tech mode of attack. Calculations have shown that this mode of attack could be orchestrated for less than $0.05 per death.[6]

In their investigation of the September 11th attacks the 9/11 Commission concluded that Al Qaeda has specific attack criteria that they evaluate when planning an attack. These six criteria are used by Al Qaeda planners to evaluate if a mission will be effective: [7]

1. An attack cannot be trivialized or marginalized.
2. The attack must be capable of international media understanding.
3. It must be relatively inexpensive.
4. It must be technologically simple.
5. Ideally, the attack should use the attributes of a capitalist society and the west against itself.
6. The attack must incorporate multiple targets simultaneously.

There is no doubt that an assault on the water supply meets all of the terrorist's criteria for an attack. It could cause mass casualties, would be inexpensive, and actually offer the terrorists a good chance of avoiding apprehension. Unfortunately, because monitoring for contamination of the water supply typically is limited to infrequent grab samples, the first indications of such an attack are likely to be casualties showing up at local hospitals. A terrorist could launch such an attack and be on a plane out of the country before the first casualty is reported.

HISTORY OF ATTACKS ON WATER

Attacks on water are not just hypothetical. The historical record of attacks on water supplies goes as far back as 1000 BC when it is reported that ancient Chinese warriors used arsenic to contaminate the water supplies of their enemies. History is replete with other examples. A few of the more recent documented incidents are listed below. There are many more.

2002 Two al Qaeda operatives were arrested in Denver with plans to poison U.S. water supplies.[8]

2004 FBI and Department of Homeland Security issue a bulletin warning that terrorists were trying to recruit workers in water plants as part of a plan to poison drinking water.[9]

2006 A water tank in Tring, England was deliberately contaminated with weed killer.[10]

2006 Strychnine was intentionally placed in a Danish reservoir.[11]

2007 In China, 201 people fell ill and one died when water intentionally contaminated with fluoroacetimide was used to make porridge.[12]

2008 In Varney, West Virginia a man was arrested with two canisters of cyanide and plans to poison the water supply.[13]

2008 The water supply of a Burmese refugee camp (pop. 30,000) in Thailand was intentionally poisoned with weed killer.[14]

2009 In the Philippines, Moro Islamic Liberation Front (MILF) rebels poisoned water sources being used by government soldiers and local residents.[15]

2010 In the Kashmir region of India, Maoist rebels intentionally poisoned a pond used by the Central Reserve Police Force, a paramilitary group, as a drinking water source.[16]

2010 In England, a father and son pair of Neo-Nazis was convicted of various terrorism charges including the manufacture of ricin and plotting with Serbian Nazis to poison water supplies used by Muslims.[17]

2011 Materials seized during the raid that resulted in the death of Osama bin Laden revealed plans to poison water supplies.[18]

2011 In Spain, a plot to poison water supplies to avenge Osama bin Laden's death was uncovered.[19]

2012 In Australia, two 5,000 liter drinking water tanks were intentionally poisoned with Diuron.[20]

2012 In Afghanistan, hundreds of school girls were sickened when the water supply at their school was intentionally poisoned.[21]

One incident of particular interest occurred in 2002 in Rome, Italy. Four men from the Salafist group "Call and Combat' which had close ties to al Qaeda were arrested in the possession of chemicals, false documents and detailed maps of the city's water supply network in and around the area of the U.S. embassy. They had already obtained access to the pipe network near the embassy and investigators believed that they were about to mount an attack on the area's water supply. After the arrest it was found that the material that the would-be terrorists had in their possession was potassium ferricyanide. This chemical is not very toxic and would be an extremely poor choice for use as an agent to contaminate water supplies. Some commentators took the use of this material to mean that the terrorists had little knowledge as to how to go about such an attack. Such opinions may have played a role in the perpetrators only being convicted of minor offenses.

But upon further reflection, this interpretation of the events doesn't make any sense. From all reports these terrorists did know what they were doing. They had done everything correctly except for their choice of chemical. So why did they choose the wrong chemical? Five minutes on the Internet and a high school education would have pointed out the mistake. While it is easy to determine what chemicals you could use to poison a water system through a simple internet search and a rudimentary understanding of chemistry, the ability to target a specific site, such as the embassy, by introducing a toxicant into the water system is more difficult. A good guess as to the would be terrorist's motives in the case of the ferricyanide is not a dismal failure in understanding toxicology, but rather a dry run of the attack scenario to determine if they had correctly mapped the hydraulics of the system.

While ferricyanide is not toxic, it does have many similar characteristics to other cyanide compounds such as sodium and potassium cyanide, which are very toxic. Its solubility and flow characteristics in water would closely mimic those of the other cyanide compounds. It is easy to obtain and inexpensive. Also, solutions of ferricyanide are a highly colored yellow. Solutions with as little a 1 mg/l present can be detected with the naked eye. Best of all, the color is so light at this level that, unless you are looking for it, it would most likely not be noticeable or would be interpreted as a slight amount of rust in the water. The terrorists could inject a small amount of the ferricyanide into the system and then send a cohort into the target on legitimate business and while they are there observe the drinking or restroom water for traces of the telltale yellow color. Once the hydraulics are verified, the real attack with a different cyanide compound could commence.[22]

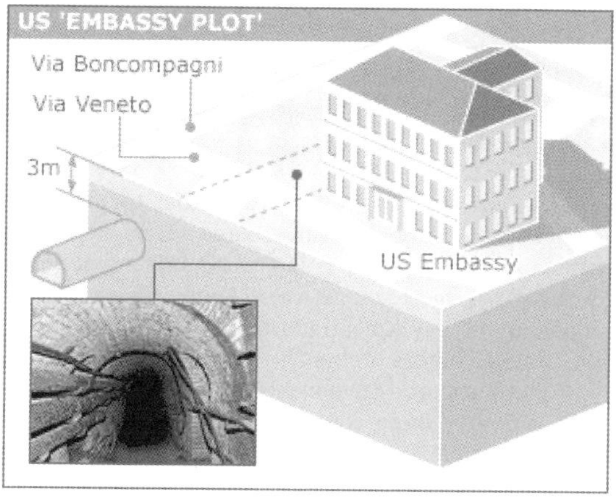

Fig. 1: Diagram of the U.S. embassy in Rome and the tunnel used to access water supply pipes.

WHERE DO THE SYSTEM VULNERABILITIES LIE?

The interest exhibited by terrorists and the potential for catastrophic damage being caused by a water attack indicates that this is indeed a vulnerability that needs attention. The forms that an attack on the drinking water supply system could take are as many and varied as the components of the system that are vulnerable. The provision of drinking water to our homes is a complex process that involves many steps. All of these steps are to some degree vulnerable to compromise by terrorist acts.

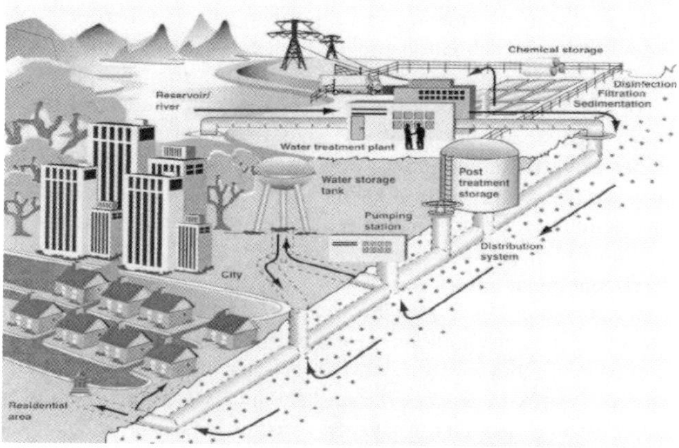

Fig. 2: A typical water supply system has many potentially vulnerable sites.

Source Water

The sources of our drinking water are diverse. Each municipality has their own source or sources. These can include rivers, lakes, streams, ground water from deep wells, ground water from wells affected by surface water, desalinization of ocean water, or even recycled sewage water. Some places rely upon a combination of different sources, for example, a river and ground water wells.

While source waters are usually not vulnerable to physical attack, they are vulnerable to contamination. As a general rule, source waters are easy to access, because it would be impossible to guard all of the points along a river that a terrorist could use as a gateway to the water. Also, most source water supplies have a built in hedge against intentional contamination. In the water industry this is known as; "The solution to pollution is dilution." Large sources are much less vulnerable than small sources. Also, the fact that many of the potential contaminants are degraded over time by exposure to sunlight and the elements give these sources a self-cleaning characteristic that makes them difficult, but not impossible, to contaminate.

Untreated Water Storage

Untreated water storage includes facilities such as dams, reservoirs, and holding tanks. They have many of the same attributes as source waters. Like source water supplies, they vary dramatically in size, and their vulnerability to attack is linked directly to that size factor. They tend to be more closely watched than source waters and do not present as large a footprint. This makes them a little harder to access, but overall, they are not difficult to approach. Many are of a dual use nature, for example, reservoirs that are used for fishing and boating as well as a supply of water. Reservoirs have limited vulnerability to contamination due to their volumes. The same constraints apply to them as to source waters, and a mass casualty contamination event seems unlikely, although, nuisance attacks are very possible.

Treatment Plants

In most cases, the treatment plant represents the last barrier between natural, accidental, or deliberate contamination and the final end user. In many cases, it is also the final point where continual routine monitoring of the chemical and physical characteristics of the water occurs. Like the previous components of the water supply system, treatment plants can vary greatly in size, treatment method and the amount of water treated per day. Different plants also use different methods of treatment and purification to make the source water potable. The method of treatment varies depending upon source water quality. Some sources require simple settling and disinfection while others require a more elaborate process.

The treatment plant offers terrorists many opportunities to inflict damage. Al Qaeda and other terrorist organizations are well aware of this and have expressed interest in recruiting workers in such facilities to orchestrate such an attack. Another important design aspect to bear in mind regarding drinking water treatment plants is that as part of the drinking water treatment process various chemicals are intentionally added to the water. These can include flocculating agents, caustics, acids, disinfectants, and fluoride among others. All of the dosing equipment is in place at the facility to feed massive quantities of solid, liquid, or gaseous chemicals into the finished water supply. If a terrorist could gain access to the plant, he could feed any of a large number of toxic substances into the system. There has also been some concern that a terrorist would not even have to gain access to the facility to mount such an attack. It may be much simpler, from the terrorist viewpoint, to infiltrate one of the companies responsible for delivery of chemicals to the treatment plant. They could then replace or contaminate the usual shipment with a toxic compound and deliver it to the plant as normal. Then, when the plant operators added the replaced or adulterated treatment chemical to the water, they would in fact be poisoning the finished water.

While the introduction of toxic chemicals from an outside source is a grave concern, in fact, it would not even be necessary to bring the toxic material in from the outside. Many of the normal treatment chemicals already present at the site can be toxic. All that would be needed to cause harm would be for the terrorist to increase the dosages of these usually benign chemicals up to toxic levels. It would be very easy for a terrorist to simply turn up the dosage on a feeder and achieve similar results. This could even be accomplished remotely via hacking into the plants Supervisory Control and Data Acquisition (SCADA) system. While disruption to many parts of the system is feasible, the treatment plant is the first location in the system where the possibility of deliberate contamination in an attempt to cause mass casualties actually becomes realistically workable due to the ease of introducing chemicals and the decreased levels of dilution that are afforded at this point in the water delivery system.

Finished Water Holding Tanks

These sites, represented by water towers and standpipes, are also locations with a limited geographic footprint so they can also benefit from some degree of physical security. They do tend to be located in remote and widely dispersed areas, so in some locations, physical monitoring can be burdensome. The volume of water involved is also greatly reduced, but there is still some potential for dilution. These tanks are post treatment plant, and if the residual disinfectant barrier is breached, there is little opportunity for contaminant neutralization before the water reaches the end user.

Fig. 3: Physical barriers to protect water storage facilities are often ineffective.

The Distribution System and Backflow Attacks

The distribution system is the network of pipes, valves, pumps stations and other accoutrements that move the water from the treatment plant to the end user. The distribution system is widely recognized as the most vulnerable component of the water supply network. At this point, the potential for dilution is vastly reduced as is the time available for attenuation.

The most likely scenario for such an attack, in which the goal is to inflict mass casualties, is to orchestrate a simple backflow contamination event. A backflow attack occurs when a pump is used to overcome the pressure gradient that is present in the distribution system's pipes. This can be easily achieved by using pumps available for rent or purchase at most home improvement stores. After the pressure gradient present in the system has been overcome and a contaminant introduced, siphoning effects act to pull the contaminant into the flowing system. Once the contaminant is present in the pipes, the normal movement of water in the system acts to disseminate the contaminant throughout the network affecting areas surrounding the introduction point. The introduction point can be anywhere in the system such as a fire hydrant, commercial building, or residence. Some areas, however, are more vulnerable than others. Access points near high flow areas and larger pipes would be favored because they would disseminate the material to a wider area more quickly; however, any point except for those at the very end of long deadhead lines could be used to effectively access the system.

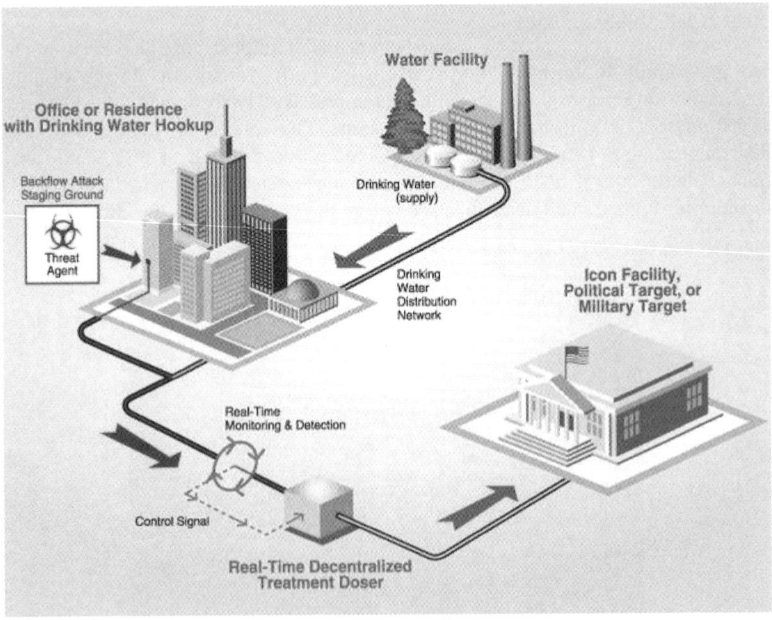

Fig. 4: Distribution systems are vulnerable to backflow attacks.

COMPOUNDS OF CONCERN FOR DRINKING WATER SECURITY

There are a large number of compounds that could be deployed by terrorists in an assault on water supplies.

Heavy Metals

Heavy metals are agents of concern due to their toxicity to humans. They are also fairly easy to obtain, and their salts tend to be readily soluble.

Herbicides

While as a general class, herbicides tend to be less detrimental to human health than some other compounds; there are some notable exceptions. This, along with the ability to easily obtain large quantities of these chemicals from agricultural supply sources, adds to the concern. Even if few fatalities resulted, the panic caused by the introduction of herbicide type compounds into a water system could be severe.

Insecticides

Insecticides tend to be more harmful to human health than herbicides. Some of the insecticides have chemical structures quite similar to some of the chemical warfare nerve agents, and there are several that are cholinesterase inhibitors. Like herbicides, insecticides are also readily available in large quantities. For some, their solubility limits their usefulness as water introduced weapons, but others are quite soluble and present more of a threat.

Nematocides and Rodenticides

Nematocides are similar to insecticides. They, with some exceptions, do tend to be more soluble than insecticides. Some nematocide compounds are also similar to chemical warfare agents in structure and mode of action. Rodenticides are of concern because they are specifically designed to be lethal to mammalian species such as humans. Both classes are readily available in large quantities.

Industrial Chemicals and Miscellaneous Agents

There are any number of industrial chemicals that could be used in an attack. Chief among these is cyanide, which is widely used in mining and other industries.

Illegal Drugs

Illegal drugs are not widely recognized as a potential threat. Street drugs, such as LSD, GHB, PCP and heroin, among others, are a mode of attack that could be employed. Some drugs, such as LSD, are easily synthesized in a home lab. Other drugs, such as heroin, are widely available. While the cost could be prohibitive for individuals working alone, supplies do exist for well-organized and funded groups. Also, it should be noted that a large portion of the illegal opiates (such as heroin) finding their way into the U.S. come from areas such as West Asia where the terrorist cells often control this trade.

Radionuclides

The use of radionuclides as a terror weapon is a distinct possibility. Even if casualties were low, the psychological impact of a radiological threat could be severe. Obtaining high purity, highly radioactive material, such as plutonium or Uranium 238, is difficult, and it is unlikely that a terrorist organization that had obtained these materials would be inclined to use them in an attack on a water system. More likely is the use of low level radioactive material or waste.

Commercial Products

While not the weapons of choice for organized terrorists, lone saboteurs, the emotionally unstable or small groups may turn to easily obtained commercial products such as bug sprays or lawn chemicals. Many of the active ingredients of these preparations are the same as the pesticides and herbicides already discussed. The difference lies in the smaller proportion of active ingredients. The vast majority of these compounds have inert ingredients listed as their main component.

Chemical Warfare Agents

Chemical warfare agents such as VX, soman, and sarin along with older type chemical weapons such as mustard gas and lewisite are not likely to be targeted against a water system due to their limited availability. If they are used, it is more likely that any assault from these weapons will be via aerosol. As the result of an aerosol attack it is possible and even likely that these agents could find their way into water supplies.

Toxins and Bioagents

There are a number of protozoa, bacteria, viruses and toxins that could be utilized in an attack. Many of these materials are extremely toxic with compounds such as botulinum toxin being some of the most toxic substances known. These types of materials are fairly easy to grow or extract from readily available sources. For example, ricin is extracted from castor beans and abrin can be obtained from rosary peas. There are several examples of the elicit production of ricin by terrorists. Bacteria can also be easily grown. For an attack on water, the production of these materials may be even simpler than for an aerosol attack as there is little need to modify the toxins to make them airborne. In fact even raw sewage could be used as a potential contaminant in a backflow or cross connection type attack.

MONITORING AS A MEANS OF DETECTING ATTACKS:

While some of the vulnerabilities inherent in the water supply system can be addressed with physical security measures, the most serious vulnerabilities cannot. The event with the highest potential of causing mass casualties is an intentional contamination incident. Source water and the water distribution system are the two components with the highest vulnerability to this form of attack and they also have the least potential for physical security being effective. Of the two, the distribution system has the highest vulnerability. Operationally there is no means to effectively prevent such an attack. Therefore, the only option is to attempt to detect such an attack as soon as possible to mitigate its effects. Numerous data inputs can be used to help detect an attack including customer complainants, syndromic surveillance and threat intelligence inputs. Possibly the most useful data input is real time water quality monitoring. Monitoring of water supplies in real time presents many difficulties.

<u>Difficulties in Monitoring Water Supplies</u>

A common misconception concerning analysis in water is that the system is stable with little variation over time or from location to location. In the real world, even after treatment, there is great diversity in the water found in distribution systems. For a parameter as simple as pH that we would expect to be around 7± 1 pH units, the diversity is much greater and can run from 3 to near 11 pH units. Also in a given system there is great heterogeneity over time in basic conditions such as pH, turbidity, conductivity, etc. Figure 5 is representative of the diversity that can be found in the real world in these types of parameters over time.

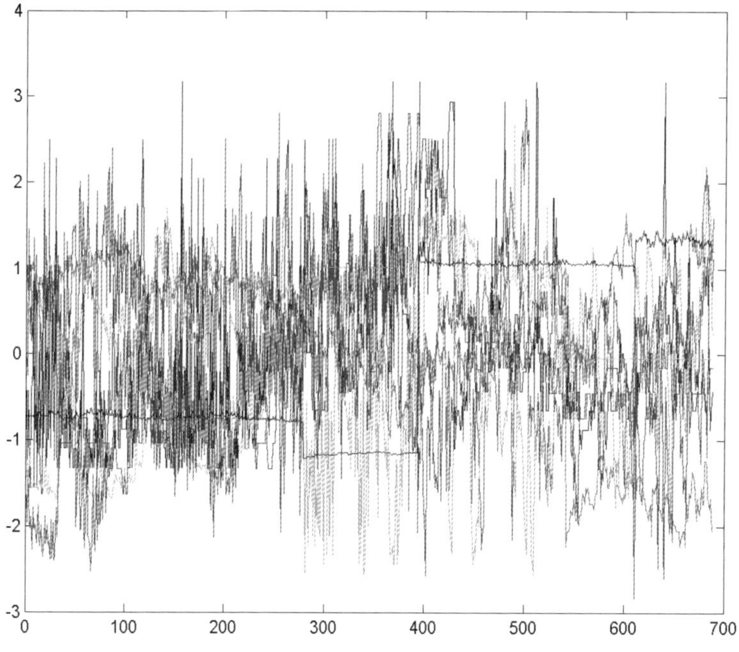

Fig. 5: Water quality changes in the real world can be very dynamic.

On top of the dynamism of water quality that may be present, the general environment is also very harsh. The water conditions may be corrosive or scaling in nature. This can lead to degradation of anything placed in the system or the formation of a coating of various types of materials. There is also present in most pipe systems something known as biofilm. This is a thin layer of bacteria and their associated slime that coats the inside of pipes and anything else present in the system. Any detection system designed to function over long periods of time must be capable of handling these harsh conditions. There is also the problem of aging infrastructure. Many of the water pipes in our major cities are over 100 years old and are occluded with rust, crumbling concrete and other debris.

Fig. 6: Conditions in water supply systems can be quite harsh.

Finally, the great diversity of potential threat agents that need to be detected presents problems of its own. This need to detect such a wide variety of diverse contaminants requires a realignment of thinking from the traditional development of a sensor for a given compound or agent. Single parameter or even multiplexed sensors can be thwarted by use of a different threat agent. To date, two main approaches have been utilized to overcome these obstacles.

Toxicity Monitoring

Toxicity is the ability of a substance to cause a living organism to undergo adverse effects upon exposure. These effects can include negative impacts on survival, growth, behavior and reproduction among others. Toxicity tests are an attempt to measure toxicity in a sample by analyzing the results that exposure produces on standard test organisms.[23]

Toxicity testing in the realm of security monitoring holds promise due to its ability to detect a wide variety of potential threats. This ability has lead to the development of a number of on-line toxicity monitoring devices as well as field verification kits that utilize a number of diverse organisms and methods to detect problems in the water supply. As a general rule, the closer an organism's physiology is to a human's the closer the organism's response will be for a given compound. Working with higher organisms can be complicated and that is why trade-offs are made and lower organisms are used for most toxicity testing. Systems for water have been developed using fish, bivalves, daphnia, algae, bacteria, photosynthesis and chemiluminescence.

There are several problems with toxicity testing. One problem is that no matter which organism or method is chosen, there will be significant differences between human and test organism response. Also, while toxicity tests are fairly adept at detecting chemical toxins, they are for the most part ineffective against biological agents such as bacteria and viruses. The largest drawback to all toxicity testing is culture maintenance. Growing and maintaining organisms in remote locations can be problematic. Another problem is that water treatment chemicals or simply common constituents of drinking water such as trace metals that are not toxic to humans may adversely affect test results. A drawback to toxicity monitoring in the distribution system is the variable environment that can come into play. Some organisms can be quite sensitive to changes in the general surroundings. Increases in vibrations, noise, air quality, and temperature could lead to false alarms. Shielding organisms from this type of upset can be costly or can severely limit options for deployment. Toxicity monitoring is often a good choice for source water monitoring where there are few monitoring points which can be easily controlled. However, for use in distribution systems, where widespread deployment is required, toxicity monitoring is usually not a viable option. For use in the distribution system multi-parameter bulk monitoring may be a better choice.

Multi-parameter Monitoring

Multi-parameter monitoring entails monitoring common water quality parameters and then looking for anomalies that may be indicative of a water contamination event. Sensors can include parameters such as chlorine residual, total organic carbon, pH, conductivity, turbidity, UV absorbance/fluorescence and others. One advantage of this approach is that it utilizes common, widely used instrumentation to make the measurements. Immediately after 9/11 the concept of deploying common sensors to act in just such a manner was investigated for water security monitoring. A number of government,[26] academic[24] and private industry studies[25] evaluated a variety of sensors to see if they would respond to the contaminants most likely to be used by a terrorist in an attack. A number of instrument manufacturers have developed multi-parameter water quality monitors for both source water and distribution system water. These systems encompass a diverse selection of different sensors and can be tailored to meet monitoring needs.

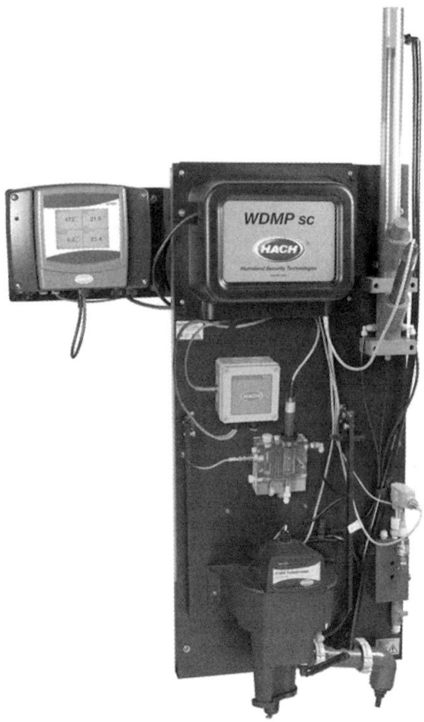

Fig. 7: One example of a multi-parameter monitoring panel.

The problem is what to do with all of this data generated by such systems. Enormous amounts of streaming data need to be processed. Another problem is the minute-to-minute variability that is present in a system. How are we to determine if alterations in water quality parameters are significant against a background of dynamic changes? Unless a full time team of statisticians is to be employed to make sense of this information, there is a need for intelligent algorithms to streamline the process. Intelligent algorithms are capable of detecting the subtle changes in bulk parameter readings that are indicative of an incursion into the system. Some are also capable of discriminating the unique pattern of responses that are elicited by different classes of agent. These differences may be enough to identify the class of an event and possibly fingerprint the most likely members of that class. The current state of bulk parameter on-line monitoring with existing instrumentation coupled with algorithms is that significant actual events are detectable. An assortment of sophisticated algorithms for interpreting on-line data and recognizing threats have been developed by a number of private and public entities.

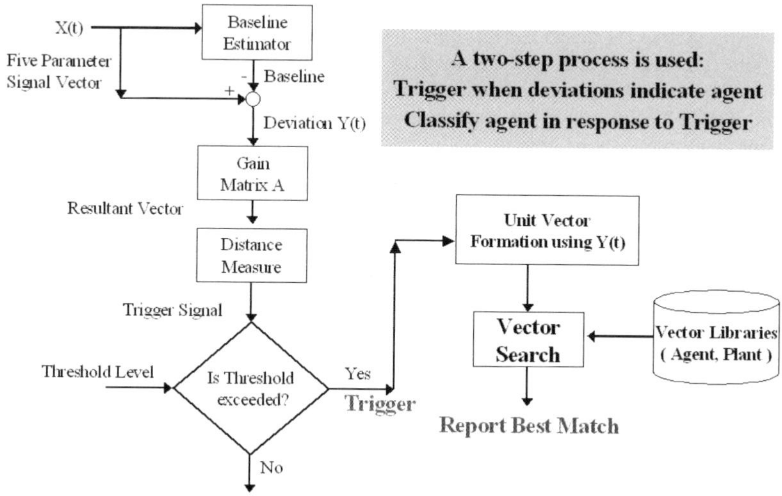

Fig. 8: Flow chart of how one such algorithm functions.

One of the largest advantages to this type of monitoring is the multi-parameter array's ability to detect a wide variety of potential threat agents from metals to organics to bio-agents. The ability to trigger on unique unknown events is also a major advantage. Some of the disadvantages are that there are some events that occur during normal operation that may trigger an unknown alarm. This however can be an advantage if the information is used to streamline operational procedures and lower costs while improving quality. None-the less, this learning phase does generate "unknown" alarms associated with normal system maintenance and requires an input of time and effort to investigate and classify these alarms so they can be placed into a plant library. Another disadvantage of such systems is that while they will detect biological events, they are not as sensitive to such events as some other methods. The majority of the detection capability comes from the growth media that may or may not be co-introduced along with the bioagents. They do not perform cell counts nor do they carry out individual bacterial identification. One bio event tends to look pretty much like another. Such systems are not likely to detect low levels of bacteria in the system. However, one of the most likely forms a bioagent attack could take would include growth media. This is due to the difficulty in removing the media and the need to overcome disinfection residuals. In these cases the instruments would respond.

Another problem has to do with deployment. Many of these instrument packages tend to be somewhat large and require a suitable site for deployment. Many also generate a waste stream that needs to be dealt with. These size and waste constraints can limit where these types of systems can be deployed. There are however options for other means of measuring these parameters than those of traditional wet chemistry and optics.

These include electrochemical and microscale devices that can be inserted directly into pipes. Microchemical based devices tend to suffer from problems of robustness. These and other electrochemical methods tend to offer less sensitivity than more traditional means of measuring bulk parameters. They may be more constrained as to what water conditions they require for proper functioning (For example electrochemical chlorine measurement may only be effective in a limited pH range) however; they may be the only option for some deployment scenarios.

Multiple parameter systems tend to be costly, however some of the cost can be justified by taking advantage of their dual use capabilities to help streamline operations and improve water quality. There are many case studies where these instruments have been deployed in real world scenarios where significant cost savings, water quality improvements, and avoidance of potential accidental contamination events have been realized.

CONCLUSION

Much progress has been made in recent years in recognizing and addressing the vulnerabilities in our water supply networks. Technological innovations have allowed us to begin monitoring the systems in real time so that we can respond appropriately if an attack actually occurs. There are, however some problems that still need to be addressed.

- Lower cost, more robust, more accurate and smaller sensor packages.
- Sensors that are reagent free and/or allow longer maintenance schedules.
- Sensors that are easily deployable in-pipe.
- Better sensors for biological detection.
- Algorithms to decrease false alarms even further.
- Coupling monitoring with response and operational procedures so that they become routine.

Progress is being made on all of these fronts, but much remains to be done.

REFERENCES

1. Nacos, Brigitte (2003) "The Terrorist Calculus behind 9/11: A Model for Future Terroism?" *Studies in Conflict & Terrorism.*
2. Jenkins, Brian. (1974) *International Terrorism: A New kind of War.* Rand Corporation. http://www.rand.org?pubs/papers/2008/P5261.pdf.
3. Jenkins, Brian (2008) *Will Terrorists Go Nuclear*. Prometheus.
4. Jenkins, John. Definitions of Terrorism. History.com. http://www.history.com/topics/terrorism.
5. Allman, T.P. (2003) "Drinking Water Distribution System Modeling for Predicting the Impact and Detection of Intentional Contamination. Masters Thesis. Dept. of Civil Engineering. Colorado State University. Fort Collins, Colorado.
6. Kroll, Dan (2004) "Mass Casulties on a Budget. Cost Effective Wtarer Terrorism" Briefing paper prepared at request of U.S. Army Corp of Engineers.

7. *"The 9/11 Comisssion Report"* (2004) W.W. Norton Company, NY, NY.
8. Cameron, Carl (2002) "Feds arrest al Qaeda suspects with plans to poison water" Fox News. July 30th. http://www.foxnews.com/story/0,2933,59055,00.html.
9. Kalil J.M., Berns D. (2004) "Drinking supply: terrorists had eyes on water". *Las Vegas Review-Journal* August 12, 2004.
10. Water Spiked With Weed Killer. http://www.hemeltoday.co.uk/ViewArticle2.aspx?SectionID=841&ArticleID=1685978
11. "Strychnine Found in Danish Reservoir." *United Press International*. October 7th, 2006.
12. "One Dead after Mass Food Poisoning in China's Habrin." *Xinhua* April 12th, 2007.
13. "Man Accused of Threatening Water Supply." *WSAZ News*. October 8th, 2008.
14. Pedersen, Daniel. "Poison Found in Refugee Camp Water Supply." http://www.mizzima.com/news/regional/2011-poison-found-in-refugee-camp-water-supply.html
15. Acosta, Rene. "Rogue MILF Rebels Poison Water Source of Soldiers, Local Residents." *Business Mirror* June 9th, 2009. http://businessmirror.com.ph/home/nation/11508-rogue-milf-rebels-poison-water-source-of-soldiers-local-residents.html.
16. *Times of India*. Nov. 10th 2010. "Maosists poison pond close to CRPF camp."
17. *BBC News*. May 14th 2010. "County Durham terror plot fater and son are jailed."
18. Roman, Thomas. *ABC News*. May 7th, 2011. "Terrorism experts talks about bin Laden evidence."
19. Silva, Daniel. *AFP*. August 20th, 2011. "Qaeda suspect plotted to poison water."
20. Hoffman, Bill. September 22nd. 2011. "Poison put in tanks." *Sunshine Coast News*.
21. Faiez, R. and Vogt, H. June 6th 2012. "Taliban poisoned school girls, say Afghanistan officials." *Associated Press*.
22. Kroll, Dan (2006) *Securing Our Water Supply. Protecting a Vulnerable Resource.* Pennwell Publishers. Tulsa, OK.
23. Kroll, Dan (2007) "The Role of Rapid Toxicity Testing in Security Breach Evaluations" *WaterWorld*. January Issue.
24. Byer, David and Kenneth H. Carlson (2005) "Real-time detection of internal chemical contamination in the distribution system." *Journal of the American Water Works Association*. 97 (1): 58-61.
25. Kroll, Dan (2002) "Results of threshold beaker testing on chemical threat agents: Is on-line water security monitoring feasible?" *Internal Hach HST Report*. September 12th, 2002.
26. EPA. (2006) Environmental Test Verification (ETV). http://www.epa.gov/etv/verifications /verification-index.html

CONTAMINATION OF WATER DISTRIBUTION SYSTEMS

WALTER M. GRAYMAN
Consulting Engineer
Cincinnati,USA

ABSTRACT

It is widely accepted that our water supply systems are susceptible to contamination—intentional or accidental. When a contaminant enters the water distribution system, depending upon the point of entry, the type and quantity of contaminant, the duration of entry, and the operation and design of the water system, it may travel short distances or may reach large portions of the distribution system. This paper discusses the following issues related to contamination of distribution systems:

- The movement and transformation of contaminants in the distribution system.
- Methods (models) for predicting the movement of contaminants and how these models can be used to mitigate the consequences.
- The use of contamination warning systems as a mechanism for responding to an event and reducing the impacts of the contamination event.
- Siting of monitors in the distribution system to detect the contamination.
- Design of distribution systems to reduce impacts if a contaminant enters the system.
- Decontamination of a contaminated distribution system.

INTRODUCTION

Water distribution systems are inherently susceptible to both intentional and accidental contamination. As shown in Figure 1, there are numerous points at which a contaminant may enter the distribution system. These range from major components such as treatment plants, pumps and tanks, to individual hydrants and customer connections. Since it is highly infeasible (or virtually impossible) to fully protect all potential points of entry, in addition to trying to lower the likelihood of contamination events, a major emphasis has been on managing the distribution system in case of contamination and trying to minimize the impacts of such events.

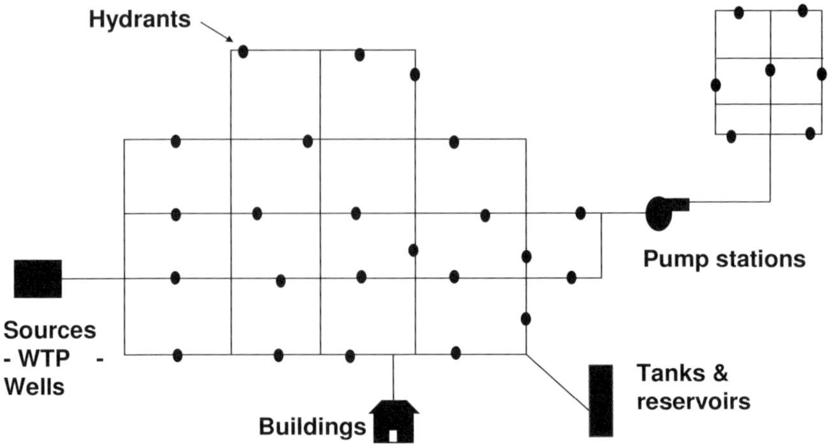

Fig. 1: Example potential entry points for contaminants into a distribution system.

FATE AND TRANSPORT OF CONTAMINANTS IN A WATER DISTRIBUTION SYSTEM

When a contaminant enters a distribution system it will mix with the water flowing in the distribution system and be transported along with the flowing water. Depending upon the particular contaminant, it may also experience chemical, physical or biological reactions and may interact with other constituents in the water and with the pipe wall. The degree to which the contaminant spreads through the distribution system and potentially impacts customers depends upon the point of entry, the type and quantity of contaminant, the duration of entry, and the operation and design of the water system. Consequently, it may travel short distances or may reach large portions of the distribution system.

Mathematical hydraulic and water quality models of distribution systems may be used to calculate (estimate) the movement of a contaminant in a distribution system (Grayman, 2005). Both public domain software [e.g., EPANET (Rossman, 2000)] and commercial packages are available. The models first calculate the flows and pressures throughout the system over the course of a day or longer periods and then calculate the movement of the contaminant and resulting concentration within the flow assuming either conservative or reactionary behavior. As with all models, the results are only approximations with the accuracy dependent upon factors such as the level of detail of the model and how well it represents the actual system. These models have been used for a variety of tasks related to the contamination of distribution systems including estimating the vulnerability of a system to contamination, design of monitoring systems, and as part of a contamination warning system for mitigating and reducing the impacts of a contamination system.

In order to demonstrate the movement of a contaminant in a distribution system, a model has been applied to an example system that serves approximately 25,000 people under two separate contamination scenarios. In the first scenario, shown in Figure 2, the clearwell at the water treatment plant is contaminated over a period of several days. As illustrated, the contaminant (highlighted in bold in the map) moves through the water distribution system relatively rapidly and after 24 hours a large majority of the system is contaminated. Eventually the entire system would be affected. In the second scenario (Figure 3), a contaminant is injected into the system along a small diameter transmission line (150 mm) serving a neighborhood in the northwest corner of the system. As illustrated, this results in the entire neighborhood being contaminated with the remainder of the system unaffected.

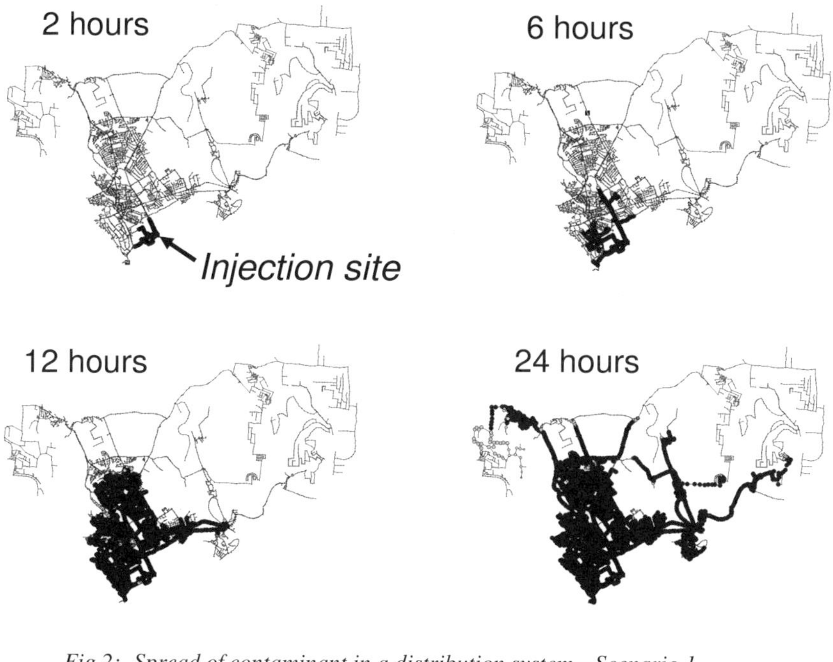

Fig 2: Spread of contaminant in a distribution system—Scenario 1.

Fig. 3: Spread of contaminant in a distribution system—Scenario 2.

The speed at which a contaminant moves through the system is largely controlled by the water demand (water consumption by customers), resulting in more rapid movement during higher demand daylight hours than at night. The concentration of the contaminant at any point within the distribution system is controlled by a set of interacting factors including the volume of contaminant that enters the system, the duration of the injection and the amount of uncontaminated water that dilutes the contaminant. Other factors that affect the movement of the contaminant and its concentration include the operation of the water system and the presence of storage tanks. The model effectively integrates all of these factors in calculating the resulting concentration.

CONTAMINATION WARNING SYSTEM

The primary mechanism that is currently available to mitigate or reduce the impacts of a contamination event in the distribution system is a contamination warning system (CWS). A contamination warning system is a combination of monitors, institutional arrangements, analysis tools, emergency protocols, and response mechanisms designed to provide early warning of contaminants in order to minimize customer exposure.

USEPA (2007) describes a conceptual model for contamination warning system operation as follows:

- Monitoring and surveillance. The basic components of online water quality monitoring, sampling and analysis, enhanced security monitoring, consumer complaint surveillance, and public health surveillance occurs on a routine basis, in near-real time until an anomaly or deviation from the baseline or base state is detected.
- Event detection and possible determination. Event detection is the process or mechanism by which an anomaly or deviation from the baseline or base state is detected.
- Credible determination. Credibility determination procedures are performed using information from all contamination warning system components as well as external resources when available and relevant. If contamination is determined to be credible, additional confirmatory and response actions are initiated.
- Confirmed determination. In this stage of consequence management, additional information is gathered and assessed to confirm drinking water contamination. Response actions initiated during credible determination are expanded and additional response activities may be implemented.
- Remediation and recovery. Once contamination has been confirmed, and the immediate crisis has been addressed through response (e.g., flushing, emergency warnings, etc.), remediation and recovery actions defined in the consequence management plan are performed to restore the system to normal operations.

On-line monitors or sensors are widely considered as the primary means of detecting a potential contamination event in a distribution system. The role of monitors in a contamination warning system and how they may be applied during a potential contamination event to detect, characterize and confirm a contaminant event is shown schematically in Figure 4 (Grayman, 2010). In order to reliably and efficiently detect potential contaminants, the monitors must be sensitive to the presence of a wide range of agents and there must be a sufficient number of appropriate monitors so that detection occurs in a timely manner. Historically, monitoring sites were selected primarily on the basis of informal selection criteria that reflected the representativeness and accessibility of the sites and the specific purpose(s) of the monitoring system.

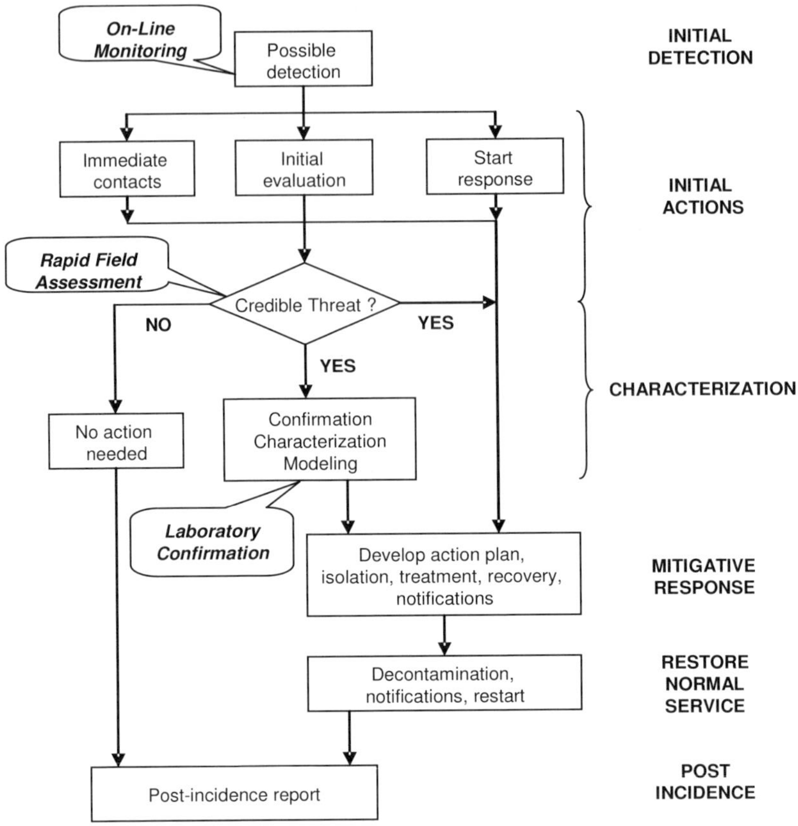

Fig. 4: Flowchart of a CWS as part of a response to a potential contamination event.

The success of a contamination warning system in reducing impacts of a contamination event is highly dependent upon the length of time of the various steps shown in Figure 4. In a study in Ann Arbor, Michigan, Skadsen et al. (2008) found that a total delay time greater than about 8 hours between the entry of the contaminant into the distribution system and the cessation of water use by customers (associated with the initial detection, initial actions, characterization and mitigative response steps in Figure 4) would significantly reduce the effectiveness of the CWS. This threshold may vary considerably in other systems. Discussions with water utility personnel suggest that in order to meet a delay time goal of less than eight hours would require a very robust on-line monitoring network and a very efficient emergency response plan.

Over the past decade, there has been extensive research on the optimal placement of a set of sensors in a distribution system (Murray and Hart, 2010; Grayman, 2010). Berry et al. (2005) provide a succinct description of the sensor placement problem for CWS design:

> "The goal of a sensor placement optimization formulation is simple: to place a limited number of sensors in a water distribution network such that the impact of an accidental or intentional injection of contaminant to public health is minimized. However, no specific, concrete formulation for sensor placement has emerged that is widely accepted by the water community. There are a wide range of alternative objectives that are also important when considering sensor placements, such as minimizing the cost of installing and maintaining the sensors, minimizing the response time to a contamination event, and minimizing the extent of contamination (which impacts the recovery costs). Additionally, it is difficult to quantify the health impact of a contamination event because human water usage is often poorly characterized, both in terms of water consumption patterns as well as how the water consumption impacts health effects. Consequently, surrogate measures like the total volume of water consumed at all sites have been used to model health impacts; this measure assumes that human water consumption is proportional to water consumption at all junctions within the network."

Many different algorithms have been proposed and tested for designing optimal sensor networks (Ostfeld, 2008). Most of these methods utilize integrated search techniques in conjunction with distribution system hydraulic/water quality models that assess the benefits associated with alternative sensor network designs. The most widely used software in this category is the Threat Ensemble Vulnerability Assessment—Sensor Placement Optimization Toolbox (TEVA-SPOT) distributed by the Environmental Protection Agency (Berry et al. 2010).

DESIGN OF WATER SYSTEMS TO REDUCE VULNERABILITY

Contamination warning systems and other methods for detecting and mitigating contamination events require active intervention in order to minimize impacts. An alternative or supplemental "passive" mechanism for reducing impacts involves the re-design of distribution systems aimed specifically at improving security (Baker, 2008). In the United States and many other countries, distribution systems are designed as looped systems composed of transmission lines and interconnected local delivery pipes used to deliver water to customers. Looped systems generally result in multiple paths that water can follow from the treatment plant to customers and provide redundancy in case of outages. However, the loops also result in multiple pathways for contaminants and greater difficulties in isolating contaminants.

Baker (2008) suggested the greater use of "zoned" distribution systems based in part on the District Meter Area (DMA) concept that is widely used in the United Kingdom and other areas. A DMA consists of an isolated water distribution system zone for which there can be both control and calculation of flow amounts. The Baker approach, shown schematically in Figure 5, extends this approach further to minimize interchange between zones and "hardening" and reduction of access to the transmission system to discourage entry of contaminants into the transmission backbone.

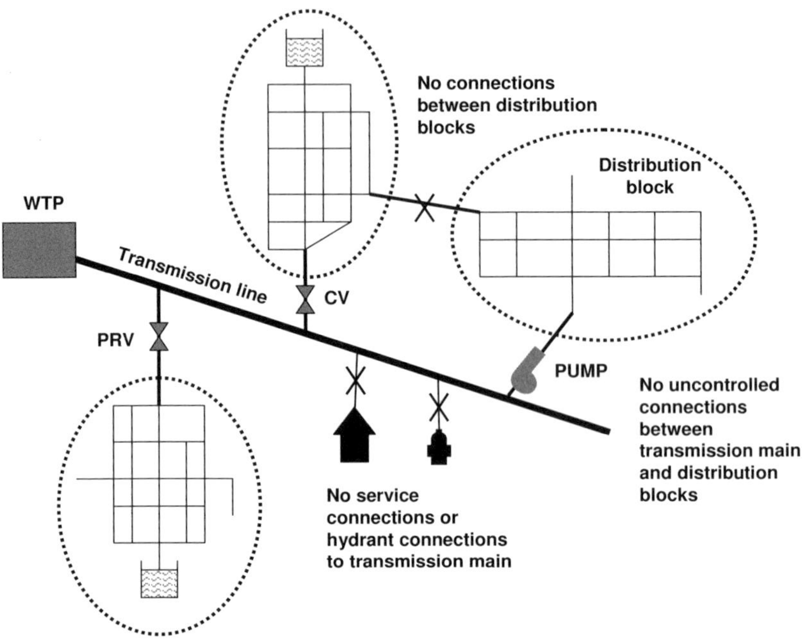

Fig. 5: Schematic representation of a zoned distribution system.

Murray et al. (2010) tested this zoned approach using mathematical modeling and simulation of several distribution systems and found that such an approach may significantly reduce the spread of contaminants with little negative impact on water age (a surrogate for water quality) and reliability. Converting a system from a looped to a zoned system would require the addition of control valves, check valves and some piping. The level of effort and costs associated with such a conversion is likely to vary considerably depending on the existing distribution system configuration.

REMEDIATION, DECONTAMINATION AND RECOVERY

When a contamination event occurs, the immediate concern is minimizing the impacts of the event on the customers of the water system. Once the immediate concern is taken care of, attention turns to the goal of remediating and decontaminating the system and then returning it to normal operation (USEPA, 2008). There are many challenges during this period when the system is being returned to normal operation including, for example:

- Providing interim water supply while the water system is not available
- Treatment of water contaminated through accidental or deliberate actions. This water needs to be treated before being released into treatment systems or the environment.
- Decontaminating pipes. Many contaminants that have been identified as likely threats to drinking water and distribution facilities can adhere to, or become embedded in, rusty or corroded pipes or biologically active layers (biofilm).
- Mitigating the "fear factor" associated with the contamination event and regaining the confidence of customers that their water supply is safe.

CONCLUSIONS

In the United States, following the events of September 11, 2001, the concerns over possible terrorist acts affected almost all aspects of life including water supply systems. Previous concerns over accidental contamination of water sources or intrusion into distribution systems took a backseat to the much greater concerns over intentional contamination of the water supply. This fostered a significant upturn in research and regulations affecting water supplies. The goal was certainly a water system that was much less vulnerable to intentional and accidental events, a robust and ubiquitous monitoring system that would immediately detect a wide range of contaminants at any point from source to tap, and a response plan that would protect customers from exposure to unwanted agents. In the years since 2001, some progress has been made in reaching for the goals of a safer water system. However, since there have been no major intentional contamination events in water systems in the U.S., one can postulate that: (1) it is difficult to fully assess the degree of protection that has resulted and; (2) the rate of research and development (and progress towards the lofty goals of 2001) has slowed down.

There are still many issues that need to be confronted in order to ensure a safe water supply. Some issues and questions that are specifically related to potential contamination of water distribution systems are outlined below:

- Given the current state of technology, can a contamination warning system be effective in protecting customers from contamination events?
- Can we afford the relatively high capital and operational costs associated with a contamination warning system to defend against what is assumed to be very rare events?
- Can we broaden the usage of the monitoring equipment associated with contamination warning systems in order to address more routine water quality issues and thus, broaden the cost basis?
- What are the prospects for future breakthroughs in monitoring capabilities that will lower costs and improve capabilities?

- Should we consider more radical systems such as point-of-use treatment or dual water delivery systems as a mechanism for improving the security of the water delivery process?

REFERENCES

1. Baker Corporation. (2008) Municipal Water Distribution System Security Study: Recommendations for Science and Technology Investments, Final Report to the U.S. Department of Homeland Security, Washington, DC.
2. Berry, J.W., Hart, W.E., Phillips, C.A., Uber, J.G., and Watson, J.P. (2005) "Validation and assessment of integer programming sensor placement models." In *Proceedings of the World Water & Environmental Resources Congress*, EWRI, ASCE, Reston, VA.
3. Berry, J. et al. (2010) User's Manual TEVA-SPOT Toolkit. Version 2.4. U.S. Environmental Protection Agency. EPA 600/R-08/041B.
4. Grayman, W.M. (2010) Designing an Optimum Water Monitoring System. Wiley Handbook of Science and Technology for Homeland Security, Ed., J.G. Voeller. John Wiley & Sons, Inc.
5. Grayman, W.M. (2006) Use of Distribution System Water Quality Models In Support Of Water Security. In Security of Water Supply Systems: From Source to Tap, Edited by J. Pollert and B. Dedus. Springer Press, Netherlands.
6. Murray, R.E., Grayman, W.M., Savic, D.A., and Farmani. (2010) Effects of DMA redesign on water distribution system performance. In Integrating Water Systems—Edited by Boxall & Maksimovi´c. Taylor & Francis Group, London.
7. Murray, R. and Hart, W.E. (2010) "Review of Sensor Placement Strategies for Contamination Warning Systems." *Journal of Water Resources Planning and Management*. American Society of Civil Engineers (ASCE), Reston, VA, 136(6):611-619.
8. Ostfeld, A., et al. (2008) "The battle of the water sensor networks (BWSN): A design challenge for engineers and algorithms." *J. Water Resources Planning and Management*. American Society of Civil Engineers (ASCE), Reston, VA. 134(6): 556–568.
9. Rossman, L.A. (2000) EPANET 2 Users Manual. U.S. Environmental Protection Agency.
10. Skadsen, J., Janke, R., Grayman, W., Samuels, W., TenBroek, M., Steglitz, B., and Bahl, S. (2008) "Distribution System Monitoring For Detecting Contamination Events And Water Quality." *J. AWWA*. 100(7): 81-94.
11. U.S. EPA. (2007) "Water security initiative: Interim guidance on planning for contamination warning system deployment." EPA-817-R-07-002, Office of Water, Office of Ground Water and Drinking Water, U.S. EPA, Washington, D.C.
12. U.S. EPA. (2008) "Decontamination and recovery planning: Water and wastewater utility case study"." EPA-817-F-08-004, Office of Water, U.S. EPA, Washington, D.C.

WATER AND CBRNE/CYBER TERRORISM: WATERBORNE THREATS AND SUSTAINABILITY OF CITIES INCLUDING UNUSUAL SECONDARY AND TERTIARY WATER CONTAMINATION LESSONS FROM FUKUSHIMA

SALLY LEIVESLEY
Newrisk Limited London, Mitigation of Terrorists Acts PMP, World Federation of Scientists, London, UK

ABSTRACT

Water may become a medium for attacks through chemical, biological, radiological, nuclear, explosives, cyber impacts (CBRNE/cyber terrorism), and psychological operations from terrorists. The objective of this paper is to discuss strategies, policy, practice and technologies that prevent, disrupt, respond, mitigate and assist recovery from waterborne threats.

It is proposed that mitigation of potential waterborne components of CBRNE/cyber terrorism is critical for the sustainability of cities and that the problem can be addressed within a wider definition of waterborne threats to cover conflict, natural hazards and accidents as well as CBRNE/cyber terrorism. Included within the scope of this discussion are radiological 'dirty bombs', improvised nuclear devices and conventional conflicts between nations. Psychological operations of terror groups may have significant impacts in attacks on water facilities. Recommendations are made for new forms of fast, near-real time, trusted and unambiguous scientific communications techniques to mitigate unnecessary fear in the population and to limit other harmful effects.

Two case studies are presented to illustrate the importance of the sustainability of cities from waterborne threats. The first case study, of the Fukushima Daiichi Nuclear Power Plant (NPP) accident, shows the capacity of Tokyo to function despite city-wide contamination from Caesium-137 and Iodine-131—although rural Japan continues to have unresolved challenges to agricultural land sustainability a year after the disaster. The second case study discusses Iran as a country under threat of conflicts which may breach underground and aboveground nuclear facilities, including a nuclear reactor. The potential for a perceived threat to fragile water resources within this region illustrates the importance of scientific communications for real-time public advice (for example on whether any incident requires shelter in-situ or evacuation) and the formulation of twelve hour plans to recover cities' water access to prevent panic and refugee movements.

Keywords: *Waterborne, terrorism, social activists, CBRNE, cyber, SCADA, dirty bombs, nuclear, chemical, biological, explosives, psychological operations, media, social media, Fukushima, Iran, Syria, recovery, sustainability, water resources, water facilities, refugees.*

WATERBORNE THREATS DEFINED

The term 'terror threats' covers a plethora of targets, tactics, motivations and shared information (including information transmitted via the internet or other technological means). In the face of uncertain threats scientists need to find generic scientific solutions to ensure the continuity of water facilities and the protection of water resources. To assist the search for solutions that will protect across many different causal factors, this paper proposes to use an analysis of waterborne threats based on a consideration of terrorism, international conflict, natural hazards and accidents.

Within this context, waterborne threats are defined as accidents or natural hazards that threaten sustainable water supplies or water resources and where water is used as a platform to deliver an attack or is a target of CBRNE/cyber terrorism.

Water has an important weighting in threat assessments because of the emotive response of populations to any threats to life from losses of sustainable water supplies and waterborne threats may become a national or international security issue.

Once there is a wide definition of waterborne threats it is then possible to allocate multidisciplinary teams to develop generic solutions to address the threat spectrum and create specifications for policy, practice and operations. For example, it may be feasible to design a monitoring system for the water inlets of nuclear power plants that registers intruders as well as accidental radiation releases. A few key factors in the analysis of waterborne threats are summarised in Table 1 below:

Table 1: Causes, Means and Consequences of Waterborne Threats.

1.	Terrorism, conflict, natural disasters and accidents are causes.
2.	Water is used as a medium or platform for attack or there is a disruption to the sustainability of water supply or water resources.
3.	Terrorists or hostile nations may deploy chemical, biological, radiological, nuclear, explosives and cyber weapons or devices (CBRNE/cyber) as well as psychological operations. Examples may include poisoning a building's water supplies and attacks on nuclear reactors so as to cause radiation emissions contaminating water resources and water supplies.
4.	The emotive consequences of waterborne threats may include panic, flight, refugees, war and low level conflict.
5.	Other consequences of waterborne incidents may include threats to the sustainability of cities and agricultural land, mass casualty emergencies, economic losses and a weakening of national security.

This paper discusses two case studies to illustrate waterborne threats and the usefulness of generic policies in providing long term mitigation. The first case study is an assessment of waterborne consequences from an accidental release of radiation from the Fukushima Daiichi NPP as this incident provides data that is relevant for identifying the consequences of a terrorist attack with a radiological distribution device (in layman's terms, a dirty bomb). The impact of the Fukushima Daiichi NPP accident was extensive and Tokyo presents an example of a modern city that experienced caesium-137, caesium-134 and iodine-131 in its water supply. Public response, government decisions and the effects on industry in Tokyo and other parts of Japan also provide valuable lessons which are applicable worldwide where there are threats of nuclear terrorism.[1] Caesium-137 has long been recognised as one of the many nuclear terror materials of interest to terrorists.[2]

The second case study is Iran—especially its nuclear programme which has attracted an increased threat of conflict which could lead to breaches of underground nuclear containments with conventional and nuclear weapons as well as threats to the Bushehr reactor. Iran has plans for two more nuclear reactors and an expansion of facilities within its nuclear programme. Within Iran and the surrounding region any perception of radiation threats to water could lead to flights of population and refugee crises unless (a) scientific clarification of any incident was immediately available and (b) there was a well-advertised plan for sustaining water supplies and water resources for cities in the region. It should be understood that it is the perceived threat to water that will create population flight if people have no threshold or benchmark which explains the health effects of radiation contamination of water. It is more difficult to prove a negative effect and to overcome general fears associated with the word 'radiation'.

The objective of this paper is therefore to open a discussion on strategies, policy, practice and technologies that will prevent, disrupt, respond, mitigate and recover from waterborne threats. The top challenge for governments is the sustainability of cities if waterborne threats cause public panic and flight from cities and disrupt the economy of rural environments. Further, the scope of any regional radiation threat from Iran's nuclear programme is one which needs quantification and consideration by the international scientific community and governments of the region.

TERROR THREATS—INTERNATIONAL TERROR GROUPS AND ACTIVISTS

An early Al Qaeda plot to destroy critical water infrastructure was discovered in 2002 through video reconnaissance film of water pipelines at the Johor-Singapore Causeway. The video material was recovered from terrorists operating in Singapore.[3] This water facility was critical to Singaporean national security as it was the primary source of water supply for Singapore.

---

[1] Waterborne issues from the Fukushima Daiichi NPP accident are summarised from data collected in a year-long analysis commencing 03/2011 of the daily risks to Tokyo from three reactors and the Reactor No 4 spent fuel pool. There was a daily threat to the continuity of business in Tokyo as well as other parts of the country. Data was gathered daily from Japanese papers, the JAIF daily reports, NHK World news reports and documentaries and scientific papers. Invaluable expert assistance in analysis and interpretation of commercial issues, economic and social impacts was provided to the Author by the commercial sector in London and Tokyo over a ten month period and by members of the WFS Permanent Monitoring Panel for the Mitigation of Terrorist Acts and the wider WFS scientific community. Additionally, the analysis in this paper is based on experience in crisis communications and media interviews during the height of the radiation crisis on Tokyo on the subject of whether Tokyo should be evacuated, the international effects on the business supply chain from Japan and whether recovery was feasible in the damaged areas.

[2] *'The Nuclear Regulatory Commission (NRC) has issued its final policy statement on the protection of sealed radiation sources containing cesium-137 chloride (CsCl). Cesium chloride sources have received special attention because the CsCl powder is highly soluble and dispersible, which makes it attractive for use in a radiological dispersal device.':* U.S. NRC Final Policy Statement on Security of Cs-137 Chloride, Sources, http://www.nti.org/analysis/articles/us-nrc-final-policy-statement-security-cs-137-chloride-sources/ Accessed 20120506.

[3] White Paper the Jemaah Islamiyah arrests and the threat of terrorism, 7 January 2003, www.mha.gov.sg/get_blob.aspx?file_id=252_complete.pdf, *see page 17, Accessed 20120605.*

Terrorists within Pakistan have demonstrated a capacity to threaten workers of nuclear sites with attacks on journeys to work using improvised explosive devices and the radicalisation of workers within power plants. In respect of insider threats, the capacity of terrorists to radicalise and persuade workers in nuclear power plants presents a critical vulnerability if insiders consequently assist in a terror attack on key areas of a nuclear facility.

There are potential threats to nuclear and conventional power plants if water cooling systems or dams are disrupted by explosives or cyber attacks on control systems. The ability to turn off or destroy safety systems using cyber penetration of control systems or insider cooperation would be extremely destructive. The attraction of nuclear power plants as a target for terrorist attacks received some media attention when, some weeks after the Fukushima Daiichi NPP accident and three days after the death of Bin Laden, five individuals of Bangladeshi origin who had travelled 300 miles from London were arrested when photographing the Sellafield MOX plant 'which has long been regarded as a major target for Islamic terrorists'.[4] The individuals were later released.

Indications of a core Al Qaeda interest in acquiring nuclear weapons and using nuclear materials have arisen from statements of intent and the disruption of ineffective plots. Dhiren Barot was arrested for planning to detonate dirty bombs.[5]

Although terrorists' targets are not definitively foreseeable it is still possible to assess the range of devices or weapons that may be used to create radiation emissions or to attack critical infrastructure through waterborne vulnerabilities. These devices may include chemical, biological, radiological, nuclear, explosives, cyber impacts (CBRNE/cyber) and an attack may be accompanied by psychological operations. Waterborne nuclear terror threats are summarised in Table 2 below.

Table 2: Examples of Waterborne Nuclear Terror Threats.

1.	Acquisition of control over and/or access to nuclear facilities through radicalisation of personnel.
2.	Backdoor entry to nuclear power plants or planting of explosives by divers in water inlets.
3.	Backdoor cyber disruption of water cooling systems or cyber penetration of plant control systems.
4.	Targeting of nuclear workers on journeys to and from work with improvised explosive devices.
5.	Dirty bomb plots to distribute radiological materials over cities.
6.	Psychological operations to enhance the threat and disrupt the economy.
7.	Activist groups and sole activists. For example, in 2001, Anders Breivik a sole Norwegian activist distributed to activist groups a 1500 page compendium with an assessment of nuclear power plant containment vulnerability to explosives and a list of European nuclear power plants.[6]

[4] Five men arrested on suspicion of terror offences at Sellafield nuclear power station, http://www.dailymail.co.uk/news/article-1383025/After-Osama-Bin-Ladens-death-5-terror-arrests-Sellafield-nuclear-power-station.html#ixzz1wwYv37dd Accessed 20120605.

[5] Prosecution case against al-Qaeda Briton, http://news.bbc.co.uk/1/hi/uk/6122270.stm, Accessed 20120605.

[6] Norway killings: Breivik's diary of a killer, http://www.telegraph.co.uk/news/worldnews/europe/norway/8663755/Norway-killings-Breiviks-diary-of-a-killer.html, Accessed 20120605.

CYBER THREATS AND WATERBORNE VULNERABILITIES

Cyber terrorism is an emerging threat and the dimensions of this threat remain poorly defined, particularly at the catastrophic level of attacks on critical infrastructure.

Cyber terrorism in this paper is defined as the use of the world-wide-web (WWW) and digital and electronic systems as a means of destroying, disrupting or controlling critical systems including communications infrastructure or as a platform for the delivery of terrorist propaganda in order to promote a terrorist objective.

The emerging pattern of terrorists' cyber capability reveals a complex family of cyber attack tools which when considered together follow a pathway similar to a military attack with reconnaissance, intelligence gathering, probing for vulnerabilities and penetration of Supervisory Control and Data Acquisition (SCADA) controls to deliver an overwhelming loss of capability to critical systems and national security. This pattern has a particular application to water systems as well as other critical infrastructure.

Cyber security within water infrastructure remains an ongoing technical challenge although the most serious risk to water systems in the near term may be insider activity that successfully bypasses encryption and firewall protections. The capacity to diagnose, defend against and manage cyber incidents may require cooperation between technical experts within industry and government. 'Communities of Interest' of private sector and government organisations to pool cyber defence capability can give early warnings of cyber attack intent, recognise threats, share data, speed up repairs of systems and provide alternate operating platforms if systems are overwhelmed by a cyber attack. Cyber terror sources are summarised in Table 3.

Table 3: Examples of Cyber Terror Sources.

1.	Activist groups and sole activists.
2.	National and international terror groups.
3.	Proxy terror groups acting under direction of or in association with a nation's state apparatus or with elements of a state apparatus that may or may not be under direction of the head of state.
4.	Nation states engaging in an undeclared war or engaging in low level conflict with or without attribution.
5.	Terrorist publications or chat forums using the WWW to share information.

The most frequent form of attack on governments and the economy in the recent past appear to have been conducted primarily by activists using denial-of-service attacks to overwhelm a computer or internet service or the more destructive distributed denial-of-service attacks where the attacker controls compromised computers, commonly described as 'bots' which form a 'botnet' to deliver a focused and overwhelming attack on a victim. In 2012 the British Government was embarrassed by activist groups which attacked a Home Office web site and perpetrated a telecommunications intercept on a police counter-terrorist hotline accompanied by interviews to the media to demonstrate their capability.

The vulnerability of water facilities to an external cyber control is a security risk but incidents have not been frequently reported in open source literature.[7] A sophisticated

[7] *'An example of an intentional cyber attack on a SCADA system was in January 2000 in Queensland Australia, when a disgruntled ex-employee of a sewerage plant covertly took control of the plant's*

attack could take control over internal systems in a facility, causing damage whilst masking the reports of damage so that a facility does not, and cannot, respond to the incident. This capability was indicated in the reported performance of the Stuxnet malware on centrifuges in the Iranian nuclear programme.[8] In addition, the Duqu virus[9] was found to have been designed to specifically target control systems for critical infrastructure and the Flame virus (suspected of deployment in 2010) was discovered by a Russian laboratory and described as a cyber weapon with widespread data gathering capability which appeared to target specific countries.

A serious proxy cyber terror plot involved FARC, a Colombian terror group backed by Iran, that planned to use students in a Mexican university laboratory[10] to mount cyber attacks on U.S. critical infrastructure. Audio and video obtained by the students at the National Autonomous University of Mexico, suggested a Venezuelan diplomat was seeking information about the servers of nuclear power plants in the U.S. The generic growth in capability of a nation state to use terrorist proxies and to mix this with organised crime proxies has been indicated in low level conflict reports about Iran in 2011 and 2012. An example was plots on Saudi and Israeli diplomats in several countries.[11]

Water facility operators and nuclear power plant operators may wish to assess specific components of cyber threats when planning for the sustainability of facilities.

---

operating systems—opening and closing valves and disrupting communications systems. The attack resulted in 264,000 gallons of raw sewerage flooding a nearby river'. Cyber Threats to SCADA systems, United Nations, http://www.unicri.it/emerging_crimes/cybercrime/explanations/scada.php. Accessed 20120605.

[8] Albright, D., et al, Stuxnet Malware and Natanz: Update of ISIS December 22, 2010 Report, http://isis-online.org/isis-reports/detail/stuxnet-malware-and-natanz-update-of-isis-december-22-2010-reportsupa-href1/, Accessed 20120605.

[9] '*Flame appears to be part of the state-sponsored campaign that spied on and eventually set back Iran's nuclear program in 2010, when a digital attack destroyed roughly a fifth of Iran's nuclear centrifuges. .. The first virus, Duqu, was a reconnaissance tool that researchers say was used to copy blueprints of Iran's nuclear program... Duqu, was a reconnaissance tool that researchers say was used to copy blueprints of Iran's nuclear program' Researchers Find Clues in Malware*, http://www.nytimes.com/2012/05/31/technology/researchers-link-flame-virus-to-stuxnet-and-duqu.html, Accessed 20120606.

[10] James R. Clapper, Director of National Intelligence, Unclassified Statement for the Record on the Worldwide Threat Assessment of the U.S. Intelligence Community for the Senate Select Committee on Intelligence January 31, 2012 http://www.dni.gov/testimonies/20120131_testimony_ata.pdf, Accessed 20120605.

'*A former computer instructor at the National Autonomous University of Mexico told Univision that he was recruited by a professor there in 2006 to organize a group of student hackers to carry out cyber attacks against the United States, initially at the behest of the Cuban Embassy.*' Más en Univision.com: http://foro.univision.com/t5/Al-Punto/U-S-probing-cyberattack-plot-by-Venezuela-Iran/td-p/435614085#ixzz1wzkxgfav Accessed 20120606.

[11] Israel Says Iran Is Behind Bombs, http://www.nytimes.com/2012/02/14/world/middleeast/israeli-embassy-officials-attacked-in-india-and-georgia.html? Accessed 20120606.

See also, '*A group of Iranians detained after explosions in Thailand's capital Bangkok were intending to target two Israeli diplomats, Thai police say..... Iran believes Israel has carried out almost identical acts in Tehran killing scientists linked to the nuclear programme.* (Statement by Gordon Corera, BBC security correspondent)'Bangkok blast suspects 'targeting Israeli diplomats', http://www.bbc.co.uk/news/world-asia-17055367, Accessed 20120606.

Cyber threats to water facilities and nuclear power plants are summarised in Table 4 below.

Table 4: Examples of Cyber Threats to Water Facilities and Nuclear Power Plants.

1.	Recruitment of organised crime and terror groups for proxy attacks by nations engaged in low level conflict with or without attribution.
2.	Insider recruitment by terror groups which may not be identified through traditional vetting of personnel where deception is used and if radicalisation occurs post-recruitment and within a very short period.
3.	Nation states deploying sophisticated cyber attack tools that have the capacity to overwhelm critical infrastructure and employ deception to achieve control and destruction. Encryption and firewall security breaches with insider assistance, spying tools and possibly espionage.
4.	Access to SCADA control systems.
5.	Cyber weapons deploying counter measures to mask a destructive attack on a facility.
6.	Denial–of-service attacks and distributed denial-of service – effective in legacy systems and where patching is inadequate or there are other human failures in systems security maintenance.
7.	Use of telephone systems and other linked communications routes to access systems and control operations including Bluetooth which may span an air gap to defeat unconnected systems.
8.	External maintenance routes into the system or external dial-in by employees on their laptops may provide a route to bypass firewalls or dual use of computers, phones or other systems that may carry unsecured non-corporate private communications routes.
9.	Externalisation of information technology services to service providers with Cloud computing may provide new routes for penetration depending on the integrity of any shared Cloud system.

Terrorists may self- recruit, go onto the internet to seek connections in terror chat room forums and attend religious meetings to attempt to link into radical elements. Recruitment of 'clean skins' by terrorists may be done within a short time so the continuous scrutiny for insider risks is a key factor in critical infrastructure. Insider recruitment could lead to breaches in encrypted systems and firewalls. Insiders may bypass encryption unless monitored and counter measures may be used to mask insider operations on a system. Breaches of database systems (as in the Wikileaks classified military database publication incident)[12] have indicated the catastrophic loss potential and threat to life and national security from insider recruitment.

Insider recruitment presents a significant threat to water utilities and other critical infrastructure. Examples of cyber tools for terrorism could be masking of poison releases into a water treatment facility or where perimeter security is breached without triggering a web cam or other sensor recognition.

Solutions to cyber attacks require a capacity to diagnose and act within very short periods of time and in practical terms losses will occur before a protection can be deployed. However, the capacity to isolate a problem and defeat a cyber attack on a water facility is critical for continuity of water distribution. As such defeat cannot be

[12] *'Manning, a 24-year-old Crescent, Okla., native, faces the possibility of life in prison if convicted of the most serious charge, aiding the enemy. He allegedly sent to the anti-secrecy website WikiLeaks hundreds of thousands of classified diplomatic cables and war logs downloaded from government computers while working as an intelligence analyst in Baghdad in late 2009 and early 2010.'* Bradley Manning WikiLeaks Case, http://www.huffingtonpost.com/2012/05/24/bradley-manning-wikileaks-case_n_1542819.html, Accessed 20120606.

guaranteed it is necessary to have an alternate supply of water available in any crisis plan for a city. Engineered solutions that replace information systems programming with hard wiring may provide a redundancy solution.

Any psychological operations launched against critical water infrastructure also require counter measures using all forms of media, social media and communications so as to encourage the public to manage during a water loss event. An aggressive communications policy will limit the potential for panic runs on water supplies or evacuation.

BIOLOGICAL

Biological threats and natural hazards from bacteria, virus or toxins entering water distribution systems are a significant issue for water facility managers. Biological threats from terrorism are not developing at the same pace as other improvised forms of attack but should be considered as a threat especially if nation states use proxy terrorism to spread fear and weaken national security. Cities have a considerable capacity to manage and monitor water supplies and with specialist epidemiological experience (from natural hazards such as Legionnaires Disease) there is a capacity to identify and track unusual outbreaks. A primary problem in the biological field is dual use technology evolving from laboratory investigation and scientific progress in health and medicine which can be used for destructive purposes by terrorists or activists. Synthetic biology involves enabling reproduction of agents with harmful effects—drugs that interfere with the central nervous system have applications to aerosolisation and distribution over crowds and the modification of immunological processes are all processes being discussed for ethical management and education[13] but which will not be safely regulated by groups with interests in death and destruction.

The worldwide experience of pandemic planning has also introduced a protection for biological attacks with increased preparedness for medical response and management of infectious disease. This has created new capabilities for early medical recognition of symptoms, public self-diagnosis and infection prevention, communication with communities and the upgrading of biological monitoring technology and pharmaceutical response. However, a lack of tertiary treatment of water systems in some countries has provided continuing problems when tackling biological threats. An unusual example is the spread of resistance to antiviral treatment of the pandemic avian flu H5N1 virus where extensive treatment of people with anti-viral drugs and untreated sewerage may have been a factor that increased a risk of virus mutations from birds using waterways with the untreated sewerage.

Summaries of biological terror threats and specific vulnerabilities of water facilities are outlined below in Table 5, Biological Terror Threats and Table 6, Water Facility Vulnerabilities to Biological Terrorism.

---

[13] Dr Amy Smithson, Professor Malcolm Dando and other colleagues discussed recent research on this subject which is leading to recommendations for widespread education and control when giving presentations from J.B. Tucker et.al., Innovation, Dual Use and Security: Managing the Risks of Emerging Biological and Chemical Technologies, MIT Press, Spring 2012. Chatham House, 21 June, 2012.

Table 5: Biological Terror Threats.[14]

1.	Conflict in Syria leading to the risk of access by terrorists to stored biological weapons some of which may be pre-loaded munitions e.g., anthrax, smallpox and plague. Al Qaeda terrorists and proxy terror groups such as Hezbollah which are supported by Iran have been named in media reports as operating within the Syrian conflict in 2012.
2.	Dual use technology that may appear to be used for other purposes; experimentation on animals to identify biological weapon effectiveness.
3.	Research programmes which mask biological weapons programmes; the use of traditional tertiary education facilities for recruitment or to insert students into the biological sciences; and the use of laboratory bench work to build weapons.
4.	'Kitchen biological capability'—home experimentation by interest groups.
5.	Market demands for rapid down-scaling in size and up-scaling in complexity of technology available for home or small laboratory use e.g., DNA analysis. This provides an opportunistic capacity for kitchen biological weapons development.
6.	Scientific publications that describe pathways for engineered mutation of the avian flu virus.[15] These will inform the terrorist 'Community of Interest'.
7.	Possible use of remote camps by terror groups for experimentation.
8.	Terror group access to laboratories for chemicals and knowledge e.g., Yemen in 2012 following capture of city of Zinjibar in the south by Al Qaeda in the Arabian Peninsula.[16]
9.	Availability of biological agents in a natural state such as ricin, anthrax, avian flu, botulism and cholera some of which could be used in waterborne attacks.
10.	Waterborne transmission potential of cholera, tularemia, anthrax, ricin and other agents.[17]
11.	Nation state recruitment of terror groups and the provision of resources and improvised bio weapons that present an effective threat.
12.	Health workers' fear of pandemic that may translate into uncertainty for medical response plans when a pandemic commences.
13.	Scientific and policy failures to identify key factors in cross infection rates or multidisciplinary solutions.
14.	Commercial issues related to pharmaceutical distribution which may limit treatment and deliver less effective measures.[18] The cost of pharmaceuticals is a limiting factor for some countries.

The structures and operation of water resources, water treatment and water distribution systems are a focus for the analysis of vulnerabilities to bio-terrorism.

---

[14] For these purposes toxins are included within the reference to biological agents. For a comprehensive overview see Shea, D.A., Small-scale Terrorist Attacks Using Chemical and Biological Agents: An Assessment Framework and Preliminary Comparisons, Small-scale Terrorist Attacks Using Chemical and Biological Agents: An Assessment Framework and Preliminary Comparisons, CRS Report for Congress , May 20, 2004, http://www.fas.org/irp/crs/RL32391.pdf.

[15] *'According to the World Health Organisation (WHO), H5N1 kills around 60 per cent of people who become infected with the virus. So the decision by scientists to conduct experiments making such a deadly virus transmit MORE easily between human beings is controversial. The knowledge that has been generated could be dangerous, especially if it falls into the wrong hands (e.g., bioterrorism).'* Publish and be damned? Sussex debates the bird flu research controversy, http://www.sussex.ac.uk/alumni/newsandevents?id=13915, Accessed 20120606.

[16] How Yemen became a centre of global terrorism, 'May 29,2011, hundreds of suspected Al-Qaeda gunmen seize control of Zinjibar and surrounding area' The Sunday Times,13052012, p.17.

[17] Chemical and Biological Agents, Nuclear Events, http://www.alerts.si.edu/docs/CommonChemicalBiologicalReferences.pdf, Accessed 20120606. See also Cholera, http://www.cdc.gov/cholera/epi.html, Accessed 20120606.

[18] This has been subject of medical health governance discussions, Chatham House 2011 and 2012, 'Competing Visions of Global Health Governance', 11 May 2012, Owain, D., Williams, The Global Governance of Access to Medicines.

Table 6 below summarises vulnerabilities and identifies a special risk issue of terrorists targeting the water supply of buildings.

Table 6: Water Facility and Buildings' Vulnerabilities to Bio-Terrorism.

1.	Cyber intervention to take over control of operating systems.
2.	Denial-of-Service cyber attack that degrades a utility operation.
3.	Unmonitored chemical treatment sites that may be the entry point for biological agents.
4.	Insider recruitment for water contamination.
5.	Hoax events using false demonstrations of capability to disrupt, destroy or impact target populations.
6.	Disruption from natural incidents mistaken for water poisoning where a cluster of people showing symptoms creates distrust of water supply.
7.	Public psychological vulnerability to water contamination which triggers panic, flight and overwhelming demand for retail purchase of water or for access to alternate water supplies.
8.	Public vulnerability to psychological operations by terror groups or hoaxers.
9.	Modern buildings may have well designed water systems with protection but building administrators may need to plan for recognising early warning signs of attempts to tamper or that indicate an interest in contamination of water supply.

Biological threats to water are possibly a class of incident which surpass other contamination incidents because of the difficulties inherent in monitoring and detection. Table 7 provides a brief summary of ways in which scientists may assist in the mitigation of biological water threats.

Table 7: Scientific Mitigation of Biological threats to Water.

1.	Public communications by scientists in response to psychological terror created by threats to a water system.
2.	Early diagnosis of the form of contamination, the lethality, health or infection risks.
3.	Self protection measures for the public and advice to operational response organisations and emergency responders.
4.	Technology to contribute to the diagnosis, or protection, of water from contamination.
5.	Expedient remediation solutions.
6.	Policy, plans and stress test design to simulate catastrophic losses at key points in a system and the consequences of mitigation controls.
7.	Research and development to mitigate residual risks of catastrophic loss of water supply or water resources through bio-threats. This would cover urban and rural consequences.

There are generic communications techniques and priorities for policies that warn the public and mitigate the effects of contamination of water distribution facilities. These are discussed in the penultimate section of this paper on 'Communications, Media, Social Media and Psychological operations'.

JAPAN CASE STUDY 1: FUKUSHIMA DAIICHI NPP INCIDENT—WATERBORNE RADIATION CONSEQUENCES?

Nuclear terrorism presents a threat to populations, water distribution and water resources. A year after the Fukushima Daiichi NPP accident there is data as to how a capital city such as Tokyo has considerable resilience and can absorb a major event releasing caesium-134, caesium-137 and iodine-131 onto land, water reticulation areas, water treatment centres and into the sea and rivers.[19]

The Fukushima Daiichi NPP incident illustrates the broad spectrum of contamination that follows a nuclear reactor accident. Although information is available from 30 years of Chernobyl data, the Japanese accident is directly applicable to scenarios of radiological devices (dirty bombs) and their effects on modern industrial cities and urban lifestyle. Furthermore there is information relevant to policy-makers on the effects on topographical features of radiation deposits, waterborne threats to rivers and marine life and unexpected dynamic effects of contaminated water where it acts as a route of secondary contamination on plants, animals and humans. Japan faces a long term effort in recovering 8% of its land area which remains subject to some level of contamination and in the management of economic and population effects.[20] The population's reluctance to restart the country's reactors a year after 3/11 shows how fear of radiation may limit lifestyle and economic development in the aftermath of an incident.[21] The government has now decided to restart a nuclear facility in Western Japan to avoid further deterioration in quality of life (resulting from energy shortages) but there is considerable public trepidation about the decision.

The mitigation of residual risk of waterborne threats from radiation following a terror attack, international conflict or accident is a significant challenge to science as the sustainability of cities may be under threat unless drinking water can be delivered to a population within a very few hours after an incident or the population can be convinced with adequate and accurate scientific communication, as to the quality of the water supply. Within Japan there was an additional burden of the long term psyche of the population 60 years after the bombing of Hiroshima and Nagasaki which gave an additional impetus to the fear of radiation effects and a concern for the next generation born into the post-Fukushima environment.

---

[19] A. Stohl et al., Xenon-133 and caesium-137 releases into the atmosphere from the Fukushima Dai-ichi nuclear power plant: determination of the source term, atmospheric dispersion, and deposition, Atmos. Chem. Phys., 12, 2313-2343, 2012, http://www.atmos-chem-phys.net/12/2313/2012/acp-12-2313-2012.html, Accessed 20120606.

[20] 'Spanning 13 prefectures, the affected area has accumulated more than 10,000 becquerels of cesium 134 and 137 per square meter, according to the science ministry. November 21, 2011
Some 8 percent of Japan's land area, or more than 30,000 square kilometers, has been contaminated with radioactive cesium from the crippled Fukushima No. 1 nuclear power plant. Spanning 13 prefectures, the affected area has accumulated more than 10,000 becquerels of cesium 134 and 137 per square meter, according to the science ministry. Ministry officials said the plumes flowed mainly via four routes between March 14 and 22 after the plant was damaged by the Great East Japan Earthquake and tsunami on March 11. Radioactive materials fell with rain and snow, particularly in the northern parts of Tochigi and Gunma prefectures....A branch of the plume moved southward ...passing through western Tokyo... The fourth plume headed southward from the night of March 21 and early March 22. It passed through northern Chiba Prefecture but largely skirted central Tokyo due to a pressure pattern, limiting contamination in Tokyo. Radioactive cesium blankets 8% of Japan's land area,
http://ajw.asahi.com/article/0311disaster/fukushima/AJ201111210014, Accessed 20120606.

[21] '*The government of Prime Minister Yoshihiko Noda has been trying to gain local consent to bring the Oi reactors back online in an attempt to beat the scorching summer heat in the coming months*', 71 percent against hastily restarting Oi nuclear plant: Mainichi poll,
http://mainichi.jp/english/english/newsselect/news/20120604p2a00m0na018000c.html, Accessed 20120606.

Tokyo is an example of a highly industrialised modern city that managed a very large deposition of radiation into its water supplies[22] and hot spots over the city.[23] The capacity of the city to manage the incident and keep operating as an important financial centre indicates that modern cities can robustly respond to radiation deposition and are not helpless in such situations. The water disruption from iodine-131 contamination was managed and hot spot caesium-137 deposits were identified, made known to the public and cleaned up.

The lesson from Tokyo is that in the event of a terrorist attack with a radiological distribution device (dirty bomb) or multiple devices there is every reason to consider that a city will not be disabled. Although there are many caveats (in relation to type of radionuclide that may be distributed in any terror attack), the fact remains that water supplies in Tokyo were significantly contaminated and the city managed the disruption without losing its coherence and capacity to deliver services. A dirty bomb in a terror attack would be more limited in its area of distribution in comparison to the accident affecting Tokyo.

However the main lesson for future terrorist incident mitigation is that water is a sensitive service in cities and it is necessary to provide a plan for fast access to independent water supply and distribution in the event of contamination of water resources. If there is not a significant immediate or long term health threat to the population it is then necessary to ensure that the public have credible scientific information. An initial speculative estimate of 12 hours may be the time taken by the public to determine whether to remain, to self-evacuate or to demand assisted evacuation for life preserving access to water.

Examples of waterborne threats from the Fukushima NPP accident are presented in Table 8 below.

---

[22] '*On Wednesday, 210 becquerels of iodine-131 per liter of water, more than double the government-set safety level of 100 for infants, were detected in water at a purification plant in Tokyo*' *http://www.japantimes.co.jp/text/nn20110325f1.html*, Friday March 25,2011. Accessed 20120606.

[23] Mainichi: 'Hot spots' found in large numbers near Tokyo, in northwest Chiba, http://enenews.com/mainichi-hot-spots-found-large-numbers-tokyo-northwest-chiba-gov-alarmed-unexpected-population-drop,

WSJ: High radioactivity has experts 'puzzled' and Tokyo-area residents alarmed '*Little attention has gone to the rainfall that started in the morning of March 21 and lasted until early hours of March 23. The news [sic] study concludes that this round of precipitation dumped relatively large doses of tainted materials to a swath of land much farther away from the plant that covered parts of Tochigi, Saitama, Chiba prefectures and Tokyo.*'http://enenews.com/wsj-high-radioactivity-experts-puzzled-tokyo-area-residents-alarmed-study-relatively-large-doses-tainted-materials-dumped-parts-tokyo.

Table 8: Fukushima Daiichi NPP Accident: Lessons for Dirty Bombs on Cities.

Crisis from iodine-131 and caesium-137 entering into the Tokyo water supply: • Bottled water ran out; [24] • Mothers reluctant to breast feed and/or provide water to infants.	
1.	Secondary contamination of cow's milk was not recognised immediately in the days following radiation emissions commencing from the Fukushima Daiichi site. In the year that followed the authorities accepted a responsibility for monitoring young people (under age 18) for thyroid symptoms and also the health of the general population in prefectures close to the Fukushima Daiichi site.
2.	Rigorous standards for radiation levels in drinking water and other ingested produce were not introduced until April 2012.[25]
3.	Temporary and less rigorous standards for ingested food and liquids were not effectively monitored for most of the first year because of limited and unregulated sampling, inadequate technology and lack of centralised control. Local authorities were left with the responsibility for radiation measurement.
4.	Ingestion of contaminated food, water and other beverages by populations close to the Fukushima Daiichi site were the result of disruption in external supplies in the weeks following 3/11 and difficulties for local authorities in (a) understanding the breadth of contaminated items and (b) doing technical measurements.
5.	Unexpected ingestion of contaminated food, water and other beverages above the temporary levels set by the government continued in Tokyo and other areas of the country throughout the first year after 3/11. Contaminated beef, rice, tea, vegetables, marine life, water, milk, infant formula, beverages and other foodstuffs were found through citizen action or retrospectively (after consumption) when the government identified a pathway of contamination. One example was rice straw from Fukushima that was fed to beef herds in other parts of the country.[26]
6.	Tokyo's water testing became rigorous once contamination was recognised some days after deposits had occurred and the authorities placed continuous data on web sites for the public which showed the public when water supplied to the city was clear.[27]
7.	Tourism was sensitive to contamination information and the Government placed data on contamination on its tourism web site to reduce the economic threat.

[24] *'In the early afternoon on Wednesday, March 23, the Tokyo Municipal Government announced that 210 Bq/liter of iodine was detected in water at a water filtration plant in north-eastern Tokyo. This level exceeded MHLW's guideline for infant intake, which suggests a threshold of 100 Bq/liter. Upon the announcement, the municipal water authority urged residents in Tokyo to use bottled water for infant formula, which created an immediate panic as Tokyo residents rushed out to buy bottled water. Beverage shelves at most supermarkets and convenient stores went empty within hours. Convenience stores put in place purchase limits, as consumers started buying everything in stock. Chief Cabinet Secretary Edano asked consumers to not hoard bottled water, stressing that Tokyo water is completely safe to drink for everybody except infants. The crisis ended on the afternoon of March 24, when the Tokyo Metropolitan Government lifted the ban on tap water for infants, and reported that the morning's iodine 131 reading at a water purification plant in Katsushika plunged sharply. The return to rational water purchasing behavior was almost immediate, though the near daily health and safety scares have obviously taken a serious psychological toll on the Japanese consumer'*, USDA Foreign Agricultural Service, Gain Report, http://gain.fas.usda.gov/Recent%20GAIN%20Publications/March%2024%20Update%20-%20Japan%20Food%20and%20Agriculture_Tokyo_Japan_4-21-2011.pdf, Accessed 20120606.

[25] NHK World news, 20111217. Contemporaneous notes. These guidelines have been introduced in a manual by the Environment Ministry.

[26]Fukushima rice straw contaminates beef, http://www.bloomberg.com/news/2011-07-15/beef-contamination-spreads-in-japan-as-cattle-eat-radiation-tainted-straw.html, Accessed 20120606.

[27] Monitored data on environmental radiation levels in Tokyo
Monitoring site: Hyakunincho, Shinjuku-ku, Tokyo, 'Radioactive material level in Water in Tokyo/day', http://monitoring.tokyo-eiken.go.jp/monitoring/index-e.html, Accessed 20120606.

Table 8: Fukushima Daiichi NPP Accident: Lessons for Dirty Bombs on Cities (continued).

8.	Contamination of agricultural land, forests, rivers and sea water was gradually recognised and the government undertook large scale studies to determine the extent to which the nation's resources were affected. The International Atomic Energy Authority (IAEA) was consulted on recovery programmes.
9.	Secondary contamination from rain and snow and concentrated contamination occurred in the first year affecting house roofs, downpipes, roadside drainage, fountains, school swimming pools[28] and swimming pool covers.
10.	Water runoff created a form of mobile contamination and caused new hot spots during seasonal weather—particularly sensitive when close to schools, road routes to schools and around the perimeter of houses (where rain runoff from roofs caused concentrations in the soil).
11.	Drainage ditches produced contaminated sludge which could only be temporarily stored as neighbourhoods rejected permanent storage, burial or incineration.
12.	There was secondary contamination of animals and humans when water was ingested.
13.	There was tertiary contamination of people from ingestion of resources grown in contaminated water or which carried radiation deposition on surface areas – for example, beef, mushrooms, rice, tea, seafood and fruit.
14.	Tertiary contamination was also identified in Tokyo domestic sewerage sludge which rose to levels rejected for landfill because the sludge concentrated in its collection. [29]
15.	The most extensive land contamination from water was caused by rain on contaminated tree canopies which formed contaminated runoff down mountains into land growing rice and other produce. There was also contamination of forest animals, plants including mushrooms and wood (which is used extensively in Japan's household products), cooking utensils and food containers and in building construction. Hilly areas fed contaminated water onto the valuable and scarce level ground where produce is grown.
16.	The extent to which forests would contribute to widespread ecological contamination was identified in the measurement of caesium-137 in cedar pollen which is a seasonal highlight in the country and appears in yellow clouds. The pollen was measured as presenting a measurable inhalation dose (but at a negligible level).[30]
17.	Effects on river fish and sea marine life were not measured in a significant way. Plugs of water moved down rivers and contaminated organic life. This was significant owing to the Japanese population's dependence on marine life in their daily food intake.
18.	Contamination of pits used for gravel and cement making in prefectures close to the Fukushima Daiichi site led to some secondary contamination of house construction which was consequently deemed unsuitable for habitation.[31]

---

[28] Fukushima University is assisting with new technology for decontamination of school pools but in a test run 300 tons of water took ten days to clean and cost $200,000. It left a 1% (by volume) residue of highly radioactive caesium sludge which the school then had to store on its grounds. The decontamination method uses ferric ferrocyanide which captures the caesium and then is taken up by an absorbent flocculent so that the sludge drops to the bottom of the water and the cleaned water is clear to detectable limits and can be released into the environment. NHK World Contemporaneous notes, 20120217.

[29] 8 Jun 2011 – Radioactive sewage sludge has since turned up at facilities in Tokyo, Kanagawa, Saitama and other prefectures http://www.asahi.com/english/TKY201106070178.html, Accessed 20120606.

[30] NHK: Extremely high radiation in cedar flowers that release pollen—Not a "great" health hazard says gov't—2,000+ grains in a cubic meter of air during season, http://enenews.com/nhk-extremely-high-radiation-cedar-flowers-release-pollen-great-health-hazard-govt-2000-grains-cubic-meter-air-during-season-video.

[31] Radioactive Construction Sites: 1,000 Sites May Have Used "Contaminated" Crushed Stones in Fukushima, http://www.infiniteunknown.net/2012/01/18/radioactive-construction-sites-1000-sites-may-have-used-contaminated-crushed-stones-in-fukushima/, Accessed 20120606.

Policies and research evolved at different stages in the first year after 3/11 in Japan. National strategies for guidance required committees and lengthy analysis and research testing before decontamination of the country could be successfully pursued and in some situations (such as forests) the method for decontaminating in a cost effective way was not identified.

A summary of the evolution of the response to the recovery of contaminated resources is in Table 9.

Table 9: Lessons for Decontamination of Dirty Bombs on Cities from the Fukushima Daiichi NPP Accident.

1.	The devolution of decontamination policy to local authorities caused a delay in national strategies and affected public communications as local authorities did not have the expertise, technology or capability to evolve strategies.
2.	University departments and central technical committees undertook scientific research on contaminated areas and government committees used research to set standards permitted for food and water. This took time to evolve.
3.	Technology was rapidly developed in the private sector for monitoring contamination in products with monitoring equipment made available in some shops for public use. Mobile phone sleeves were designed over the course of the year so that the public could check local radiation and affected produce. (The accuracy of public measurement using these devices is still to be assessed).
4.	Large scale decontamination of land resources remained the most difficult problem a year after the accident.
5.	Clean up programmes of urban areas and rural land moved caesium-137 from one point to another and remained ineffective—with central government slow to set priorities. The IAEA made visits and offered scientific advice.
6.	At the Fukushima Daiichi site contaminated water management required international innovation and cooperation for decontamination of the large volumes of cooling water required each day to stabilise the damaged reactors. Constant problems with leaks from water piping, damaged reactor containments and leaking drains delayed attempts to stabilise the site. Seawater contamination around the plant outlets was successfully managed with silt dams which contained outflows and allowed decontamination within the dammed area.[32]

In the Fukushima Daiichi NPP incident there were evacuation related deaths from the lack of medical resources, shelter, food and inadequate facilities for personal hygiene.[33] Local authorities were unable to obtain adequate supplies of clean water, food or medicine from outside the contaminated region during the crisis and there were many difficulties with distribution. In the year following the accident permanent evacuation has been more noticeable and there are depopulated areas. There are difficulties for local communities in relocating back into reopened zones because of disruption to industry, fear of radiation and slow progress in compensation payments. There is also a delay in

[32]Information on Status of Nuclear Power Plants in Fukushima, http://www.neimagazine.com/journals/Power/NEI/July_2011/attachments/ENGNEWS01_1307247746P.pdf, Accessed 20120606.

[33] *'Of the 634 deaths in the Fukushima evacuation zone's population since March 11, 573 have been certified as "disaster-related". A disaster-related death certificate is issued when a death is not directly caused by a tragedy, but by fatigue or the aggravation of a chronic disease due to the disaster. All deaths came from 13 municipalities in the no-entry, emergency evacuation preparation or expanded evacuation zones around the nuclear plant.'* http://www.hiroshimasyndrome.com/fukushima-accident-updates/fukushima-25.html, Accessed 20120606.

rebuilding as many communities need to relocate to higher sites free from tsunami risk. These experiences in Japan provide lessons that would guide planning for dirty bombs or other nuclear effects in other countries.

Rural areas in Japan have been subjected to extensive scientific measurement, mapping and experiments in crop growing and decontamination. This has produced a clearer picture of water engineering solutions that may be required for management of extensive areas of contamination. The radiation threat is dynamic because of movements in water and subject to cyclical changes from weather patterns. Water borne radiation thus presents a continuous and spreading threat to rural livelihoods under these contamination conditions.

Crop contamination has proved to be a problem that may not be overcome by removal of topsoil as this has also removed essential elements of the soil necessary for crop growing. Water flows off mountain slopes are a seasonal problem and trees, rain and snow may contribute to the spread of contamination across agricultural land. One solution identified in the Japanese scientific work was to capture of groundwater flows into 'sacrificed' land areas close to the mountain slopes in order to divert run off from agricultural areas closer to the coast.

The lessons from Fukushima provide evidence of the value of generic solutions to water contamination from radiation as seen in Table 10 below.

Table 10: Generic Solutions to Water Contamination with Radiation.

1.	Use of social media for local, trusted information distribution that is fast and reassuring.
2.	Encouragement of the public to use hand held monitors, such as adapted mobile phones to understand the levels of contamination in food, water and the environment.
3.	Publication of a 12 hour response plan for drinking water replacement to hot spots in a city to supplement retail water supplies in a crisis. As discussed above, 12 hours is an estimated figure of a reasonable time for household access to additional water supplies.
4.	Public private partnerships and contractual arrangements for flexible integration of public resources and private commercial structures. This may deliver a long term capacity for water distribution following incidents.
5.	Planning and construction of national water grids as a long term, protective measure.
6.	Publication of governance procedures to ensure public trust in emergency water resources and to psychologically reduce risk of panic and flight by populations.

After 3/11 the Japanese Government used established figures (such as the Cabinet Secretary) who spoke with calm authority to provide continuous information on radiation effects. This message was diluted by conflicting local information, media and advice from foreign governments that persuaded foreign nationals to consider evacuation from Tokyo. The Japanese government eventually resolved much of the governance issues by displaying visual water and background radiation measurements for local communities on the WWW and in print media.

IRAN—CASE STUDY 2: RADIATION THREATS AND TO THE REGION AND WATERBORNE ISSUES?

The conflict between Israel and Iran raises many threats one of which is waterborne radiation.[34] The sites involved in Iran's nuclear programme are situated across the length of the country in an arc from the far north-west to the south and some are located close to Tehran's population.[35] Subject to the prevailing meteorological conditions when a nuclear containment is breached, the region's fragile water infrastructure and natural resources may be at risk. This risk may not be significant but nonetheless it requires study and quantification.

On May 6, 2012 there was an announcement in Israel of plans to hold an early election so as to provide a mandate *'on whether to bomb Iran's nuclear facilities'*.[36] Following this announcement another solution was found for strengthening the Israeli Government—by way of a political coalition to present a united front on policies. The capacity for Israel to undertake an all-out bombing programme was also openly discussed in May 2012[37] and there are numerous broadsheet media discussions on the possible involvement of the USA with Israel in a bombing campaign. On 6 June 2012, The Times (UK) stated: *'If the imminent nuclear talks fail, there will be only a five month window for a strike on Iran with U.S. support.'*[38]

Despite beliefs in underground containments retaining hazardous by-products of explosions it may be useful for some clarification of the potential for radiological emissions or nuclear fallout to impact on the region. Underground facilities may be constructed with deceptive features and contain many pathways by which a confined explosion would reach the external atmosphere. Additionally any conventional attack on a reactor such as the Bushehr 1,000-megawatt nuclear power plant would have consequences depending on the operational status of the reactor at the time. Consequences of accidents with this reactor will have been documented during its construction as part of the site risk management.

There are key sustainability issues in relation to water and populations which arise from the risk of conflict with Iran and it may be important for the international scientific community to quantify the waterborne challenges. A waterborne challenge from radiation contamination within a region which has fragile water resources requires scientific study and discussion, particularly to determine whether pre-emptive planning for protection of the region is required. Some of these challenges are set out in Table 11 below. It is to be noted that waterborne risks are one category of a much broader range of effects that would require scientific advice including critical population protection in relation to

---

[34] Data for analysis of waterborne radiation threats for Iran and the Region comes from papers and discussion in annual workshops and Plenary meetings of Permanent Monitoring Panel for the Mitigation of Terror Acts (PMP-MPA) and daily analysis of news media.

[35] Cordesman, A.H. and Burke, A.A., The new IAEAreport and Iran's evolving nuclear and missile forces, November8, 2011, http://csis.org/files/publication/111108_irans_evolving_nuclear_forces.pdf., Accessed 20120606.

[36] Mandate to Bomb Iran, Uzi Mahnaimi, Sunday Times, 20120506, p.11.

[37] *'A major Israel TV station on Sunday night broadcast a detailed report on how Israel will go about attacking Iran's nuclear facilities in the event that diplomacy and sanctions fail and Israel decides to carry out a military strike'*. http://www.timesofisrael.com/iaf-plans-for-iran-attack/ Accessed 20120506.

[38] 'What Israel fears is seeing Obama re-elected', The Times, 20120604, p.19.

airborne radiation and the more immediate and dangerous effects of explosions. Adequate shelter and warning systems are key components but these will not be discussed within the framework of this paper.

Table 11: Waterborne Threats from Breaches of Iranian Nuclear facilities.

1.	Vulnerability of the Bushehr Nuclear Power Plant to a breach of containment whilst the facility is operating.
2.	Nuclear programme sites above ground in Iran—atmospheric consequences.
3.	Nuclear programme sites underground -breached by nuclear or conventional munitions.
4.	Potential nuclear weapons' effects on containments of nuclear materials—for example, the Fordow Fuel Enrichment Plant, an underground facility near the city of Qom.
5.	Proximity of some sites to dense urban populations.
6.	Rural impacts from contamination.
7.	Meteorological conditions moving radiation beyond Iran's borders across water and land to the region.
8.	Effects on water resources in Iran and the region.
9.	Effects on water distribution and water quality in Iran and the region.
10.	Effects on seawater and local marine ecology.
11.	Waterborne impacts on the population, animals and agriculture and secondary contamination effects on the population.
12.	Immediate risks of population flight, especially from high density urban areas.
13.	Refugee issues across borders.
14.	The immediate and life-long consequences to young persons and adults from various levels of contamination of water resources and water distribution systems.
15.	The psychological consequences of a perceived threat to water and impacts on decisions to self evacuate from rural and urban areas.

It is to be noted that the secondary and tertiary contamination of plants, livestock, food, water and of manufactured products such as infant formula, milk and processed foods caused alarm in different parts of Japan in the first year post 3/11.

Some of the scientific solutions to pre-empt a regional impact from breaches of Iranian nuclear sites are summarised in Table 12.

Table 12: Mitigation of Possible Regional Contamination from Breaching Iranian Nuclear facilities.

1.	Scientific information to inform and protect populations from unnecessary fear.
2.	A crisis water supply plan.
3.	Policies for evacuation based on radiation thresholds that have international agreement. Threshold radiation levels with consequences for immediate and long term health will assist the public in decisions to stay in shelter, decontaminate themselves if they have been exposed in the open, limit contamination transfer into shelter or to evacuate to an uncontaminated zone. The importance of threshold risk information is to reduce unnecessary consequences of unwarranted evacuation, allow evacuation to take place with time for a prepared evacuation[39] or to stimulate immediate life protecting movements out of a hazardous zone.
4.	Radiation exposure creates such concern in populations that some evacuation decisions may be expedient if people perceive there is even a long term health risk. In emergency planning evacuation has generally been based on immediate acute dose effects because of the ambiguity of long term consequences.
5.	Provision of rational information to the public that is both timely and based on plume deposition assumptions and measurement but which may commence with assumptions biased towards public protection where discrete data cannot be immediately collected.
6.	Regional level planning for water, food, shelter, medical assessments, communications and continuous monitoring of the environment in the event of a catastrophic release of radiation from breaches of nuclear sites.
7.	Planning for provision of long term health monitoring services for the region .

See Table 13 below for an overview of multidisciplinary scientific challenges. The two case studies of Fukushima and the effects of any breach of Iranian nuclear facilities illustrate that generic responses can be designed for waterborne contamination threats.

Table 13: Generic Scientific Contributions to Waterborne Contamination Threats following a Radiation Incident.

1.	Methods of mitigation to take into account cultural and economic factors and agreed international thresholds for immediate or long term harm.
2.	Communications in near-real time for protection so that the public understand the significance of radiation in water and whether or not there is a health hazard.
3.	Assessment of the residual risks to a country and international risks from waterborne threats arising from an incident.
4.	Replacement of critical water distribution to a population within a time period that is less than the assumed time it will take for a refugee flow to become established from a city.
5.	Arrangements for a fast inflow of international assistance to strengthen local efforts if a country agrees to have international assistance and there is cultural acceptance of assistance.

Distribution of clean water to populations when there is radiation contamination is a criticality for the sustainability of cities as well as rural economies. This was seen in Japan when there were overwhelming disruptions from the earthquake and tsunami after 3/11 and there was not an easy solution for the supply of water to the most affected prefectures. This occurred despite Japan being pre-eminent in efficient supply chain planning.

---

[39] Professor Richard Wilson has outlined in discussions an option of delaying evacuation and using non-emergency planned movements of populations in zones where dosage is calculated to remain below thresholds for immediate or long term harm: http://www.physics.harvard.edu/~wilson/

However, the management of water resources is a challenge for all countries and waterborne radiation threats may have consequences outside the country of origin of any incident. In the case of 3/11 the seaborne radiation from deposits from the Fukushima Daiichi nuclear power plant emissions created long term problems in the 30 kilometre seawater zone which were difficult to monitor but which also spread contaminated fish for thousands of kilometres by way of sea currents.[40] Although dilution in water removes the risk, the contamination of fish and marine products (which are staple diets for the region) remained an issue in the first year after 3/11 and local fisheries were unable to operate. Reputation damage has remained a long term effect for fisheries and producers in the Fukushima area.

Where water resources and supply systems are fragile—as in the Middle East—populations have little buffer in terms of alternate sources of supply in the event of an incident. This means that it is essential that there is a policy for sustainability of populations. Water sustainability may be most quickly effected through regional co-operation and planning especially where the resources of the affected country may be limited. A 12 hour re-supply plan (as an estimated time) may be important in providing stability to a country and mitigating catastrophic effects. The significant lesson from Japan's experience in 3/11 is that the level of technological sophistication and joined up planning does not reduce the requirement for international assistance in the first few days, weeks or months.

COMMUNICATIONS, MEDIA, SOCIAL MEDIA AND PSYCHOLOGICAL OPERATIONS

Terrorists and social activists' use of psychological operations is widening in scope. Communications are a primary form of extending the scope, motivation, area and scale of operations and power of terror groups. Sophisticated online publications by terror groups provide an information medium for followers across the world. Such publications may be accessible, sophisticated and instructive for members of terror networks. Publications distributed to international terrorists through the internet, such as the 'Inspire' magazine produced by Al Qaeda in the Arabian Peninsula, are a formidable psychological weapon of terror. Inspire has the online appearance of a glossy magazine with a message that attracts followers, promotes commitment and details past and future plots and methods of killing.

Terror groups also use mainstream print and television news media as a vehicle for their organisational messages. There are frequent statements sent to mainstream media as part of terrorist psychological operations that support attacks on critical infrastructure and the public in many countries.

Psychological operations enhance the terror effect and are an important component to consider in any incident involving water contamination. Water is a particular emotive trigger for the public and incidents may cause panic runs on drinking water supplies. It is therefore sensible for cities to consider a 12 hour plan (as an

[40] Bluefin tuna spawned off Japan's coast have been found to have carried ... from Fukushima to waters 6000 miles (9650km) away off California in the U.S., http://www.metro.co.uk/news/900484-tuna-contaminated-with-fukushima-radiation-discovered-off-us-coast, Accessed 20120606.

estimated time) to re-supply a city's water stocks. This may be needed in the event of hoax incidents as well as any attack actually affecting water distribution, or after a disaster. There are very few incidents which genuinely panic populations as shown by the extreme resilience of the Japanese public to the 3/11 earthquake and tsunami disasters and the Fukushima Daiichi NPP accident. In response to the extreme threats of the earthquake and tsunami there were many cases where lives were sacrificed by members of communities to help save their families and communities. However, much of the survival in Japan was culturally underpinned by planning and practising (even by very young children accustomed by drills) and such plans held communities together, maximising social cohesion and cooperation.

Hoaxes are a special tool deployed by social activist groups and can usually be filtered out by police methods. However, at times hoaxes are a device used by terrorists to assess strengths in security systems or to cause uncertainty and economic impact.

Activist groups and 'sole activists' have a capacity to carry out plots that may focus on water facilities. However the more serious threats may come from terror groups and proxy terror groups because of technical effectiveness, a capacity to sustain an attack or a capability to perpetrate attacks at multiple points. Poisonings of a water treatment facility or the water supply of a target building or a swimming pool may be a potential threat from terrorists or activists.

In the event of a crisis, the ability of scientists to communicate directly to the public through media and social media and to communicate persuasively and effectively to political leaders requires a sound understanding of media communications and social media. The speed of international news media may require scientific information to be broadcast within a two minute interview which essentially may give the scientist 3-4 sentences to explain the cause, effects and solutions. This time may even be the limit of time available for a briefing interview to media who will then edit and broadcast from a collection of information from several sources.

A two minute span of time may be further contracted in live or pre-recorded scientific interviews to 40 seconds for widespread distribution through media as a 'sound bite' within a package. A 'package' will typically be a brief item on an incident which may be distributed widely through the media and played as often as every 15 minutes within the programmed intervals for news broadcasts.

In social media the 140 character limit of Twitter is a model for preparation of messages that will be taken up and have very wide distribution through trusted networks if the form and content of the information is credible and helpful to people. YouTube is another potent media vehicle which uses vision and grabs of information. Successful visual or descriptive information may be copied and distributed from mainstream media or social media sources and this is a natural cascade that follows the public's interest or appetite for information on the subject. The organic movement of information then translates into person to person communication within trusted networks and other contact groups- i.e., the scope for global escalation is now very brief in time and almost unbounded in the numbers of people who will receive information on a threat.

Further, the directions for creating media messages (as described above) are relevant only at the time of writing and may change owing to the rapid evolution of social media and international broadcast communications. A possible trend will be a move of most emergency messages to mobile media (for example, media accessible by mobile phone or any device carried by the person) because of the worldwide penetration of this technology. Successful continuity of communications however will depend on highly distributed systems of communications to the public to attract take up into the informal social media and person to person communication. Planning must also account for feedback effects onto the communications platforms and in some instances this may require a capacity for many millions of persons to communicate as well as to receive communications. At present the only available solution for this remains international grids (such as that conceived in CERN) as these grids have the capacity and continuity for national or international communications loads under disaster conditions. Cloud computing may develop to the extent that it can manage messaging for a national level crisis and this platform is rapidly evolving with a capacity for the commercial sector to escalate services because of arrangements for quick access and contractual arrangements.

Important psychological factors in message content during a catastrophic threat are the reduction of perceptions of vulnerability to realistic levels by clarifying what people will sense (see, hear, taste, smell and touch) and clarifying what can be done for self-protection. Once personal perception of vulnerability is reduced the potential for panic is reduced. The objective of mass communications in a catastrophic event is to empower the general public, industry and workplaces to instantly respond to a threat.

Psychological fear creates physiological responses that prompt the body into survival mode and this will limit attention to information (tunnel vision and auditory acuity is reduced) so it is important for messages to reach the public very quickly before there is a sense of helplessness and vulnerability. If the communications reduce fear, public perception of the threat becomes normalised. The content of the communication needs to explain the context of the incident, describe cues in the environment that will help survival and demonstrate how people are not helpless or in a hopeless situation. Once there is a growth in the perception that people will survive then the fear reaction will reduce, physiological survival mechanisms will subside and people can both listen and respond without going into an instinctive flight response.

Streaming information via social media, broadcast media or websites for the public presents real time visual and auditory messages and reduces uncertainty and fear. Governments can admit to the public that there may be much uncertainty about an incident as long as they stream information on actions being taken to get a situation under control and, in the interim, how the public can act to protect themselves. Where people are clustered in large numbers in confined spaces i.e., buildings, sports arenas and underground railways then there is an enhanced capacity for social media to normalise behaviour because the digital messages to handheld mobile phones will trigger person to person communications which give a secondary reinforcement or another 'cue' that there are responses that will assist survival. This will reinforce compliance with messages such as 'stay in situ' or 'evacuate'.

Historically 'the Golden Hour' has been assumed as the time for government communications to analyse and political leaders or civil servants to make public statements. Now there is a requirement for near real-time streaming of information through many different media formats. This has an effect of creating a partnership with the public whereby information is interactive as there are many routes for public feedback and trust has been developed in many channels of communication for two way communications. Social media provides a new resource for intelligence on incidents and when fused with scientific data and other facts the ongoing situation awareness information may be fed back to the public and business to improve protection. Global or national or local intelligence on social media now presents an opportunity for fast, accurate and highly penetrating messages that will protect the public within the context of their local threat environment.

A few of the issues to be considered by scientists when developing effective communication policies for application to any event triggering a panic over waterborne threats are set out in Table 14.

Table 14: Communications to the Public in the event of Catastrophic Threats.

To be prepared to deliver information within 5 minutes of breaking news.	
1.	Real-time information stream following the progress of the event.
2.	Twitter "tweets" of 140 characters length.
3.	Message ambiguity is limited through a constant assessment of content to minimize uncertainty and panic.
4.	Using and fusing visual information with scientific explanations for adaptive strategies as an incident progresses.
5.	Content of the information describes self protection from weapons, tactics, and shock effects.
6.	Content describes danger cues and expedient survival.
7.	Communications platforms are used with a capacity for high speed, timely two way communications with the public.
8.	The communications medium uses sensor reporting that allows geographic and statistical collations in real-time so that messages align with the real environment and experiences of the public.

CONCLUSIONS AND RECOMMENDATIONS: CONTINUITY OF CITIES UNDER WATERBORNE THREAT—IS THERE A 12 HOUR RE-SUPPLY AND RECOVERY PLAN?

Japan's experience with the Fukushima Daiichi NPP incident provides a series of lessons in relation to the limits of threats of dirty bombs to the sustainability of major cities—and it is good news for the rest of the world. This does not remove more extreme threats from evolution of radiological devices which could deny parts or a city but the Tokyo experience provides the basis for sustainable planning and minimising disruption.

Chemical and biological threats arising as a result of attacks on water facilities may be dependent upon facility design but in an incident, mitigation will occur if there has been planning for medical monitoring, reporting and response and real-time communications to the public. Any symptoms and signs of poisoning or distress within clusters of the population downstream will be a major incident for medical services and other first responders. Medical response plans already cover communications through media, social media and develop comprehensive intelligence by plotting public information on the WWW e.g., requests in web searches as indicators of spread of symptoms faster than people will attend doctor's surgeries.

Attacks on critical facilities with explosives (as per the Singapore plot described earlier in this paper) pose a significant threat but one which may be more easily managed by water engineers in relation to repairs and diversions around the damaged location. The difficulty for all water facility managers is identifying critical points that may be of interest to terrorists and this requires an integrated approach to intelligence gathering whereby the industry works in partnership with police and security services.

Other unusual forms of waterborne threat require effective physical security of critical points in water facilities and also consideration of waterways being used for backdoor terror attack entry points to facilities or targets. Finally there was a most unusual example of a waterborne incident in the disruption of the Oxford and Cambridge Boat Race in April 2012 by an activist swimmer in wet suit who rose up in the river unexpectedly between two fast moving boats. The world viewing the boat race saw in real-time the spectacle of unprepared athletes who remained focused on the race and without evident understanding of underwater threats to iconic events. At the most extreme level and in different circumstances, underwater divers can deliver devastating explosive attacks.

Recommendations in this paper apply to policy and operations by water engineers, hydrologists and scientists from many disciplines. It is proposed that further discussion be developed on the use of multidisciplinary work to consider waterborne threats from natural disasters, terrorism and conflict over short, medium and long term periods in the lifetime of critical infrastructure.

Short term threats have been discussed in the paper in relation to a threat to breach nuclear containments in Iran and in the event of future nuclear reactor incidents. Fortunately the lessons from Fukushima provide a significant data set applicable to for rural areas and city populations and Japan has produced many scientific innovations to deal with the effects of contamination. Transparency and governance have been a point of contention inside and outside Japan because of cultural factors which have limited the sharing of information and providing quantifications. However Japan's solution of putting up data in real-time for public scrutiny overcame some of these difficulties.

The scope of cyber threats to water facilities is still emerging and will remain less accessible to quantification and mitigation for various reasons. There are examples in recent times of the capacity of opponents to devise cyber weapons which could be applied to cause catastrophic effects on control systems of water facilities. Additionally there is a capacity to mask the damage to a facility and this creates new challenges in infrastructure design to guarantee recognition of an incident

Waterborne biological threats from terrorism could replicate cholera or typhoid but in developed countries the capacity to identify these incidents from clusters of symptomatic patients is built into medical monitoring and thus there is a speedy capacity to identify and limit such an attack. However the biological dimension remains one that needs to be flagged by countries for constant monitoring and scientific cooperation, especially in relation to the monitoring of terrorist interest in scientific information. Waterborne biological threats may have the greatest impact across the world if an infectious agent resulted in a pandemic. An evolving dual use technology is one of the more potent threats from the biological sciences as it is rapidly escalating the methods and routes of entry into the biological sphere for the purpose of harming populations.

Table 15 contains a set of recommendations that summarise the generic findings from the analysis of CBRNE/cyber threats to water and the consequences of natural disasters or accidents.

Table 15: Recommendations for Contributions of Science to Policy and Practice for Generic Protection from Waterborne Threats.

1.	Briefing papers for governments to describe waterborne threats and to detail mitigation measures covering a wide spectrum of natural disaster, terror and emerging conflict threats.
2.	Water engineers and hydrologists to engage with a wider scientific community to pool many solutions and capabilities so as to manage emerging waterborne threats.
3.	Solutions to mitigation of contamination of water facilities and water resources from dirty bombs or nuclear power plant accidents to include lessons from the Fukushima Daiichi NPP incident.
4.	Solutions to be based on an analysis of primary, secondary and tertiary contamination effects through water following a catastrophic incident.
5.	Solutions to consider a comprehensive water resource protection strategy.
6.	Cyber terrorism to be assessed as a core threat to water resources and water distribution systems that can be mitigated with multidisciplinary scientific teams.
7.	Cyber generated Denial of Service attacks to be assessed and mitigated along with consequences on critical infrastructure such as water.
8.	Sophisticated nation state cyber attacks to be mitigated possibly with new design engineering to bypass programmed systems and where necessary a back up with reversion to hard wired systems to be considered as a solution to deny cyber threats.
9.	Recognition of the importance of waterborne threats in national security policy.
10.	Definition of scientific research and development requirements for waterborne threats.
11.	Development of communications skills and a plethora of media platforms for crisis communications in real-time during water crises.
12.	Communications capacity to include fast responses to major crisis within five minutes and for scientific information to be encapsulated in social media length communications.

In conclusion, the scope of waterborne threats has been opened for discussion with a preliminary analysis across the threats posed by conflicts, terrorism, natural disasters and accidents. A number of tables have been presented in the paper and may operate as a check list for waterborne events and to assist with mitigation planning. The challenge is to utilise international scientific cooperation to promote solutions to waterborne challenges for the benefit of local and regional populations and the environment.

STRIVING FOR GLOBAL HEALTH AND WATER SECURITY: ACHIEVABLE OR NOT?

ANNETTE SOBEL
University of Missouri, Senior Advisor for Bio-security Engagement, Office of the Secretary of Defense/Nuclear, Chemical, Biological, Columbia, USA

BACKGROUND

Current events underscore the need for robust critical infrastructure to ensure economic growth and regional stability in emerging nations. India's recent massive power failure (July 30-August 1, 2012) spotlighted the challenges and opportunities due to an asymmetry of infrastructure investment to population growth, yet was a testament to the wonderfully peaceful culture of the Indian people (*The Economist*, p.35-36, August 4, 2012). Power, water, food, and health security policy enable the lifeblood flow of every economy.

All-source research performed for the U.S. Department of State (Multiple Sources, Bureau of Intelligence and Research, DoS, October 2011) strongly supports the conclusions that water policy is linked to long-term impacts in economic development, regional stability and vulnerability to terrorist actions. Current research typically selects 2040 as the endpoint for high confidence predictive analysis. A longstanding medical truism adopts low (less than one per cent) perinatal mortality as one of the most predictive indicators of a country's emergence and regional stability.

Global health security (the intersection of public health and biosecurity) necessarily includes water, food and power security and the associated critical infrastructure and virtual information security enablers. Similarly, human security implies sufficient access to clean food, water, power, and infrastructure including shelter, health care, and other essential services necessary to sustain healthy living. This paper concludes with a summary of recommendations for a roadmap of globally and regionally interconnected infrastructure based on One Health-One Science-One Security.

Assumptions in this paper include: (1) lack of significant technology deployment, and implementation of research advances in emerging nations will occur in the next three decades to significantly impact global health security; (2) investments in critical infrastructure will be insufficient given population growth, human and environmental factors, and other issues; (3) social networking and crowd sourcing integrated with access to other information technologies will continue to impact disaster mitigation due to natural and manmade events; 4) policy will continue to be driven primarily by cultural norms, independent of regional instabilities. On a more optimistic note, this author is highly confident that a science-based response, if communicated efficiently and effectively to decision makers and the public sector, is highly effective during complex, high uncertainty events such as disasters. Most notable within the category of complex disasters are those impacting infrastructure and resulting in second and third order effects to health, security and confidence in government. These events are hybrid by their very nature, and cause asymmetric threats to life, property, critical infrastructure (e.g., water, power, health, food, logistics, and information systems) and government leadership/continuity. Ultimately, mitigation strategies may be impacted through the perfect storm of inadequate critical infrastructure investment, security, scientific knowledge, and public policy resulting in unconstrained health catastrophes such as pandemic, (re-) emerging infectious disease outbreaks, and/or biological attacks.

THE TRIAD OF ONE HEALTH-ONE SCIENCE-ONE SECURITY

The motivation for One Health (http://onehealthinitiative.com) stems from a global virtual professional community dedicated to recognition that "...human health, animal health and ecosystem health are inextricably linked..."described in Figure 1. Figure 1 represents the dynamic interrelationships of multiple entities and also provides a framework for the insertion of One Science into this paradigm. One Science provides a mechanism for understanding the complexity of the relationships of these interdependent systems and a deeper understanding of vulnerabilities and measurable impact(s) and limitations of interventions.

Emerging infectious diseases (EID) are now recognized as elements of adaptive, dynamic systems. The mathematical modeling work of Kenkre et al. from the University of New Mexico (BTR Proceedings 2005: Unified Science and Technology for Reducing Biological Threats & Countering Terrorism, A.L. Sobel, Editor, Albuquerque, NM, March 16-18, 2005) represents some of the earliest validated work reflecting the re-emergence of a significant pathogen (Hantavirus) given the optimal set of "boundary conditions" in humans, animals, plants, precipitation, climate and the entire ecosystem. This work underscores the flaws in logic and presumed high confidence associated with use of heuristic models in isolation to predict disease outbreaks, and especially when hybrid (i.e., natural/manmade) threats are to be considered. In addition, these models represent some of the earliest field validated work supporting the use of combined linear analysis, simulations and complexity, and suggest strategies for mitigation and containment of a biological or hybrid threat. [V.M. Kenkre, "Modern Challenges in Statistical Mechanics: Patterns, Noise, and the Interplay of Nonlinearity and Complexity, AIP Proc. Vol. 658 (2003) 63.]

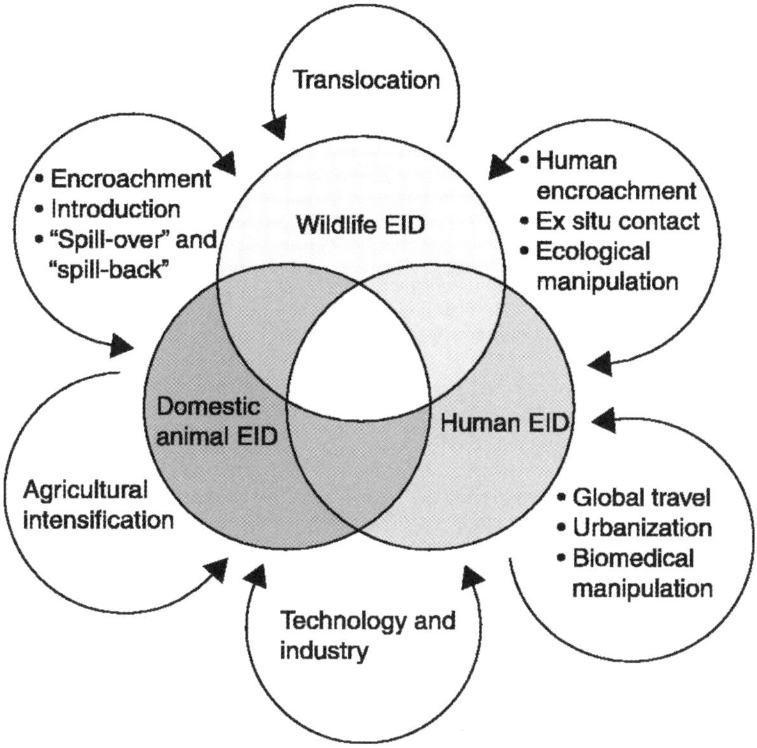

Fig. 1. P. Daszak et al. Science *287, 443-449 (2000).*

This paper also posits that economic stability and emergence of nations are highly interdependent, and recognizes the relationship to health and water security and the naturally extrapolation to the broader strategic framework of One Health-One Science-One Security.

METHODOLOGY: A SUGGESTED OPERATIONAL FRAMEWORK FOR ONE SCIENCE-ONE HEALTH-ONE SECURITY

Recent Haitian disasters (2010) have reinforced the value of technology and shared information systems in complex emergencies. When well designed and mindful of security vulnerabilities, such platforms can assist with timely flow of actionable information, vulnerability assessments of critical infrastructure, prioritization of resource deployment, and capacity building. The over-whelming human neediness is compounded by access and logistical challenges and may be mitigated through use of real-time information and advanced imagery collection techniques.

The recently forward-deployed system Medical and Public Health Information Sharing Environment (MPHISE) used in Haiti met with limited operational success and acceptance (http://haiti.mphise.net). Unfortunately, sustainability with limited infrastructure investment remains an enormous challenge. This information platform was modeled after a successful White House pilot project, and subsequently adopted by the Pan American Health Organization (Data submitted by Kathy Gilbeaux, www.new.paho.org , July 2011). The final pilot project included 1300 representatives of organizations and social networks, all with a common goal: to improve human and animal health, water and food security in Haiti. The system's mission included: situation awareness; identification of mission critical gaps; employment of smart swarms for data-mining, information retrieval and intelligent modeling. The system was built on a task server solely purposed to converge smart swarms to resolve gaps in actionable information and recommend courses of action (personal communication, Dr. Jim Wilson, December 7, 2011).

Information fusion or integration using a set of pre-determined 'rules', i.e., parameters, is most helpful when the goal is to form a 'common operational picture' or COP during complex emergencies. All-source surveillance data (including epidemiologic and imagery, critical infrastructure data, logistics, and other relevant operational data) will result in a translational pathway from data to (actionable) information and knowledge. All-source information systems, such as demonstrated in Haiti, include executable elements that assist with prioritization and optimization of resource allocation and local/regional distribution networks, and support favorable, measurable outcomes for all sectors of society involved: government, public, and private. When combined with social network analysis, a system for education, capacity building, and sustainability may be achieved using a 'bottom-up' approach to humanitarian assistance and disaster management through actively engaging community leaders.

The American Association for the Advancement of Science (AAAS) and the Association of Public and Land-Grant Universities (APLU) recently published a sentinel report entitled 'Bridging Science and Security for Biological Research: A Dialogue between Universities and the Federal Bureau of Investigation' (AAAS and APLU, February 21-22, 2012, Washington, DC). In summary, this report stresses the need for open communication between national security, government, private, and scientific communities regarding mitigation of risks associated with biological research and dual-use applications. Adequate biosecurity is defined operationally by optimization of security for the given entity/situation, understanding vulnerabilities to accidental and deliberate exposures, and use of risk mitigation strategies across critical infrastructure research areas. This systematic approach addresses the need to build collaborative global frameworks for critical infrastructure public policy that ensures pursuit of peaceful, collaborative, and yet potentially dual-use biological and life sciences research. The scientific community is charged with educating policy makers and non-scientists to enable informed decision-making.

The decision to operationalize day-to-day seamless sharing of science and technology resides with local, regional and national communities. During unstable and rapidly evolving disasters, reciprocity of information sharing from the field leads to a "tipping point" as rare events begin to repeatedly occur and attempts are made at analysis and pattern recognition. This area of computational dynamics and Big Data analysis offers many exciting opportunities for science and technology to contribution to real-world problem-solving.

Traditionally, many academic institutions are driven by the search for fundamental knowledge and applications-driven science is a new phenomenon for many researchers. As applied science research becomes increasingly more common, we must strive to streamline the processes, polices, and compliance requirements to enable true scientific 'breakthroughs' and contributions. Over the past 30-40 years, concerns about ethics, safety and security have begun to trump operational needs. Post 9/11, growth in regulatory requirements has constrained the academic research and innovation milieu in the United States. Despite protest from the academic community as a whole to these expanding regulations and controls, (American Association of Universities, Association of Public and Land Grant Universities, to name a few entities) regulatory requirements have continued to escalate and approval processes have become increasingly steeped in bureaucracy. Notably, research and development by terrorist organizations, unilateral terrorists, and state-sponsored terrorist organizations are not constrained in a similar manner. As a result, new asymmetries have begun to emerge that favor terrorism, and are amplified and accelerated by the unbridled science and technology investment being made in proliferating nations.

CONCLUSION: A SUGGESTED ROADMAP EMPLOYING THE FRAMEWORK OF ONE HEALTH-ONE SCIENCE-ONE SECURITY

In a world in which security and stability are often the most fleeting, yet desirable, end-states, the ability to create and sustain the seamless flow of validated information takes on a new dimension of urgency. When faced with the operational challenges of a hybrid threat, the ability to make efficient, scientifically sound decisions and share information within an appropriate security environment is paramount to mitigation, containment, and recovery. Unconventional disasters are characterized by uncertainty and rarity and create challenges and opportunities for decision making that may be enhanced through use of science and technology, and innovative analytic techniques.

A political-social-financially sound roadmap intended to prevent and mitigate acts of terrorism may be charted through information and knowledge integration in a system empowered through continuous feedback from the operational environment. In other words, connecting scientists to policy makers and the operational community describes an ideal feedback loop in complex emergencies. A notional, holistic One Science-One Health-One Security roadmap must include input from the scientific, operational, private sector and policy communities. Only through these types of partnerships may the global community addressing the most challenging issues of energy, water, food, and health security.

As governments and senior decision-makers begin to recognize the emerging power of social networks, so must these leaders strive for balanced information analysis that is science-based. Science cannot 'de-risk' decision making, rather, careful analysis may suggest a compelling plan for risk mitigation.

This paper has attempted to extrapolate from the lessons-learned in a real-world complex disaster and encourage the greatest scientific minds and innovators to engage. In conclusion, I have attempted to re-emphasize the need for continuous dialogue between the health, science, security and political leadership communities as we traverse the continuing threats of terrorism and high uncertainty of outcomes they pose.

ACKNOWLEDGEMENTS

I wish to thank Dr. Jim Wilson, colleague, and Dr. Gary Simpson, colleague and mentor, for their unfaltering efforts in One Health and their selfless work to help improve life for us all.

SESSION 8

BRAIN AGING & BEHAVIOR

BRAIN AGING: THREATS AND OPPORTUNITIES

MARY GANGULI
University of Pittsburgh School of Medicine, Division of Geriatrics and Neuropsychiatry, Pittsburgh, USA

Census data reveal that life expectancy is increasing across the planet, with steeper increases in the developing or low-and-middle-income (LMIC) countries than in the established economies. This increase reflects improved living standards, including factors such as nutrition, sanitation, and health care, especially prevention and cure of many acute infectious diseases. As a consequence, premature mortality has been reduced and lifespan has been greatly extended. Currently older adults thus have new opportunities for productive and satisfying lives that were not enjoyed by previous generations.

However, older adults today also face new threats from increasing chronic disease burden, coupled with the challenges of physiologic aging. Together, these factors conspire to limit the older person's ability to continue being productive and contributing to society. They also heighten the challenges to older people of functioning independently in their own environments, especially as traditional extended family support systems become fragmented. Thus, longevity is not universally celebrated, and old age is often regarded as a burden both by individuals and by society. Yet, since people today are living so much longer than their parents did, we have relatively little knowledge from previous generations to guide us in predicting what the future holds for an aging world.

Physiological aging is a gradual process and is not tightly correlated with chronological aging. Brain aging parallels the aging of the rest of the body. The brain itself shrinks in weight and volume, neural circuits become less "plastic," certain important brain chemicals (neurotransmitters) are produced in lesser quantities, and processes such as inflammation and oxidation take their toll. But the actual number of brain cells seem to be largely preserved in healthy older adults who do not have brain diseases (e.g., degenerative diseases like Alzheimer's or Parkinson's, or cerebrovascular disease with or without over strokes). Further, many old people remained cognitively intact throughout their lives and yet are found at death to have degenerative brain pathology that looks like Alzheimer's disease. This consistent finding suggests that the brain has some kind of reserve capacity which allows the individual to develop and engage in compensatory strategies to carry out life's functions, despite the pathology that may be present in his or her brain. One theory suggests that the adaptive brain engages in compensatory "scaffolding" to respond to the challenges posed by declining neural structures and function. Across the lifespan, scaffolding involves use and development of complementary, alternative neural circuits to achieve a particular cognitive goal, thus protecting cognitive function.

The most clear-cut changes that normal aging produces in cognition are a slowing in the speed with which we process information, an increase in reaction time, and a reduction in our ability to sustain attention. Generally, language functions like vocabulary remain well-preserved, whereas other functions, like the ability to acquire and retain new information, working memory, and inhibitory functions, tends to decline with age. By the time these changes become substantial, the demands of most older people's lives have also changed with regard to their need for speed, multitasking, and acquisition of new knowledge. They are thus able to continue functioning simply by pacing themselves.

Aging is not exclusively a negative process. Normal human development takes place across the lifespan, with different developmental tasks and challenges at different stages of life. Healthy older adults usually compensate for their diminishing abilities in some areas (e.g., in processing speed) because of enhanced abilities in others. For example, accumulated experience and knowledge over many years allows older individuals to be more adept at pattern recognition, and thus be better able to deal with uncertainty, than they may have been at earlier ages. Many older adults compensate in different ways for declining abilities, and maintain or even enhance their levels of social engagement. There is emerging evidence that cognitive functioning, and the "scaffolding" mechanism, are strengthened by both physical and mental exercise (cognitive engagement).

Not necessarily synchronous with cognitive aging is emotional aging, and here it appears that aging may confer advantages. Although personality traits remain stable with age, cross-cultural studies have demonstrated declines in neuroticism, extraversion, and openness with aging along with increases in agreeableness and conscientiousness. These attributes are consistent with our views of maturation. Factors that seem to predict diminished psychological well-being in later life are poor emotional regulation, failure to develop "wisdom," and inadequate self-protective processes. While "maturity" is not synonymous with chronological age, the characteristics of less mature and more mature thinkers have been compared as in the following table (*adapted from Cozolino, 2008*).

Less Mature Thinkers	**More Mature Thinkers**
• Need for certainty. • Need for control. • Inability to tolerate opposites. • Lack of openness to unconscious processes. • Belief that all the important information needed to make a decision is apparent. • Limited empathic ability. • Greater use of coping strategies of denial, blame, and projection. • Lesser use of coping strategies of humor, sublimation, and suppression.	• Ability to tolerate own ignorance • Lesser concern with being in control • Recognition of complexity • Ability to incorporate opposites • Understanding that not all important information is apparent. • Openness to unconscious processes • Tolerance for time needed to gather and process information. • Realistic expectations and forgiveness of others • Empathy and connectedness • Greater use of coping strategies of humor, sublimation, and suppression • Lesser use of coping strategies of denial, blame, and projection.

Some authorities in lifespan psychology distinguish between the "mechanics" and the "pragmatics" of accomplishing a given task. The mechanics are the physical infrastructure (in the present context, brain structure) that controls the ability to complete the task, and are usually superior at younger ages. The pragmatics are the sophisticated strategies acquired through effortful practice. Wisdom has been defined as an expert knowledge system concerning the fundamental pragmatics of life. (Baltes and Stoudinger, 2000). These are further categorized as factual knowledge about life, procedural knowledge about life, life-span contextualism, value relativism, awareness and management of uncertainty. These attributes appear to be facilitated by age and experience, in the absence of cognitive impairment. Efforts are now being made to study whether wisdom can be activated and transmitted between generations, e.g., by older adults in mentoring young children.

While wisdom cannot entirely compensate for the structural and physiological losses of aging, there are some key strategies that older experts use to maintain performance in the face of normal aging. These have been described as selection, optimization, and compensation. Selection refers to specialization, such that the older person selects certain areas in which to maintain expert performance while letting others slide. Optimization refers to pragmatics; the effortful practice of the selected, reduced, repertoire. Compensation refers to using one's strengths in key areas to compensate for weaknesses or losses in other areas.

One perspective on optimization includes the concept of "control" which has been categorized as primary and secondary. Primary control is the ability to manipulate the world around oneself to make things happen. Secondary control is the ability to manipulate one's own mind, i.e., to adapt psychologically to one's inability to manipulate the world. The motivation for primary control is described as being constant across the lifespan, while secondary control changes with age. Adapting to health problems, role changes, approaching mortality, etc., require appropriate active control strategies. Studies are now showing that the use of active control strategies buffers the effect of physical illness and elevated blood cortisol ("stress hormone") in older adults, resulting in reduced functional disability.

As lifespan increases across the world, we have expanding opportunities to understand the threats and opportunities associated with brain aging. The ultimate goal is to utilize this new understanding to improve the ways in which the aging human organism adapts to its changing internal and external environment.

Age and treachery will overcome youth and skill. (source unknown)

REFERENCES

1. Baltes, P.B., Staudinger, U.M., Lindenberger, U. (1999) "Lifespan psychology: theory and application to intellectual functioning." *Annual Review of Psychology*, 50: 471-507.
2. Cozolino, L. (2008) *The Healthy Aging Brain. Sustaining Attachment, Attaining Wisdom*. Norton, New York.
3. He, W., Muenchrath, M.N., Kowal, P. (2012) *Shades of Gray: A Cross-Country Study of Health and Well-Being of the Older Populations in SAGE Countries, 2007–2010. International Population Reports.* U.S. Census Bureau, www.census.gov. Issued May 2012.

4. Kinsella, K., He, W. (2009) *An Aging World 2008.* U.S. CENSUS BUREAU. *International Population Reports*. Issued June 2009.
5. McRae, R.R. et al. (1999) "Age differences in personality across the lifespan: Parallels in five cultures." *Developmental Psychology*, 35: 466-477.
6. Park, D.C., Reuter-Lorenz, P. (2009) "The adaptive brain; aging and neurocognitive scaffolding." *Annual Review of Psychology*, 60:173-96.
7. Salthouse, T. (2012) "Consequences of age-related cognitive decline." *Annual Review of Psychology*, 63:201-226.
8. Vaillant, G.E. (2002) *Aging Well.* New York, Little Brown.
9. Wrosch, C. Schulz, R. (2008) "Health-engagement control strategies and 2-year changes in older adults' physical health." *Psychological Science,* 19:537-541.
10. Wrosch, C., Miller, G.E., Schulz, R. (2009) "Cortisol secretion and functional disabilities in old age; importance of using adaptive control strategies." *Psychosomatic Medicine,* 71: 996-1003.

FRONTOTEMPORAL DEMENTIA (FTD): INSIGHTS INTO WISDOM AND ALTRUISM

BRUCE L. MILLER
Memory and Aging Center, University of California
San Francisco, USA

ABSTRACT

FTD attacks the frontoinsular brain circuits that mediate wisdom and altruism. By contrast, Alzheimer's disease (AD) attacks posterior brain regions that facilitate memory, navigation, reading and writing. These circuits function in a reciprocal fashion—turning on the frontal circuit turns off posterior brain regions, while turning on posterior circuits turns off frontal brain regions. The symptoms associated with FTD include profound deficits in behavior, social cognition, moral reasoning and empathy. Conversely, navigation, writing and drawing often remain normal or become enhanced in these same patients. Visual creativity is seen in many patients with left frontal or temporal degeneration. In AD, some individuals show enhanced kindness and generosity in the early stages of the illness. The phenomenology of these changes is described, and the aspects of brain aging that protect the future of humanity are noted.

BEHAVIORAL VARIANT FRONTOTEMPORAL DEMENTIA (BVFTD)

Patients with behavioral variant frontotemporal dementia (bvFTD) typically present with late-onset behavioral and personality changes that can be mistaken for psychiatric disorders.[1] The earliest behavioral changes include disinhibition, socially inappropriate behavior, poor impulse control, hyperorality, diminishing empathy and increasing apathy.[2] These behaviors create a personality that is socially disconnected from other people, shows little concern for any difficulties they cause others, has little insight to the situation and shows poor decision-making skills. Many of these patients first show up in a neurologist's office after shoplifting, losing large amounts of money, being swindled or other behaviors resulting from diminished executive functioning. Social wisdom has disappeared. The degeneration of these frontoinsular circuits has stripped the patient of his social cognitive abilities.

By contrast, memory, language and visuospatial skills are initially spared, and cognitive testing may look reasonably normal despite decreased functionality in daily life. Structurally, the imaging shows focal anterior cingulate cortex (ACC) and orbital frontoinsula (FI) degeneration, especially in the right hemisphere.[3] Disinhibition is strongly linked to atrophy in the right orbital-frontal region and apathy correlates with dysfunction in the right cingulate and medial frontal lobes.[4]

ALZHEIMER'S DISEASE (AD)

Typical Alzheimer's disease (AD) begins with memory problems, including impairment in the learning and retrieving of recently learned information.[5] Most AD patients maintain appropriate social behavior, maintain the expected eye contact with their loved one in an experimental laboratory environment and can become gentler and sweeter at disease onset.[6] These skills help them maintain relationships that help and support them, even if specific identities and the details of the relationship are lost. The personality of a typical AD patient presents as someone who is able to maintain an emotional connection and relationship with someone, even if they cannot tell you who the person is and why they know them. In AD, brain atrophy the posterior hippocampal-cingulo-temporal-parietal network involved in the recall of episodic memory.[7,8]

NORMAL AGING

Healthy old age brings the growth of emotional control, social flexibility, creativity, wisdom, optimism and resilience. Older adults become more adept at mitigating or avoiding negative emotions by focusing on positive stimuli and a positive interpretation of situations and suppressing reactions to the negative.[9,10] Over a 30-day evaluation, older individuals were more likely than younger participants to report never experiencing guilt, shame or hostility within the time period.[11] In essence, older adults create stable positive experiences and experience fewer and more transient negative experiences.[12] Older adults show a greater increase in prefrontal cortex (PFC) activity than younger adults when viewing emotional stimuli and may recruit a variety of PFC regions (ventromedial, dorsal and lateral) to compensate for any age-related loss in the area when faced with negative stimuli.[13]

Socially, older adults improve at recognizing emotions when tested with dynamic, naturalistic stimuli (i.e., emotionally laden, unrehearsed, interactions between real married couples).[14] This age-related increase in emotion recognition helps older people form stronger social networks for support and enjoyment, which in turn can help promote cognitive and physical wellbeing through lowered blood pressure, stress, and loneliness as well as mental engagement and stimulation. Older adults are also tend to forgive more readily and quickly,[15] which may further improve their relationships.

In support of the common sense that wisdom increases with age, a large, randomly selected sample of older adults showed an increased composite score for wisdom in studies of societal and interpersonal conflict, despite a concurrent age-related decline in Wechsler Adult Intelligence Score IQ.[16] These gains seem to occur mainly between 60 and 80 years old.

RECIPROCAL NETWORKS

Human brain activity occurs within spatially distinct, functionally related groups of cortical and subcortical regions.[17,18,19,20,21] These functional and structural networks can be imaged on fMRI and monitored for change due to activity or disease. Neurodegenerative diseases such as bvFTD and AD disrupt specific, separate networks—leading to the distinct clinical syndromes. AD affects the default mode network involved in episodic memory while FTD interrupts the emotional salience processing network.[22] The AD and FTD circuits also function in a reciprocal fashion.[23] Activity in the frontal circuit inhibits posterior brain regions, while turning on posterior circuits turns off frontal brain regions. This reciprocity provides some insight into the clinical strengths and deficits seen in patients with these diseases.

DISCUSSION

Neurodegenerative diseases like bvFTD and AD offer insights into how the brain functions due to the selective destruction of specific networks. We see skills diminished in isolation while other skills thrive, based on the connectivity patterns between neurons. These changes raise interesting questions about who we are and what is our capacity for self determination. They also foster compassion, with increased understanding of the patient's experience and perspective. These networks may be able to serve as early markers of disease and disease progression. Ultimately, these networks may be able to spread therapeutics along a vulnerable track and protect or restore function.

Popular culture today tends to idolize youth and fear, even demonize aging. Aging is seen as a process of steady destruction and demise. While yes, there are physical and cognitive challenges associated with aging, this view misses the joys and strengths of aging: improved emotional well-being, improved ability to resolve highly emotional problems like interpersonal conflicts, the accumulation of wisdom and experience, increased resilience to bounce back from difficulties, creativity, generosity and concerns for future generations. These skills are critical to the survival and evolution of human society. Many of our great artists, leaders and humanitarians barely got started until their golden years. Appreciating and finding the appropriate ways for elderly citizens to contribute to a community is a winning strategy for any society.

REFERENCES

1. Woolley, J.D., Khan, B.K., Murthy, N.K., Miller, B.L., Rankin, K.P. (2011) "The diagnostic challenge of psychiatric symptoms in neurodegenerative disease: Rates of and risk factors for prior psychiatric diagnosis in patients with early neurodegenerative disease." *J Clin Psychiatry*. 2011 Feb.72(2):126-33.
2. Rascovsky, K., Hodges, J.R., Knopman, D., Mendez, M.F., Kramer, J.H., Neuhaus, J., van Swieten, J.C., Seelaar, H., Dopper, E.G., Onyike, C.U., Hillis, A.E., Josephs, K.A., Boeve, B.F., Kertesz, A., Seeley, W.W., Rankin, K.P., Johnson, J.K., Gorno-Tempini, M.L., Rosen, H., Prioleau-Latham, C.E., Lee, A., Kipps, C.M., Lillo, P., Piguet, O., Rohrer, J.D., Rossor, M.N., Warren, J.D., Fox, N.C., Galasko, D., Salmon, D.P., Black, S.E., Mesulam, M., Weintraub, S., Dickerson, B.C., Diehl-Schmid, J., Pasquier, F., Deramecourt, V., Lebert, F., Pijnenburg, Y., Chow, T.W., Manes, F., Grafman, J., Cappa, S.F., Freedman, M., Grossman, M., Miller, B.L. (2011) "Sensitivity of revised diagnostic criteria for the behavioural variant of frontotemporal dementia." *Brain*. 2011 Sep.134 (Pt 9):2456-77.
3. Seeley, W.W., Allman, J.M., Carlin, D.A., Crawford, R.K., Macedo, M.N., Greicius, M.D., Dearmond, S.J., Miller, B.L. (2007) "Divergent social functioning in behavioral variant frontotemporal dementia and Alzheimer disease: reciprocal networks and neuronal evolution." *Alzheimer Dis Assoc Disord*. 2007 Oct-Dec. 21(4):S50-7.
4. Rosen, H.J., Allison, S.C., Schauer, G.F., Gorno-Tempini, M.L., Weiner, M.W., Miller, B.L. (2005) "Neuroanatomical correlates of behavioural disorders in dementia." *Brain*. 2005 Nov.128 (Pt 11):2612-25.

5. McKhann, G.M., Knopman, D.S., Chertkow, H., Hyman, B.T., Jack, C.R. Jr, Kawas, C.H., Klunk, W.E., Koroshetz, W.J., Manly, J.J., Mayeux, R., Mohs, R.C., Morris, J.C., Rossor, M.N., Scheltens, P., Carrillo, M.C., Thies, B., Weintraub, S., Phelps, C.H. (2011) "The diagnosis of dementia due to Alzheimer's disease: recommendations from the National Institute on Aging-Alzheimer's Association workgroups on diagnostic guidelines for Alzheimer's disease." *Alzheimers Dement.* 2011 May. 7(3):263-9.
6. Sturm, V.E., McCarthy, M.E., Yun, I., Madan, A., Yuan, J.W., Holley, S.R., Ascher, E.A., Boxer, A.L., Miller, B.L., Levenson, R.W. (2011) "Mutual gaze in Alzheimer's disease, frontotemporal and semantic dementia couples." *Soc Cogn Affect Neurosci.* 2011 Jun. 6(3):359-67.
7. Seeley, W.W., Allman, J.M., Carlin, D.A., Crawford, R.K., Macedo, M.N., Greicius, M.D., Dearmond, S.J., Miller, B.L. (2007) "Divergent social functioning in behavioral variant frontotemporal dementia and Alzheimer disease: reciprocal networks and neuronal evolution." *Alzheimer Dis Assoc Disord.* 2007 Oct-Dec. 21(4):S50-7.
8. Braak, H., Braak, E. (1995) "Staging of Alzheimer's disease-related neurofibrillary changes." *Neurobiol Aging.* 1995 May-Jun. 16(3):271-8; discussion 278-84.
9. Charles, S.T. (2010) "Strength and vulnerability integration: a model of emotional well-being across adulthood." *Psychol Bull.* 2010 Nov. 136(6):1068-91.
10. Mather, M. (2012) "The emotion paradox in the aging brain." *Ann N Y Acad Sci.* 2012 Mar.1251:33-49.
11. Hay, E.L., Diehl, M. (2011) "Emotion complexity and emotion regulation across adulthood." *Eur J Ageing.* 2011 Sep. 8(3):157-168.
12. Carstensen, L.L., Pasupathi, M., Mayr, U., Nesselroade, J.R. (2000) "Emotional experience in everyday life across the adult life span." *J Pers Soc Psychol.* 2000 Oct. 79(4):644-55.
13. Mather, M. (2012) "The emotion paradox in the aging brain." *Ann N Y Acad Sci.* 2012 Mar. 1251:33-49.
14. Sze, J.A., Goodkind, M.S., Gyurak, A., Levenson, R.W. (2012) "Aging and Emotion Recognition: Not Just a Losing Matter." *Psychol Aging.* 2012 Jul 23. [Epub ahead of print]
15. Cheng, S.T., Yim, Y.K. (2008) "Age differences in forgiveness: the role of future time perspective." *Psychol Aging.* 2008 Sep. 23(3):676-80.
16. Grossmann, I., Na, J., Varnum, M.E., Park, D.C., Kitayama, S., Nisbett, R.E. (2010) "Reasoning about social conflicts improves into old age." *Proc Natl Acad Sci U S A.* 2010 Apr 20. 107(16):7246-50.
17. Beckmann, C.F., DeLuca, M., Devlin, J.T., Smith, S.M. (2005) "Investigations into resting-state connectivity using independent component analysis." *Philos Trans R Soc Lond B Biol Sci.* 2005 May 29. 360(1457):1001-13.
18. Seeley, W.W., Menon, V., Schatzberg, A.F., Keller, J., Glove,r G.H., Kenna, H., Reiss, A.L., Greicius, M.D. (2007) "Dissociable intrinsic connectivity networks for salience processing and executive control." *J Neurosci.* 2007 Feb 28. 27(9):2349-56.

19. Vincent, J.L., Patel, G.H., Fox, M.D., Snyder, A.Z., Baker, J.T., Van Essen, D.C., Zempel, J.M., Snyder, L.H., Corbetta, M., Raichle, M.E. (2007) "Intrinsic functional architecture in the anaesthetized monkey brain." *Nature*. 2007 May 3. 447(7140):83-6.
20. Seeley, W.W., Crawford, R.K., Zhou, J., Miller, B.L., Greicius, M.D. (2009) "Neurodegenerative diseases target large-scale human brain networks." *Neuron*. 2009 Apr 16. 62(1):42-52.
21. Zhou, J., Gennatas, E.D., Kramer, J.H., Miller, B.L., Seeley, W.W. (2012) "Predicting regional neurodegeneration from the healthy brain functional connectome." *Neuron*. 2012 Mar 22. 73(6):1216-27.
22. Seeley, W.W., Crawford, R.K., Zhou, J., Miller, B.L., Greicius, M.D. (2009) "Neurodegenerative diseases target large-scale human brain networks." *Neuron*. 2009 Apr 16. 62(1):42-52.
23. Zhou, J., Greicius, M.D., Gennatas, E.D., Growdon, M.E., Jang, J.Y., Rabinovici, G.D., Kramer, J.H., Weiner, M., Miller, B.L., Seeley, W.W. (2010) "Divergent network connectivity changes in behavioural variant frontotemporal dementia and Alzheimer's disease." *Brain*. 2010 May. 133(Pt 5):1352-67.

SESSION 9

FRONTIERS IN FAST COMPUTING & INFORMATICS

COMPUTING FOR THE LARGE HADRON COLLIDER AT CERN

HELGE MEINHARD
CERN IT-PES
Geneva, Switzerland

INTRODUCTION

Since 2009, the four detectors at CERN's Large Hadron Collider (LHC), ALICE, ATLAS, CMS and LHCb, have been recording very high data rates that have put high demands in terms of storing, processing and distributing the data. Handling the data successfully has, next to the superb performances of the collider and the detectors, been a cornerstone in the analysis leading to the announcement, on July 4th, 2012, of a new bosonic particle with a mass of about 125 GeV/c^2 that is consistent with the standard model Higgs particle. All three elements: the accelerator, the detectors, and the computing, have been truly global efforts.

CERN

CERN stands for Organisation Européenne pour la Recherche Nucléaire, but is commonly referred to as the European Laboratory for Particle Physics. The laboratory is located near Geneva, Switzerland; its installations straddle across the Swiss-French border. It was founded in 1954 by 12 European States, and is hence older than what is now known as the European Union. Since then the number of member states has increased to 20, with a number of more countries in different stages of the membership accession process. Some 2,300 staff and 980 other paid personnel provide services to the 10,000 registered scientists from universities and research labs all around the world that use CERN's facilities. The annual budget of about 1 billion Swiss francs is covered by the member states according to their gross national product.

THE LARGE HADRON COLLIDER

The Large Hadron Collider (LHC) is a proton-proton collider installed in a tunnel of 27 km circumference. Protons have been accelerated up to 4 TeraelectronVolt (TeV) so far, about 4,000 times their rest mass; the design energy of 7 TeV per beam will be reached in 2015. A total of more than 1,000 superconducting dipole magnets produce the strong magnetic field of 8 Tesla necessary to keep the protons on their path, making the LHC the largest cryogenic installation world-wide. The machine can also be run in a mode where one or both proton beams are replaced by a beam of heavy ions (typically lead).

At 4 TeV beam energy, protons have been packed into 1400 bunches per beam of 100 billion protons each. CERN aims for doubling the number of bunches per beam in 2015. The instantaneous luminosity, a measure for the instantaneous collision rate, has exceeded $7 * 10^{33}$ cm^{-2} s^{-1} several times in 2012; the integrated luminosity in 2012 exceeded 23 fb^{-1}, more than four times the integrated luminosity achieved in 2011. This performance was key to the discovery of the Higgs-like particle in July 2012; the reason is that only one out of ten thousand billion events is relevant for the Higgs search.

THE EXPERIMENTS

At the four places around the circumference of the LHC, where the proton beams are brought into collision, large detectors have been built, the task of which is to record the traces of the particles emerging from the collisions with high precision. Two detectors, ATLAS and CMS, have been designed as multi-purpose detectors covering the widest possible range of physics questions, while LHCb and ALICE have each been optimised for specific domains of physics. ATLAS and CMS weigh about 10,000 tons each and feature about 150 million active sensor elements each, positioned with a spatial accuracy of down to a 100th of a millimetre close to the collision points. The detectors, like the LHC machine, require very significant effort; the ATLAS experiment, for example, has been built and is being exploited by almost 3,000 physicists from more than 150 institutes from more than 30 countries all around the world.

DATA PROCESSING REQUIREMENTS

The 150 million active cells result in data rates of the order of Petabytes (PB; 1 PB = 1,000 Terabytes = 1,000,000 Gigabytes = 1,000,000,000 Megabytes) per second. These data are reduced on-line in several steps in order to filter out noise as well as signals from collisions that have already been perfectly understood; the resulting data rate from the experiment for off-line processing is in the order of one or a few Gigabytes (GB) per second.

Even this reduced data volume per second comes from a large number of collisions, each of which contributes a few Megabytes—about the size of a digital photograph. These collision data are often referred to as “event” data. An important characteristic of these data is that one event is entirely independent from any other one; hence the data can be processed in a trivial parallel fashion. Single-core performance is usually sufficient; there is no need for the complexity of large SMP servers and/or fast interconnects between them.

Another important characteristic is that most of the calculations require integer rather than floating-point processing power. Typically, applications scale to 90% with integer-centric benchmarks.

With these two characteristics, data processing for particle physics is quite unique with respect to other compute-intensive sciences.

Typical steps of data processing include reconstructing events from the collision data (i.e., finding particle trajectories and properties from the signals of the sensor elements) and physics analysis (applying selections on event data and running statistical analysis over a large number of events). In order to compare with known models and find agreement or deviation, the known models form the basis for simulating collisions and detector responses; these simulated data are then running through reconstruction and analysis as well. There are a number of ancillary processes, for example for calibrating detectors for future reconstruction, or for verifying the data integrity and quality.

Initial expectations were that the data from the four LHC experiments (not taking into account any simulated data) would amount to 15 PB per year, corresponding to a tower of more than 20 km height of CDs each fully written with LHC data. In reality, the total volume amounted to 22 PB and 30 PB in 2011 and 2012, respectively. These very large data volumes require correspondingly large data processing facilities estimated at 250'000 processors and many tens of PBs of disk space.

THE WORLD-WIDE LHC COMPUTING GRID (WLCG)

For a variety of reasons, it would not have been feasible to implement a data processing facility of the size needed for the LHC data at CERN. Hence a distributed computing infrastructure with a multitude of data centres scattered all around the world was a natural choice; this infrastructure is known as the worldwide LHC computing grid (WLCG).

It is operated by a collaboration between the LHC experiments and the participating data centres. Functionally (but not managerially), it is organised in a hierarchical fashion, with the CERN data centre as the Tier-0, 11 major centres around the world acting as Tier-1 centres, and almost 150 Tier-2 centres. The functions are different per Tier; the Tier-0 looks after data recording, the initial reconstruction as well as the data distribution; the duty of the Tier-1 centres is to provide capacity for data-heavy analysis, reprocessing and permanent storage; the Tier-2 centres run simulation and end-user analysis. The centres are interconnected via wide-area high-speed links sized and configured appropriately so that all data traffic is handled by the networks. (For example, the links between the Tier-0 and each one of the Tier-1s are dedicated, permanent 10 Gigabit per second circuits.) While some of the Tiers, including the Tier-0, are supporting all four experiments, some others may be supporting a subset only. The rights and obligations from participating in this collaboration are described in a Memorandum of Understanding that has been signed by 49 parties (for most countries, multiple data centres are covered by one funding agency only that serves as signatory).

In order for the user not to worry about specificities of the data centre that happens to run his application, a software layer has been deployed that implements the concept of a computing grid (grid middleware). With this software layer, the entirety of the resources appears to the user like a large virtual compute and storage facility.

The WLCG collaboration's record of achievement is very positive. Already for the first physics results presented at conferences in summer 2010, data were used that had been taken by the experiments just days before; they had gone through the full cycle of Tier-0/Tier-1/Tier-2 in order to produce the results presented. At the end of 2010, the processing capacities of all Tiers were fully used. The processing capacity is now about twice what it was in summer 2010, and daily job counts of 1.5 million and more are normal. CERN as Tier-0 contributes about 20% to this capacity; about 35% are provided by the Tier-1 centres, and the Tier-2 centres are responsible of a total of about 45%. For this distributed data processing to work, very high rates of data transfers on the wide-area networks are required; aggregate transfer rates of about 10 Gigabytes per second are typical.

FUTURE OF WLCG

Despite this great success, active work is needed to ensure future sustainability of the system. The current setups, involving a large number of different grid services, are rather complex, and require significant effort to maintain them. At the same time, technology has moved on and is now offering options that did not exist when WLCG was designed initially. Experiments and sites are working together to plan for significantly simplifying the setup.

As part of the simplification, WLCG is moving from a strictly hierarchical networking model, where Tier-2s connect to their "parent" Tier-1 only, and the Tier-1s connect to the Tier-0 only, to a mesh where multiple connections between Tier-1 sites and Tier-2 sites are integrated into the model. This will allow for a more flexible location of data. Distributing grid workload to the individual centres is another area of active consolidation and simplification.

SOCIETAL IMPACT

Even though grid computing has had a major impact on data processing for particle physics, it has neither been invented for it, nor are its impacts limited to it. Other sciences such as archaeology, astronomy, astroparticle physics, civil protection, computational chemistry, earth sciences, nuclear fusion, geophysics, life sciences, material sciences and others have benefited from the increased data processing potential thanks to grid computing. Practical commercial applications in the area of finance and multimedia are known as well.

COMPUTING CHALLENGES FOR THE FUTURE

Grid computing is only one way of addressing large-scale computing problems. For a long time, large supercomputers have been on the market as well. More recently, clouds have emerged, and the role of volunteer computing (for example SETI@home, Einstein@home or LHC@home) has increased. These approaches are neither exclusive nor strictly competitive; they provide solutions to different problems. Hence a solid understanding of the problem is required in order to choose the right solution.

For particle physics, supercomputers are not very competitive due to their high price point and the fact that the applications can make only little use of the advantages of supercomputers, for example high-speed interconnects of processors. Volunteer computing is being used at a rather low level; due to its distributed nature with unpredictable network performance, it is particularly suited for applications requiring no or very little data transfer, but more significant processing resources, which is typical for simulations.

Grid computing and clouds are again addressing different problems. The former provides a computing service integrating distributed resources. The latter can be viewed as a large virtual remote computer centre. They are complementary and are expected to both have an important role for the future of WLCG.

More generally, the question emerges whether the world-wide growth of IT resources is sustainable. The demands for computing resources are growing exponentially, as seen by the growth of data centres both in size and in number. Their energy consumption has already well exceeded 2% of the world's total energy consumption. Demands, technologies and infrastructure are all evolving at very different paces, making matches over longer time periods very difficult. In the same sense, hardware, operating software and tools, and application software are all evolving at different pace, giving rise to a similar problem. Finally, in terms of computing architectures, the last 15…20 years have been rather stable—x86-architecture CPUs at the centre of systems with some RAM, disk space and I/O capacity have been a commodity solution well adapted to our needs for single-threaded performance, allowing particle physics to benefit from regular cost decreases. However, the future of this architecture is uncertain; the per-core performance growth has basically stopped, newer systems provide more (perhaps many more) cores than previously, which is difficult to exploit for particle physics applications. Adapting those to properly support multi-threaded execution environments requires very significant development effort. Lastly, next to the x86 architecture, there are other processor architectures, most notably ARM that is typically used in smartphones, tablets and similar devices. The popularity and growth of this platform has already resulted in small server solutions; it is currently unclear what the mainstream computer server platform will look like in a few years' time.

CONCLUSIONS

Computing for the LHC experiments has presented unprecedented challenges, which have been addressed very successfully by a distributed computing model. Despite the rapid pace of change in computing industry, a stable platform for the requirements of the LHC experiments has been provided, which was one of the cornerstones for discovering a Higgs-like particle at 125 GeV/c^2 as announced on July 4th, 2012. However, adapting the facility to the increasing requirements of the experiments and to the rapidly changing markets and technologies will remain a very high-profile challenge for the years and decennials ahead. Change will be the only constant that will always be there.

THE PLANETARY EMERGENCIES: A CHALLENGE FOR CHANGE

ANNAMARIA CANCELLIERI
Ministro degli Interni della Repubblica Italiana
Rome, Italy

I feel very honoured to be here today, at Erice, in such an outstanding research centre, in the presence of such a highly qualified assembly of renowned scientists, and I thank Professor Zichichi, to whom I'm tied by a long relationship of esteem and friendship, for allowing me to make some considerations at the end of today's working session.

The project of setting up a Study and Research Centre for Planetary Emergencies, which your Foundation has developed, is an initiative of great value not only from the techno-scientific perspective, but also and above all from a cultural point of view.

It is an initiative Professor Zichichi already talked me about last May, when we inaugurated, also thanks to his effort and resolve, the renovations works of the study centre devoted to Enrico Fermi, the place in which he made his important discoveries, and that is located within the Viminale, the present site of the Ministry of the Interior.

Therefore it is with great pleasure that today I see materialising the start of this new and stimulating endeavour and consider with particular sympathy and satisfaction the outlook that some of the activity of that Centre for the Study of Planetary Emergencies can have its site within the Fermi Centre, as a seal of a renewed alliance between the world of science and that of institutions and culture.

I think that the subject of planetary emergencies recalls apocalyptical scenarios for each of us, and when I say "us" I refer to all those who, like me, consider science and scientists as a separated and unattainable world.

I confess that I shiver when I learn, as I learnt today, that we have to face 63 planetary emergencies.

Yet, as Professor Zichichi said in his opening statement, if we lived in the Science Era, if scientific culture had really entered into the heart of modern culture, in practice such emergencies wouldn't exist.

Hence the theme of planetary emergencies is first of all the issue of recomposition of two worlds: the one of the so called humanistic culture as opposed to scientific culture. These two worlds have evolved separately for to many centuries, looking at each other sometimes with a sort of detachment, but more often with condescension or presumed mutual superiority.

According to common opinion and for most of us—non-experts—the word "science" immediately evokes the idea of precision, certainty, infallibility. "Scientific" methods, "scientific" demonstrations, "scientific" evidence are expressions that in our common language—or in the media language which often amplifies it—suggest the concept of an authority without weaknesses and appeals. A supreme, undisputable authority of last instance.

All this—which could appear as a sign of an extraordinary recognition and prestige—is indeed, in my humble opinion, the mark of an essentially anti-scientific mentality. Even, perhaps, the mark of a magical concept of science, its methods and its results.

We look with admiration and naivety the images of exulting scientists at the CERN in Geneva, and hear obscure words such as "boson", "quantum", Standard Model"; we look with admiration the images of the NASA explorer landing on Mars to tell us something about life and its origins. And we are happy—as Italians—and indeed proud of learning that hundreds of Italian researchers, under the guide of an eminent physician of our country, made a fundamental contribution to the experiment in Geneva; We are happy, and indeed proud, to learn that a chip designed and developed in our country contributes to the functioning of computers on board of the explorer named *CURIOSITY*.

This news makes us happy, gives us relief and makes us reflect. Yet as long as we go on considering science as the place of acquired certainties and indisputable authorities, I think we will show—besides our ignorance—our essentially anti-scientific mentality.

Only when we learn that science is the reign of hypotheses and counter-hypotheses, of conjectures and confutations, of reiterated attempts and reflexive control exerted by the intellectual community over itself, with patience and honesty, with neutrality and far-sightedness, with courage and sacrifice; when we finally learn that science is not the reign of certainty but the reign of doubt, not the reign of infallibility, but the reign of essential fallibility, not the reign of indisputable authority, but of the permanent criticism and self-criticism, perhaps only then we, even non-experts, will be able to understand something about science.

And in our country—as we all know very well—it so easy to slip into an anti-scientific mentality, and the irrational veneration for science is only its specular and symmetric other side. Italy, which is a young country, has not only experienced a prolonged technological backwardness, but has also suffered the supremacy of prejudices who led generations of students, scholars and intellectuals to experience what Gramsci, a social thinker and not a scientist, had considered as a permanent "scientific denutrition". We know these kind of prejudices, both religious and lay: we know how our Illuminists and Positivists were fragile and marginal; we know that Croce's idealism, as well as the school reform elaborated by Gentile have established the subaltern character of techno-scientific disciplines, confining the science role into a merely subsidiary and ancillary knowledge.

In our country, still nowadays, the fact of ignoring Dante or Leopardi is considered a symptom of unculture (and it is rightly so!). On the contrary, it is considered an unimportant sin to ignore mathematics, chemistry or physics, and knowing nothing about Einstein or Bohr. I told a "venial sin": but for somebody—we cannot deny it—is even a cause for haughtiness and a kind of snobbism. Of course, things are changing, even though slowly. Nowadays techno-scientific schools are no more considered as a secondary choice as compared to the supreme "grammar school". Today the level of the research carried out in our universities is no lower than the one of other countries, both European and worldwide, despite of painful government financial cuts and due to an economic policy that does not invest in research and development as much as in other states. Today, at least, science is no longer subordinate to humanities, even if students enrolled in scientific courses—according to statistical data—are far fewer than those enrolled into humanities courses or in professionalizing courses. Something is changing, even though slowly. Yet, if it seems that the "two-cultures" pendulum, as Snow called it, is now swaying towards the so called "hard" or "exact" sciences, it is not granted that this can really help us overcome our prejudices and our harmful anti-scientific heritage; it is not granted this help us working out a correct idea of "science" and drawing from it the appropriate legislative and financial consequences.

Now I would like to repeat the words used by Ivano Dionigi, the Rector of Bologna University, and an expert in the field of history of ideas about the Past-Present comparison:

"The new ways of science (Chemistry, Physics, Genetics, Information Science, Biotechnologies labs) announce us a radical transformation of life and environment: a sort of change of the concept itself of man and his destiny in the world. In the presence of this epistemological and cultural turn it is far too easy to foresee the resurgence of ancient *querelles,* with the risk of division between anti-modernists (who distrust science due to its obnoxious applications and fearful of future scenarios, equally disquieting) and techno-supporters (who on the contrary see the realm of machines as a new and unlimited golden age). In the presence of these extreme attitudes, i.e., obscurantist fear and scientist fideism, it is worth recalling some basic principles: that science is essential to the knowledge and perception of the world, to technical and industrial progress and change, and to freedom and independence of a nation; that scientific research, aiming at tomorrow and at the "whys" of *natura rerum* is completely different from its applications aimed at immediate objectives and practical advantages depending on economics and market; that scientific research—which by nature is experimental, anti-dogmatic and fallible-continuously questions its aims, results and methods. We know well that science is neither the supreme value nor asks itself about the ultimate questions; and that beside the "technical knowledge" subject to be worn out, aiming at external progress, there is the "civil knowledge", which is not subject to be worn out, and aims at internal progress. From this point of view, not only scientists—as Snow assumed—but also humanists have "the future in their blood".[1]

I think that this invitation should be accepted. In a world which science, with difficulty but unstoppably, is revolutionizing, it is our duty as citizens and men:

1. To know and to know how to teach that science is not acquired certainty, but continuously evolving knowledge;
2. To demand that, exactly for these reasons, science be always free, not conditioned by conservatism constraints or economy imperatives;
3. To be convinced that science is not an enemy but an ally of the traditional humanistic knowledge, and that right now- in the face of dilemmas always derived from scientific progress and in the presence of the extraordinary opportunities they provide- it is more necessary than ever to achieve a co-operation between scientific knowledge and historical, ethical, legal knowledge. And it is also necessary to be convinced that any opposition centred on the two cultures model is negative for both of them: to the humanists as well as to scientists. *If scientists- thanks to their discoveries- have the fundamental and unique task of giving the right answers to the present challenges (and you have gathered here for this purpose), humanists—thanks to the lessons of history—have the equally fundamental*

[1] Ivano Dionigi, *I classici e la scienza. Gli antichi, i moderni, noi*, Milano, BUR-Rizzoli, 2007, p. 14.

> *and unique task of asking the right questions. The paradigm of obliviousness of the first ones must be balanced with the paradigm of the memory of the latter.*

Above all in a country like Italy, which is so rooted in its past and so courageously oriented, thanks to the excellence of its research and despite the current difficulties, towards the future.

We do hope—also in line with the recent warning by the President of the Republic—that the present government as well as the future government will manage to find sufficient resources in the field of innovation, research and training; moreover, we do hope that Italy will invest both in the preservation of the traditional knowledge, to which it owes so much of its grandeur and glory, and in the development of the new branches of knowledge, forming the changing and lively panorama of "science".

We also hope that one day ignoring both Dante and Einstein will be considered not a venial sin, and nonetheless a cause for haughtiness, but a grave fault.

The science is a challenge to the thought, a cultural conquest, an economic advantage; but, do not forget it, also a constitutional obligation (art. 9 "the Republic promotes the development of culture and the scientific and technical research) and, I would add, a deeply ethical need. As a master of classic wisdom taught us: "What is good?" asked Seneca and answered "Knowledge". "And what is evil? Ignorance". I repeat it in Latin, which is even more understandable: "Quid est ergo bonum?rerum scientia. Quid malum est? rerum imperitia" (letter 31, 6).

SESSION 10

INFORMATION SECURITY

REGULATING CYBER BEHAVIOR: SOME INITIAL REFLECTIONS ON CODES OF CONDUCT AND CONFIDENCE-BUILDING MEASURES

HENNING WEGENER
Ambassador of Germany (Ret.), Information Security Permanent Monitoring Panel World Federation of Scientists, Madrid, Spain

This Permanent Monitoring Panel has from its inception geared its work to the overriding issues of harnessing cyber conflict, and to the organizational challenge of creating a universal order of cyberspace. We were convinced that the cyber sphere is, if not a lawless space, at least one that lacks a comprehensive legal framework to manage and control the all-pervasive, infinite potential of digital technologies. Consequently, there is little or no ability to effectively control the escalation of cyber conflict, and there is no common understanding of how the existing norms of international law would apply —if at all. A dangerous, precarious state of affairs, which our group has never ceased to analyse and denounce. From the beginning of our activity, we have argued that a comprehensive international legal framework is required, to fill this void, establishing norms and rules for adequate, responsible state behaviour in order to ensure and guarantee the peaceful use of cyber space. Our first major document, "Toward a Universal Order of Cyberspace: Managing Threats from Cybercrime to Cyberwar"[1] reflects that approach. At that time, we ambitiously called for an UN-led effort to create a Comprehensive Law of Cyberspace. Our erstwhile colleague Ambassador Kamal has later fleshed out this project in a book "The Law of Cyber-Space: An Invitation to the Table of Negotiations".[2] Yet, we were not the first ones to embrace such a complex endeavor. At the United Nations, Russia has as of 1998, in a series of UN draft resolutions, advocated a Cyber Treaty, proposing detailed, although conflictual and probably unimplementable contents.[3] These draft resolutions became an annual exercise in frustration: the Russian initiative was for many years rejected by a number of Western countries, but yet had the undoubted merit of keeping the argument alive that a major normative effort was required. ITU Secretary General Hamadoun Touré has early on, and then repeatedly, called for a "Global Cyber Treaty",[4] laying out some essential provisions. He has thus added an authoritative UN voice to the quest for a normative framework. Many other advocates for comprehensive treaty-making could be cited.[5]

But however attractive the concept of a comprehensive Convention on Cyber Space—comparable to the UN Convention on the Law of the Sea of 1982—the obstacles

---

[1] Document WSIS-03/GENEVA/CONTR/6-E, www.itu.int/dms_pub/itu-s/md/.../S03-WSIS-C-0006!! PDF-E.pdf, also at www.unibw.de/infosecur.
[2] UNITAR, Geneva 2005, www.in.int/kamal/the_law_of_cyber_space.
[3] Res. A/53/70, annually up to A/65/41.
[4] "UN Chief proposes int'l accord to prevent cyber war," 31 Jan. 2010, www.thepoc.net/breaking-news/ world/ 3930-un-chief-proposes-intl-a.
[5] The Federation of Russia has, in 2011, developed a new draft for a comprehensive treaty, although apparently it has not been officially submitted to the UN (http://www.mid.ru/bdomp/ns-osndoc.nsf/ e2f289bea62097f9c325787a0034c255/6912ce36aa5f1e92c32579250035bebd!OpenDocument).

to such an instrument and its creation were increasingly recognized as overwhelming. Cyberspace might be even more complex than the ocean world. Digital technologies and their uses are still evolving at a rapid pace. Universal treaty-making would be beset by still bigger cleavages in individual nations' views. Treaty negotiation would be a lengthy process, and national ratification procedures would not proceed on a time scale even marginally in keeping with the urgency to fill the legal void and the growing, shared perception among governments that the threat of cyberconflict and unmanageable cyber damage is escalating out of control. Binding commitments to avoid attack or hostile actions and corresponding sanctions would be unworkable. Important definitional issues would probably be unresolvable in treaty language. Thus, while a Universal Treaty/Law on Cyberspace remains a preferred objective, a target concept, practical reasons require an alternative approach.

If a one-shot cyber treaty thus is not a realistic option, or is not timely—or if early concentration on the treaty idea was perhaps even a wrong start—neither does a simple extension by analogy of the laws of armed conflict or humanitarian international law into cyberspace do the trick. There are areas of ambiguity, including the scale and nature of damage from cyber attack that could qualify as the use of force which, in a rapidly evolving digital environment, would defy treaty-making or treaty analogies.

Fortunately, collective thinking about the necessary processes of cyber strategy has notably evolved. To make a long story short, a new age of cyber diplomacy has begun around 2008 with a manifestly emerging international consensus to concentrate efforts on an alternative to formal treaty-making: the elaboration of confidence-building measures or codes of conduct as normative tools. We may be witnessing a turning point in cybersecurity diplomacy.

Both of them, CBMs and codes, are time-honored techniques of international diplomacy. Confidence-building measures have been a major focus in reshaping the East-West relationship in the CSCE/OSCE and in military affairs (taking their methodological guidance in particular from the Guidelines for Confidence-Building Measures adopted by the UN General Assembly in 1986[6]), and codes of conduct are constructed in many other areas of the social environment including the private sector. In the UN, there have been negotiations on a code of conduct for the transfer of technology, and a code of conduct of behaviour in Outer Space is presently in process.[7] In our context, the two methods are frequently mentioned together, and indeed, they may be somewhat interchangeable: both aim at regulating behaviour; CBMs look more to the intended result, the term code of conduct reflects the notion that an ample, coherent body of norms is being negotiated.

The prevailing view is now that CBMs open a window of opportunity to make real progress towards common definitions and behavioural standards. CBMs have the potential to reduce threat, enhance transparency, make State behaviour predictable, are flexible, voluntary, and offer a variable geometry in terms of participants—it is possible to include non-State actors—and follow-up: contrary to coherent treaty-making, participants are free to adopt partial solutions and enact them without delay and independently or with other like-minded stake-holders. CBMs which States embrace do

[6] Reprinted in UN document A/S-15/3.
[7] See also the Montreux Document on Operations of Private Military and Security Companies During Armed Conflict, www.icrc.org.

not require ratification; they invite emulation, and are at most—and at best—politically binding. They are thus uniquely suited to foment international consensus-building on an evolutionary scale. A well negotiated package of CBMs with a critical mass of participants may set in motion a process of further incremental change and heightened sensitivity. Clarification of behavioural standards may provide an incentive for going for more.

One catalyst for this new campaign may well have been the terse report of the second UN Group of Governmental Experts 2010[8] (the first having ended inconclusively) recommending further international discussion on norms, confidence-building, stability and risk reduction measures. A natural result was the convocation of a third GGE, this time with the concrete mandate to define "cooperative measures including norms, rules of principles of responsible behaviour of States and confidence-building measures with regard to information space".[9] The Group will convene in August, coinciding in time with this Seminar, under the presidency of an Australian diplomat. Governments have provided numerous inputs to the Group at the request of the UN Secretary General.[10] Their views have strongly supported the idea of identifying CBMs. In a short time, flurries of other national statements to the same effect have surfaced: Australia, the UK, Germany, at least by implication the U.S., among others.[11] An authoritative academic voice from India has joined the concert.[12] China, Russia, Tajikistan and Uzbekistan, reflecting work within the Shanghai Cooperation Council, submitted to the UN Secretary General, in September 2011, a draft international code of conduct for information security[13]. Although the document, by virtue of its authors, did not seem to display a sufficient flavour of political correctness, the catalogue of commitments, offered to voluntary subscription by States, should not be disdained. In the meantime, member countries have organized prestigious international conferences, where the CBM idea has been ventilated, and more or less detailed catalogues of CBM contents have figured in the Conference summaries (London,[14] Berlin,[15] Beijing, Vienna,[16] with other conferences forthcoming in Budapest in October 2012 and South Korea in 2013). Apart from the on-going UN exercise, regional organizations are also getting into the act. The OSCE is intensively working on a draft code of conduct (see "A Comprehensive Approach to Cyber Security", and the Astana Declaration of 2010) and APEC, as well as the aforementioned Shanghai Council are also working on regional arrangements. The Council of Europe, famous for its contribution to a world penal law on cyber crime

[8] A/65/201.
[9] A/Res/66/24 of 13 December 2011.
[10] A/66/152 and A/66/152 Add.1.
[11] See previous footnote. For Germany, see also "Challenges in Cyber Security: Risks, Strategies and Conference Building, Conference Report, December 13 and 14, 2011, Berlin, http://www.auswaertiges-amt.de/DE/Aussenpolitik/Friedenspolitik/Abruestung/Projekte/Cybersicherheit.html. The German Federal Foreign Office, in addition, supports a 2012 UNIDIR project on International Cybersecurity and CBMs
[12] Arvind Gupta, CBMs in Cyber Space: What Should Be India's Approach?, IDSA, Institute for Defence Studies and Analysis, June 27, 2012.
[13] A/66/359. See also the Agreement between the Governments of the Member States of the Shanghai Cooperation Organization on Cooperation in the Field of International Information Security, signed in Yekaterinburg on 15 June 2009.
[14]See www.fco.gov.uk/en/global-issues/london-conference-cyberspace/.
[15] See footnote 9.
[16] An OSCE Conference, May 9 and 10 2012.

through the Convention on Cybercrime, has adopted 10 Principles on Internet Governance.[17] UNIDIR helps to provide the academic underpinning for these endeavours.[18] NGOs in the cyber area, as the ICT4Peace,[19] or individual researchers like the Estonian scientist Eneken Tikk,[20] provide catalogues of rules of conduct of their own. These catalogues can obviously not be reproduced or analysed here—I will revert to the task waiting for us there—but provide effective tools for spurring the debate and facilitating CBM negotiations. I hardly need to remind you that our own Erice Declaration on Principles for Cyber Stability and Cyber Peace of August 2009[21] also urges a common code for cyber conduct.

Just to give the audience the flavour of where codes of conduct may lead us, I will provide subsequently an indicative list of important provisos which CBMs or a code of conduct negotiation should consider. The list comprises the tenets I personally have been working with, but also reflect the principles Secretary Hamadoun Touré and the ITU consider essential, and some ideas from other catalogues. The list is not only indicative, but also very general, as CBM negotiations, other than treaty-making, only lend themselves to very basic commitments.

The list would read as follows:

- Binding agreement on the fundamental principle that a cyber attack against another State, direct or through hired perpetrators, constitutes a violation of international law.
- Commitment of all States not to practice first use of cyber weapons against another State, as long it has not already undergone an attack by conventional weapons.
- States subscribe, nationally and internationally, to a policy of prevention of cyber conflicts, placing the emphasis on cyber defence, and will therefore, jointly with the private sector, secure their systems and networks through maximum robustness and potential of resilience, built-in redundancies, net segmentation, "cyber hygiene", efficient management, and similar measures. States will to that effect collaborate in international standard-setting.
- In case of a cyber attack or grave disturbance of their net structures in process, States will aim at the restoration of intact networks and of stable peaceful communication structures at the earliest point in time.
- States are responsible for the protection of critical infrastructures on their national territory, particularly of those ensuring energy availability, banking services, public health, and installations serving other humanitarian purposes, and to ensure the inviolability of transnational digital net structures. Attacks on these structures are prohibited.

---

[17] www.coe.int/media, Press Release DCO/79(2011).
[18] UNIDIR organizes conferences and participates in others. See also the articles in DISARMAMENT FORUM.
[19] http://ict4peace.org.
[20] "Ten Rules for Cyber Security", SURVIVAL vol.53 no.3, June-July 2011.
[21] Reprinted in "The Quest for Cyber Peace" www.itu.int/S-GEN/WFS,01-2011-PDF-E, , p.110.

- States are obliged, in the interest of a uniform or harmonized international protective legal framework, to introduce comprehensive penal legislation for the persecution of cyber crime—be it by adhering to the Convention of Cybercrime of the Council of Europe, or by adopting comparably comprehensive arrangements—and to ensure the efficiency of law enforcement and the requisite international cooperation for that purpose.
- States have the duty to protect their national citizens in cyberspace.
- States must not tolerate cyber terrorists or cyber criminals operating in or from their territory without due punishment. States are obliged, within available technological means (for example, through deep packet inspection of the principal fibre circuits of their ISPs) to impede and reveal the use of their territory by cyber criminals.
- The use of botnets and other irregular cyber war practices is forbidden, and States are obliged to implement this prohibition on their national territory.
- Neutrality continues to be valid in the cyber age, and cyber attacks must not be perpetrated through the net structures of neutral States.
- States must commit themselves to mutual assistance in the reconnaissance and detection of cyber crimes, particularly if the attack has issued from their territory. States are to participate fully in the existing information and early warning systems, by adhering to the agreements on 24/7 contact points and international CERT networks, among others.
- States are invited to conclude, in addition to multilateral agreements, bilateral accords containing reciprocal non-aggression commitments, arrangements for the joint defence against cyber attacks, and mutual assistance in case of damage inflicted. Unilateral declarations aimed at renouncing any cyber activity that might cause large-scale harm to civilian critical infrastructures are also desirable.
- States will ensure an open Internet that allows individuals access to content and services with only such restrictions as are permitted under international legal obligations, while protecting users against abuse, especially children. Government restrictions on the freedom of the Internet will be duly incorporated into law and subject to independent judicial review.

That such catalogues of suggestions for future norm-making now appear realistic is certainly a welcome development. Yet, the road ahead is far from clear or without obstacles. Current activities in the field of cyber diplomacy take place in an environment of prevailing deep distrust, high secrecy especially as regards military cyber applications, wide disparities of views among States, insufficient research on important regulatory issues, and lack of a common vision of how a future order of cyberspace would look: most of the suggested lists are unsorted, with unclear overlaps. Some crucial issues may not lend themselves to a CBM negotiation on broad principles at all: many definitions, detailed rule-making on what constitutes "armed attack" or what responses would be justified, what rules of engagement should prevail, etc. would probably need long-drawn supplementary negotiations, to be outsourced to specialized bodies. (NATO and some governments like the U.S. are already working on these issues). Important as general principles are, this may well prove to be the most difficult part.

Then there is the general question of venue. Where should negotiations on CBMs or codes be located? (I will revert to that issue). What are the ground rules to be? Will participation be inclusive, allowing stakeholders other than States, to join? If there are to be several parallel negotiations, for instance by regional international organizations—the OSZE, APEC, the Shanghai group, NATO—how can the results be synthesized to generate uniform rules and a combined impact? As no one universal draft treaty is aimed at, and involvement is voluntary, participants should agree that each one has the right to pick individual measures or code elements and enact them early and separately; there must not necessarily be packages. Could there, then, be a sequence of stages? What would be the time frame?

What we see at the present juncture is no more than work in progress. As it shapes up, problems of structure and content will constantly emerge. They need to be identified and researched. This is where contributions from outside, including the academic community and dedicated NGOs need to come in to accompany and help shape the process. I see a definite role for our PMP. In our work we can contribute to a stock-taking of the multiple efforts presently under way, make an independent assessment of the ideas proffered by Governments and explore their consensus potential, It may sound ambitious, but this may be the opportune moment for a cool, objective assessment of each stage of the evolving process, and for helping to put matters into a coherent and clear focus. A group as ours can help to prioritize the various planks of the negotiation, to make for more transparency in the underlying doctrines, and to press toward better mechanisms of international cooperation and crisis management. We may also be able to make a well-reflected input into the pending issues of developing international law for the requirements of the cyber age, as we have already begun to do in several of our documents. In short, a creative input of the PMP could be a very apposite contribution in the furtherance of our concept of cyberpeace.

An issue, already touched upon and needing penetrating thought, and also discussion within our Group, is the identification of an optimum venue to host the (hopefully) forthcoming negotiations on CBMs or Codes of Conduct. I do not want to preempt that debate, but from the universal perspective of our PMP, a single forum offering such universal perspective would certainly be the most adequate. On the other hand, preliminary regional endeavours are already under way, and their dynamics should be used. The OSCE, although it has no recent treaty-negotiating record, has the advantage of comprising States on the East-West axis, some of which have distinct cybersecurity problems. Confidence is particularly needed in such environment. If a regional approach prevails, some coordinating mechanism should be developed to avoid contrasting or contradictory standard-setting. If one wishes to avoid the heavy UN machinery, an autonomous Conference of States, perhaps serviced by the UN, would be a good bet. I have never hidden my preference for the ITU as the principal venue and coordinating instance in the cybersecurity domain.[22] But in other contexts I have also thrown out the idea of entrusting the Conference on Disarmament with its perennial negotiating experience in the security domain. And has anybody probed the suitability of the Council of Europe? The political implications and acceptance potential of any of these options

[22] Henning Wegener, Harnessing the perils in cyberspace: who is in charge? UNIDIR, DISARMAMENT FORUM 3/2007.

have to be weighed carefully, and experts on international machinery are invited to provide their input.

CYBER GOVERNANCE FOR CYBER CONFIDENCE*

JODY R. WESTBY[1]
Global Cyber Risk LLC, Information Security PMP, World Federation of Scientists, Washington, USA

DEFINITION OF CYBER GOVERNANCE[2]

A well-accepted definition of *enterprise governance* is set forth by the International Federation of Accountants (IFAC) and the Information Systems Audit and Control Association (ISACA):

> Enterprise governance is the set of responsibilities and practices exercised by the board and executive management with the goal of providing strategic direction, ensuring that objectives are achieved, ascertaining that risks are managed appropriately and verifying that the organization's resources are used responsibly.[3]

The U.S. Business Roundtable, comprised of America's leading officers and directors, has determined that effective enterprise governance includes:

- Selecting a chief executive and overseeing the CEO and senior management in the ethical operation of the corporation
- Setting the managerial tone of the organization to create culture of legal compliance and integrity and overseeing the ethical and effective operation of the corporation.
- Determining the strategic direction and appropriate risk profile of the organization and overseeing the management of risks
- Producing fairly representative financial statements and providing timely information that investors need regarding the business operations, risk management, and financial condition of the corporation.

---

[1] Jody R. Westby is CEO of Global Cyber Risk LLC, based in Washington, DC and serves as Adjunct Professor to the Georgia Institute of Technology School of Computer Science and Distinguished Fellow to Carnegie Mellon CyLab. She is co-chair of the World Federation of Scientists' Permanent Monitoring Panel on Information Security and chairs the American Bar Association's Privacy & Computer Crime Committee. She led the development of the *International Toolkit for Cybercrime Legislation* and served as a member of the ITU Secretary-General's High Level Experts Group on cyber security. She has co-authored and edited four books on privacy, security, cybercrime, and enterprise security programs, as well as the UN publication, *The Quest for Cyber Peace*.

[2] This section of the paper is adapted from: Jody R. Westby and Julia H. Allen, Governing for Enterprise Security Implementation Guide, Carnegie Mellon University, Software Engineering Institute, 2007, www.cert.org/archive/pdf/07tn020.pdf.

[3] *Enterprise Governance: Getting the Balance Right,* Int'l Federation of Accountants, Professional Accountants in Business Committee, 2004, http://www.ifac.org/publications-resources/enterprise-governance-getting-balance-right.

- Shaping the governance and leadership of the organization and assessing the background, skills, and expertise of the board
- Producing financial statements that fairly present the conditions and results of operations and making timely disclosures
- Aligning compensation with performance and incentivizing long-term gain for shareholders without excessive risk
- Engaging with shareholders on issues and concerns and exemplifying the highest standards of corporate citizenship.[4]

Governance extends to the management of an organization's use of information technology (IT). The IT Governance Institute declares that:

> *IT governance* is the responsibility of the board of directors and executive management. It is an integral part of enterprise governance and consists of the leadership and organizational structures and processes that ensure that the organization's IT sustains and extends the organization's strategies and objectives.[5]

In its publication, *Information Security Handbook: A Guide for Managers*, the U.S. National Institute of Standards and Technology (NIST) extends this definition for *information security governance* as follows:

> Information security governance can be defined as the process of establishing and maintaining a framework and supporting management structure and processes to provide assurance that information security strategies are aligned with and support business objectives, are consistent with applicable laws and regulations through adherence to policies and internal controls, and provide assignment of responsibility, all in an effort to manage risk.[6]

Although implicit in all of these definitions, cyber risk management involves three layers of consideration: the management of risks to networks, applications and operating platforms, and data. If adequate attention is not given to all three areas, it is impossible to manage the cyber risks associated with business operations and processes and compliance requirements.[7]

---

[4] "Principles of Corporate Governance 2012," Business Roundtable, 2012, http://www.scribd.com/doc/98420021/BRT-Principles-of-Corporate-Governance-2012.

[5] *Board Briefing on IT Governance*, IT Governance Institute, 2nd Ed., 2003, http://www.isaca.org/Knowledge-Center/Research/ResearchDeliverables/Pages/Board-Briefing-on-IT-Governance-2nd-Edition.aspx.

[6] Pauline Bowen, Joan Hash, and Mark Wilson, *Information Security Handbook: A Guide for Managers*, National Institute of Standards and Technology, NIST Special Publication 800-100, 2006, http://csrc.nist.gov/publications/nistpubs/800-100/SP800-100-Mar07-2007.pdf.

[7] Jody R. Westby and Julia H. Allen, Governing for Enterprise Security Implementation Guide, Carnegie Mellon University, Software Engineering Institute, 2007, www.cert.org/archive/pdf/07tn020.pdf.

FIDUCIARY DUTY OF DIRECTORS AND OFFICERS[8]

It has long been recognized that corporate directors and officers (D&Os) have a fiduciary duty to protect the assets of their organizations.[9] Today, this duty extends to "digital assets"—information, applications, and networks. This duty has been expanded by the enactment of state and federal laws and regulations that impose specific privacy and security requirements on targeted industry sectors and types of data.

The Sarbanes-Oxley Act requires both management and external auditors to attest to the effectiveness of internal controls that provide meaningful assurance about the security of information assets.[10] In late 2011, the Securities and Exchange Commission (SEC) issued guidelines that require public companies to disclose the risk of cyber incidents if they materially affect a registrant's products, services, relationships with customers or suppliers, or competitive conditions, or if they make an investment in the company speculative or risky.[11]

The pressure on critical infrastructure industry sectors to secure their systems according to best practices and standards persists, with the U.S. energy sector already subject to regulations.[12] Today, the tone in Washington has moved from persuasive to compulsory, with numerous bills pending in Congress that mandate security measures for corporate systems if enacted.[13]

In addition, the reputational and financial consequences of a breach can be significant. When a company is the victim of an attack on its information systems—whether from an insider or an outside bad actor—studies have shown that this can result in a lack of confidence in the company and even a drop in the company stock price.[14]

---

[8] This section of the paper from: Jody R. Westby, Governance of Enterprise Security: CyLab 2012 Report, Carnegie Mellon CyLab, May 16, 2012 at 11-12, www.rsa.com/innovation/docs/CMU-GOVERNANCE-RPT-2012-FINAL.pdf (hereinafter "2012 CyLab Governance Report").

[9] *See* Jody R. Westby, Testimony before the House Committee on Government Reform, Subcommittee on Technology, Information Policy, Intergovernmental Relations and the Census, Sept. 22, 2004, http://www.cccure.org/Documents/Governance/westby1.pdf . For a discussion regarding the fiduciary duty of boards and officers and the extension of that duty to protect the digital assets of their organizations, *see* Jody R. Westby, ed., *International Guide to Cyber Security,* American Bar Assn., Privacy & Computer Crime Committee, 2004 at 189-93.

[10] Sarbanes-Oxley Act of 2002, Pub. Law 107-204, Sections 302, 404, http://news.findlaw.com/hdocs/docs/gwbush/sarbanesoxley072302.pdf. The SEC has taken a narrow interpretation of Sarbanes-Oxley to the point that information security and risk management pertain only to the financial statements of a company. The Federal Reserve has countered this by saying a broader interpretation is needed to include all of the operational risks since there are many aspects that can impact the financial standing of an organization that can affect the integrity and accuracy of the financials.

[11] "CF Disclosure Guidance: Topic 2, Cyber security," Securities and Exchange Commission, Division of Corporate Finance, Oct. 13, 2011, http://www.sec.gov/divisions/corpfin/guidance/cfguidance-topic2.htm.

[12] "Legal Resources," Critical Energy Infrastructure Information (CEII) Regulations, Federal Energy Regulatory Commission, http://www.ferc.gov/legal/maj-ord-reg/land-docs/ceii-rule.asp.

[13] *See, e.g.,* Jody Westby, "Cyber Legislation Will Cost Businesses and Hurt Economy," Feb. 27, 2012, *Forbes.com, http://www.forbes.com/sites/jodywestby/2012/02/27/cyber-legislation-will-cost-businesses-and-hurt-economy/.*

[14] Kevin Coleman, "Battle Damage Increases From Widespread Attacks," *Defense Systems,* Aug. 1, 2011, http://defensesystems.com/Articles/2011/07/18/Digital-Conflict-cyberattacks-economy-security.aspx?Page=1; Brian Cashell, William D. Jackson, Mark Jickling, Baird Webel, *The Economic Impact of*

Breaches of personally identifiable information (PII) are expensive and frequently result in civil and class action lawsuits and investigation by state attorneys general or the Federal Trade Commission. The 2011 *U.S. Cost of a Data Breach Study*, conducted by Symantec and the Ponemon Institute, calculated that data breaches cost companies an average of USD5.5 million per incident.[15] Another recent Ponemon survey found that brand and reputation can decline 17-31% after a breach, and that it may take an organization more than a year to recover its corporate image.[16]

Corporate data is at a higher risk of theft or misuse than ever before, and the systemic nature of recent attacks has alarmed both industry leaders and government officials around the world. *Managing these cyber risks now requires active oversight by boards and senior executives.*

Failure to properly govern cyber security and privacy may result in shareholder derivative suits against D&Os for breach of fiduciary duty as a result of losses on stock price, decrease in market share, or damage to brand caused by inadequate attention to the security of the company's data, applications, and networks. Although Delaware case law provides strong protections to D&Os under the business judgment rule and recent case law,[17] harm caused by security breaches *may* receive stricter scrutiny because:

- Security best practices and standards are well-developed, harmonized, and widely available;
- Many privacy and security laws require organizations to have an enterprise security program that is regularly reviewed and tested;
- The Council of Europe (CoE) Convention on Cybercrime,[18] which has been signed by 47 countries and ratified by 33 (including the U.S.), holds companies civilly, administratively, or criminally liable for cybercrimes that benefit the company and were made possible due to the lack of supervision or control by someone in a senior management position, such as an officer or director. Article 9 of the European Union's (EU) Council Framework

---

Cyber-Attacks," Congressional Research Service, RL32331, Apr. 1, 2004, www.cisco.com/warp/public/779/.../images/CRS_Cyber_Attacks.pdf; A. Marshall Acuff, Jr., "Information Security Impacting Securities Valuations: Information Technology and the Internet Changing the Face of Business," Salomon Smith Barney, 2000, at 3-4.

[15] *2011 Cost of Data Breach Study: United States,* Ponemon Research Institute & Symantec Corp., Mar. 2012, http://www.symantec.com/about/news/release/article.jsp?prid=20120320_02.

[16] *Reputation Impact of a Data Breach: U.S. Study of Executives & Managers,* Ponemon Research Institute & Experian Corp., Nov. 2011, http://www.experian.com/blogs/data-breach/2012/01/17/how-data-breaches-harm-reputations/.

[17] In re *Citigroup Inc. Shareholder Derivative Action*, No. 3338-CC, 2009 WL 481906 (Del. Ch. Feb. 24, 2009), http://www.delawarelitigation.com/uploads/file/int99(1).pdf ; *Stone v. Ritter*, 911 A.2d 362, 366–67 (Del. 2006), http://caselaw.lp.findlaw.com/data2/delawarestatecases/93-2006.pdf.

[18] Council of Europe *Convention on Cybercrime* – Budapest, 23.XI.2001 (ETS No. 185) (2002), http://conventions.coe.int/Treaty/Commun/QueVoulezVous.asp?NT=185&CL=ENG, Council of Europe *Convention on Cybercrime Explanatory Report*, Nov. 8, 2001, http://conventions.coe.int/Treaty/en/Reports/Html/185.htm.

Decision on attacks against information systems,[19] which applies to all 27 EU member countries, mirrors the CoE language.

Thus, D&O duties with respect to privacy and security may be more prescribed in this area and negligence more easily proven. There are also situations where higher standards apply to directors and officers, such as acquisitions, takeovers, responses to shareholder suits, and distribution of assets to shareholders in preference over creditors. In these circumstances, directors and officers are required to obtain professional assistance or perform adequate analyses to mitigate the risks that ordinarily accompany these activities. Some information assurance experts assert that a "higher degree of care will also be required of Directors and Officers regarding the complex nature of issues involved in information assurance."[20]

In addition, securities laws and regulations also require public corporations to adequately disclose the risks relevant to the corporation and its assets in their public filings. The *Independent Director* put this in the context of information systems by reporting that: Management of information risk is central to the success of any organization operating today. For Directors, this means that Board performance is increasingly being judged by how well their company measures up to internationally accepted codes and guidelines on preferred Information Assurance practice.[21]

Clearly, directors and officers need to undertake a certain level of involvement and oversight in ensuring that the organization is properly secured and data is protected.

Fortunately, boards and senior executives have access to standards and best practices that guide them in fulfilling their governance responsibilities. The IT Governance Institute has an excellent collection of materials, as does ISACA, and Carnegie Mellon University. In addition, the International Organization for Standardization (ISO) has released ISO 38500, the international standard for corporate governance of IT, and the National Institute of Standards and Technology (NIST) has produced world-class materials on privacy and security best practices and guidance—including risk management—that are available at no cost.

ANALYSIS OF CURRENT STATE OF CYBER GOVERNANCE[22]

The Carnegie Mellon CyLab Governance Surveys, conducted biennially in 2008, 2010, and 2012, were intended to measure the degree of governance, or oversight, exercised by boards of directors and senior management regarding the privacy and security of their

[19] *Proposal for a Council Framework Decision on attacks against information systems*, Commission of the European Communities, Article 9, Apr. 19, 2002, COM(2002) 173 final, 2002/0086 (CNS), http://europa.eu.int/eur-lex/en/com/pdf/2002/com2002_0173en01.pdf; *see also Proposal for a Directive of the European Parliament and of the Council on attacks against information systems and repealing Council Framework Decision 2005/222/JHA*, European Commission, COM(2010) 517, http://europa.eu/rapid/pressReleasesAction.do?reference=MEMO/10/463.

[20] John H. Nugent, "Corporate Officer and Director Information Assurance (IA) Liability Issues: A Layman's Perspective," Dec. 15, 2002.

[21] *Id.* (citing Dr. Andrew Rathmell, Chairman of the Information Assurance Advisory Council, "Information Assurance: Protecting your Key Asset," http://www.iaac.ac.uk).

[22] This section of the paper is from the 2012 CyLab Governance Report at 5-7.

organizations' digital assets. Unlike the two previous surveys, which focused a U.S. survey population, the 2012 survey was the first global survey to be conducted, which enabled a comparison of responses between geographic regions and industry sectors.

The CyLab 2012 survey was based upon results received from 108 respondents at the board or senior executive level from Forbes Global 2000 companies. Half of the respondents was board members, and the other half was non-director senior executives. Twenty-four percent (24%) of the respondents were board chairs and 44% are on board Audit, Governance, or Risk Committees. Seventy-five percent (75%) of the respondents were from critical infrastructure companies.

For the third time, the survey revealed that boards are not actively addressing cyber risk management. While placing high importance on risk management generally, there is still a gap in understanding the linkage between IT risks and enterprise risk management. Although there have been some measureable improvements since the 2008 and 2010 surveys, boards still are not undertaking key oversight activities related to cyber risks, such as reviewing budgets, security program assessments, and top-level policies; assigning roles and responsibilities for privacy and security; and receiving regular reports on breaches and IT risks. Involvement in these areas would help them manage reputational and financial risks associated with the theft of confidential and proprietary data and security breaches of personal information.

Improvements in governance over the three surveys are largely organizational. There has been a noticeable increase in the number of boards with Risk Committees responsible for privacy and security risks (48% in 2012 compared with 8% in 2008) and in the number of companies that have established cross-organizational teams to manage privacy and security risks (72% in 2012 compared with 17% in 2008).

Boards and senior management are lagging, however, in establishing key positions for privacy and security and appropriately assigning responsibilities. *Less than two-thirds of the Forbes Global 2000 companies responding to the survey have full-time personnel in key roles for privacy and security (CISO/CSO, CPO, CRO) in a manner that is consistent with internationally accepted best practices and standards.*[23] An amazing 82% of the respondents indicated that they did not have a CPO. In addition, the majority of CISOs (58%) and 47% of CSOs are assigned responsibility for both privacy and security and tend to report to the CIO, creating segregation of duties (SOD) issues that are against best practices.

Despite these organizational improvements, respondents indicated that Audit Committees and full boards still are mostly responsible for oversight of risk. The report highlights the SOD issues that arise when Audit Committees both oversee the development of security programs and also audit the controls and effectiveness of such programs.

Although most boards (89%) review risk assessments, less than half of them hire outside expertise to assist with risk management. Only 16% of board Risk Committees and 16% of IT Committees hire outside experts. Despite 91% of the respondents indicating that risk management was being actively addressed by the board, the issues that received the least attention were IT operations (29%), computer and information security (33%), and vendor management (13%). The continuing low scores in these areas

[23] CISO is an acronym for chief information security officer, CSO is an acronym for chief security officer, CPO is an acronym for chief privacy office, and CRO is an acronym for chief risk officer.

indicate that boards do not understand that, today, all business operations are supported by computer systems and digital data, and that risks in these areas can undermine operations. The low response for vendor management is concerning because it indicates that the privacy and security of data at cloud and software providers and outsource vendors are receiving little oversight.

Another positive sign from the survey was the importance that boards are placing upon IT and security/risk expertise in board recruitment. Results indicated that IT expertise was very important or important for 37% of the respondents and somewhat important for 42%. Risk and security expertise was even more encouraging, with 64% of the respondents indicating that it was very important or important and 27% indicating it was somewhat important.

Industry Sector Comparisons

Industry sector and regional comparisons from the survey provide interesting insights into how privacy and security risks are managed among critical infrastructure industry sectors and across geographical regions. *The survey confirmed the belief among security experts that, overall, the financial sector has better privacy and security practices than other industry sectors.*

Respondents indicated that the financial sector paid more attention to IT and security issues and was more engaged in best practice activities, such as budget reviews, roles and responsibilities, and top-level policies, than the energy/utilities, IT/telecom, and industrials industry sectors. The financial sector also has a higher rate of (1) board IT/Technology Committees, and (2) Risk Committees separate from the Audit Committee that have responsibility for privacy and security.

The IT/telecom sector tends not to establish board IT/Technology Committees and assigns privacy and security oversight responsibilities to their Audit Committees. The energy/utilities and industrials sector respondents each indicated that their boards never (0%) address vendor management issues, whereas the financial and IT/telecom respondents said they do (28% and 15%, respectively). Energy/utilities respondents also ranked the lowest in establishing board Risk Committees separate from the Audit Committee, but indicated that when they do form a Risk Committee, they assign it responsibility for privacy and security. Only half of the energy/utilities and infrastructure sectors indicated that they have cross-organizational committees.

Respondents indicated that all industry sectors surveyed are not properly assigning privacy responsibilities to CPOs. None of the IT/telecom respondents (0%) indicated that they had a CPO, even though they have some of the most stringent privacy and security compliance requirements, and only seven percent (7%) of the energy/utilities respondents said they had a CPO. Just 13% of industrials sector respondents said they had a CPO, and 17% of the financial sector respondents said they did.

Interestingly, none of the energy/utilities sector respondents (0%) indicated that they have a CRO even though their risks are high. The energy/utilities sector also places a much lower value on board member IT experience than the other sectors, which is puzzling since their operations are so dependent upon complex supervisory control and data acquisition (SCADA) systems.

The energy/utilities and IT/telecom sector boards also are not reviewing cyber insurance coverage (79% and 77%, respectively) compared to the financial sector (52% not reviewing) and industrials sector (44% not reviewing) boards. The industrials sector respondents indicated that they exclusively (100%) rely upon insurance brokers to provide outside risk expertise, while the energy/utilities and IT/telecom sectors never do (0% for each). The financial sector respondents indicated that they seldom use insurance brokers for this purpose.

Geographic Regional Comparisons

Although Europe leads globally in privacy regulation and enforcement, few European organizations have a CPO (3%), with Asia only slightly ahead at five percent (5%) and North America at 23%. European companies, however, are more likely to have CISOs and CSOs (72%) than North American or Asian organizations (58% and 52%, respectively).

North American boards lag behind European and Asian boards in undertaking key activities associated with privacy and security governance. European boards, however, pay less attention to IT operations and computer and information security (19%) than North American and Asian boards (40% and 38%, respectively).

Boards across geographical regions were even in their neglect to review cyber insurance coverage (56-58%).

Asian boards (76%) are much more likely to have a board Risk Committee responsible for privacy and security than North American and European boards (35% and 41%, respectively). Asian organizations (82%) are much more likely to have privacy responsibilities assigned to security personnel than North American and European organizations (44% and 48%, respectively). Asian organizations are less likely to have the CISO/CSO report to the CIO, however.

2012 SURVEY CONCLUSIONS

Even though the top management of the Forbes Global 2000 companies appears to give risk management serious consideration, the 2012 survey indicates that cyber risk management requires more attention at the board and executive level. This will likely require raising the awareness of this audience on the sophistication of cyber threats and the dangers that they pose to corporate operations. Absent greater involvement, top management will not be prepared to make necessary decisions at the time of an event because they will lack the context of the risk.

This lack of attention at the board and executive level to digital risks actually increases the risks facing an organization because they expose the organization to compliance fines and penalties, legal liabilities from lawsuits and breaches of legal obligations, reputational harm, and operational and financial risks. Security incidents are commonly investigated by state authorities and/or regulatory agencies and media coverage about incidents raises the risk of declines in stock price and market share, even if temporarily. Additionally, directors and officers themselves are at risk of derivative lawsuits for such drops in market capitalization and market share resulting from security incidents.

The perception among some that regulation is required to improve corporate security is put in question by the survey results. Highly regulated industry sectors, such as utilities/energy, actually indicated that they had the weakest governance of privacy and security risks. Even the financial sector, which fared better in the survey results than other industry sectors, did not have stellar results. Indeed, there is the need for better security in all industry sectors, but it is doubtful that regulations are the solution. More analysis in this area is necessary.

An additional risk is that at the time of an incident, boards and executives may not receive full and complete information about the situation. When risk is not properly aligned across an organization, with roles and responsibilities for privacy and security a component of job descriptions for business unit executives and their personnel, general counsels, chief financial officers, and procurement personnel, all the risk is disproportionately placed on the security personnel. Thus, at the time of an incident—especially in hard economic times when jobs are hard to come by, the CISO may be fearful of revealing just how bad an incident really is. Therefore, it is critical that boards and senior executives receive independent advice regarding cyber risk management, which may in part come from annual assessments by auditors.

Lastly, culture may play a role in the degree of cyber governance that is exercised. It is worth wondering if the high percentages in the results for the Asia/Pacific region's respondents may in part be due to their cultural emphasis on compliance and security.

On a more specific level, the following conclusions can be drawn from the findings of the 2012 CyLab Governance Survey:[24]

- Boards are actively addressing risk management, but there is still a gap in understanding the linkage between cyber security risks and enterprise risk management.
- Boards are not undertaking key governance activities that would help protect their organizations from some of the highest risks: the reputational and financial losses flowing from theft of confidential and proprietary data or security breaches involving personally identifiable information.
- Organizationally, improvements are seen in (1) the increased number of boards with Risk Committees responsible for privacy and security risks, and (2) the high percentage of companies that have established cross-organizational committees to focus on privacy and security risks.
- Although most boards hire outside expertise, less than half hire it for assistance with risk assessments and risk management. There is a higher reliance upon IT security experts than risk services firms.
- The majority of boards are not evaluating the adequacy of their organizations' insurance coverage for cyber risks.
- Boards are recognizing that IT security and risk expertise are important skills when recruiting board members.

[24] This section of the paper is from the 2012 CyLab Governance Report at 24-26.

- Less than two-thirds of the Forbes Global 2000 companies responding to the survey have full-time personnel in key roles responsible for privacy and security in a manner that is consistent with internationally accepted best practices and standards. For organizations that do have these roles assigned, there is a serious lack of functional separation of privacy and security responsibilities.
- CISO/CSOs still tend to report to CIOs more than to CEOs or CFOs.

Regional Conclusions

- European boards pay less attention to IT operations and computer and information security than North American and Asian boards.
- Other than receiving privacy/security and breach/data loss reports, North American boards lag behind European and Asian boards in undertaking key activities associated with privacy and security governance.
- Asian boards are much more likely to have board Risk Committees responsible for privacy and security than North American and European boards.
- Across all regions, 56-58% of boards are not reviewing their organization's cyber insurance coverage.
- North American boards are much more reliant upon insurance broker risk expertise than risk services firms or IT security experts when seeking assistance with risk assessments and risk management.
- North American and Asian boards value board member IT experience much more highly than European boards. All geographical regions value risk and security expertise.
- Although Europe leads globally in privacy regulation and enforcement, few European organizations have a CPO (3%), with Asia only slightly ahead (5%). European companies, however, have a higher percentage of CISOs and CSOs than North American or Asian organizations.
- Asian organizations (82%) are much more likely to have privacy and security responsibilities assigned to key personnel than North American and European organizations (44% and 48%, respectively). Asians are less likely, however, to have the CISO/CSO report to the CIO.

Industry Sector Conclusions

- The financial sector has better privacy and security governance practices than the energy/utilities, IT/telecom, and industrials industry sectors. It also has a high rate of board IT/Technology Committees and Risk Committees separate from the Audit Committee, which are assigned oversight of privacy and security.
- Unlike the financial sector, the IT/telecom sector tends to assign the Audit Committee responsibility for cyber security and privacy risks.
- The IT/telecom industry sector respondents indicated that none of their organizations have a board IT/Technology Committee.
- The energy/utilities and IT/telecom respondents indicated that their organizations never (0%) rely upon insurance brokers to provide outside risk

expertise, while the industrials sector relies upon them 100%. The financial sector seldom does.

- Energy/utilities and IT/telecom sector boards are not adequately reviewing cyber insurance coverage.
- The energy/utilities sector places a much lower value on board member IT experience than financial, IT/telecom, and industrials industry sectors.
- Zero percent (0%) of the IT/telecom industry sector said they have CPOs, even though they have some of the most stringent privacy compliance requirements. Likewise, none (0%) of the respondents from the energy/utilities and industrials sectors indicated they have CROs.

In conclusion, cyber governance is a critical component of confidence in the security of networks, systems, and data. Although organizations may believe that they have good risk management plans in place, they actually have injected new risks in an organization through their failure to exercise appropriate oversight of the privacy and security of its digital assets.

PRIVACY IN INTERNET TRANSACTIONS: ECONOMIC AND SECURITY ASPECTS

UDO HELMBRECHT
European Network and Information Security Agency
Heraklion, Greece

One of the priorities in the work of the European Network and Information Security Agency (ENISA) is the identification and assessment of challenges and risks of new technologies, particularly regarding information security as well as privacy and data protection. At the same time, ENISA provides support and input to a number of policy initiatives of the European Commission (EC) and the European Union (EU) Member States. In this respect, ENISA is making contributions aiming towards *'a pan-European approach in the area of NIS and Privacy is necessary for the EU'* since we are convinced that it:

- Constitutes an obligation of the policy maker to protect the citizens of EU, but also because,
- Such an approach enables the deployment of an industrial policy in ICT. This is particularly important also in creating a level playing field to EU industry in areas where the market is currently dominated by non-EU players (e.g., Social Networks, online services).

In this talk we will provide a brief overview of the top level challenges in NIS from an EU perspective. In this context we group challenges into the following categories:

<u>Technology and R&D challenges</u>

The EU as well as EU industry are investing a considerable amount of resources to collaborative R&D in ICT technologies including NIS. Although considerable progress is achieved as a result of this investment in terms of technologies developed Europe lacks strategic research initiatives that put forward high level ambitious objectives and federate large amounts of resources under them (often called as the 'AIRBUS project/initiative'). At the same time, from ENISA's experience working together industry/research and policy makers a clear gap is observed between the advances made by the research community and the relevant policy priorities set at EU level.

<u>Economic challenges</u>

A typical argument used by NIS experts is that senior industry managers usually underestimate the necessity of investing on NIS until an incident occurs. Although, on some occasions this happens to be to the case, one should not underestimate that business models used by the ICT industry often are in conflict with policy initiatives. Recently the EC put forward its proposal concerning the review of the DP framework. A key aspect of the EC proposal is that personal data constitute a 'new form of currency'. Nevertheless, the business model used by established online services depends on the ability of the provider to create profiles of its users that then exploits (e.g., target-based ads, etc.).

EU legal framework and governance

It is widely acknowledged that Europe lacks a uniform legal system in the EU. As examples one may consider the cases of the Data Protection and ePrivacy Directives or Data Retention Directive and how it is transposed at the level of MS. One should also not ignore that at the EU level there is a multiplicity of actors that contribute to this debate, i.e., DPA, NRA, Art29, EDPS, DG JUS, DG INFSO, DG HOME.

Standardisation

For a number of years the role of standardisation in the new ICT landscape is being discussed by International Standardisation Organisations (ISO). It is still unclear though what this new role of ISO should be.

Users perspective

Research in this area indicates that the main priorities of users of online services are convenience/ease of use and low price (preferably free). These requirements often contradict with rigorous security settings and privacy preserving mechanisms. At the same time the perception of average users on the value of their security varies significantly depending on whether the question is asked before or after a security incident affecting them occurs (usually is much higher after the incident). Finally, there are many concerns whether educating users is the answer to all these problems. Certainly increasing awareness of the risks involved is desirable nevertheless users expect that the services will just work seamlessly and assume that their security and privacy will be preserved by the service providers.

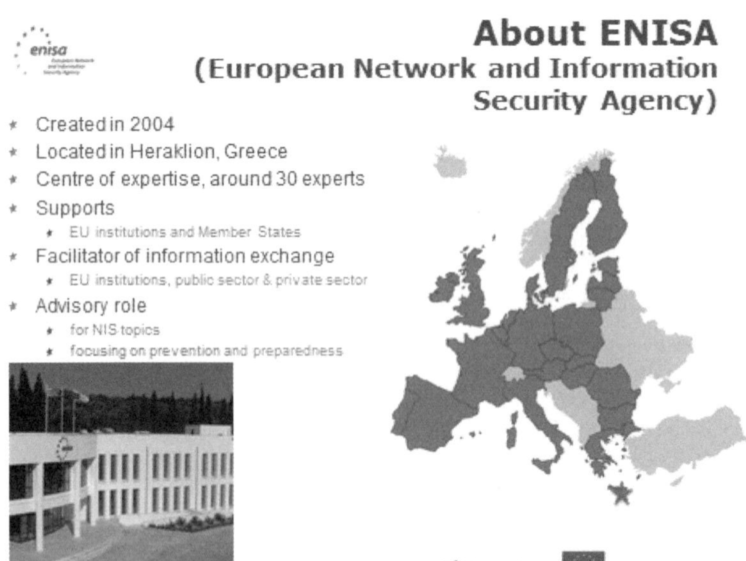

Development of attacks

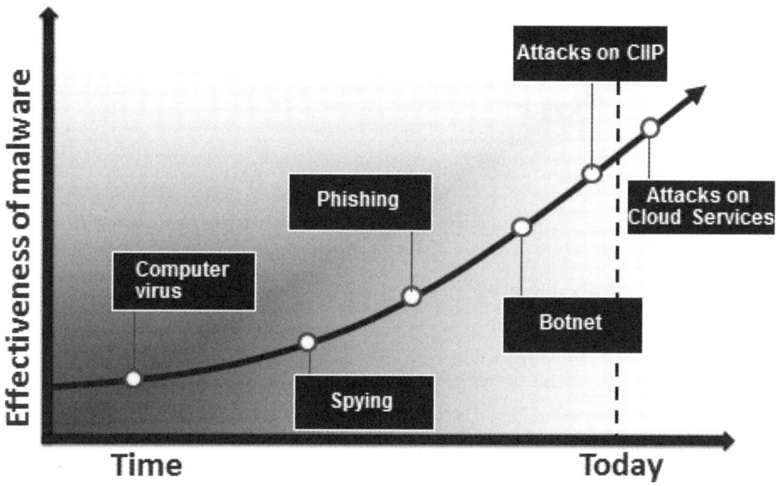

New virtual world (I)

- No national borders
- No uniform legal system
- Divergent approaches
  - Personal data protection vs. data retention
- Difference of perception across countries/regions
  - Privacy – human right in EU or consumer right in US
- Not a level playing field for EU online service providers
- Policy framework reactive instead of proactive

New virtual world (II)

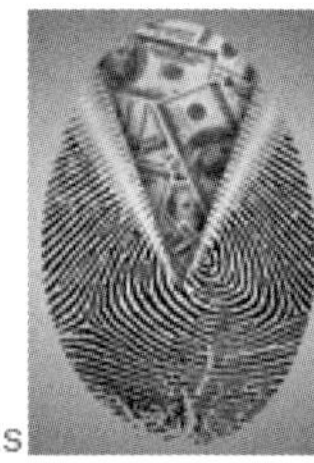

- A new currency: personal data
- Contradictory expectations and practice
  - Privacy - fundamental human right in the EU
  - Users concerned about privacy
    - 93% of participants in recent ENISA study
  - Users wiling to disclose more personal data for discounts
    - up to 87% of participants, in some cases, for 0.5 € discount in the study
- New business value chains
  - What is a fair price of a user's profile?
  - How this price may vary if considered as part of a group ('bargaining power of individual')?
  - What is the effect on consumers & competition?

http://www.enisa.europa.eu/activities/identity-and-trust/library/deliverables/monetising-privacy

www.enisa.europa.eu

Challenges in NIS from EU perspective

- Technology and R&D challenges
  - Clear gap between research and EU policy priorities
- Economic challenges
  - Business models used by online services depends on profiling
- EU legal framework and governance
  - Lack of uniform legal system
- Standardisation
  - Current approaches are ineffective on EU level. New role for SDOs?
- User perspective
  - Requirements of end users contradict rigorous security/privacy preserving mechanisms

www.enisa.europa.eu

Privacy as human right

- "*Everyone has the right to respect for his private and family life, his home and his correspondence.*"
    - Article 8 of The European Convention on Human Rights
        - adopted by states member of The Council of Europe
- "*Everyone has the right to the protection of personal data concerning them".*
    - Article 16, The Treaty of Lisbon, The Treaty on the Functioning of the European Union states
- "*Everyone has the right to the protection of personal data concerning him or her" [..] "Such data must be processed fairly for specified purposes and on the basis of the consent of the person concerned or some other legitimate basis laid down by law. Everyone has the right of access to data which has been collected concerning him or her, and the right to have it rectified.*"
    - Article 8, the Charter of Fundamental Rights of the European Union

www.enisa.europa.eu

Attitudes on Data Protection and Electronic Identity in the EU

- Special Eurobarometer (data end 2010, published 2011)

QB2. Which of the following types of information and data that are related to you do you consider as personal?

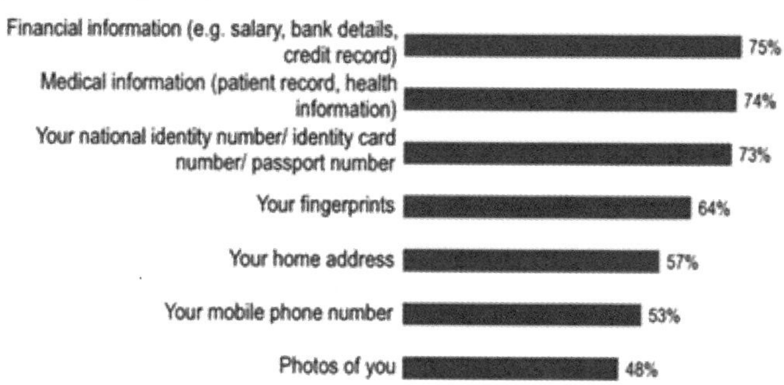

www.enisa.europa.eu

Privacy & Trust – context

- End user attitudes
- EU legislation reform
  - Data Protection
    - Directive 95/46/EC to be replaced
    - Proposal of a new regulation
      - Personal data protection by design and by default
      - Right to be forgotten
      - Sanctions
      - Personal data as 'new currency'
  - Data Retention reform
  - Electronic identification and trust services
    - Proposal of a new regulation

www.enisa.europa.eu

Privacy & Trust – paradoxes

- Necessity of a pan-EU approach in the area of Privacy
  - EU policy makers have to protect the citizens
  - Market currently dominated by non-EU players
  - Instrument of industrial policy should be developed
- Necessity of giving up some of MS national interests
  - Negative examples from the past
    - Exclusion of IT from critical infrastructure (DE)
    - No operational tasks for ENISA (SE, DK)
- Lack of a uniform legal system in the EU
  - Examples: Data Protection and ePrivacy directives
  - But: review of Data Protection framework
- Business model of the successful web services depends on profiling

www.enisa.europa.eu

Monetising privacy
An economic model for pricing personal information (I)

- Questions
  - Do some people pay a premium for privacy?
  - What is the effect on consumers & competition?
  - Does product personalization lead to lock-in?
  - Does it pay for firms to differentiate along the privacy concern of individuals?
- Model developed and experiments during 2011
  - Lab experiment, 440+ participants ; field experiment and hybrid experiment
    - the largest laboratory experiment in the field of privacy economics to date
    - 2,300+ participants with 139 transactions and observations in the field & hybrid experiments

www.enisa.europa.eu

Monetising privacy
An economic model for pricing personal information (II)

- Findings
  - Privacy-friendly firms get a greater share of market if no price differentiation
  - Privacy-unfriendly firms obtain greater share market if they practice lower prices, but privacy friendly firms stay above 10%
- Recommendations
  - Users should be provided with options that allow them to disclose less personal data
  - Easy-to-understand comparison of the data protection practices is needed
  - Portability of profiles for consumers required to reduce lock-in
- http://www.enisa.europa.eu/act/it/library

www.enisa.europa.eu

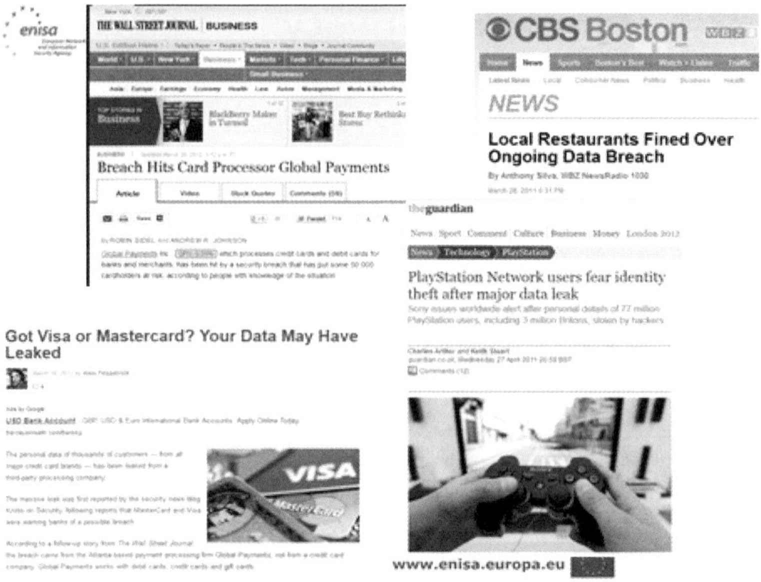

Data breach notifications

- Review of ePrivacy Directive (2002/58/EC)
  - Article 4

  *In the case of a **personal data breach**, the provider of publicly available electronic communications services shall, **without undue delay**, notify the personal data breach to the **competent national authority**.*

  *When the personal data breach is likely to adversely affect the **personal data or privacy** of a subscriber or individual, the provider shall also notify the **subscriber** or individual of the breach **without undue delay**.*

  *Notification of a personal data breach to a subscriber or individual concerned **shall not be required** if the provider has demonstrated to the satisfaction of the competent authority that it has implemented **appropriate technological protection** measures, and that those measures were applied to the data concerned by the security breach. Such technological protection measures shall render the **data unintelligible** to any person who is not authorised to access it.*
- Proposal of Regulation on Data Protection
- Proposal of Regulation on electronic identification

www.enisa.europa.eu

Implementing the notifications

* Differences between Member States
* Concerns of competent authorities
  * Lack of resources
  * Stress on education and prevention
* Concerns of industry
  * Confidence in internal procedures
  * Need of support in interpretation of law
  * Feeling of injustice in the telecom sector
* Impact of Data Breach Notifications
  * Short term: DBN will ensure information is given and actions taken
  * Long term: contribution in data protection in general
* Problems are not country-specific (cloud!)
* Recommendations for technical implementation: ENISA 2011/12

www.enisa.europa.eu

Towards acceptable governance and security

* The challenges extend beyond MS borders, hence...
  * MSs and the EU need close collaboration with industry and research
* A **gap** is observed between
  * what is possible at technological level
  * what is available at market place and proposed by policy makers
* PPP can be such an instrument although we are still at early phases of establishing effective PPP
* The role of standardisation is still not clear
* Users are primarily interested for
  * Convenience, ease of use
  * Price (preferably free)
* ENISA is trying to address this gap

www.enisa.europa.eu

Some 'going-away' messages

- Users should be provided with options that allow them to disclose less personal data
  - Free vs subscription-based services, with different privacy preserving conditions?
- Easy-to-understand comparison of the data protection practices is needed
- Portability of profiles for consumers required to reduce lock-in
- Right to be forgotten: **<u>How</u>** can it be implemented in practice?
- Minimal disclosure: can it be considered as an alternative to the notion of 'the right to be forgotten'?

www.enisa.europa.eu

Annual Privacy Forum 2012

'closing the loop from research to policy'

10-11 October 2012, Cyprus

- Organisers
  - ENISA, DG INFSO and University of Cyprus
- General Co-chairs
  - Jesus Villasante, European Commission, DG Information Society and Media (INFSO)
  - Demosthenes Ikonomou, European Network and Information Security Agency (ENISA)
  - Prof. Marios Dikaiakos, University of Cyprus
- Program Committee chair
  - Bart Preneel, KU Leuven
- www.privacyforum.eu

Annual Privacy Forum

www.enisa.europa.eu

SESSION 11

MESSAGGIO DEL PRESIDENTE DELLA REPUBBLICA ITALIANA: LE EMERGENZE PLANETAIRE E IL PROGETTO WFS

THE PLANETARY EMERGENCIES AND THE PROJECT WFS

DONATO MARRA
Segretario Generale, della Presidenza della Repubblica Italiana
Rome, Italy

I have the great pleasure and honour to open the closing plenary session of this important seminar on planetary emergencies, organised by the Research Centre named after the great physicist Ettore Majorana and directed with passion and outstanding competence by Professor Antonino Zichichi.

Firstly, I would like to extend President Giorgio Napolitano's most cordial and sincere greetings to the many leading scientists and specialists of so many different fields and to the prominent political figures from over forty countries, to give an essential contribution to the advancement of scientific knowledge and also to the affirmation of a more responsible form of politics, one that is also aware of the contribution that science can offer it.

President Napolitano, who knows very well the other Head of State, H.E. Vaclav Klaus, President of the Czech Republic, who gave an important *lectio magistralis* here and to whom he addresses a special greeting, has long been following closely and appreciating the activity that this Study and Research Centre has been carrying out for over 40 years not only to develop scientific culture, but also—and above all, I should say —to promote the widest and closest collaboration between scientists from all over the world, trying successfully to overcome divisions based on ideological stances, religious faith and ethnic backgrounds.

He was therefore very willing and fully convinced about granting his High Patronage to this 45th Session on the role of science in the third millennium. Indeed it is thanks to the perseverance and tenacity with which the Centre has developed its own initiatives that the premises have been laid for scientists to transmit to the various States their willingness to discuss and collaborate, so as to bring scientific and political culture closer together.

In her very fascinating speech yesterday, Minister Cancellieri rightly pointed out the artificiality of the distinctions between humanistic and scientific culture: allow me to point out that there must be no opposition between science and politics either. This is an essential condition in order for politics to be able to recover in full its necessary credibility and morality. It is indispensable to become fully aware that the only way to address and solve emergencies that put the fate of the entire planet and of future generations at risk is by pooling the limited resources that are available to each State and using them with the rigour proper of the scientific approach, without being conditioned by the subjection of science to special interests or by the pursuit of easy but short-lived consensus.

Hence the importance of the WFS (World Federation of Scientists) project, born of an even more structured collaboration among scientists from all over the world, which aims to provide a decisive contribution to the study and solution of the problems related to the real planetary emergencies, ensuring a targeted use of resources, without the waste and distortions caused by a partial and therefore wrong identification of these problems.

All this takes place in the context of a genuine alliance between science and politics, in which politics doesn't seek to enslave scientific knowledge for its own purposes, but intends to use the free contribution of scientists to solve problems that transcend the responsibilities as well as the capabilities of the individual states. This is the only way that we will be able to overcome the most treacherous enemy for humanity and Democracy itself: ignorance and misinformation.

This is another important aspect of the need for new forms of governance in the world that know how to deliver to the new generations not only models of sustainable economic development at the social and environmental level, but also a healthy planet with all of its rich original resources.

The focus on the new generations, who represent the future of all States and to whom the further development of political, cultural and social processes is particularly entrusted, is indeed the main preoccupation of President Giorgio Napolitano, whose heartfelt message of appreciation and encouragement I had the privilege of bringing you.

For President Giorgio Napolitano, the implementation of the project which sees the engagement of the WFS scientists is a significant contribution to ensure to the younger generations the legacy of a planet in which Planetary Emergencies, which have their roots in the tragic errors occurred in the twentieth century, are effectively addressed and quickly resolved.

In this same spirit President Napolitano thus constantly underlined the need to promote investment on research, innovation and training, by ensuring the allocation of adequate resources: we will thus be able to offer to young people the possibility of giving a decisive contribution to developing their knowledge and building their future.

WFS GENERAL MEETING

PMP REPORTS

CLIMATE PMP REPORT, AUGUST 2012

CHRISTOPHER ESSEX
Department of Applied Mathematics, University of Western Ontario
London, Canada

INTRODUCTION

Climate modeling has been among my many research pursuits as a professor of applied mathematics. I began working on it in the 1970's, including being part of the Canadian General Circulation Model (GCM) group of the Canadian Climate Centre. I was not satisfied with models on physical grounds thirty years ago, so I returned to more fundamental work that was needed before they could be made better. The very best of them were then, and are still now, not good enough for what has become expected of them. It's a little-known fact that all climate models are inherently didactic and not prognostic. This is obscured by the unusual usage of the term "model" in climate research. A friend in astrophysics asked me about this. After he develops a model for what goes on with some star or some nebula he, or an opponent, looks for observations that contradict his model; that is, his model could be shown to be wrong. But climate models are not falsifiable in the sense of Karl Popper.

Many of my skeptical friends attempt to catch climate models in false predictions. But this is a romantic notion. There is no observation or experiment that can be performed that will invalidate a climate model. Any deviation between observations and model behaviour can be simply fixed by tuning the empirical parameterizations without necessarily gaining any deeper understanding, assuming such deviations are not just dismissed outright. The models don't even agree with each other in many important features, such as clouds for example.

But even if you can get them to agree, one empirical treatment may do as well as another, since none of them need to fully respect the actual physics. Moreover, without future climate data, how do you know whether they are all agreeing on the wrong thing or not? One has no data to check against future climates because we have not experienced them. But it is a modern conceit that the same climate dependent empiricism of today will work as well in a different future climate. With that thought in mind, it makes no sense to look at the average across models, although, perhaps out of desperation, that has been and is being done.

If we reject empiricism and we use the basic physics to do proper computing, as we teach it to our students with all structure above the grid scale, then the computation time becomes much longer than the age of the Universe! So empirical models are the only possibility for anyone, no matter how clever. This does not mean that climate models have no value. It does not mean that they cannot provide useful insights into the science of climate. It does not mean that climate models are not the best we have or the state-of-the-art—they are the state-of-the art! But the state-of-the-art can't tell us what the future is going to be. I note here that there are many people who have made full careers making specific predictions about climate, ostensibly backed up by what climate models actually cannot do. This is a warning sign that something is wrong.

In case you think I am the only one saying this, here is the consensus position from the full scientific component of the third assessment report of the IPCC.

> "In climate research and modeling, we should recognize that we are dealing with a coupled non-linear chaotic system, and therefore that long-term prediction of future climate states is not possible."[1]

You may not have heard this extremely important quote from the so called consensus position of the IPCC. If that is so, why?

CLIMATE, COMPLEXITY, AND CULTURE

There is indeed something wrong both with climate science and how our culture deals with it. The IPCC is part of the problem certainly, but in spite of its problems (as Dick Lindzen rightly points out) the worst of what we have to deal with is not the IPCC scientific reports themselves. Many aspects of them are not too bad, and even the most egregious parts of them (and there are some) are small transgressions compared to the crazy cultural and political reactions to them (a 2012 example: melted streetlights in Oklahoma were attributed to global warming). There is a relentless torrent of this kind of nonsense.

Climate dynamics viewed from a meteorological standpoint is governed by a mathematical complex system. Loosely speaking, such complexity refers to mathematics, or properties of the natural world, that defy all of our best tricks to make them simple. Is there always a simple rule that can get you a particular value of a solution of a differential equation without integrating to it from an initial condition? The answer is probably no, and the solutions of chaotic dynamics are among the best known examples. If you can prove that the answer is indeed no, by the way, this is the famous P—NP problem, which is one of the Clay Millennium problems of fundamental mathematics. There is a million dollar prize for a solution.

Simple chaotic dynamics is child's play compared to the climate problem. So then why has such a fiendishly complex, unsolved problem of physics and computation become regarded as solved with such certainty? Some reasons are political for sure. Certainty as a slogan is a byproduct of the "orchestration" of scientists for policymakers. But the real reason why is that people do not understand what "complexity" means to modern science. Everyone prefers a simple world to comprehend, but for complex systems you can't have it. This was a stunning discovery that the scientific community grappled with throughout the 1980's and 1990's. It was a scientific revolution, which we are not quite over. The significance of complexity and nonlinearity of the climate system is an even more challenging idea for the public at large.

No matter how good the explanation of complexity is, it is hard to find a forum to explain it in, and even if you do, there is considerable resistance because of false intuition. The "shoe box" model for climate, illustrated in the cartoon, provides this intuition. The

[1] The Intergovernmental Panel on Climate Change (IPCC), third Assessment Report (2001), Section 14.2.2.2, page 774.

scene is in a school yard because plastic wrap over a shoe box is actually used widely in schools to explain the climate problem itself to children.

But the shoe box is not how climate works. The shoe box delivers simplicity instead of complexity, in addition to other infectious misunderstandings. The powerful subtext for the children, and the grown-ups they become, is one of anticomplexity. The real-world complexity has been swept away by making the problem purely radiative, which it most assuredly is not. This is a metaphor that has replaced the thing itself. While such things are not unheard of, the enormity of it, in connection with the climate problem, is unprecedented.

If this is the secret picture in the back of your mind for how climate works, juxtapose it with the IPCC quotation. That is another warning sign that something is very wrong.

BROKEN SCIENCE AND POLITICS

You may know that some scientists have been distancing themselves from their former positions on the climate change canon. Consider the 2012 remarks of James Lovelock, a guru of the environmental movement, and the originator of the Gaia idea. This is from an interview with the American MSNBC channel:

> "The problem is we don't know what the climate is doing. We thought we knew 20 years ago. That led to some alarmist books—mine included—because it looked clear-cut, but it hasn't happened...We were supposed to be halfway toward a frying world now."

He words it like a recent discovery, but 20 years ago we did know that we didn't know. It was never clear cut! But those of us who tried to point this out then had a great deal of trouble getting anyone to listen. Why?

A century from now, historians may still be arguing about this. But what we do know is that there has been a sustained assault on simple basic principles of science and it originates from the world of politics. Some of it, but not all of it, may even have been unintentional collateral damage from other political actions because politicians don't understand or care about scientific ideals. But both scientists and the general public have become extremely confused about where science ends and politics begins as a result.

There is a popular, but naive, idea that science has no internal controversies of its own. But disagreements are inevitable, because if there is no room for doubts or questions, then what starts as science degenerates into dogma and ceases to be science at all. Skepticism is among the most basic components of a scientific mindset. Where skepticism is under attack there is no science.

But for climate work, skepticism itself has become depicted as wrong. This is political thinking, which leads to all manner of thoughtless actions, dirty tricks and even actual threats. I won't dwell on such things here. But the Climategate emails proved what many of us already knew. This political thinking had grown into a large scale phenomenon, even penetrating into the scientific world with many participants cooperating to manipulate, intimidate, bully, and conceal undesirable results.

Erice was not insulated from this. I am proud to say that there is an email that complains about Skeptics, with a capital "S," at Erice, as if Skeptics were a political party. The author makes himself sound like a political lobbyist. I am proud of this because, while a skeptic is a person, skepticism is a trait that all scientists need to do their work. It is not an ideology. It is a virtue and a talent, not something to be railed and struggled against. All scientists should be skeptical. Only fortunetellers and scoundrels rail against "skeptics."

It is only necessary here that you understand that we have a significant problem and that we have a chance to fix it before it is too late.

THE PMP MEETING

The climate PMP meeting was set against this background, including ten presentations and discussion sessions. We were most honored to have Dr. Vaclav Klaus in attendance, present as a private citizen, for much of the meeting.

Three special themes were put forward:

1. Disentangling politics and science.
2. Developing a better alternative to the precautionary principle.
3. Fostering good science hurt by current policies

Points arising from the meeting:

Climatology is in the unusual position that a politically-appointed organization, the UN Intergovernmental Panel on Climate Change (IPCC), has been given the authority to review ongoing work and, at regular intervals, issue judgments about the credibility of published scientific findings, decide among rival hypotheses and declare a consensus of experts. There was extensive discussion about problems arising from the existence and institutional dominance of the IPCC, its questionable review procedures and its role in polarizing the field of climatology. The following suggestions arose:

1. Criticisms of the IPCC were made by the InterAcademy Council in a 2010 report. These focused on potentially serious problems of lack of transparency, especially in author selection, and deficiencies in the peer review process, among other things. Even though the IAC was itself compromised as not being truly independent of the IPCC and not taking sufficient evidence from IPCC critics, its report nevertheless identified some genuine flaws and made important recommendations for reform. Regrettably the IPCC has dismissed most of the recommendations out of hand, and indeed has enacted new procedures that decrease transparency further.
2. There is no compelling rationale for the existence of Working Groups II and III within the IPCC. They focus on issues of policy and politics, making it more likely that the entire organization will subordinate science to policy advocacy. Support was expressed for abolishing both Working Groups and bringing the IPCC's focus back specifically to objective questions in physical science.
3. Increasing polarization of the scientific community is exacerbated by academic societies issuing position statements on behalf of their members, without giving members the right to sign or not as individuals. This phenomenon stifles free inquiry and is antithetical to the principle that scientists should think and speak for themselves, and not be forced to conform their opinions to an administratively-imposed dogma. This principle is already formally adopted by, for example, the American Economic Association. It also ought to be the logical consequence of the motto, *Nullius in Verba*, of the Royal Society. But such enlightened policy has not been realized by that organization.

Scientific issues discussed included climate sensitivity, which included a very interesting discussion on sensitivity from Dick Lindzen suggesting an unexpected dynamical consequence to the magnitude of climate sensitivity. Also notable was Henrick Svensmark's latest results on cosmic ray induced aerosol formation. He has been able to close one of the most crucial gaps in the aerosol idea—a very exciting scientific result involving the radiation induced catalysis by sulphuric acid.

CONCRETE PROJECT SUGGESTIONS DISCUSSED IN THE PMP

1. Robustness and Resilience.

You may recall that this topic escaped embargo with some enthusiasm in Monday's session.

This project suggestion followed from theme 2 (a replacement for the precautionary principle). The heart of the notion is that uncertainty and nescience are qualitatively different (recall the sheep farmer who went to the Falkland Islands to avoid war—that's nescience). These two things imply different policy responses. The goal would be to use expertise with, for example, modern dynamical systems theory, and the theory of risk, to formulate simple (not complicated) principles or guidelines for formulating policy.

2. The Climate Forecasting Prize

Ross McKitrick noted in his Monday presentation that climate modeling is not "incentivized" for forecasting. He proposed a taxation approach that might affect modeling positively. But the presumption that good forecasting already exists is inherent in the very premise of the climate change issue. Such underserved confidence has even induced calls for offices of climate forecasting to be instituted!

However, in the PMP meeting, data from individual climate models showed, as we would expect, that they are far from being able to do what many people believe they can do or would wish them to do. They really are more focused on the needs of policymakers instead of how well they can actually forecast with money at stake. But rather than argue endlessly over whether such forecasts are even possible or not, there is a clever way to push the discussion to a more pragmatic level: give a prize for actually doing successfully what is in doubt. But is it possible to define an intelligent criterion for what an accurate long-term forecast would be? Perhaps it would involve some climate version of meteorological skill. A substantial prize would be offered to the first person or team able to achieve it.

We are thinking in terms of more than $1 million over, say, 10 years. But it would take some time to elaborate and the idea needs some development to confirm that the prize plan is possible to do and to do well.

2012 REPORT FROM THE PERMANENT MONITORING PANEL—MITIGATION OF TERRORIST ACTS (PMP-MTA)

ANNETTE SOBEL
University of Missouri, Senior Advisor for Bio-security Engagement, Office of the Secretary of Defense/Nuclear, Chemical, Biological, Columbia, USA

SALLY LEIVESLEY
Newrisk Limited London, Mitigation of Terrorists Acts PMP, World Federation of Scientists, London, UK

PROGRAM INTRODUCTION

This year's program began the long-term process of developing a more comprehensive and strategic approach to the science of terrorist threats, public policy, and implementation of science-based solutions with cultural, social and technical appropriateness. We chose to partner with the Pollution PMP, chaired by Dr. Lorne Everett. This decision was based on the ubiquitous nature of water, the significant challenge of the science, and the ability to recommend actions within the realms of policy, social and technical spheres. There is also a historical precedent for terrorist threats to water and the use of water as a weapon, a notable contextual backdrop for this area of study.

The program focused on the panoply of threats, to include the broader entity of "hybrid" threats such as chemical, nuclear, radiation, biological and cyber, etc. (presenters Grayman, Kroll, vom Saal, Leivesley, Sobel, von Hippel, Parker, Moore). In all cases, we attempted to select challenges that held promising technical solutions and public policy options and were representative of global concerns.

The technical presentations were selected to stimulate thought in the areas of threat characterization, modeling and simulation, technical water surveillance and analysis, countermeasures (especially, EU initiatives, presenter, Steinhausler), dual-use technologies, human factors of terrorism and response/recovery, economic impacts (Rajaraman), and future threats. A robust discussion of nuclear and radiologic concerns was also presented to encourage further deliberation on lessons-learned from the Fukushima disaster (presenter, Wilson).

A special emphasis was placed on introduction of the topic of situation awareness and understanding of public perceptions of terrorism, with a focus on IRA atrocities and personalized stories, as introduced by a documentary film presented by the Lord John Alderdice and selected for presentation to the United Nations in the near future. This film's theme was further expanded in PMP discussions and a presentation on the topic of terrorism change agents and public policy (Rajaraman). A continuing dialogue stressing evolution of the complexities of current day Russian policy in the context of global terrorism was pursued, and potential areas for further PMP action were identified (Krivokhizha).

The PMP-MTA believes greater understanding of the social science and human factors issues impacting terrorism motivation, intent, and response directly impacts community resilience and is an area worthy of further study and quantification.

DISCUSSION (DICK GARWIN AND PMP INPUT)

The science of terrorism stems from an understanding of the physical and social sciences and the significant interactions between the two disciplines. Only through continued trans-disciplinary work will true breakthrough science occur.

It may be useful to look back not only on the work of the PMP-MTA, but especially on human experience in recent decades with terrorism of mass impact. They are linked by the kind of "filter" we have applied, in choosing to concentrate on individual acts with massive impact, rather than on assassinations, vehicle bombs, IEDs, and the like.

Indeed, there has been a transformation of terrorist mode over the past ten years, from suicide terrorists being a rarity to their being the usual approach.

There has also been a transition to an increased number of female terrorists. This facilitates terrorism and makes it more operationally reliable, since it minimizes much of the difficult planning that would be required for the terrorist himself or herself to escape and avoid interrogation. Female and male terrorists also have different characteristic signatures that may avert detection. The Institute for Counter-Terrorism, Herzilya, Israel has several scholars studying this phenomenology in depth.

In a way, suicide terrorism applies management and market approaches, since the actual perpetrator (driver of the vehicle; wearer of the explosive vest) is rarely the plotter of the act and is rewarded in some fashion. PMP-MTA has not concentrated on the motivations or means of recruitment of the actual suicide terrorists, but is cognizant that there are many "soft factors" at work. Clearly, this is a matter of important leverage in understanding and preventing such an activity. PMP-MTA is beginning to study the applicability of modeling and simulation to further understand these "soft factors" (presenter, Duncan).

We have, for the most part, concentrated on potential terrorist acts with major individual consequences—the use of biological weapons to spread disease—even communicable and highly lethal disease such as smallpox. Or, attack on nuclear reactors or even chemical plants or shipments of toxic chemicals, attack on which might kill thousands or tens of thousands of people (Bhopal) and cause billions of dollars of damage. We have looked at radiological threats from the dissemination of radioactive materials such as Co-60 or other radioisotopes used in commerce and industry, and have found that the direct consequences to life and health are small, although economic loss can be very great, according to current standards of environmental protection.

We have begun to transition to more in-depth exploration of terrorist attacks on water in collaboration with the Pollution PMP, and will include other critical infrastructure. In parallel, we plan to expand our work into an understanding of the complexity of terrorism in cities through cooperation with the Energy-Cities PMP.

When Dr. Richard Garwin wrote about mass-impact terrorism in 2001,[1] former national intelligence officer for the Near East and South Asia of CIA Paul Pillar judged that terrorists would confine themselves to the use of explosives rather than engage in the more complex approaches of acquiring and using radiological, biological, or even chemical threats. This trend is evolving.

---

[1] "The Many Threats of Terror" and an Epilogue by R.L. Garwin in a New York Review Book *Striking Terror: America's New War*, edited by R.B. Silvers and B. Epstein, March 2002, pp. 235-256. To be found at http://www.fas.org/rlg/020300-strikingterror.pdf.

For the most part, Pillar must be judged right, but that is no guarantee for the future, and we must begin to unravel the possibilities of hybrid threats to include combinations of chemical, cyber, biological, radiologic, nuclear, with explosive (C2BRNE) agents. The science of countermeasures development and early detection could certainly benefit from a great understanding of this new class of threats and weapon delivery systems.

NEXT YEAR'S TOPICS AND INTEGRATED ONE SCIENCE SOLUTIONS

Dick Garwin Strategic Recommendations with Concurrence from the PMP-MTA Membership are as follows:

- A more systematic study of vulnerabilities of water systems and other Critical Infrastructure;
- An integrated approach to understanding the threats and vulnerabilities of cities and opportunities for mitigation of terrorist acts;
- Further exploration of the interface of the social sciences and terrorist actions, to include motivation, intent, public policy, response and recovery; *this PMP does not believe pre-emption/prevention is a realistic strategy without improved human resources and understanding of "soft factors and further study is planned;*
- This PMP has reviewed data supporting concerns that...beyond relatively low-tech and sensible actions, and given the state of today's technology, we may not be able to afford to significantly lower the risk of contamination events, psychology and economic impacts of terrorism; we plan to further explore this interdisciplinary area and make further recommendations;
- The EU has developed and is implementing a number of programs that may produce global "lessons-learned" and capabilities that should continue to be tracked by this PMP;
- U.S.-Russian joint workshops addressing new approaches to strategic policy and MTA are worthy of pursuit; insights to future science and technology investment are likely to result from these workshops.

ENDANGERED SUSTAINABILITY OF A MEGALOPOLITAN REGION: THE UNEXTINCTED ACTIVITY OF THE *POPOCATEPETL* VOLCANO, MEXICO

ALBERTO GONZÁLEZ POZO
Theory and Analysis Department Universidad Autónoma Metropolitana
Mexico D.F., Mexico

The notion of sustainability has been described mostly as a critical influence of human activities trying to extract and transform natural resources beyond a limit that does not assure their availability for later generations. But announced or unexpected natural hazards like earthquakes, or impacts of cosmic objects (two issues already shown in the Erice Seminars during the last decade) may have similar, even worse consequences. That is the case of the unextincted *Popocatepetl* volcano, because it arises in the middle of several converging metropolitan areas that form together a greater megalopolis in the highlands of Central Mexico.

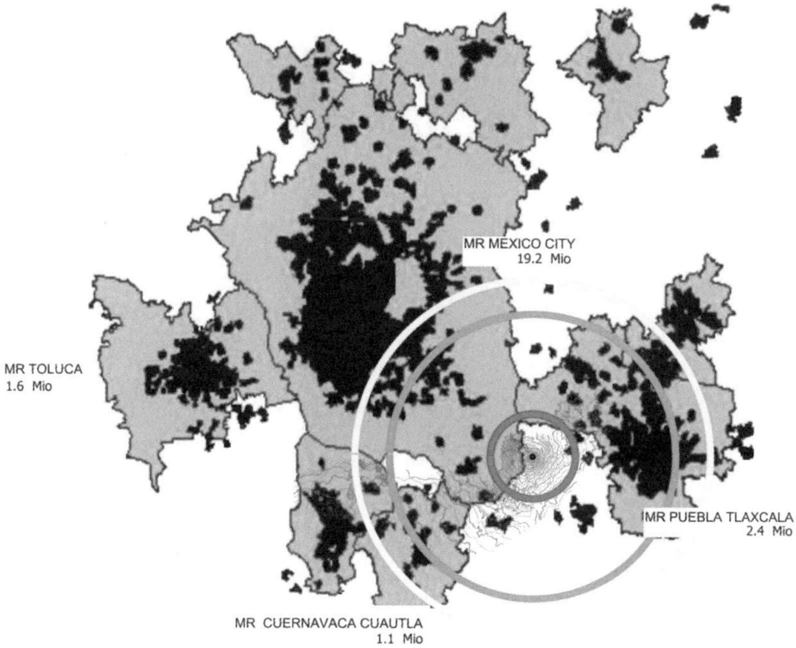

Fig. 1: Four Metropolitan regions around the Popocatepetl volcano in Mexico form a Megalopolis of 24.3 million inhabitants. The circles describe risk areas discussed in this article. (Sources: PUEM-UAM, 2000, Macías Vázquez et al. 1995 and INEGI, 2000).

I have shown in other seminars the spectacular growing of the urbanized giant that arose from Mexico City, with less than half a million inhabitants starting in the 20th century and reached more than 20 million to the closing of the same period, swallowing ancient towns and villages around it (González Pozo, 2000 and 2003). As it continues to grow, and as in other neighboring valleys with their own central cities happens the same, it gives an good example of a diminishing urban sustainability, not only because of the serious practical problems these giant settlements have, related to the provision of drinkable water, or the disposal and evacuation of rainwater (because one of the metropolises is a closed basin), or the increasing traffic and air-pollution problems, but also because of the latent danger to human lives and assets potentially exposed to a great eruption of a volcano.

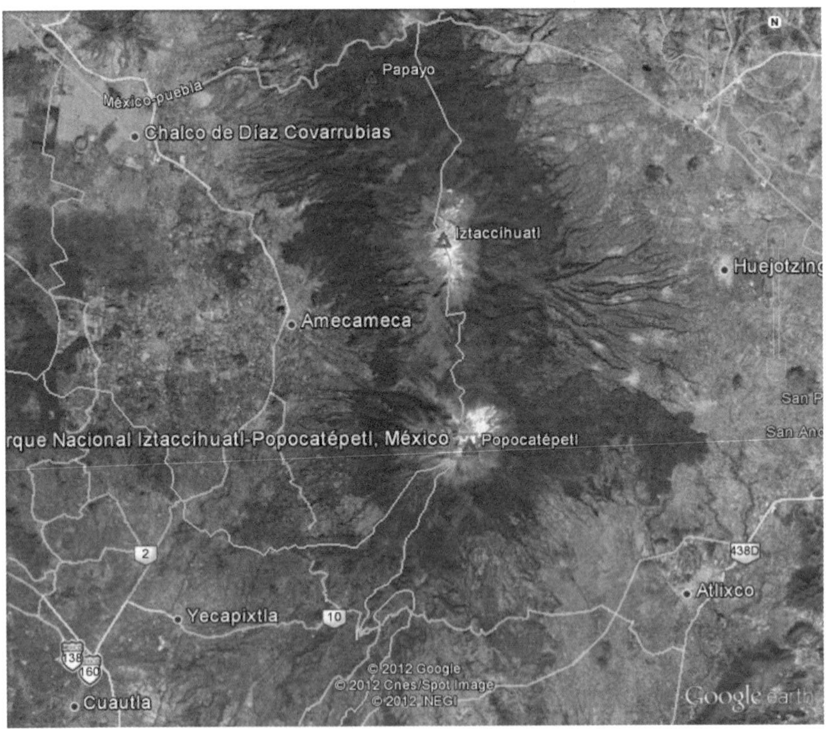

Fig. 2: Eastern limit of the Valley of Mexico with the Iztaccihuatl and the Popocatepetl volcanoes (Googlearth).

This presentation gives some facts about the present situation, as well as the measures taken to face and mitigate the threat of a sudden increase of the risk level.

A COUNTRY DIVIDED BY A VOLCANIC BELT

Volcanic activity seems to coincide with border regions between continental or oceanic plates, and Mexico is no exception. But when one sees more closely the distribution of active volcanoes in our region, it is clear that most of them are in a band along the 19°N latitude that crosses the whole country from the Pacific Ocean to the Gulf of Mexico. It is called the "Trans-Mexican Volcanic Belt" (also neovolcanic axis) and was probably formed between Pleistocene and Holocene times. (Santoyo Villa et al. 2005) The Popocatepetl volcano occupies there a central position, next to the biggest human agglomeration of the Mexican central high plateau.

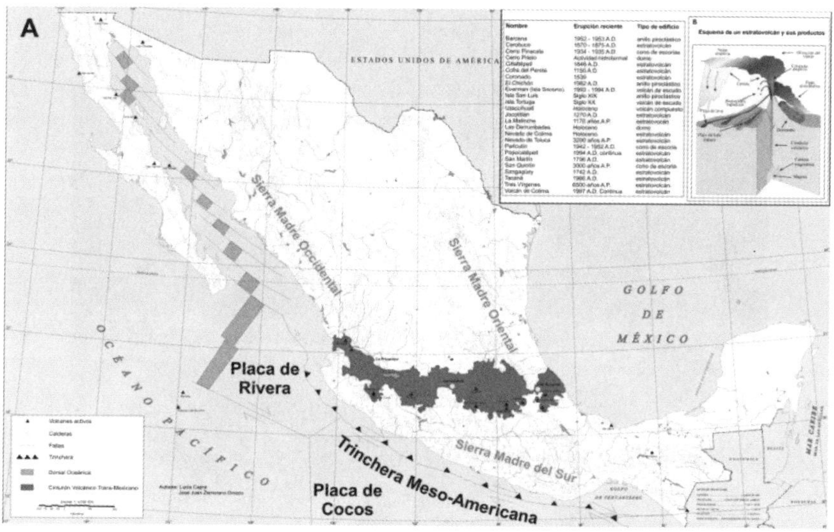

Fig. 3: "Trans-Mexican Volcanic Belt". Mexico City and the Popocatepetl volcano are in the middle (Source: Instituto de Geofísica, UNAM).

The fact that the giant Stratovolcano (5,452 m. over sea level) is active since long ago explains its ancient Aztec name. Popocatepetl means "smoking mountain", and the inhabitants of the Mexican capital consider it and its neighbor, the Iztaccihuatl (in Aztec "sleeping lady", 5,285 m), as important and typical features of the surrounding landscape that had been permanently covered by heavy snowcaps, thick enough to supply important amounts of freshwater to the surrounding valleys.

Fig. 4: Two snow giants: Popocatepetl seen from Iztaccihuatl, ca. 1960 (Photo: Virgilio Gonzalez Pozo).

But since the last 15 years, two remarkable changes have gradually taken place: the first one was the unusual awakening of eruptive activity inside the crater of the Popocatepetl, followed by sudden explosions throwing outside solid fragments or exhaling giant "plumes" or columns of smoke with gases and volcanic ashes; and the second one is that the once called "permanent" snowcaps have gradually disappeared, due to the climatic changes that affect now the whole world. Depending on the climatic conditions, those and other Stratovolcanoes in Mexico may recover their snowcaps for several weeks or months, but most of the year show a dry image in the landscape.

Fig. 5: Left: Thermographic infrared image of a strong smoke column after a crater explosion, 2000 (Source, CENAPRED, 2000).
Right: Big smoke column in daylight. In foreground: a colonial church on the top of a pyramid. (Source, Cronica Digital, 2012), México)

Depending on the height of the plume (up to 8 kilometers) and the prevailing winds (especially those from east to west or the opposite) the ashes are deposited in the surrounding territory, traveling distances of 10 to 50 kilometers. Until now, the ashes are a minor disturbance even if the city cars suddenly are covered by a thin layer of a fine and abrasive powder. But sometimes, their presence is heavy enough to stop for several hours or some days the aircraft activity in the International Airport of Mexico City.

Types and effects of volcanic emissions. (Translated after Reyna de la Cruz, 2004)

Material expelled	Associated risk	Speed	Distance range	More frequent consequence
Liquid lava	Lava flows	Low	Short	Ground destruction
Ashes	Rain of ashes	Medium	Intermediate	Ashes accumulation
Several types of fragmens	Pyroclastic flows	Very high	Short to intermediate	Devastation
Ashes	Rain of ashes	Medium	Long to very long	Ashes accumulation, sewage blockage
Mud (fragments and wáter)	Mud flow (lahar)	Medium to high	Intermediate to long	Devastation
Ground slipping or falling	Rubble avalanche	Hign to very high	Intermediate to long	Devastation

The fear under such new situation is now present, but only in the population of small villages and hamlets next to the volcano. They are aware that if the hazard is strong enough they will be compelled to evacuate their homes for security reasons.

THE LONG RECORD OF VIOLENT EPISODES OF A SEEMINGLY HARMLESS GIANT

The oscillations between weak and strong activity periods in the Popocatepetl since 1994 is only a warning sign that the volcano is reawakening. Today, we know better its long history, where violent eruptions took place between lapses of some centuries or some millennia of calm. In strong eruptions, the volcano expelled flows of lava or volcanic mud that contributed to build its typical conic shape by successive layers. But it could also throw pyroclasts (rocks or smaller fragments) that covered larger areas. Finally, crater explosions could expel giant smoke columns already described.

Short Eruptive History of the Popocatepetl Volcano, Mexico. (Shortened, after CENAPRED)

Between 23,000-14,000 years before present	At least two large Plinian eruptions, one of them, St. Helens-type, destroys previous volcanic edifice.
Between 14,000 years before present and 800 A.C.	Seven large eruptions and several minor eruptions. In the first Cent. B.C., another small volcano called **Xitle** in the southern part of the Valley of Mexico, covers with lava the ancient pre-columbian town of Cuicuilco, rediscovered in the first half of the 20th Century after hard excavation in the rocky lava.
800 A.C. to 1927 A.C.	About 1200 years, with at least 17 moderate or mild eruptions. Most of them produce ash and pumice emissions that fall at different distances (some reach 40 and 50 kilometers away from the crater) and form soil layers between few to 30 centimeters thick. All these episodes restart fumarolic activity. Some of them are historically recorded and described.
1927	Moderate eruption. Besides episodic explosions inside the crater that produce ash and pumice emissions, a small lava dome grows in the crater floor. Several people die inside the crater during sulphur extraction works. The next six decades the volcano seems to rest, with remnants of fumarolic activity and is covered by permanent icecaps, as it was since ancient times.
1994-1997-2001	Moderate eruptions. Episodic explosions produce ash and pumice emissions. A small lava dome grows on the crater floor and fills up to 20% of the crater capacity. 5 people were killed close to the crater rim during an explosion in May, 1995. The former icecaps start to disappear.
2001-2012	Moderate eruptions, some produce very high smoke columns, up to 8 kilometers high, that carry volcanic sand (silica and pumice). One of them causes the closing of the International Airport of Mexico City. The presence of icecaps is rather rare. They return but melt again after several months or weeks.

Fig. 6: Left: Nocturnal explosion inside the crater, 2000. Pyroclastic material runs down outside and gasses and ashes mix with the atmosphere. Right: Interior view of the crater in a "resting" period before 1994. (Sources: Instituto de Geofísica, UNAM and H. Delgado, UNAM).

RENEWED CONSCIENCE OF THE VOLCANIC RISK AND THE MITIGATION PROVISIONS

The recent reawakening of the Popocatepetl is now closely followed by experts and the media, thus creating a renewed conscience of the risk and the vulnerability of the whole involved region. The creation of a Federal Government Agency specialized in disasters, the National Centre for Disasters Prevention (CENAPRED for Centro Nacional de Prevención de Desastres) within the Secretary of Government, which has the highest rank among the Cabinet. Its physical location within the main Campus of the National Autonomous University of Mexico (UNAM), allows a close relationship with research experts of several convergent disciplines like geophysicists and vulcanologists, engineers, planners and social scientists. Local governments of the three Federal States involved, as well as city authorities of the most important or smaller urban settlements (especially those closer to the Stratovolcano), also watch the volcano's behavior.

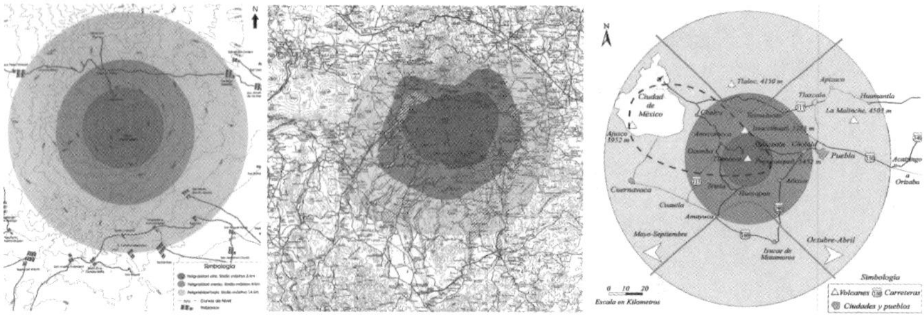

A B C

Fig. 7: Three different risk maps related to the Popocatepetl hazards: (Source: Macìas Vázquez et al., 1995. Instituto de Geofísica, UNAM).
A. Describes possible falls of volcanic fragments in eruptions of three intensities between a radius of 28 Km.
B. Irregular area where lava and volcanic mud could flow over slopes and river-beds, with maximum distances of—and—Km respectively.
C Possible falls of ashes in a broader area with maximum radius of 50 Km that reach Mexico City. Dotted line shows the area covered in a Plinian eruption. The central area between diagonals represents predominant wind directions either east-west or the opposite.

With their participation, several maps have been produced showing the hazard areas for different situations. One of them shows the possible fall of tephra (pyroclastic material) around the mountain flanks. Another, the superficial flow of lava or hot volcanic mud that quickly descends following the gorges and topographical valleys. Another, more ominous, considers the case of violent explosive eruptions, showing the fall of ashes and volcanic sand once the prevailing winds disperses them from the big smoke plumes and transport them to short or long distances depending from the violence of the explosions, some of them reaching the Metropolitan Areas around the volcano.

The menacing giant is now surrounded by sensors and recording devices that follow constantly its behavior, forming a network that alerts the scientists and eventually triggers a sort of volcanic "traffic light" put in public and visible places within a distance of ca. 10 km of the volcano crater, where the green light means that the volcano is resting, the yellow light indicates some activity and the need to watch constantly the news and prepare for an eventual evacuation, and the red light the imminent evacuation measures and the movement to provisional shelters.

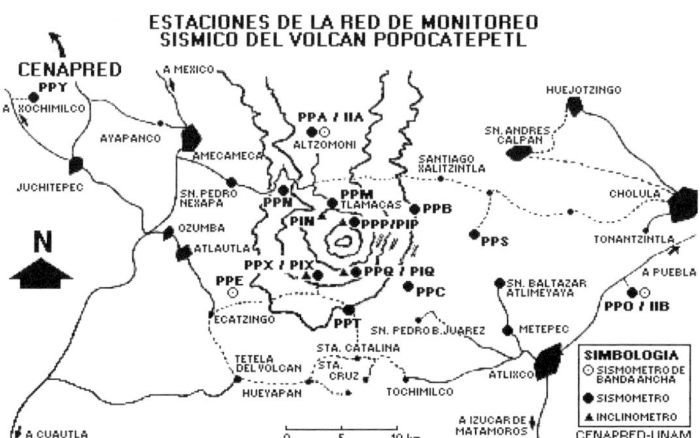

Fig. 8: The volcano is constantly watched with different monitoring devices from several places. (Source, CENAPRED-UNAM, 2002 and 1991)

Up to now, the mitigation measures are concentrated upon the most vulnerable settlements within the smaller area of 14 Km of radius, but this does not exclude the possibility of another strong or violent eruption that may cause damages in a much broader area of 50 Km radius.

And meanwhile, the megalopolitan region continues to expand in each of the four metropolitan areas around the volcano, approaching slowly but continuously their urbanized zones to the smoking colossus. This of course, represents a crucial question to its future sustainability.

SOURCES

1. CENAPRED, in http://cenapred.gob.mx/es/Instrumentacion/InstVolcanica/MVolcan/HistoriaVolcan.
2. Delgado, H. in http://earth-of-fire.over-blog.com/article-les-volcans-mexicains-tmvb-le-popocatepetl-75508714.html.
3. Espíndola, Juan Manuel and Macías Vázquez, José Luis (1996) "El vulcanismo", *CIENCIAS,* 41, January-March 1996, UNAM.
4. González Pozo, Alberto (2000) "Water use, abuse and waste: limits to sustainable development in the Metropolitan Area of Mexico City", in *International Seminar on Nuclear War and Planetary Emergencies,* 25th. Session, Singapore.
5. González Pozo, Alberto (2003) "Urban Mobility in the Mexican Metropolis" in *International Seminar on Nuclear War and Planetary Emergencies,* 30th. Session, Singapore.
6. Macías Vázquez, José Luis et al. (1995) *Mapa de Peligros del Volcán Popocatépetl* INSTITUTO DE GEOFÍSICA, UNAM-CENAPRED, June1995.
7. Reyna de la Cruz, Servando (2004) *Volcanes, Peligro y Riesgo Volcánico en México,* Serie Fascículos, CENAPRED, México.
8. Santoyo Villa E., Obando Shelley, E., Mooser, F. and Leon Plata, E. (2005, *Síntesis Geotecnica de la Cuenca del Valle de México*, TGC, México.

POLLUTION PERMANENT MONITORING PANEL — 2012 ANNUAL REPORT

LORNE G. EVERETT
Chancellor Emeritus, Lakehead University, Thunder Bay, Canada and L. Everett & Associates, LLC, Santa Barbara, USA

THE POLLUTION PERMANENT MONITORING PANEL ACTIVITIES FOR 2012 INCLUDED:

- Pollution PMP August 19 Meeting
- 4 Pollution PMP Papers and Reports
- 4 Centre Proposals and 3 Session themes, 2013
- Joint PMP Meeting with Terrorism Panel on Water
- Joint Plenary Session with Terrorism Panel: Water, Pollution and Terrorism

THE FOUR NEW CENTRE PROPOSALS UNDER PREPARATION FOR 2012 INCLUDE:

- Hardening Water Infrastructure Against Terrorism
- Hydraulic Fracturing Soil Gas Impacts
- Encouraging Global Cooperation on Spent Fuel Management
- A Non-Proliferating Fuel Cycle Approach

IN 2013 THE PROPOSED POLLUTION PANEL SESSIONS INCLUDE:

- Innovative Cleanup Technologies
- Nanotechnologies Environmental Applications
- Endocrine Disrupting Chemicals (EDCs) and Drugs in Water: A Terrorism Concern

HARDENING WATER INFRASTRUCTURE AGAINST TERRORISM PROPOSAL

Terrorist attacks on water infrastructure play only a minor, but important, role in global water problems. Far more lives can be saved with the provision of clean water and clean sanitary facilities than by frustrating attacks on water infrastructure. Equally, equitable division of available water can resolve political and social problems to a greater extent than avoidance of terrorist attacks. We have discussed these issues in earlier meetings of the World Federation of Scientists. However, terrorist attacks on water infrastructures are important and can cause severe public reaction and chaos.

"Those who know do not predict. Those who predict do not know"—Lao Tzu-6th century BC. However, other options, such as black boxes or crones on three legged stools at Delphi, are less useful than predictions. How to prepare for terrorist attacks on water infrastructure systems? In the Black Swan, Nassim N. Taleb suggests that we prepare to repair the effects of unexpected events as we cannot predict what these events are and

when they will occur. However, we all know that it is better to prevent such potential events. That is a difficult task.

We do know that terrorists want to instill fear. They want to see blood and bodies. Most predictable attacks on water infrastructures do not do this. This does not mean that such attacks will not occur. We can also assume that such attacks will be on major cities and/or iconic facilities such as the White House, Pentagon, etc. Therefore, the targets that need to be hardened are limited. We also know that we could limit damage to water supplies in iconic building by such relatively easy, but not necessarily inexpensive measures as providing holding tanks at such sites and testing the water before it is allowed into the building. One could also limit the size of such tanks by installing drinking water only pipelines.

Natural and human generated disruptions of water supplies have occurred such as in Katrina and Fukushima. In the event of a disruption in service in major cities, one could take advantage of the large stock of bottled water in the homes, stores, warehouses etc., until clean water supplies can be restored. Rationing of the supplies could result in a more equitable distribution of supplies.

Studies have been made of the cities and buildings most likely to be attacked. However, to our knowledge, the engineering needed and the costs to harden water supplies for iconic buildings and to provide drinking water only pipes have not been studied. Further, the externalities such as the value of having secured water supplies in iconic buildings have not been studied. Such studies would be valuable in determining whether hardening of water supplies is feasible.

Various attack scenarios as noted below were presented at the meetings.

Various Attack Scenarios

Agent Type	LD-50 70 Kg male in 1 L	Cost LD-50 Dose	Total cost to contaminate 10^6 gallons	Cost per casualty 50% Lethality	Amount Required
Plant Extract	1 Unit	$0.002	$7500	$1.50 – 0.04	0.8 Kg
Plant Extract	7 units	$0.0012	$4500	$0.90 – 0.04	7.9 Kg
Street Drug	0.0001 mg	$0.0075	$28125	$2.81	0.375 Kg
Pesticide	140 mg	NA	NA	$0.04	525 Kg
Insecticide	62 mg	$0.001	$3750	$0.75 – 0.04	281.25 Kg
Industrial Chemical	300 mg	$0.0039	$14625.00	$2.93-0.04	1125 Kg

Inferential monitoring as noted below appears to be the most viable choice to protect buildings, however these indirect measurements are deemed to provide only limited protection at this stage.

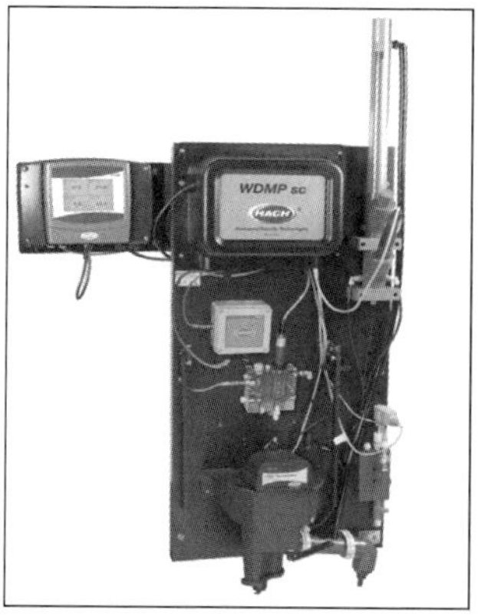

Toxicity Monitoring *Multi-Parameter Monitoring*

Examples, as noted below, were given relative to how fast an attack on a city's water supply will kill 1000's of people:

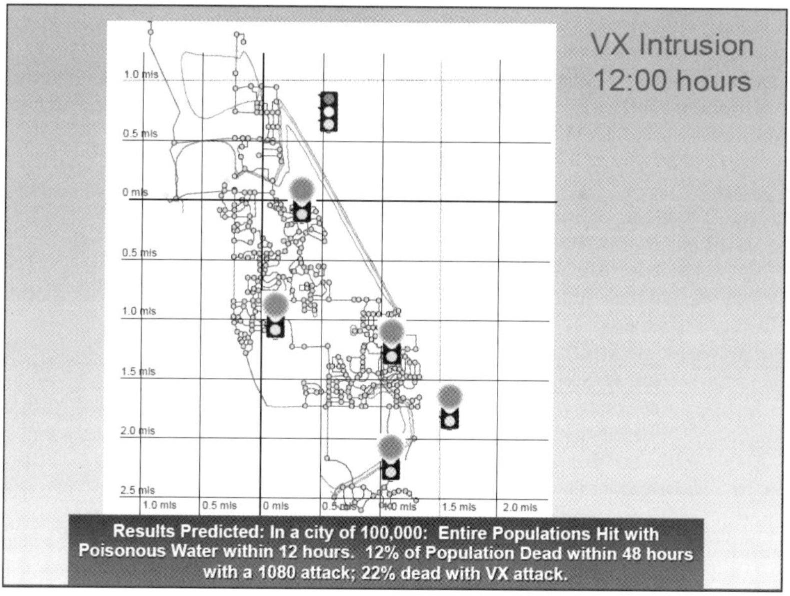

A PROPOSAL FOR SOIL GAS MONITORING IN SHALE-GAS REGIONS IN EUROPE

Background conditions:

- Soil gas anomalies due to geologic structure
- Shale-Gas Exploration
- Relation between groundwater dissolved chemistry and soil gas

Methodology—Situate soil gas monitoring above yet close to the groundwater table. Monitor soil gas and environmental parameters over time and relate to environmental or human variables.

<u>Site selections</u>

- Shale-Gas regions—To Be Determined
  Background sites (5-10)
  - No historic or active shale-gas drilling
  - No near future drilling plans (NY state, or other locations with moratoria or bans)
- Geologic Structure (2-5)
  - Fault zones
  - Anticline/syncline regions (highly fractured)
  - Where exploration is presently permitted
  - No existing evidence of soil gas anomalies
- Shale-gas exploration (4-8)
  - Where exploration is presently permitted
  - No existing evidence of soil gas anomalies
  - No Existing wells, however wells planed in near horizon (1-3 years)

- Soil Gases of Interest:
  - Methane (CH_4)
  - Ethane (C_2H_6)
  - Propane (C_3H_8)
  - Butane (C_4H_{10})
  - Hydrogen Sulfide (H_2S)
  - Carbon Dioxide (CO_2)
  - Nitrogen (N_2)
- Other monitoring variables (may need equipment in addition to Gas Clams):
  - Barometric pressure
  - Atmospheric pressure at monitoring location
  - Groundwater level
  - Temperature
  - Precipitation
- Soil gas monitoring equipment:
  - Gas Clam
- Soil gas data Collection and storage:
  - Waiora
- Groundwater sampling and analysis—Data Analysis:
  - Simple parametric and non-parametric comparisons
  - Principal component analysis

A legacy Image to Overcome!

A PROPOSAL WAS SUBMITTED ON SAFE AND SECURE MANAGEMENT OF SPENT FUEL MANAGEMENT

In all developing countries, there is a large and growing demand for energy, and specifically for electricity. In developed countries the demand is also growing for affordable and clean electricity. Renewables such as solar, wind, geothermal and hydropower will not be able to satisfy these needs in the foreseeable future. The facts that increased use of nuclear power can not alone solve the problem, but that it must be part of the solution are also recognized. The severe reactor accident at Fukushima has led to a few nations proposing to phase out nuclear power. However, in the majority of countries, the argument for carbon free energy and the over-riding global need for secure energy sources have together maintained global interest in expanding nuclear power. The so-called "nuclear renaissance" was slowed but not stopped by Fukushima.

But extensive nuclear growth will remain wishful thinking unless some crucial requirements are satisfied. Nuclear energy production must be safe, secure and economic both at the front-end (from the mining of uranium through its enrichment to its burn up in nuclear reactors) and also at the back-end, i.e., in the waste management area. The front end issues tend to dominate the nuclear debate, most recently because of the concerns about enrichment capabilities bringing nations closer to weapons capability. However, the back end must not be neglected.

Proposal—A multiple year project at the Centre for Planetary Emergencies could contribute to ensuring that nuclear power is recognized as a sustainable energy source, whose extensive use can enhance global security and safety. A small team based at the Centre could provide invaluable scientific support and advice to those responsible for radioactive waste management in particular in the large number of developing countries that are intent on introducing nuclear power.

Each country should have a structured national strategy for managing such wastes. The Centre can provide advice and training for national experts, using input from developed nuclear nations. Some countries will decide to implement purely national facilities for managing their wastes. Even if they do so, it is a global concern that these facilities are well engineered and operated in a way to minimized hazards. Experience has shown that nuclear accidents of any kind, in any land—even if the direct consequences are localized—can have a global impact on the acceptability of nuclear power, and hence a global impact on the environment. The expertise available at the Centre can help ensure that national facilities are indeed state-of-the-art.

Ultimately, however, highly hazardous spent nuclear fuel containing fissile plutonium should not end up in numerous, scattered locations around the globe as more and more nations, both large and small, contemplate expanding or introducing nuclear power. Fewer, safely constructed and well secured storage and disposal facilities must be the goal. The key challenge in this regard is the siting and construction of deep geological repositories for long-lived radioactive wastes. These repositories are expensive; even the smallest state-of-the-art deep facilities for high level radioactive wastes (HLW) or spent fuel will cost several billion dollars.

Many small nuclear programs, or countries starting out in nuclear, do not have the technical and/or financial resources to implement a national repository in a timely fashion. They will have to keep their spent fuel in interim storage facilities; this could result in numerous sites all around the world, at each of which hazardous materials will be stored for decades to hundreds of years. One safer and more secure option would be for nuclear fuel suppliers to take back the spent fuel under a leasing arrangement and add it to their own larger stocks, which would be stored for later reprocessing and recycling into new fuels. However, although there is fierce competition among nuclear suppliers to provide reactors, fuels and reprocessing services, there are as yet no offers to take back fuel. The "take-back" options that have offered by Russia are exclusively for Russian fuel—and the concept in any case covers only new fuel supplied and not the extensive further inventories of radioactive waste that must be disposed of in geological repositories.

The most promising option that remains open for small and new nuclear power programs is to collaborate with similarly positioned countries in efforts to implement shared, multi-national repositories. Most credible is the cooperation of geographically contiguous or close nations in the scope of regional repository projects. The national advantages in sharing technology and in benefiting financially due to the economies of scale in repository implementation are obvious. The global safety and security benefits in helping all nations have earlier access to state-of-the-art repositories are also clear. The big challenge, of course, is in achieving public and political acceptance in the repository host countries.

Over the past few years, significant progress in this direction has been made in the SAPIERR project (acronym for "Strategic Action Plan for Implementation of European Regional Repositories"). The Project, originally conceived by the not-for-profit Arius Association and funded by the European Commission, has carried out a range of studies that lay the groundwork for serious multinational negotiations on the establishment of one or more shared repositories in Europe. The studies (all available on the web-site www.sapierr.net) have looked at legal and liability issues, organizational forms, economic aspects, safety and security issues and public involvement challenges. Based on these studies, a Working Group with representatives of around 10 EU Member States have established of a Working Group for a European Repository Development Organization (ERDO-WG). The European Commission has issued legislation which makes feasible the implementation of shared European repositories and proposals for setting up a formal ERDO, domiciled in an EU Member State.

By combining their resources in this way, the partners in ERDO can also demonstrate to other regions of the world the feasibility of enhancing safety and security whilst increasing the economic attractiveness of nuclear power, even for small countries. The ERDO could act as a role model for regional groupings elsewhere. The Arab States have recently made clear that they intend to introduce nuclear power, and that the will to do so collaboratively. Other world regions with potential regional groupings are in Central and South America, Asia and Africa. Several countries in such regions have expressed interest in the regional repository concept and some have already attended relevant meetings on the topic. With funding from US charitable Foundations, Arius is currently promoting regional storage and disposal concepts in the Arabian Gulf and in South East Asia.

A Centre for the Study of Planetary Emergencies could include a project on the safe and secure long-term management of nuclear materials from nuclear applications. A program extending over around 5 years, utilizing some staff based at the Centre and drafting in external experts for specific studies could help advance global technical and societal preparations for extended use of nuclear power.

An inaugural workshop on the topic could attract participants from the ERDO-WG countries, further European States and also from the UAE, Jordan, Egypt, Morocco, Algeria, Iran, Taiwan, South Korea, Vietnam, Malaysia, Philippines, Thailand, South Africa, Namibia, Nigeria, Ghana, Mexico, Argentina, Brazil and Chile. All of these are interested in introducing or expanding nuclear power programs.

Further interest in such a Workshop could be expected from the large nuclear supplier countries. These should be concerned that back-end solutions are made available to small countries where they are promoting nuclear power and selling nuclear materials. The Workshop could perhaps initiate dialogue between small nuclear user countries that are being asked to forego certain rights that they have under the NPT and the potential large service providers that may be prepared to help with customers with back-end solutions if this increases commercial opportunities without raising concerns about global security.

Subsequently, the Centre could support studies and dialogue workshops on the safety, security, economic and political risks associated with managing global inventories of hazardous radioactive materials that must be stored for long times before ultimately being disposed of in deep geological repositories—a step which is agreed to be the only feasible permanent solution. The major lessons learned from past attempts at resolving these issues made it clear that repository site locations and technologies such as represented below will not be a part of the discussions.

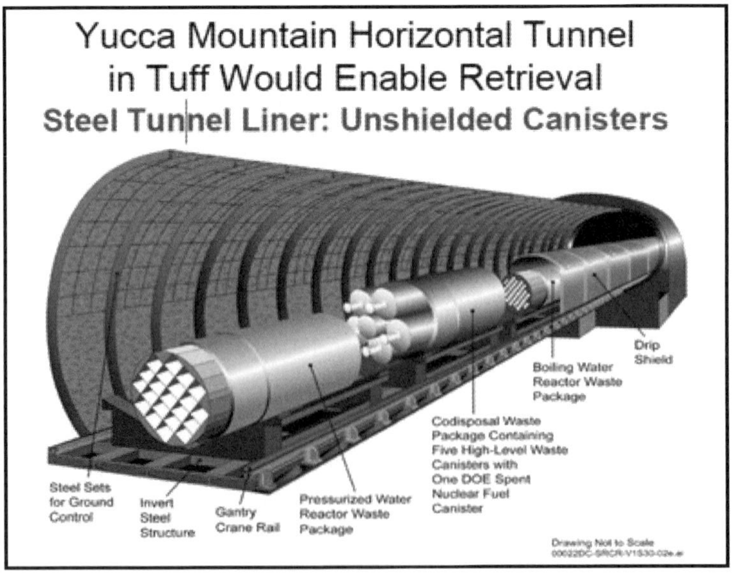

THE PCE CHALLENGE

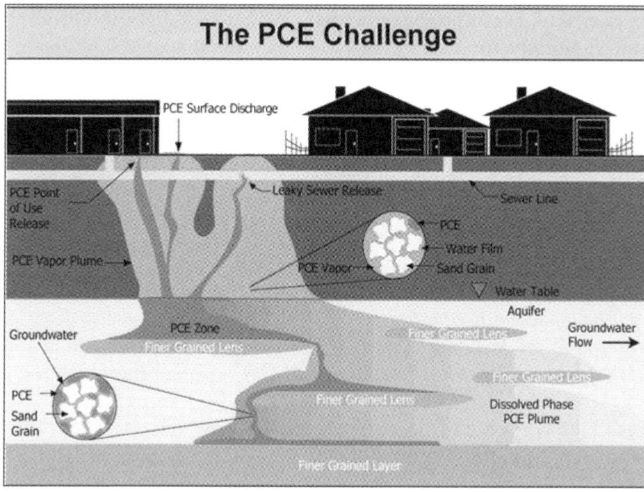

The First Pollution PMP Contaminant (see below) Discussed was Tetrachloroethylene (PCE) 24 years ago in Erice. This year we wish to recommend that in 2013 the Pollution Panel focus on "Innovative Cleanup Technologies".

When I first came to Erice we felt that DNAPL (Dense Non Aqueous Phase Liquids/PCE) could not be cleaned up. However as has been shown at the Interagency Government Test site at the Kennedy Space Centre below, DNAPL sites can now be remediated by several new technologies.

LAUCH COMPLEX 34, CAPE CANAVERAL, FLORIDA

One of the emerging areas of environmental need is soil gas investigations relative to vapor intrusion risk as noted below:

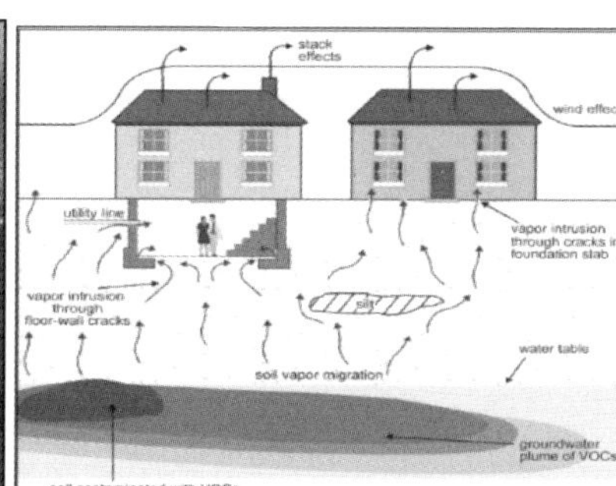

As Chairman of ASTM International D 18.21.02, I have been very active in developing soil gas monitoring systems for the past 20 years. The list of the soil gas standards developed to date is provided below.

American Society for Testing and Materials International Chairman, L. Everett, ASTM D18.21.02

- *D5314-92 (2006) Standard Guide for Soil Gas Monitoring in the Vadose Zone*
- *D7758 (2012) Practice For Passive Soil Gas Sampling in the Vadose Zone for Source Identification, Spatial Variability Assessment, Monitoring, and Vapor Intrusion Evaluations*
- *D7648 (2012) Practice For Active Soil Gas Sampling for Direct Push or Manual-Driven Hand-Sampling Equipment*
- *D7663 (2012) Practice for Active Soil Gas Sampling in the Vadose Zone for Vapor Intrusion Evaluations*
- *International Dynamic Soil Gas Monitoring Symposium including Methane risk, Jacksonville, Florida January 2013. Chairman Lorne Everett*

One of the proposed innovative technologies for 2013 is Environmental Nanotechnology Applications. The session will include discussions of:

- Remediation of contaminated groundwater using nanoparticles containing zero-valent iron is one of the most prominent examples of a rapidly emerging technology with considerable potential benefits.
- One of the main environmental applications of nanotechnology is in the water sector. As freshwater sources become increasingly scarce due to over consumption and contamination, scientists have begun to consider seawater as another source for drinking water. The majority of the world's water supply has too much salt for human consumption and desalination is an option but expensive method for removing the salt to create new sources of drinking water. Carbon nanotube membranes have the potential to reduce desalination costs.
- Nanofilters can be used to remediate or clean up ground water or surface water contaminated with chemicals and hazardous substances. Nanosensors could be developed to detect waterborne contaminants

The operating size range of these nano particles is provided in the figure below and shows that the range is approximately the same as the pollution molecule.

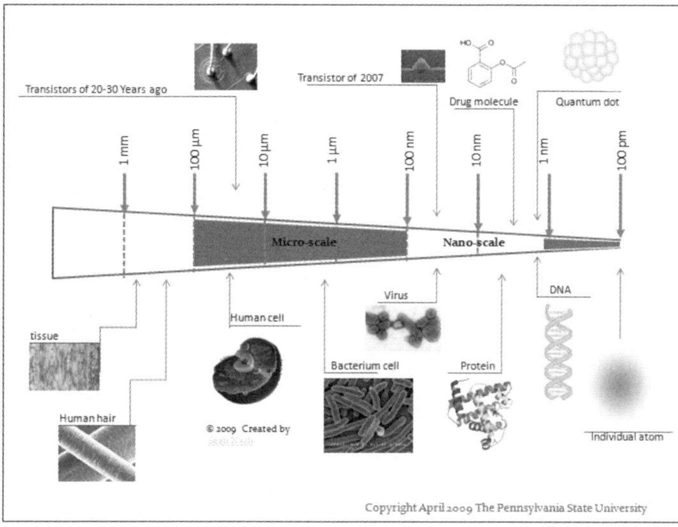

ENDOCRINE DISRUPTING CHEMICALS (EDCS) AND DRUGS IN WATER

Another of the themes for 2013 includes endocrine disrupting chemicals (EDCs) and drugs in water: A Bioterrorism Concern, 2013. Topics to be addressed include:

- EDCs act via receptors at very low doses (at and below one part per trillion).
- EDCs are easily obtained by terrorists.
- They are not being routinely monitored.
- Remediation technology is not yet in place.
- The cost of responding to the loss of a major source of drinking water would be enormous.
- Many diseases are increased by exposure to very low doses of EDCs, and damage is transmitted to future generations.

The final proposal submitted for consideration in 2013 is a feasibility study to address a Non-Proliferating Fuel Cycle-No Enrichment, Reprocessing or Accessible Spent Fuel. The topic will include discussions of:

- The non-proliferating system of uranium from the sea, (no mining or milling), use of Candu type reactors (no enrichment) and no reprocessing—infinite supplies of uranium from the oceans (no plutonium or high level waste streams) and no land based disposal of radioactive wastes (disposed of in sub-seabed sediments) removes many objections. CO_2 now allowed to be disposed in sub-seabed sediments. Spent fuel would be sent by barge or ship to U.S. Naval Reactor Shipyards for storage and staging for disposal. Such sites have most of the equipment and security needed. All of these systems have been carried out at commercial or pilot stage facilities.

- Of course, all these comments are assertions. Modeling and pilot testing need to be carried out to verify these claims and determine the costs. In addition, the environmental impact, including embedded emissions associated with making, installing, maintaining and decommissioning the facilities and equipment must also be calculated. Then, the feasibility of the non-proliferating fuel cycle can be determined.

THE 2012 POLLUTION PMP MEETING PARTICIPANTS AND CONTRIBUTORS INCLUDED THE FOLLOWING:

Dr. Lorne G. Everett—L. Everett & Associates, USA
Distinguished Professor Frank L. Parker—Vanderbilt Univ. USA
Professor Fred vom Saal—University of Missouri, USA
Professor Stephano Parmiagiano—University of Parma, Italy
Dr. Franco M. Buonaguro—Istituto Nazionale dei Tumori, Napoli, Italy
Dr. Dan J. Kroll—Hach Homeland Security Technologies, USA
Dr. Walter M. Grayman—Consulting Engineer, USA
Dr. Richard Ragaini—LLNL-Retired, USA
Dr. Gina Marie Calderone—ECC, USA
Dr. Charles McCombie—AEGIS, Switzerland
Dr. Thomas Ballestero—UNH, USA

REPORT OF THE PERMANENT MONITORING PANEL ON ENERGY, 2012

CARMINE DIFIGLIO
Office of Policy and International Affairs, U.S. Department of Energy
Washington, USA

ROBERT SCHOCK
World Energy Council, Senior Advisor
London, UK

WILLIAM BARLETTA[1]
U.S. Particle Accelerator School, Department of Physics Massachusetts Institute of Technology, Cambridge, USA

The Energy PMP organized eleven plenary-session presentations in two topics: "Global Nuclear Energy Issues" and "Energy and Sustainability in Cities".

GLOBAL NUCLEAR ENERGY ISSUES

A lively plenary morning session on the 21st August covered Global Nuclear Energy Issues. The outlook of nuclear power in the USA, Japan, China, Europe and the Middle East was each covered in five separate papers; also an IAEA paper on "World Outlook for Nuclear Power after Fukushima" was presented. The following debate considered that the Fukushima event did deliver a serious blow to the future development of nuclear power in the European Union, also the U.S., where the public opinion has turned hostile to nuclear development and the cost of the facilities has significantly increased due to more stringent safety and regulatory requirements. Contribution of existing plants in Japan and Germany was severely restricted because of this event. However, nuclear prospects in China, Korea, India, Russia and other rapidly developing economies in Asia and the Middle East are not likely to be negatively affected. These countries are pursuing their nuclear power agenda with the same past vigour. Their investment costs are less than those expected in OECD countries. Still it is predicted that the present global contribution of nuclear to electric power, presently 15%, is going to be gradual and will slightly decline over the years in the foreseeable future. The session also discussed the economics of "Nuclear Power in a Liberalised Power Sector" with the outlook that the private sector is not going to seriously consider investing in nuclear power, not only due to its cost and long implementation time, but also due to regulatory and policy uncertainties. While the substantial capital requirements for nuclear power are not easy to mobilise, the outlook for nuclear power is promising in many Asian countries where nuclear reactors are estimated to be more competitive than in the U.S., Europe or Japan and where private financing is less important. The increasing supplies of shale gas, in the US, and potential increase of natural gas supplies in many other countries are

---

[1] The summaries of the 2012 Erice plenary sessions on *Global Nuclear Energy Issues* and *Energy and Sustainability in Cities* were provided by Hisham Khatib.

encouraging the development of combined-cycle gas turbine plants by private and public investors. It is reckoned that without state involvement and funds, nuclear power is not likely to expand.

ENERGY AND SUSTAINABILITY IN CITIES

A special afternoon session was devoted to the subject of Energy and Sustainability in Cities. Urbanisation is increasing rapidly in almost every country of the world, but especially Asia and Africa. Large cities with over one million are becoming common and the number of megacities, of over ten million inhabitants, is increasing rapidly. They demand enormous amounts of energy and other resources and services. Cities are complex systems and require sophisticated planning not only in their growing energy systems, but also in other vital and related services like water, sanitation and control of pollution. A Seminar Workshop took place in the morning of the 24th in which this complex issue of meeting energy requirements and emissions reduction in cities was discussed. This subject is likely to be pursed in the future seminars.

ENERGY PMP MEETING, AUGUST 2012

The annual meeting of the Energy PMP was organized and chaired by Bruce Stram. It consisted of 9 presentations with discussion, revisions to the Energy PMP website and priority setting for 2013.

<u>Presentations</u>

• Fusion Update	Jef Ongena
• Japanese Energy Research Update	Jef Ongena for Akira Miyahara
• Energy Subsidies	Hisham Khatib and Carmine Difiglio
• Potential for Renewable Technologies	Hisham Khatib
• Potential Oil Recovery from Existing Fields	Adnan Shihab-Eldin
• Energy Democracy in Japan	Tatsuo Masuda
• China Automotive Energy Outlook	Zhang Xilang
• Insurance Value of Energy R&D	Robert Schock
• Japanese Energy Decisions	Richard Wilson

ENERGY PMP WEBSITE

Robert Schock and Carmine Difiglio provided an update on the revised Energy PMP website. The Energy PMP now has a webpage on the World Federation of Scientists' PMP page. This new Energy PMP page contains general information about the Energy PMP and past activities. It contains a link, now not functioning, to EnergyPMP.org that had been maintained by Richard Wilson. We are now in a transition stage transferring the considerable content that had been contained in EnergyPMP.org to a new page that will be reorganized and maintained by Robert Schock with small annual financial support from the Energy PMP members. We aim to have all the Energy PMP plenary papers from the 2012 Seminars and future seminars on the new website as well as many previous papers that we can assemble from World Federation of Scientists archives. Discussions are continuing as to how the webpage will be organized. We intend to provide a webpage that will build on the work of Professor Richard Wilson and expand the considerable readership that the past webpage enjoyed.

2013 ENERGY PMP PRIORITIES

The Energy PMP meeting concluded with a discussion of its 2013 priorities. Members proposed the following possible activities for 2013:

- Energy and cities
- Insurance as the real climate change issue
- New technologies for reducing GHG emissions
- Biofuels: new technologies to use bio sources—higher efficiency—prepare proposal to use new technologies to be more economic
- How do we integrate new sources of electricity into the grid?
- End use efficiency improvements: what's possible—lighting, motors—meet the gap, the cost to replace of old technologies
- Safer, smaller, more economic nuclear reactors—assessment of those future nuclear reactor technologies
- International nuclear safety regime—with international action team to support response to nuclear accidents
- Scientific limits of renewable energy and renewable technologies
- Economics of advanced energy technology deployment

In addition, it was agreed to contribute to Professor Zichichi's Project for Man, a Centre to Study Planetary Emergencies. It was announced that Professor Zichichi would ask the WFS members and, in particular, the PMPs for ambitious Centre proposals. Resources would be available for permanent staff and significant research in areas that would mitigate the likelihood of planetary emergencies in line with the principals of the Ettore Majorana Foundation and Centre for Scientific Culture, the World Federation of Scientists and the World Laboratory. The Energy PMP has defined the energy emergency as follows:

Energy supply affects economic growth, pollution, climate change and political tensions among nations. Oil prices have been, and are likely to remain a major factor limiting global economic recovery. Almost 20% of the world's population (1.3 billion people) still lack electricity and basic energy services, without which, education, economic progress and political self-realization are barely possible. While most OECD countries have vastly improved energy's environmental footprint, energy production and use remain a major source of air and water pollution in many countries, severely compromising health and well-being. Lastly, the burning of fossil fuels is the primary anthropogenic contributor to increased greenhouse gas concentrations. Consequently, the Energy PMP monitors scientific results on the following subjects:

- Energy efficiency technologies and policies.
- Renewable energy technologies and policies.
- Nuclear power technologies and policies.

- The international oil and gas markets.
- Energy services for the poor.
- Energy and cities.
- Technologies and policies to limit greenhouse gas emissions.

It was agreed that the Energy PMP would work closely with colleagues in other PMPs and develop significant proposals for the Centre.

UPDATE

Two Centre proposals have been submitted by the Energy PMP:

- Safety of Nuclear Power.
- Energy, Cities and Global Sustainability.

MOTHER AND CHILD HEALTH PMP

NATHALIE CHARPAK
Kangaroo Foundation
Bogotá, Colombia

MOTHER AND CHILD PMP MANIFESTO (ERICE 2002) ***Maternal and Child mortality is a planetary emergency***

REMINDER OF THE MISSION OF THE MCH PMP

- Recognize new and monitor existing tools that decrease maternal and infant mortality and morbidity.
- Highlight the impact of the other planetary emergencies on Maternal and Infant mortality and morbidity and therefore contribute to the enhancement of the quality of the present life and future generations.

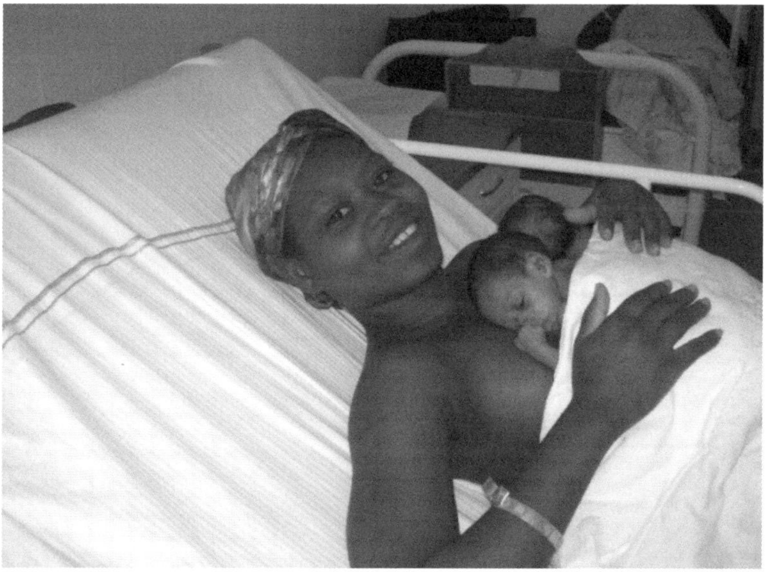

WHAT IS THE KMC METHOD?

Kangaroo Mother Care (KMC) is a multifaceted intervention involving mothers and preterm or low birth weight infants. Its short and middle term positive effects on survival, neurodevelopment, breastfeeding and quality of mother-infant bonding have been already demonstrated. At least 20 million infants per year (90% born in low and middle income countries) are candidates to take advantage of the KMC method, nevertheless diffusion is still slow and coverage is insufficient even if it is each day progressing.

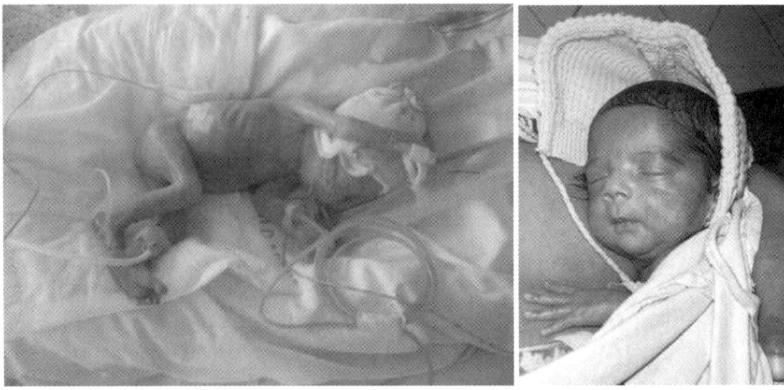

1. Kangaroo Position: Skin-to-skin contact attached to mother's chest, 24 hours a day, upright position.

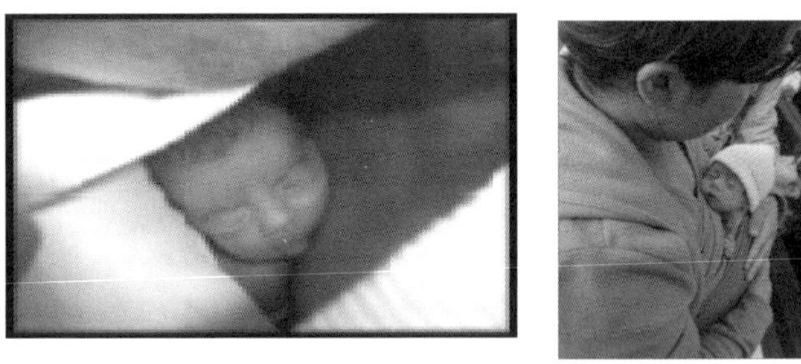

2. Kangaroo Nutrition: Exclusive or nearly exclusive breast feeding.

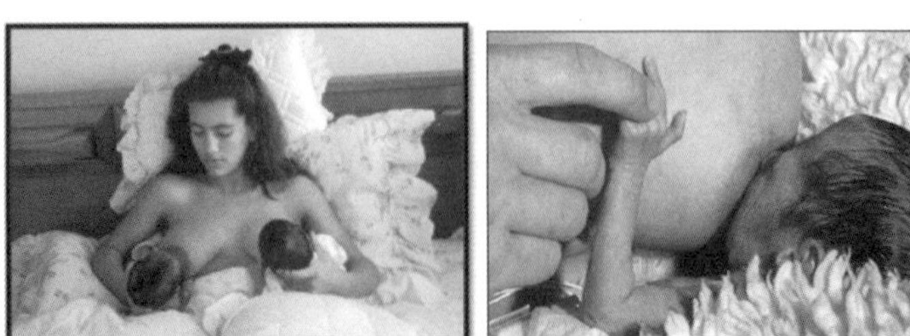

3. Early Discharge home (instead of minimal care at hospital) while in Kangaroo position with frequent and strict follow-up (Kangaroo Clinic).

RESEARCH: FOCUS ON PREMATURE INFANTS—FOLLOW-UP STUDIES

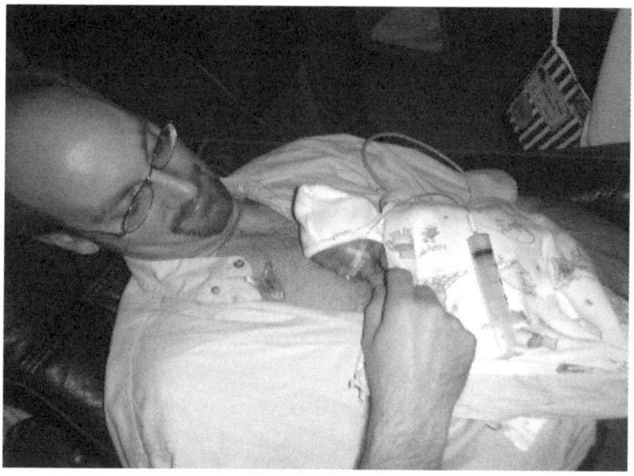

- Key focus is to examine the incidence of major disabilities: CP, Mental Retardation, cognition, Learning Disabilities and behaviour.
- The prevalence of CP in VPT children ranges between 8-20%; CP is often viewed as a marker of the quality of neonatal intensive care.
- However, it is well-known that even among infants *with no major disability*, a high prevalence of low severity motor deficits become apparent at pre-school age.
- Service-based follow-up is generally conducted up to age 2-3 years, and may miss the onset of motor problems.

(Saroj Saigal Courtesy)

Neuro-Sensorial Problems among 323,958 Children at 3y-Old
Van Naarden & Decouflé. Pediatrics 1999;104:905

Weight at birth	**RR ratio(95%IC)**
> 3999 g	1.1 (0.6-2.0)
3000-3999 g	1.0 (reference)
2500-2999 g	1.8 (1.2-2.7)
1500-2499 g	3.5 (2.2-5.4)
< 1500 g	**13.9 (8.2-23.4)**

Academic Achievement

Reference	**Measures**	**Groups**	**Résults**
Hack 94, USA	Need special class at 6 years	<1500g Term	25% 14%
Botting 98, UK	Need special support at 12 years	<1500g Peers	35% 14%
Saigal 2000, Canada	Special support or repeated grade at 14 years	<1000g Term	58% 13%
Lefebvre 2005, Canada	Repeated grade before 18 years	<1001g Term	56% 15%

Studies Reporting Motor Outcomes (Dcd) in Preterm

Author	**N**	**Birth Weight/ Gestational Age**	**Year of birth**	**Age**	**Score DCD**	**DCD% (minor neurological disorders)**
Powls '85	47	<1251g	1980-81	12-13	<15th	51%
Jongmans '97	156	<35 wks	1984-86	6	<15th	44%
Holsti '02	73	<800g	1982-87	9	<1 SD	51%
Evensen '09	54	<1500g	1988	14	<5th	23%
Foulder-Hughes '03	280	<32 wks	1991-92	7-8	<5th	31%
Davis '07	255	<1000g	1991-92	8-9	<5th	10 %
Goyen '09	50	<1000g	1992-95	8	<1 SD	42%
Marlow '07	180	<25 wks	1995	6	<1 SD	—
Epipage '07	1237	<33 wks	1997	5	-	44 % MND
Wocaldo '08	323	<30 wks	1987-97	8	<15th	31%

OUR STRATEGY—THE SCIENTIFIC RESEARCH: A COLLABORATIVE AND MULTI-FACETED QUÉBEC-BOGOTÁ FOLLOW-UP OF 15 YEAR-OLD PRETERM CHILDREN HAVING EXPERIENCED THE KANGAROO MOTHER CARE IN COLOMBIA A PILOT STUDY DURING 2010-2012

- Characteristics of the sample at 15 y-o

	No KMC (18) (Mean)	No KMC (21) (Mean)	Term group (9) (Mean)
Ballard (Weeks)	30.6	30.8	39.13
Birth Weight (g)	1486	1429	2995
Neonatal stay (Days)	27	17	1

- The Transcranial magnetic stimulation (TMS) as a new technology to test brain function in prematurity and KMC: a noninvasive brain stimulation to test brain motor excitability and corticospinal function.

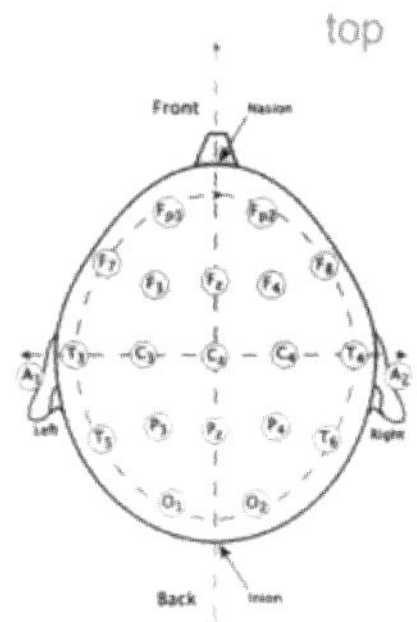

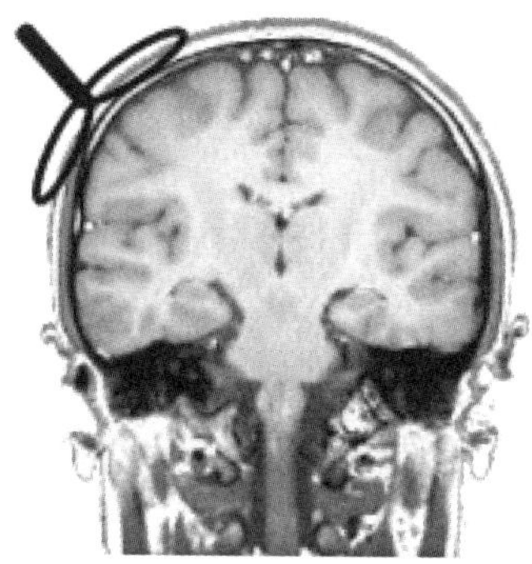

Neuroembryology

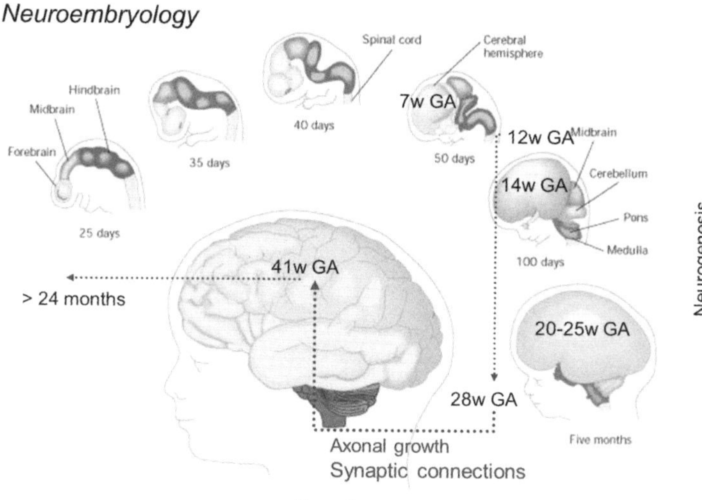

Biological risk = interrupted synaptogenesis and faulty networks (Ex.: CC)

TMS: NON INVASIVE BRAIN STIMULATION
(testing between-hemisphere function = integrity of corpus callosum)

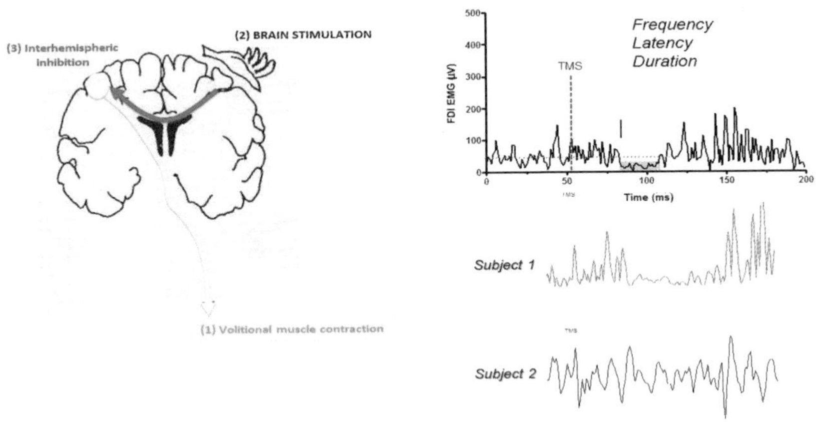

Remarks

- Results from this pilot, hypotheses-generating exercise allow us to formulate the hypothesis that KMC intervention could be associated with better long term outcomes and might help to protect the immature brain and to promote better ex-utero maturation.

ACTA PÆDIATRICA
NURTURING THE CHILD

Acta Pædiatrica ISSN 0803-

REGULAR ARTICLE

Cerebral motor function in very premature-at-birth adolescents: a brain stimulation exploration of kangaroo mother care effects

Cyril Schneider (cyril.schneider@rea.ulaval.ca)[1,2], Nathalie Charpak[3], Juan G Ruiz-Peláez[4], Réjean Tessier[5]

1.Clinical neuroscience and neurostimulation laboratory, Centre de recherche du CHUQ, Québec, QC, Canada
2.Department Rehabilitation, Faculty of Medicine, Université Laval, Québec, QC, Canada
3.Kangaroo Foundation, Bogotá, DC, Colombia
4.Department Clinical Epidemiology and Biostatistics, Faculty of Medicine, Javeriana University, Bogotá, DC, Colombia
5.School of Psychology, Centre de recherche du CHUQ et CIRRIS Université Laval, Québec, QC, Canada

ABSTRACT

Aim: Given that prematurity has deleterious effects on brain networking development beyond childhood, the study explored whether an early intervention such as Kangaroo Mother Care (KMC) in very preterm preemies could have influenced brain motor function up to adolescence.

Methods: Transcranial magnetic stimulation (TMS) was applied over the primary motor cortex (M1) of 39 adolescents born very prematurely (<33 weeks' gestational age, 21 having received KMC after birth, 18 Controls with no KMC) and nine adolescents born at term (>37 weeks' gestational age, >2500 g) to assess the functional integrity of motor circuits in each hemisphere (motor planning) and between hemispheres (callosal function).

Results: All TMS outcomes were similar between KMC and term adolescents, with typical values as in healthy adults, and better than in Controls. KMC adolescents presented faster conduction times revealing more efficient M1 cell synchronization ($p < 0.05$) and interhemispheric transfer time ($p < 0.0001$), more frequent inhibitory processes with a better control between hemispheres ($p < 0.0001$).

Conclusion: The enhanced synchronization, conduction times and connectivity of cerebral motor pathways in the KMC group suggests that the Kangaroo Mother Care positively influenced the premature brain networks and synaptic efficacy up to adolescence.

- The Future. If clearly demonstrated, we could recommend the universal use of KMC in preterm and low birth weight infants worldwide, as part of their proper basic health care, as a practice that protects and rescue this brain development.
- We will extend during the next 2 years this evaluation to 60% of subjects (450) who participated in the 1993-1997 KMC RCT we realized with the support of the World Laboratory. This Colombian cohort is unique, is still accessible and offers an exceptional opportunity to assess the long term impact of a complex, and promising intervention such as KMC. We received a grant from Grand Challenge Canada to realize this new research (2012-2014).

OTHER RESEARCH PROJECTS 2011-2012

- Cost effectiveness of KMC: a collaborative study with the U Javeriana and the U de los Andes.
- Validation of heamospect device and Validation of bilispect device: collaborative study with German team.

DOCUMENTS AND PAPERS 2011-2012

- •Evidence based National guidelines for the preterm: Health Ministery of Colombia
- •One chapter in Collection Le savoir Suisse EPFL Lausanne
- •KMC Multimedia kit as a tool for trainers: Health Ministery of Colombia and World Food Pregram
- •Acta paediatrica In press. Results of the pilot study.
- •Education and traning in KMC collaborative paper with the International KMC network

CARLOS SLIM AWARD 2011

Kangaroo foundation and KMC

DIFFUSION: 15 TRAINING OF NEW KMC CENTERS DURING THE LAST 2 YEARS

Diffusion of KMC

2011-2012

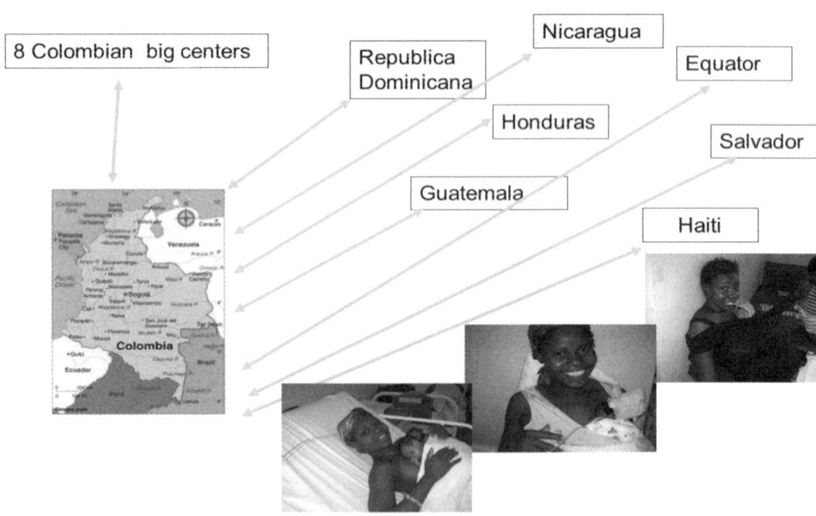

KANGAROO MOTHER CARE
Philippine Edition
Socorro De Leon-Mendoza, M.D.
BT-KMC Foundation Phil., Inc.
Erice 2011

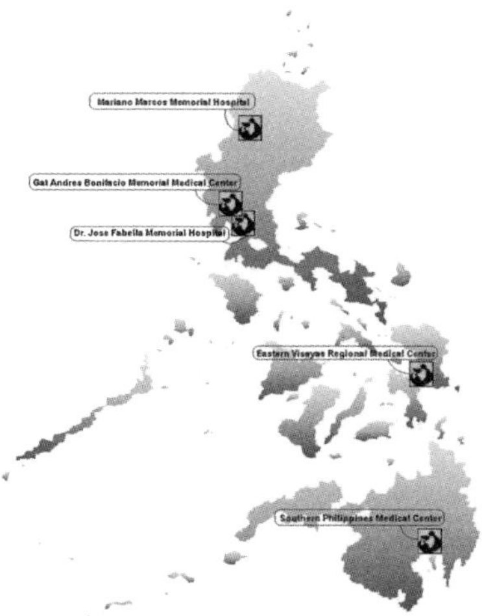

KMC MAP OF THE PHILIPPINES 2012

- Achieving Targets as of July 2012
  - Standard KMC Training of Trainers' manual in use
  - One (1) additional KMC center in City of Manila
  - Accredited KMC center in the North now conducting their own KMC training in the region
  - Two (2) Foundation-funded research completed and presented in the 18th congress of the Federation of Asia Oceania Perinatal Societies (FAOPS) in Sydney Australia
  - Conducted KMC workshop in Ulanbator, Mongolia in collaboration with the FAOPS

- Partnership: Perinatal Association of the Philippines "Adopt-a-KMC-Site" Project
  - Training and initiation of the KMC program at Gat Andres Memorial Medical Centre in Manila and furnished and inaugurated the KMC room at the same hospital
  - Trained a 7-member group of the association to serve as the core trainers who will cascade KMC training to other regions of the Philippines

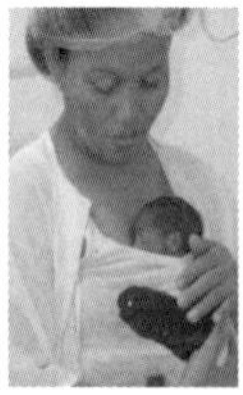
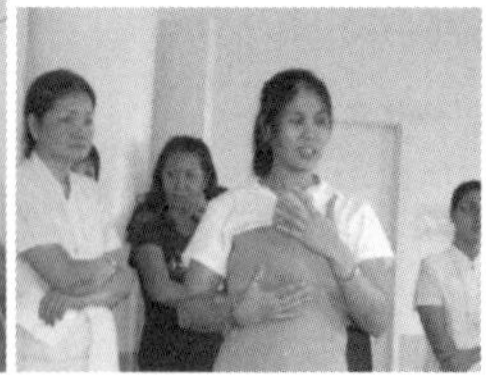

- Continuing Challenges…
  - Establishing local networking in the community by the regional centers for ambulatory KMC
  - Maintaining commitment from established centers even without government support (administrative and logistic)
  - Increasing the demand for the KMC program from other regional health facilities
  - Promoting the intervention in the private sector
  - Generating local evidence
  - Assessing the impact through multi-center research and development
  - Creating a Research Endowment Fund
  - Pushing for KMC as a national health policy covered by a Republic Act from Congress

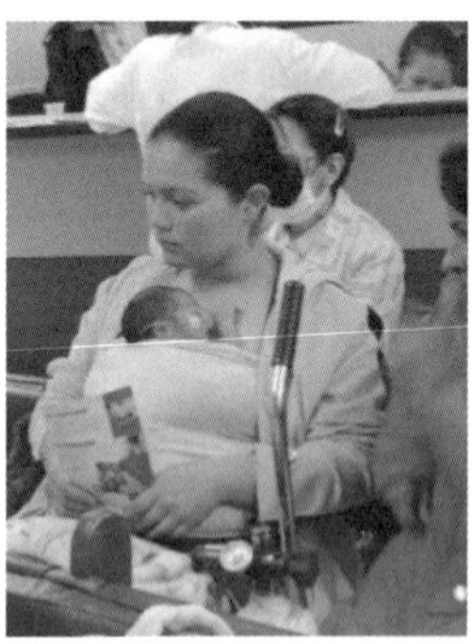
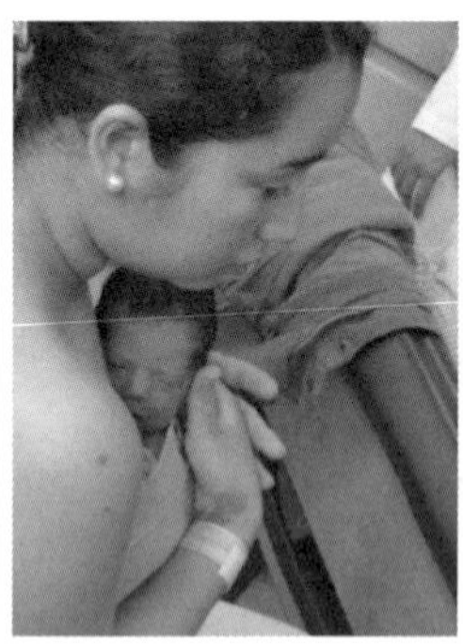
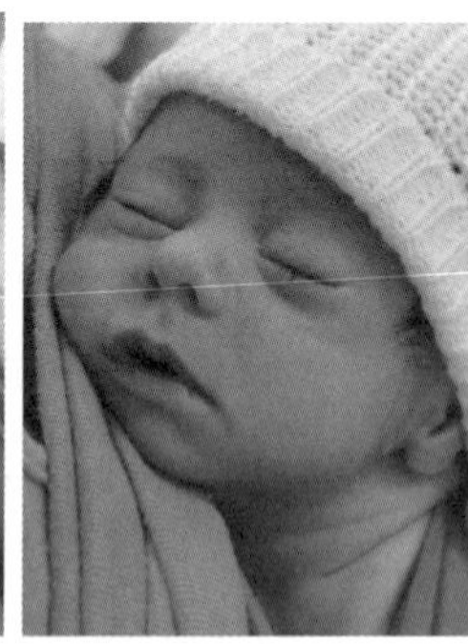

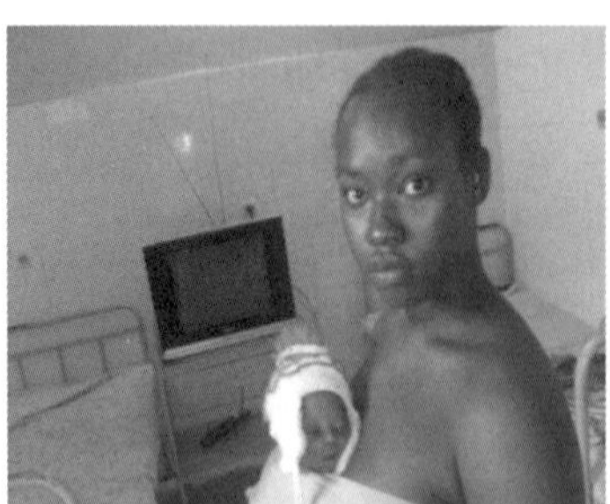
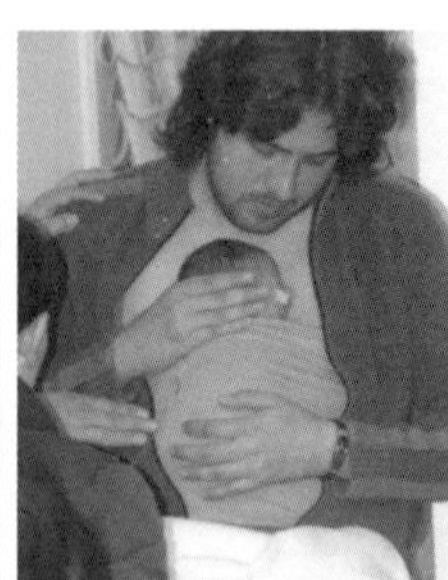

AIDS AND INFECTIOUS DISEASES (AID) PMP 2012 REPORT

FRANCO M. BUONAGURO
Istituto Nazionale dei Tumori, "Fondazione G. Pascale"
Napoli, Italy

In the last few years besides monitoring AIDS disease (with participation to International Conferences, including AIDS Vaccine Meeting), the AIDS and Infectious Diseases PMP focused on emerging diseases.

Since the establishment of the AIDS and Infectious Disease PMP, the focus has been on pathogens of risk group 3 and 4. Those pathogens are dangerous and exotic agents (to be handled at BSL-3/4) that pose a high individual risk of aerosol-transmitted laboratory infections, cause severe to fatal disease in humans for which vaccines or other treatments are not available, such as HIV, Bolivian and Argentine hemorrhagic fevers, Marburg virus, Ebola virus, Lassa virus, Crimean-Congo hemorrhagic fever, smallpox, and various other hemorrhagic diseases. In 1999, a statement by the WFS PMP on AIDS and Infectious Diseases on the urgency of an AIDS Vaccine was elaborated and signed by the most prominent scientists in the field, including Bob Gallo and Luc Montainger. Since 2004 the AID PMP has been monitoring the Avian flu epidemic and a specific session was held at Erice within the context of the 2005 Planetary Emergency Seminars.

More recently the PMP focused on the 2009 Flu Pandemic, contributing to put the H9 Pandemic flu in the right prospective. The 2009 epidemic, in fact, has been very mild considering the total number of infected individuals and the level of morbidity and mortality. It is however needed to verify the global preparedness to prevent a pandemic and the possibility to develop a broadly protective vaccine. To this matter a specific session has been dedicated within the European Congress of Virology, organized by the Italian Society of Virology (whose secretary is Dr. Buonaguro) at Cernobbio-IT, 7-11 April, 2010. New promising developments have been obtained from the crystallographic-based capacity to identify selective epitopes (F.M. Buonaguro et al. "Updates and Achievements in Virology," *Expert Review of Vaccines*, 2010) and the possibility of producing large quantity of vaccine on systems not based on eggs (cell cultures as well as tobacco plant systems, F.M. Buonaguro "Plant-derived Vaccines: Technology and Applications," *Future Medicine*, 2012).

Finally, members of the AID PMP are exploring further approaches to expand the WFS monitoring activity in AIDS and infectious diseases, including emerging diseases, participating to:

1. The AORTIC Meeting on Cancers in Africa, held in Cairo on 30 Nov-3 Dic, 2011;
2. The ASICON Meeting organized by the India AIDS Society, Lucknow, 16-18 December 2011;
3. The Global Virus Network (GVN), whose 3rd Meeting has been held in Naples, 7-9 June 2012;
4. The Burkitt Lymphoma 50 Anniversary, Abramson Cancer Center, Philadelphia, July 30-31, 2012

5. The Infectious Agents and Cancer (IAC) online scientific journal, published by BioMed Central, tracked by Medline, Scopus and Index Medicus. The journal is currently one of the fastest growing journals in epidemiology, oncology and infectious diseases categories, with an unofficial impact factor (I.F.) of 2.33.

The AID PMP proposes a Workshop on Infectious Agents and Vaccine Strategies for the Erice Planetary Emergencies Session of 2013. In particular three new members have been invited to join the group: Ishawar Gilada, Unison Medicare, Mumbai, India; Sam Mbulaiteye, NCI, Bethesda, USA and UVRI, Uganda; and Dave Pauza, IHV, Baltimore, USA. With the contribution of the newcomers, the AID PMP is planning to have the day before the 2013 Planetary Emergencies' Seminars a Workshop on innovative approaches to Infectious Agents monitoring and Preventive strategies.

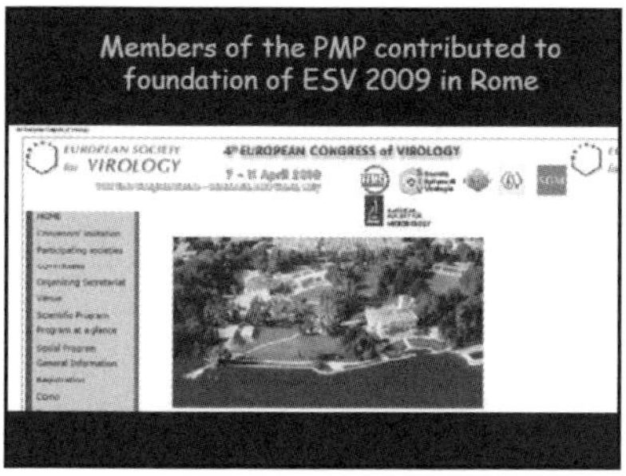
Members of the PMP contributed to foundation of ESV 2009 in Rome
EUROPEAN SOCIETY for VIROLOGY
4th EUROPEAN CONGRESS of VIROLOGY
7 - 11 April 2010

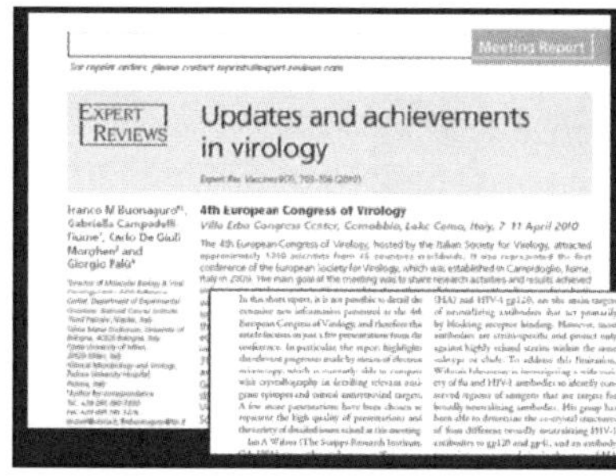
Meeting Report
EXPERT REVIEWS
Updates and achievements in virology
4th European Congress of Virology
Villa Erba Congress Center, Cernobbio, Lake Como, Italy, 7–11 April 2010

The INSTITUTE of HUMAN VIROLOGY
Program Information and Acknowledgements
Welcome
12th Annual International Meeting of the Institute of Human Virology
of the University of Maryland School of Medicine
By
Robert C. Gallo, Founder and Director, Institute of Human Virology
and
Giovanni Latorre, President, University of Calabria

INSTITUTE OF HUMAN VIROLOGY
GLOBAL VIRUS NETWORK (GVN)

Virion
EU Foundation Established for GVN-EU
THIRD GVN MEETING SET FOR NAPLES
NAPLES MEETING AGENDA

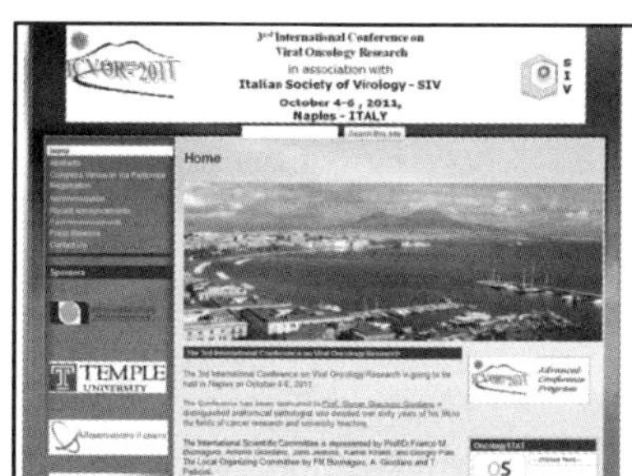
3rd International Conference on Viral Oncology Research
in association with
Italian Society of Virology - SIV
October 4-6, 2011,
Naples - ITALY
Home
TEMPLE UNIVERSITY

CONFERENCES
AORTIC

National Conference of
AIDS Society of India
ASICON
2011 LUCKNOW

Global Health Burkitt Lymphoma Meeting
Abramson Cancer Center, Philadelphi
July 30-31

Penn Medicine
Penn Medicine

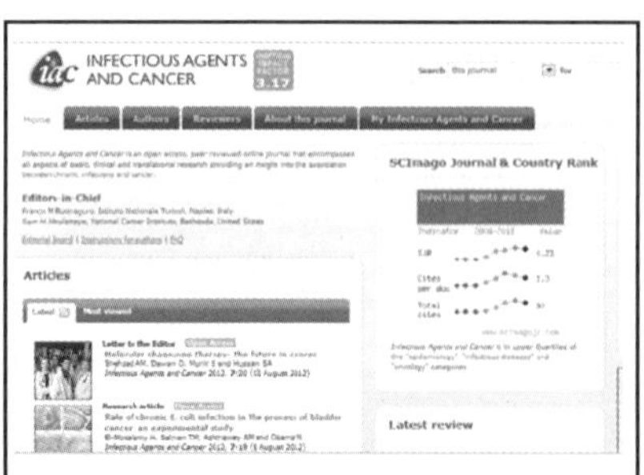
INFECTIOUS AGENTS
AND CANCER
Articles
SCImago Journal & Country Rank
Latest review

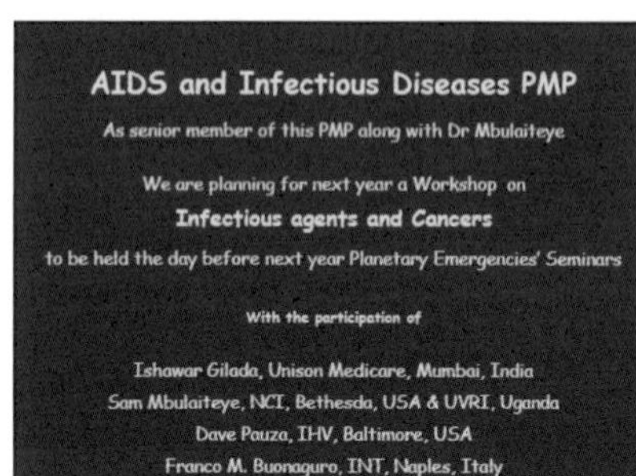
AIDS and Infectious Diseases PMP
As senior member of this PMP along with Dr Mbulaiteye
We are planning for next year a Workshop on
Infectious agents and Cancers
to be held the day before next year Planetary Emergencies' Seminars
With the participation of
Ishawar Gilada, Unison Medicare, Mumbai, India
Sam Mbulaiteye, NCI, Bethesda, USA & UVRI, Uganda
Dave Pauza, IHV, Baltimore, USA
Franco M. Buonaguro, INT, Naples, Italy

INFORMATION SECURITY PMP REPORT: AUGUST 23, 2012

HENNING WEGENER
Ambassador of Germany (ret.), Information Security Permanent Monitoring Panel
World Federation of Scientists, Madrid, Spain

JODY R. WESTBY, ESQ.
CEO, Global Cyber Risk LLC
Washington, USA

Continuing its work from the session in August 2011, the PMP continued to promote its joint publication with the Secretary General of the International Telecommunications Union (ITU), Dr. Hamadoun Touré, *The Quest for Cyber Peace*. The book was published in January 2011 in six languages and is available free of charge on the Internet at www.itu.int/pub/S-GEN-WFS.01-1-2011. As a UN publication, the book has been distributed on a global scale, and serves as an important reference of the PMP's work for a wider audience.

The success of the *Cyber Peace* book encouraged the PMP to discuss a follow-on joint publication, *The Quest for Cyber Confidence.* Over the past six months, the PMP worked with the Secretary-General and his staff to develop a detailed outline for the publication. The majority of this year's PMP session in Erice was spent reviewing the draft outline and finalizing it. Drafting assignments were agreed upon and a mid-year drafting meeting is planned for late January at the ITU in Geneva. Release of the publication in three languages is planned for October 2013 at the World ITU Telecom convention. Translation into three additional languages will be completed shortly thereafter.

The PMP also discussed its Cybersecurity Hard Problems List, which is a living document and is being updated to incorporate new issues identified in recent U.S. government cybersecurity R&D strategies. The PMP has approached the Georgia Institute of Technology's Information Security Center about posting the Hard Problems List on their website to make it available to the global R&D community. The PMP will provide links to this site from WFS and ITU websites. The goal is to have the Hard Problems list serve as a central resource point regarding who is conducting research on specific problems and what types of data they are being used in testing solutions.

The PMP continued its participation in the World Summit on the Information Society through PMP co-chair Henning Wegener's attendance at the May 2012 session in Geneva.

Throughout the year, PMP members highlighted the work of the PMP through their speaking engagements, conference participation, published work, and media interviews.

The PMP also discussed proposals for the planned Center for the Study of Planetary Emergencies and identified five possible project areas: (1) Amplifying and co-branding the new book project undertaken jointly with the Secretary General of the ITU (2) "Cyberworld 2025"—a periodic forecast of digital developments and their societal impact, (3) Early Warning: Methods of Cybersecurity Forecast, to concentrate on the area of critical infrastructures, (4) Quantitative Analysis of Cyber Resilience Strategies:

Methods and Simulation Tools to Ensure Cyber Resilience, (5) Cyber Conflict: An Extended Analysis of Conflict Space. An abstract of these five project ideas will be handed to the Secretary General. It also contains a preliminary estimate of funding needs and calculates the time requirement as at least two years.

At its working session on August 19, 2012, the PMP also:

- Discussed reports on recent cybersecurity developments, which included complexity of systems, cybersecurity economics, security issues associated with clouds, social media, and ubiquitous devices, and the sophisticated nature of cybercrime.
- Discussed PMP membership and the engagement of potential new members.
- Discussed the need to raise awareness within the WFS regarding the cyber threat and the realities of cyber warfare.

PROPOSAL TO SET-UP A PERMANENT MONITORING PANEL (PMP) ON: "FAST COMPUTING AND INFORMATICS FRONTIERS AND ECOLOGICAL SCIENCE"

PROFESSOR HORST WENNINGER
Former Director for Technology Transfer and Scientific Computing, CERN, Geneva, Switzerland

This note outlines a proposal to set-up in the framework of the "**Strategic Centre Program for Culture, Technology and Science"** a Permanent Monitoring Panel on *Fast Computing and Informatics Frontiers and Ecological Science*.

Indeed the monitoring of Planetary Emergencies is helped by real-time collection and distribution of information. Such information might help prevention, and will allow effective response and possible recovery from various types of Emergencies.

Basic Science and Technology is one of those programs, dedicated to the fight against Planetary Emergencies by promoting knowledge transfer and education.

The CERN research programs combined with an intensive R&D program, such as the LAA project—R&D works for experiments and technology—had great impact on advanced computing and information technologies. Hundreds of bright young scientists were given training in CERN, to help elaborating software, mid-ware, and hardware for the LHC experiments and related Grid computing.

Today—in the times of distributed computing with GRIDS and CLOUDS—training can be organized remotely but requires highly developed informatics infrastructures and competent areal or local teams in the institutes. That all this can be achieved has been demonstrated by the outstanding success of the LHC experimental teams of students, working in many countries and institutions around the world, when considering the speed with which the groups identified the Higgs signal within the Tsunami of data received via the computer grids.

Following examples of others,[1] an agreement should be proposed by the World Federation of Scientists to include CERN in the *Strategic Centre program for Culture, Technology and Science* allowing to use CERN's computing and informatics infrastructure.

[1] *CERN has indeed contracts with industries in the framework of the CERN Open-Lab Project and with several important Organizations, e.g., UNITAR performing large scale satellite observations to visualize emergencies around the world.*

CLOSING REMARKS ON THE CULTURE OF OUR TIME AND THE WFS PROJECT

ALFREDO MANTOVANO
Camera dei Deputati
Rome, Italy

First of all, a heartfelt thanks to Professor Antonino Zichichi for the invitation. I am honoured because it also allows me to personally follow the activities of the WFS in Erice and around the world. You, WFS scientists, are committed to providing a real contribution to the solution of problems that are barely mentioned even though they are key issues for the Earth, real Emergencies at the global level.

What you have been able to create is extremely significant for the politics and culture of our time: a culture which, as Professor Zichichi has been saying for years, calls itself modern, but actually seems to follow a course that pre-dates Aristotelian logic. If today's prevailing cultural trends followed the achievements of Science, even non experts would be fully aware of the problems to solve in order to bequeath to future generations a planet with its vital properties not upset or endangered.

I recall an example that surprised me when Professor Zichichi told me about it years ago. So-called modern Culture allowed the demonization of the greenhouse effect and carbon dioxide. Without these two "things" there would be no life. The greenhouse effect and carbon dioxide existed before the appearance of life on this satellite of the Sun of which we are privileged passengers, not absolute masters. This spacecraft might not have contained that tiny amount of carbon dioxide that the dominant Culture demonizes as a poison, and which is actually food for plants. Studies conducted by the WFS show that if you took away that tiny amount of carbon dioxide from Earth's atmosphere, plant life would disappear. This means that we would not be here if it is true that animal life presupposes plant life. Although I do not have the skills to go further into these fascinating problems, I would like to recall that the most eminent scholars of the subject known as "The problem of minimal life" gathered in this same hall three years ago to conclude that Science is far from admitting that it has understood how one goes from inert matter to living matter or, as Professor Zichichi says in very simple terms, how one goes from stone to swallow.

As a person involved in politics, I care about the problems of what is true and what newspapers and the media are full of. One thing is certain: if it were not for the activities of the WFS, Planetary Emergencies would be absent from the cultural debates of our time in the Culture of our time, even despite the fact that the WFS is committed to achieving 100 pilot projects in 50 Third World countries, including China in recent decades, and despite WFS results showing that—if there is political will—it is possible to address and solve problems linked to Planetary Emergencies. That is why, as a politician, I would like to emphasize the fact that the WFS can count on my commitment to make "The Project for Mankind" a reality.

It is a cutting edge battle which science and politics are called to fight together on the front line: it includes the protection and promotion of life, which has a practical tool in the implementation of your project to combat Planetary Emergencies. We must be realistic and admit unhesitatingly that the essence of totalitarianism coincides with the transformation of lies into truth thanks to the power that a man exerts over another to the point of changing or even taking his life. If the boundary between truth and mystification is not clear and insurmountable, if it is not identified as an objective fact to be recognized and respected, but is something variable depending on the subjective views or choices of a majority, you cannot avoid repeating the tragedies of the twentieth century.

After the horrors of the twentieth century, the quantity and quality of which remained little known for too long, the new century does not seem to be as open as it should be. Indeed, it continues to place science and technology at the service of death with tools of mass destruction, but also—at the same time—in aseptic rooms, away from cameras and uncomfortable witnesses. And it is a paradoxical time: science and its consequent technological inventions have brought us advantages and comforts. Just think of the increased life expectancy (80 years for men, 85 for women). An incredible, unprecedented result, which is accompanied, however, by the danger of a nuclear holocaust—which no one talks about anymore—and the risk of environmental holocaust. Can we trust science? In the past here, in Erice, you managed to organize some wonderful things: at the height of the East-West tension, Soviet and American scientists conversed and studied here together; scientists from the so-called free world and scientists from behind the Iron Curtain, united by their love for humanity, aimed at avoiding its self-destruction. Where politics divided you, real science brought you together. Ideologies of death, which disguised lies as the truth, led governments towards a mad arms race. Sure, it was perhaps necessary for us Westerners not to be left behind, but the price paid was and is high: an enormous amount of resources were wasted in order to fill the world with nuclear weapons and prepare environmental disasters, which resulted in real planetary emergencies, from water to soil, from air to energy.

But the leap in quality, in terms of science and in terms of politics, can only be achieved by clarifying the relationship between means and ends, i.e., if we consciously decide to retire a gentleman named Nicolò Machiavelli. Please allow me to quote Jerome Lejeune, the great geneticist who died a few years ago. He liked remembering an actual historical fact, which is a far better example than any theoretical disquisition: sixty years ago Dr. Thiersch, who used to carry out abortions in the U.S., had the idea of employing in his practice the toxic effect of aminopterin, a chemical that inhibits the metabolism of folic acid, thus blocking cell division: since embryo cells divide actively, the use of that substance, according to Thiersch, would infallibly kill the baby in the womb. However, infallibility was not so complete, because although its application was immediately lethal for some fetuses, for others death came after some time, due to severe abnormalities in the nervous system, from spina bifida to anencephaly.

Thirty years later, two doctors from different backgrounds, Smithells and Lawrence, found that folic acid administered to the mother, in special controlled doses, early in pregnancy protects babies against those very malformations of the central nervous system derived from a distorted use of the substance. Why did I mention the relationship between means and end? Because if Thiersch's goal had been to fight a disease and not to assault unborn children in the womb, the prevention of spina bifida or hydrocephalus would probably have been made possible three decades earlier, and tens of thousands of children around the world would have benefited. If we could change today's cultural perspective that seems almost uniquely and dogmatically oriented toward death (of the unborn, the elderly, the seriously sick, and the enemy), we would be open to new scientific discoveries respectful of man, the communities he belongs to, and the international community. We would thus realise the truth of Professor Zichichi's extraordinary intuition. Therefore, I would like to thank him again for the opportunity he has given me, which has also enabled us to enjoy for decades extraordinary experiences like the ones we have witnessed in these last few days.

We are all very conscious of the most unjust trial ever held: the one that took place 2,000 years ago against the Man from Nazareth. The biggest anomaly of that trial was that the judge, Pontius Pilate, the governor of Palestine, recognised the objective goal that was before him, and even spelled out the possible reasons for a judgment of acquittal, if it is true that he repeated three times "I have found no fault in this Man". He clearly recognised the substantive and procedural truth, but the operative part of the judgement did not follow that reasoning. Why? What happened? Where was the short circuit?

If we read the "court papers" in their entirety, Pilate first proceeded to interrogate the accused. During the interrogation Pilate addressed this crucial question to the defendant: "Who are you?", and received this reply: "I came into the world to testify to the truth". This answer could have been the start of a long dialogue, such as the Man was accustomed to, when talking seriously to someone. But Pilate abruptly interrupted his speech with a sentence—*quid est veritas?*—which is the synthesis of every sceptical and relativistic position when faced with reality. *Quid est veritas?* Truth does not exist, it is an opinion, it is what remains after a talk show, it is the result of a balance of power, and it is one aspect of convenience.

It is the opportunity of the moment, which can be given a semblance of democratic choice, regardless of the truth: when he asked the crowd to vote between freeing Jesus Christ or Barabbas, Pilate put the law of numbers above any other consideration. He adopted a perspective of absolute relativism.

Besides being a great philosopher, St. Augustine enjoyed making anagrams. After reading Pilate's words, *quid est veritas*, he turned them into an anagram: *est Vir qui adest!* And this allowed him to turn to Pilate and tell him: the truth is before you, if only you decide not to close our eyes to reality! Dear man of the twenty-first century, you can still make it if you grasp the truth not in virtual convenience, but in the reality that is perceived by your senses and ordered by your reason. Dear scientist, this is especially true for you and for your mission of putting your research to the service of the knowledge of the truth about man. Erice has been giving a decisive contribution in this direction for decades!

That is why I think it is the duty of every politician—it is certainly mine!—to support the "Science, Technology and Culture in the Third Millennium: A Project for Mankind", and carry it forward as far as my political-cultural competence is concerned. It is the right Project to address and solve problems that our Culture does not talk about much, despite their enormous importance.

In one of the most moving passages of *The Lord of the Rings* by John Ronald Reuel Tolkien V, a book that continues to fill the hearts of those who seek in it a sense of honour and love for the land, courage and contempt for difficulties, the exaltation of friendship and a hierarchy of values—the old Gandalf turns to his fellow adventurers, on the eve of the decisive battle against the evil armies, with these words: "It is not our part to master all the tides of the world, but to do what is in us for the succour of those years wherein we are set, uprooting the evil in the fields that we know, so that those who live after may have clean earth to till".

I am honoured to have participated today in a gathering of scientists that gives substance, quality and perspective to such healthy realism.

SEMINAR PARTICIPANTS

Lord John Alderdice	House of Lords London, UK
Professor Mikhail J. Antonovsky	Carbon Dioxide Division Institute of Global Climate and Ecology Moscow, Russia
Professor William A. Barletta	U.S. Particle Accelerator School Department of Physics Massachusetts Institute of Technology Cambridge, USA
Professor Luís A.M. Bettencourt	Santa Fe Institute Santa Fe, USA
Professor Richard L. Breyer	Documentary Film and History Syracuse University, S.I. Newhouse School of Public Communications, Syracuse, USA
Dr. Vladimir B. Britkov	Institute for Systems Analysis Russian Academy of Sciences Moscow, Russia
Dr. Jiri Brodsky	Deputy-Director of the Foreign Office Dept. Office of the President Prague, Czech Republic
Dr. Robert J. Budnitz	Lawrence Berkeley National Laboratory Earth Sciences Division Berkeley, USA
Dr. Franco Buonaguro	Istituto Nazionale dei Tumori "Fondazione G. Pascale" Napoli, Italy
Dr. Emanuele Buratti	Molecular Pathology Group International Centre for Genetic Engineering and Biotechnology, AREA Science Park Trieste, Italy
His Excellency Ambassador Petr Burianek	Ambassador of the Czech Republic Rome, Italy

Her Excellency Dr. Anna Maria Cancellieri	Ministro degli Interni della Repubblica Italiana, Rome, Italy
Dr. Nathalie Charpak	Kangaroo Foundation Bogotá, Colombia
Professor Luisa Cifarelli	Department of Physics University of Bologna Bologna, Italy
Professor Andrea Contin	Department of Physics University of Bologna Bologna, Italy
Professor Pierre Darriulat	Vietnam Auger Training Laboratory Institute for Nuclear Science & Technology Hanoi, Vietnam
Dr. Socorro de Leon-Mendoza	Jose Fabella Memorial Hospital Neonatology Unit Manila, Philippines
Dr. Carmine Difiglio	Office of Policy and International Affairs U.S. Department of Energy Washington, USA
Dr. Mbareck Diop	Science & Technology Advisor to the President of Senegal Dakar, Senegal
Professor Gheorghe Duca	The Academy of Sciences of Moldova Chisinau, Republic of Moldova
Professor Robert Duncan	University of Missouri Vice Chancellor for Research Columbia, USA
Professor Christopher Essex	Department of Applied Mathematics University of Western Ontario London, Canada
Dr. Lorne Everett	Chancellor Emeritus, Lakehead University Thunder Bay, Canada and L. Everett & Associates, LLC, Santa Barbara, USA

Dr. Michela Gallagher	Department of Psychological and Brain Sciences, The Johns Hopkins University Baltimore, USA
Dr. Bertil Galland	Writer and Historian Buxy, France
Professor Mary Ganguli	University of Pittsburgh School of Medicine Division of Geriatrics and Neuropsychiatry Pittsburgh, USA
Professor Richard L. Garwin	Thomas J. Watson Research Center IBM Research Division Yorktown Heights, USA
Professor Bernardino Ghetti	Dept. of Pathology & Laboratory Medicine Indiana University Indianapolis, USA
Professor Alberto González-Pozo	Theory and Analysis Department Universidad Autónoma Metropolitana Mexico D.F., Mexico
Dr. Walter M. Grayman	Consulting Engineer Cincinnati,USA
Professor Ara Grigoryan	Yerevan Physics Institute Yerevan, Armenia
Dr. Balamurugan Gurusamy	ERE Consulting Group Sdn Bhd Subang Jaya, Malaysia
Dr. Adalat Hasanov	National Academy of Sciences Baku, Azerbaijan
Dr. Udo Helmbrecht	European Network and Information Security Agency Heraklion, Greece
Professor Michael Jefferson	International Business and Sustainability Center, London Metropolitan Business School, London, UK

Dr. Dannie Jost	Science Advisor World Trade Institute, University of Bern Bern, Switzerland
Lady Barbara Judge	Emeritus Chairman UK Atomic Energy Authority London, UK
Mr. Anand Kamalakar	Trilok Fusion Media Brooklyn, USA
Dr. Hisham Khatib	World Energy Council Amman, Jordan
Dr. Do-Won Kim	CERN Ph-UAI Division Geneva, Switzerland
Dr. William Kininmonth	Australasian Climate Research Kew Victoria, Australia
His Excellency President Vaclav Klaus	President of the Czech Republic Prague, Czech Republic
Dr. Erwin Knupp	TOTH SA Geneva Geneva, Switzerland
Dr. Edward Kobal	Slovenian Scientific Foundation Ljubljana, Slovenia
Dr. Vasiliy Krivokhizha	International Department Federal Assembly of the Russian Federation Moscow, Russia
Dr. Dan J. Kroll	Hach Homeland Security Technologies Loveland, USA
Dr. Valery P. Kukhar	Ukrainian Academy of Sciences Laboratory of Fine Organic Synthesis Kiev, Ukraine
Professor Askar Kutanov	National Academy of Sciences Former Ambassador of the Kyrgyz Republic Bishkek, Kyrgyzstan

Lord Nigel Lawson	House of Lords London, UK
Professor Axel Lehmann	Institute for Technical Computer Sciences German Armed Forces University Munich Neubiberg, Germany
Dr. Sally Leivesley	Newrisk Limited London Mitigation of Terrorists Acts PMP World Federation of Scientists London, UK
Dr. Giovanni Levi	National Scientific Research Centre Evolution of Endocrine Regulation Paris, France
Professor Richard S. Lindzen	Earth, Atmospheric & Planetary Sciences Massachusetts Institute of Technology Cambridge, USA
Hon. Alfredo Mantovano	Camera dei Deputati Rome, Italy
His Excellency Professor Donato Marra	Segretario Generale della Presidenza della Repubblica Italiana Rome, Italy
Professor Sergio Martellucci	Faculty of Engineering Università degli Studi di Roma "Tor Vergata", Rome, Italy
Professor Tatsuo Masuda	The Graduate School Nagoya University Commerce and Business Tokyo, Japan
Professor Sam Mbulaiteye	Infections and Immunoepidemiology Branch Division of Cancer Epidemiology and Genetics, National Cancer Institute, NIH, DHHS, Rockville MD, USA
Professor Stephen McIntyre	University of Guelph Toronto, Canada
Dr. Ross McKitrick	University of Guelph Toronto, Canada

Dr. Helge Meinhard	CERN IT-PES Geneva, Switzerland
Dr. Bruce L. Miller	Memory & Aging Center University of California San Francisco, USA
Dr. Dong-Pil Min	Ambassador of Science and Technology Cooperation of Korea Seoul National University, Korea
Lord Christopher Monckton	The Viscount Monckton of Brenchley Monckton Enterprises Ltd. Carie, Rannoch, UK
Dr. Alan Leigh Moore	Research Institute Von Braun Research Hall The University of Alabama in Huntsville Huntsville, USA
Dr. Jill R. Murrell	Indiana University School of Medicine Department of Pathology and Laboratory Science, Indianapolis, USA
Dr. André Naudi	Director of Finance and Human Resources (Former) CERN, Geneva, Switzerland
Professor Joseph Niemela	International Centre for Theoretical Physics Trieste, Italy
Professor SunKun Oh	Department of Physics Konkuk University Seoul, Korea
Dr. Jef Ongena	Plasmaphysics Laboratory Ecole Royale Militaire Brussels, Belgium
The Honorable Alessandro Pagano	Chamber of Deputies, Italian Parliament Rome, Italy
Professor Garth W. Paltridge	University of Tasmania Hobart, TAS, Australia

Professor Frank Leon Parker	Environmental Engineering Vanderbilt University Nashville, USA
Professor Stefano Parmigiani	Evolutive and Functional Biology Università di Parma Parma, Italy
Dr. Robert Petroski	Nuclear Innovation Engineer TerraPower LLC Bellevue, USA
Ms. Renata Petroušová	Protocol Officer Czech Republic Embassy in Rome Rome, Italy
Dr. Daniele Pinzani	Senior Programme Advisor SPARK Mitrovica, Kosovo
Professor Nusrat Rajabov	Department of Mechanics and Mathematics Tajik State University Dushanbe, Tajikistan
Professor Ramamurti Rajaraman	School of Physical Sciences Jawaharlal Nehru University New Delhi, India
Dr. Anil Razdan	PTC India Ltd., New Delhi, India
Dr. Hans-Holger Rogner	Planning and Economics Studies Section Department of Nuclear Energy International Atomic Energy Agency Vienna, Austria
Dr. Juan Ruiz	Department of Pediatrics San Ignacio Hospital Santafé de Bogotá, Colombia
Dr. Andrew J. Saykin	IU Center for Neuroimaging Division of Imaging Sciences Department of Radiology Indianapolis, USA

Dr. Robert N. Schock	World Energy Council Senior Advisor London, UK
Professor Hiltmar Schubert	Fraunhofer-Institut für Chemische Technologie, ICT Pfinztal, Germany
Dr. Giusto Sciarabba	Third World Academy of Sciences, (TWAS) Trieste, Italy
Dr. Nir J. Shaviv	Hebrew University of Jerusalem Mount Scopus Campus Jerusalem, Israel
Dr. Adnan Shihab-Eldin	Kuwait Foundation for the Advancement of Science, Safat, Kuwait
Professor Annette Sobel	University of Missouri Senior Advisor for Bio-security Engagement Office of the Secretary of Defense/Nuclear, Chemical, Biological Columbia, USA
Dr. Gwen Sprehn	Indiana University School of Medicine Department of Neurology Indianapolis, USA
Professor Friedrich Steinhäusler	Physics and Biophysics Division University of Salzburg Salzburg, Austria
Dr. Bruce Stram	Element Markets Houston, USA
Professor Gheorghe Stratan	National Institute of Physics and Nuclear Engineering Bucharest-Magurele, Romania
Professor Nicolae Stratan	The Academy of Sciences of Moldova Center of Optoelectronics of IAP of ASM Chisinau, Republic of Moldova
Professor Yuliang Sun	INET, Tsinghua University Beijing, P.R. China

Professor Henrik Svensmark	Center for Sun-Climate Research Danish National Space Center Technical University of Denmark Copenhagen, Denmark
Professor Kyle Swanson	Department of Mathematical Sciences University of Wisconsin-Milwaukee Milwaukee, USA
Professor Michael J. Tannenbaum	Physics Department Brookhaven National Laboratory Upton, USA
Professor Abdel Nasser Tawfik	ECTP Founding Director MTI University Cairo, Egypt
Dr. Hamadoun I. Touré	Secretary General International Telecommunications Union Geneva, Switzerland
Professor Anastasios A. Tsonis	Department of Mathematical Sciences University of Wisconsin-Milwaukee Milwaukee, USA
Professor Zeblon Vilakazi	iThemba Laboratory for Accelerator Based Science Somerset West, South Africa
Dr. Jandouwe Villinger	International Centre of Insect Physiology and Ecology (ICIPE) Nairobi, Kenya
Professor Frederick vom Saal	University of Missouri Division of Biological Sciences Columbia, USA
Dr. Hans von der Schmitt	ATLAS Computing Coordination CERN, Switzerland and Max-Planck-Institut für Physik Munich, Germany
Professor Frank N. von Hippel	Woodrow Wilson School Princeton University Princeton, USA

Dr. Henning Wegener	Ambassador of Germany (ret.) Information Security Permanent Monitoring Panel World Federation of Scientists Madrid, Spain
Dr. Horst Wenninger	Former Director for Technology Transfer and Scientific Computing CERN, Geneva, Switzerland
Dr. Jody Westby	Global Cyber Risk LLC Information Security PMP World Federation of Scientists Washington, USA
Dr. Crispin Williams	Advanced Technology for Young Scientists from Developing Countries CERN, Geneva, Switzerland
Professor Richard Wilson	Department of Physics Harvard University Cambridge, USA
Dr. Lowell Wood	Hoover Institution, Stanford University University of California *and* Lawrence Livermore National Laboratory Stanford, USA
Professor Xiliang Zhang	Institute of Energy, Environment & Economy, Tsinghua University Beijing, P.R. China
Dr. Yiyan Zhang	Second Secretary Chinese Embassy in Italy Rome, Italy
Professor Antonino Zichichi	CERN, Geneva, Switzerland and University of Bologna, and Centro Enrico Fermi, Rome, Italy
Ms Hana Zvonková	Deputy Head of Mission Embassy of the Czech Republic Rome, Italy